Analysis of Shells
and Plates

Analysis of Shells and Plates

Phillip L. Gould

Prentice Hall
Upper Saddle River, New Jersey 07458

Library of Congress Cataloging-in-Publication Data

Gould, Phillip L.

   Analysis of shells and plates / Phillip L. Gould.

      p. cm.

   Includes bibliographical references and index.

   ISBN 0–13–374950–9

   1. Shells (Engineering) 2. Plates (Engineering) I. Title.

TA660.S5G644 1998

 624.1'776--dc21   98–16289

                                      CIP

Acquisitions editor: Bill Stenquist

Production Editor: Patty Donovan

Managing Editor: Eileen Clark

Editor-in-Chief: Marcia Horton

Vice-President of Production and Manufacturing: David W. Riccardi

Cover Art Director: Jayne Conte

Cover Designer: Bruce Kenslaar

Manufacturing/Full Service Coordinator: Donna M. Sullivan

Editorial Assistant: Meg Weist

Composition: Pine Tree Composition, Inc.

Reprinted with corrections June, 1999

Printed in the United States of America

10 9 8 7 6 5 4 3 2

ISBN: 0-13-374950-9

To my wife, Deborah

Contents

Preface

The study of three-dimensional continua has been a traditional part of graduate education in solid mechanics for some time. With rational simplifications to the three-dimensional theory of elasticity, the engineering theories of medium-thin plates and of thin shells may be derived and applied to a large class of engineering structures distinguished by a characteristically small dimension in one direction. Often, these two theories, plates and shells, are developed somewhat independently because of their distinctive geometrical and load-resistance characteristics. On the other hand, the two systems share a common basis and might be unified under the classification of *surface structures* after the German term *Flächentragwerke*. This common basis is fully exploited in this book.

A substantial portion of many traditional approaches to this subject has been devoted to constructing classical and approximate solutions to the governing equations of the system in order to proceed with applications. Within the context of analytical, as opposed to numerical, approaches, the limited generality of many such solutions has been a formidable obstacle to applications involving complex geometry, material properties, and/or loading. It is now relatively routine to obtain computer-based solutions to quite complicated situations. However, the choice of the proper problem to solve, through the selection of the mathematical model, remains a human rather than a machine task and requires a basis in the theory of the subject.

With the requirement of a strong grounding in the mechanics of shells and plates remaining firm, this book presents a unified development with emphasis on the fundamental engineering aspects. The basic material is designed to be covered in a single semester graduate course or through equivalent self-study; also, ample enrichment is provided for further independent study. Initially, the geometrical relationships are developed on a somewhat general level, with specific applications to frequently encountered forms. Following the geometric description of the surface and the consideration of equilibrium, we introduce a first logical simplification, the membrane theory of shells. After a further theoretical exposition of linear deformations, constitutive relationships, and energy principles, we present the flexural theory of plates and the bending theory of shells, including elastic instability. Although this sequence postpones the introduction of the complete theory of plates until much of the theory of shells is covered, we believe it is logical and consistent with the unifying objective of the text. Illustrative examples are contained throughout the text and instructive exercises are provided at the end of each chapter.

In the presentation, the fundamental geometric and static relationships are considered from an integrated mathematical and physical point of view. Orthogonal curvilinear coordinates and vector calculus are used to provide a concise general derivation of the field equations. Early on, however, we present the physical resolution and interpretation of forces and deformations for specialized geometric forms, such as rotational shells and flat plates. We believe that the physical notions are more meaningful in the specialized geometric context, whereas the mathematical formulation yields a conciseness not attainable with a strictly physical viewpoint. Also, we stress the energy aspects of the formulations because of the importance of energy methods in modern computational techniques.

With regard to applications, our focus is on classical examples that illustrate the basic resistance mechanisms of selected configurations without undue mathematical complications. The availability of numerically-based computer-implemented solution algorithms which are capable of performing complete nonlinear analyses of complex structures has diminished the need to rely on cumbersome, sometimes oversimplified, analytical solutions for such problems; hence, they are paid scant attention in this text. Rather, we emphasize the essential aspects of equilibrium and compatibility, as they apply to the resistance of surface loading and the satisfaction of boundary constraints. We believe that a firm grasp of these principles is necessary to develop the appropriate mathematical models and to perform the vital critical interpretation required of the analyst when computer solutions are employed.

This book is dedicated to developing in the engineer the physical and mathematical understanding required to perform analysis and design in an interactive computer-assisted environment. The book also provides a foundation of the subject which will enable the interested reader to progress to more advanced texts and technical papers.

Regarding the background of the reader, this book is written primarily for advanced students and practitioners in civil, mechanical, and aeronautical engineering and engineering mechanics. We presume that the reader has an elementary knowledge of vector calculus and matrix algebra, and some familiarity with the linear theory of elasticity.

Phillip L. Gould

Acknowledgments

This book is based on previous volumes, *Static Analysis of Shells* (D.C. Heath, 1974) and *Analysis of Shells and Plates* (Springer-Verlag, 1988). The author is indebted to the many contributors to these volumes.

In the past decades, the author has been influenced by a number of contemporary engineers and scientists. Among those whose contributions directly impacted this book are, in alphabetical order: Steven M. Baldridge, Consulting Engineer; Prof. P. Bergan, Norway; Prof. D. Billington, Princeton University; Dr. D. Bushnell, Lockheed-Palo Alto; Dr. J. Bobrowski, Consulting Engineer U.K.; Prof. C. R. Calladine, University of Cambridge; Prof. J. G. A. Croll, University College, London; Prof. Tom Farris, Purdue University; Prof. K. J. Han, University of Houston; Prof. S. Kato, Toyohashi Institute of Technology; Dr. M. Ketchum, Consulting Engineer; Prof. W. Krätzig, Ruhr-Universität-Bochum; Prof. D. Pecknold, University of Illinois at Urbana-Champaign; Prof. E. Reissner (deceased), University of California, San Diego; Prof. J. M. Rotter, University of Edinburgh; Prof. A. Scordelis, University of California-Berkeley; Prof. W. Schnobrich, University of Illinois at Urbana-Champaign; Prof. S. Simmons, University of Alberta; Prof. U. Wittek, University of Kaiserslautern; Chris Wise, Consulting Engineer; Prof J. K. Wu, Peking University.

The author is especially grateful to Prof. E. Ramm, University of Stuttgart, for acquainting him with the outstanding achievements of the great Russian engineer V. G. Shukov.

The author is also indebted to current and recent students B. J. Lee, J. S. Lin, Robert Elkin, Michael Williams, Michael Wendl, Manahar Kollegal, R. V. Ravichandran, Hidajat Harintho, X. J. Ma, and Nathan Gould for careful proofreading and suggestions. Also, Mr. Sakul Pochanart, Dr. Y. X. Cai, and Prof. Jerry Craig assisted in improving the illustrations and preparing many of the drawings. Ms. Tesha Harris and Ms. Linda Buckingham carefully revised and retyped much of the manuscript.

Finally, the author is appreciative of the academic atmosphere provided by Washington University, and of the patience and cooperation of his colleagues throughout the manuscript preparation process.

Analysis of Shells and Plates

1

Introduction

1.1 ROLE OF THE THEORY OF ELASTICITY

The theory of elasticity is the basis for several engineering theories which, in turn, are applied to mechanical and structural design. The basic components of the elasticity problem are often designated as *equilibrium, compatibility,* and a *constitutive law.* The equilibrium equations represent a statement of Newton's laws, which are restricted here to the static case. The compatibility conditions express the kinematic relationships between strains and displacements, and the constitutive law embodies the stress-strain behavior of the material which is presumed here to be linear elastic. In general, this set of basic components may be collected as a set of differential equations or as an energy principle. The simultaneous satisfaction of each component of the elasticity problem is often foreboding from the mathematical standpoint; therefore, engineers have naturally looked to simplifications and approximations.

One common simplification often imposed on elasticity problems is to formulate less restrictive theories based on distinctive geometric characteristics. Among the relaxed theories, we find: (1) the *theory of beams,* which is concerned with flexural members having one dimension, the span, characteristically far greater than the other two which define the cross-section; (2) the *theory of plates,* which treats initially flat components having the two plan dimensions far greater than the thickness; and (3) the *theory of shells,* which deals with curved bodies which are relatively thin. Within these three theories are a variety of subtheories. For the purposes of this introduction, it is sufficient to refer to the most common subtheory for each case; for example, (1) *shallow* beam theory; (2) *medium-thin* plate theory, and (3) *thin* shell theory.

Although a beam may be considered a one-dimensional member, whereas plates and shells are two-dimensional, the similarities in the theories are numerous, and the beam serves as a useful analogue for the exposition of the higher theories.

1.2 ENGINEERING THEORIES

Sections of a beam, a plate, and a shell are shown in Figure 1–1. On each figure, the characteristic dimensions in the transverse and lateral directions are denoted as h and l, respectively. Also, a reference position is identified in the transverse direction, midway between the boundaries for symmetric cross-sections. We call this reference position the *neutral* or *middle axis* for a beam, the *middle plane* for a plate, and the *middle surface* for a shell. In the ensuing treatment, the terms *plane* and *plate,* and *surface* and *shell,* are used synonymously

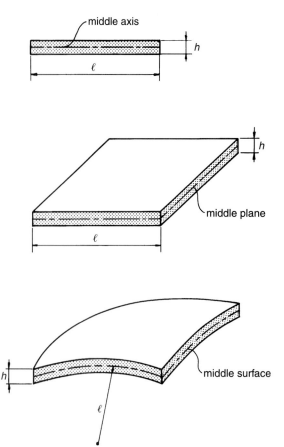

FIGURE 1–1 Characteristic Dimensions of Structural Forms.

and interchangeably to avoid repetitive distinction between the mathematical and physical objects.[a,1]

The initial step in the derivation of each of the simplified theories usually consists of a set of assumptions with respect to the ratio of the characteristic dimensions, the relative magnitude of the deflection under the applied loading, the rotation of a normal to the undeformed reference position, and the stresses in the transverse direction. The statement and justification of these assumptions are largely attributable to several distinguished mathematicians and scientists of the eighteenth and nineteenth centuries. Thus, we have the Navier hypothesis and the Bernoulli-Euler theory for beams, the Kirchhoff theory of plates, and Love's first approximation to the theory of shells.[b,2]

These assumptions are collected in Table 1-1. A number of comments with respect to the assumptions and justifications are appropriate:

[1] This assumption may be regarded as the most fundamental since it clearly delineates the class of problems with which we are concerned from a physical standpoint. Also, [1] is the justification for [3] and [4]. The bounds on h/l are only approximate and are subject to considerable latitude depending on loading, geometry, etc.

[2] This assumption is independent, although the cases for which it would be violated might well coincide with the lower range of h/l stated in [1]. If [2] is not justified, geometrically nonlinear theories can be formulated retaining the remaining assumptions. (An illustration of such a formulation is presented in Section 11.3.) We may also accommodate material nonlinearities within the range of [2]. However, as the magnitude of the admissible displacements increases, the possibility of exceeding the limits of linear elastic material behavior increases proportionally.

[3] This assumption is developed in detail within the treatment of deformations in Chapter 7. When only classical solution techniques were available, the suppression of transverse shearing strains permitted many otherwise intractable problems to be approached. With powerful numerical procedures now well developed, this necessity has diminished, although it is still popular. It has been suggested in an independent observation that this strategy, originally conceived to facilitate analytical solutions, indeed often complicates numerically based solutions.[3] Relaxation of [3] enables the upper limit on h/l in [1] to be extended in many cases.

[4] Situations for which assumption would not be justified, apart from the immediate vicinity of concentrated loads, would coincide with the upper limits of [1].

The assumptions and consequences collected in Table 1-1 are of utmost importance in what follows and are referred to frequently.

[a]Translations of the notebooks of Leonardo da Vinci (Ladislao Reti, ed. *The Unknown Leonardo,* Switzerland) reveal that he discovered that the extreme fibers of a cross-section shorten in compression and elongate in tension. Therefore, the fibers in the center do not change length. This occurred perhaps two centuries earlier than frequently cited references. [1]

[b]Readers interested in the historical development of solid mechanics and in the distinguished personalities who contributed to this development are referred to Todhunter and Pearson, A. E. H. Love, S. A. Timoshenko, and H. M. Westergaard. [2]

TABLE 1–1   Basic Assumptions

Assumption	Theory	Consequence	Justification
[1] Transverse characteristic dimension is small in comparison to lateral characteristic dimension	Shallow beam Medium-thin plate Thin shell	Beam is shallow in comparison to length, $h < l$ Plate is thin in comparison to lateral dimension, $h < l$ Shell is thin in comparison to minimum radius of curvature, $h < l$	$0.01 < h/l \leq 0.5$ cable deep beam $0.001 \leq h/l \leq 0.4$ membrane thick plate $0.001 \leq h/l \leq 0.05$ curved thick membrane shell
[2] Displacements are small in comparison to transverse characteristic dimension	All	Equilibrium may be formulated with respect to the initial undeformed geometry. Products of deformation parameters may be neglected. The system may be described by a system of geometrically linear equations.	Validity may be established by calculation in the course of the solution
[3] Transverse shearing strains which act on planes parallel to middle (section, plane, surface) are neglected	Shallow beam Medium-thin plate Thin shell	Plane sections before deformation remain plane after deformation Straight fibers perpendicular to the middle plane before deformation remain perpendicular to the middle plane after deformation Straight fibers perpendicular to to the middle surface before deformation remain perpendicular to the middle surface after deformation	$h < l$ $h < l$ $h < l$
[4] Normal stresses acting on planes parallel to middle (section, plane, surface) are neglected	Shallow beam Medium-thin plate Thin shell	Beam depth does not change during deformation Plate thickness does not change during deformation Shell thickness does not change during deformation	$h < l$ $h < l$ $h < l$

1.3   LOAD-RESISTANCE MECHANISMS

The common basis of shallow beam, medium-thin plate, and thin shell theories is illustrated by the unified set of underlying assumptions. However, the means for resisting applied loading among these structural forms may be quite different. Idealized free-body diagrams of the

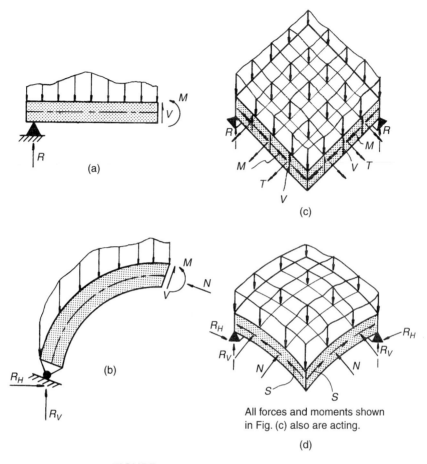

FIGURE 1–2   Means of Load Resistance.

three cases, along with an additional form, the arch, are shown in Figure 1–2. For simplicity, only a vertical loading is shown in each case, but the following observations are generally valid for any transverse loading.

The beam, being straight, depends on the shear V to resist transverse loading. In turn, V, acting together with the loading and reaction R, requires a moment M for equilibrium. The beam may be classified as a *one-dimensional flexural* member.

The arch, because it is curved, can develop a thrust N to resist the applied loading, in addition to the shear V. Although V and M are still present in the general case, the efficiency of the arch form lies primarily in resisting the loading with N and minimizing V and M. The arch may be called a *one-dimensional extensional* member.

The plate, being flat, relies on the transverse shear V to resist transverse loading in the same manner as the beam. The two-dimensional configuration results in bending moments M

and, additionally, twisting moments T on each internal face. Because the loading is generally carried in both directions and because the twisting rigidity in isotropic plates is quite significant, a plate is considerably stiffer than a beam of comparable span and thickness. The plate may be considered a *two-dimensional flexural* member.

The shell, being curved, can develop thrusts N to form the primary resistance mechanism in addition to those forces and moments present in the plate. Also, the two-dimensional curved configuration mobilizes an in-plane shear S in each direction. Although V, M, and T are still present in the general case, the efficiency of the shell form rests with the reliance on N and S as the primary means of resistance, with V, M, and T minimized. The shell may be termed a *two-dimensional extensional* member. Furthermore, whereas an arch of a specific shape can resist only one loading pattern in a purely extensional state, a shell can remain virtually momentless for a variety of loadings. The classification of these forms is summarized in Table 1–2. As mentioned in the Preface, the term *surface structures* is used in this text to denote both plates and shells. While uncommon in the English language literature, this term emphasizes the common basis of the engineering theories for these structural forms, which is the thrust of this book.

Of course, there is some overlap within the classification in Table 1–2. One common situation is the presence of axial loading in a beam and, correspondingly, in-plane loading in a plate. Within the limitations of assumption [2], these effects, called *rod* and *diaphragm* action, respectively, are uncoupled from the primary action and may be combined with the flexural behavior by superposition. The study of instability, however, involves a coupling between extensional and flexural effects.

Another possibility with respect to the beam or plate is the presence of axial or in-plane forces which develop a transverse component through curvature induced by the flexural action. This case may be treated only with a relaxation of assumption [2].

A third combination, which is of considerable practical importance, is a form curved in only one direction, such as a cylinder or cone. Such a surface may be formed from an initially flat sheet without distortion, e.g., rolling up a piece of paper to make a cylinder. Although these so called *developable* surfaces are generally regarded as shells, the resistance mechanism in the uncurved direction is basically flexural, whereas in the curved direction, it may be primarily extensional.

As a further introductory comment, it should be pointed out that structural materials are generally far more efficient in an extensional rather than in a flexural mode, making the arch and shell preferable over the beam and plate. The extensional mode, within the scope of small deformations, can be mobilized only through initial curvature; this limits the applica-

TABLE 1–2   Classification of Structural Forms

Primary Resistance Mechanism	Configuration	
	1-Dimensional	*2-Dimensional*
Flexural	Beam	Plate
Extensional	Arch	Shell

tion of arches and shells both from a fabrication and from a utilization standpoint. The structural form is subject to many constraints apart from the most efficient use of the material and will be considered as specified *a priori* in most of this text. However, the appropriate design and construction of a shell structure is perhaps the most eloquent demonstration of the art of structural engineering. As attributed to E. Torroja, one of the pioneers in the field of reinforced concrete shells, "The best work is that which is supported by its shape and not by the hidden strength of its material."[4]

It is of present, as well as historical, interest to recognize that significant structures were erected utilizing the efficient doubly curved form of the dome long before the development of modern engineering analysis.[5] Several of these structures survive today. The Pantheon of ancient Rome (Figure 1–3), attributed to Marcus Agrippa and Emperor Hadrian and constructed of a cementitious material, has stood for about two thousand years; the Hagia Sophia in Istanbul (Figure 1–4), originally completed by Isidorus, Jr. has epitomized Byzantine architecture for fifteen centuries. Beautifully tile-covered mosques from the Persian empire (Figure 1–5) survive in Iran. Increased geometrical refinement is exhibited in the Renaissance cathedrals of Santa Maria del Fiore in Florence (Figure 1–6) constructed without shoring by Brunelleschi, and St. Peter's in Rome (Figure 1–7) designed by Michelangelo. Also, Sir Christopher Wren's St. Paul's Cathedral (Figure 1–8) remains to grace the London skyline.

FIGURE 1–3a   Pantheon, Rome, Italy, Dome Span = 43.4 m; Dome Rise = 21.6 m.

FIGURE 1–3b   The Interior of the Pantheon; National Gallery of Art, Washington; Samuel H. Kress Collection.

FIGURE 1–4   Hagia Sophia, Istanbul, Turkey. Dome Span = 31.9 m; Dome Rise = 13.8 m. (Courtey Dr. I. Mungan)

FIGURE 1–5   Dome of Shah's Mosque, Iran.

FIGURE 1–6a S. Maria Del Fiore, Florence, Italy. Dome Span = 42.4 m; Dome Rise = 36.6 m.

Of these ancient domes, the Pantheon and Santa Maria are regarded as among the greatest construction achievements of all time.[6,7] In addition to the obviously spectacular clear span, the roof of the Pantheon is formed with intersecting ribs (Figure 1–3b) which provide greater stiffness and stability than an equivalent amount of material of uniform thickness, and is reinforced with bands of vitrified tile. Further weight reduction is achieved by incorporating three grades of lightweight aggregate in the concrete.[6] Brunelleschi's dome, documented in considerable detail by Parsens,[7] marks a deliberate departure from the then-traditional hemisphere to a more efficient pointed (circular) arch profile. The cross-section is composed of a separated double wall (Figure 1–6b), tied together by the meridional arches and a secondary system of horizontal arches. Also incorporated was a timber chain circling the dome to resist the somewhat vaguely understood outward thrust resulting from the arched

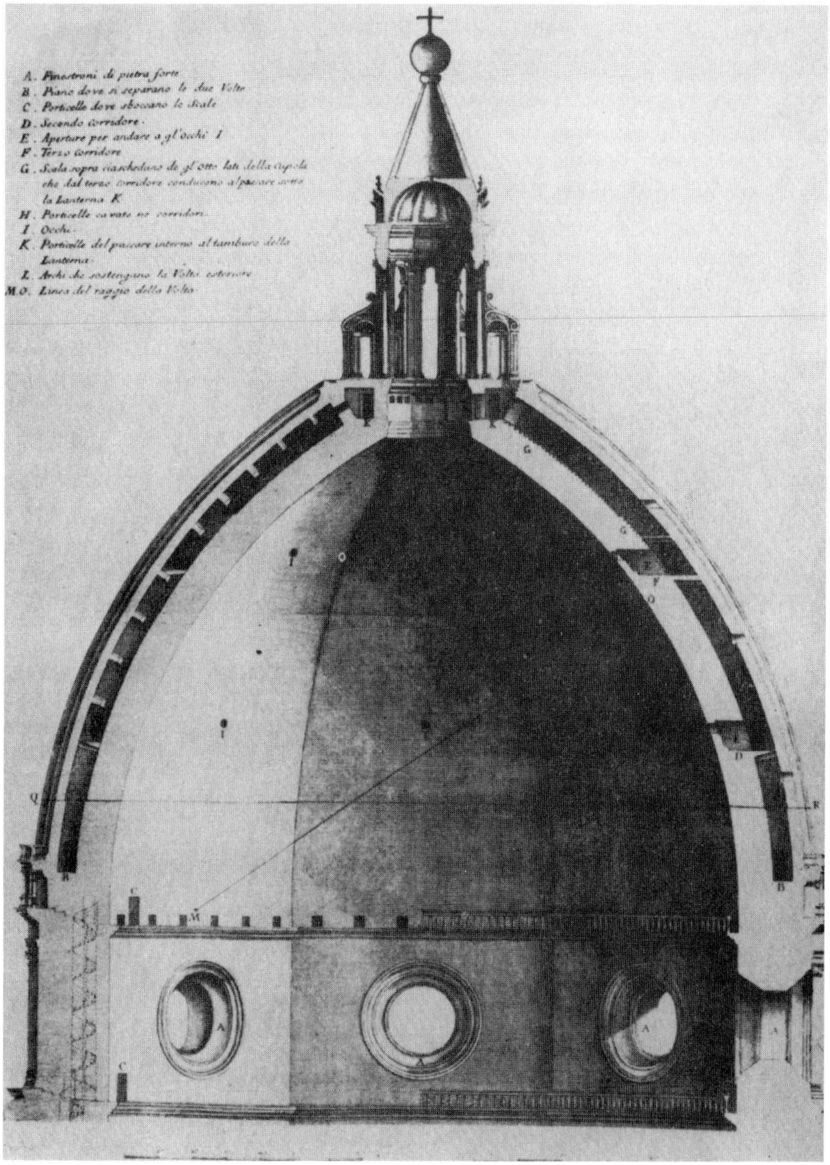

FIGURE 1–6b   Cross-Section of Dome of Santa Maria del Fiore from Sgrilli's "Descrizione e studi dell' insigne fabbrica di S. Maria del Fiore" (Florence, 1733). (From W.B. Parsons *Engineers and Engineering in the Renaissance*, Williams and Wilkins Co., Baltimore, © 1939)

FIGURE 1–7   St. Peter's, Rome, Italy. Dome Span = 41.6 m; Dome Rise = 35.1 m.

meridional profile. Remarkably, the two techniques used for increasing the structural effi-ciency of the walls of these ancient structures have modern counterparts, in other words, stiffened and multilayered shells as described in Chapter 8.

Although the ancient domes were not thin or engineered in the modern sense, they ex-hibit the unique capability of the curved surface to bridge considerable space without inter-mediate supports utilizing construction materials capable of resisting only compressive forces. Modern techniques of structural analysis, most recently computer-based finite ele-ment modeling, have revealed the intuitive understanding of structural mechanics exhibited by the Romans and by the men of the Renaissance in designing and building their spectacu-lar domes.

Perhaps one reason that several shells remain from antiquity is the ability of surface structures to survive extreme loading. It was reported that a cooling tower shell was among the few surviving structures in the Tangshun, China, earthquake of 1976.[8] The hyperbolic paraboloid shown in Figure 1–9 resisted the 1985 Mexico City earthquake without apparent

FIGURE 1–8 St. Paul's, London, England, Dome Span = 30.8 m; Dome Rise = 33.5 m.

structural damage amid totally destroyed conventional structures.[9] Surely, the ancient domes from the Persian empire such as that in Figure 1–5 have withstood many earthquakes. These examples inspire confidence in the toughness of well-designed and constructed surface structures, and the remainder of this book is an exposition of the basic principles by which such structures resist the forces of nature and man.

The connection between the modern developments in thin-shell technology and corresponding significant scientific events in the post-industrial revolution period are documented in Chapter 1 of Fung and Sechler.[10] The remaining chapters of that modern collection of papers (which is devoted to the unification of theory, experiment, and design of thin-shell structures) is probably best appreciated after the fundamental material presented in this book has been mastered.

FIGURE 1–9   Hyperbolic Paraboloid, Mexico City, After 1985 Earthquake. (Courtesy M. Celebi)

REFERENCES

1. K. Griffin, The Art of Engineering; Leonardo da Vinci—A Master Mind of Engineering, *Structural Engineering Forum* (Logan, Utah: Engineering Publications, Inc., June-July 1994): 28–32.

2. I. Todhunter and K. Pearson, *A History of the Theory of Elasticity and of the Strength of Materials.* 2 vols. (Cambridge: Cambridge University Press, 1886 and 1893). Also A. E. H. Love, *A Treastise on the Mathematical Theory of Elasticity,* 4th ed. (New York: Dover Publications, 1994), Introduction; S. A. Timoshenko, *History of Strength of Materials* (New York: McGraw-Hill, 1953); and H. M. Westergaard, *Theory of Elasticity and Plasticity* (New York: Dover Publications, 1964), Chap. 2.

3. J. T. Oden, *Finite Elements of Nonlinear Continua* (New York: McGraw-Hill, 1972): 144–145.

4. P. A. Andres and N. F. Ortega, An extension of Gaudi's funicular technique to the conception and generation of structural surface, *Bulletin of the IASS, 35,* No. 3 (December 1994): 161–172.

5. C. T. Grimm, Brick masonry shells, *Journal of the Structural Division,* ASCE 101, no. ST1 (January 1975): 79–95; discussion by K. Anadol and E. Arioylu, no. ST11 (November 1975): 2451–2455.

6. J. Bobrowski, Design philosophy for long spans in building and bridges, *The Structural Engineer,* 64A, no. 1 (January 1986): 5–12.

7. W. B. Parsens, *Engineers and Engineering in the Renaissance* (Baltimore, MD: Williams and Wilkins, 1939): 587–607.

8. J. K. Wu, private communication.

9. M. Celebi, private communication.

10. Y. C. Fung and E. E. Sechler, eds. *Thin Shell Structures* (Englewood Cliffs, NJ: Prentice-Hall, 1971).

EXERCISES

The numerical problems in this book are given without specific units to allow English metric, or Sl unit dimensions to be selected.

1.1. It has been claimed that structural materials are generally more efficient in an extensional rather than a bending mode. To illustrate this, consider the two cases shown in Figure 1–10. Both pinned end structures span the same distance 2 *L*, carry the same load *P*, and contain the same amount of material. For a span of *L* = 40, compare the maximum fiber stress in each case.

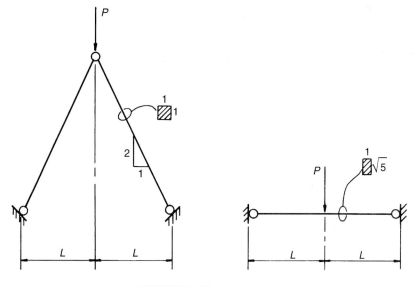

FIGURE 1–10   Exercise 1.1

2

Geometry

2.1 CURVILINEAR COORDINATES

Consider a portion of the middle surface of a shell as shown in Figure 2–1.

The surface is defined with respect to the X-Y-Z global Cartesian coordinates by

$$Z = f(X,Y) \tag{2.1}$$

Then, a set of coordinates α and β are selected which are related to the Cartesian system by

$$X = f_1(\alpha,\beta) \tag{a}$$

$$Y = f_2(\alpha,\beta) \tag{b} \quad (2.2)$$

$$Z = f_3(\alpha,\beta) \tag{c}$$

where f_1, f_2, and f_3 are continuous, single-valued functions. Because each ordered pair (α, β) corresponds to only one point on the surface, the surface is uniquely described in terms of α and β, which are called curvilinear coordinates. If one of the coordinates, for example, β, is incremented $\beta = \beta_1$, $\beta = \beta_2$, $\beta = \beta_n$, then we define a series of parametric curves on the surface, along which only α varies. These curves are termed the s_α coordinate lines Similarly, if α takes on the values $\alpha = \alpha_1$, $\alpha = \alpha_2$, $\alpha = \alpha_m$, we get the s_β coordinate lines. The coordinate lines are shown in Figure 2–1.

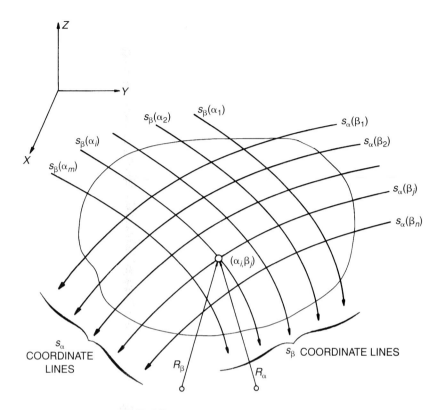

FIGURE 2–1   Middle Surface of a Shell.

If the coordinate lines s_α and s_β are mutually perpendicular at all points on the surface, the curvilinear coordinates are said to be orthogonal. Orthogonal curvilinear coordinates are used exclusively in this book.

2.2   MIDDLE-SURFACE GEOMETRY

Equation (2.1), which describes the middle surface, may be written in terms of a position vector emanating from the origin as shown in Figure 2–2:

$$\mathbf{r} = X\mathbf{u} + Y\mathbf{v} + Z\mathbf{w} \tag{2.3}$$

in which $\mathbf{u}$, $\mathbf{v}$, and $\mathbf{w}$ are unit vectors along the X, Y, and Z axes, respectively.

Substituting equation (2.2) into (2.3), the position vector is defined in terms of the curvilinear coordinates by

$$\mathbf{r}(\alpha,\beta) = f_1(\alpha,\beta)\mathbf{u} + f_2(\alpha,\beta)\mathbf{v} + f_3(\alpha,\beta)\mathbf{w} \tag{2.4}$$

The derivatives of $\mathbf{r}$ with respect to the curvilinear coordinates are considered next:

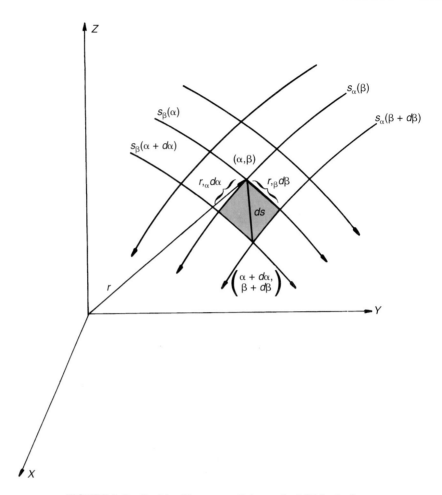

FIGURE 2–2 Position Vector to a Point on the Middle Surface.

$$\frac{\partial \mathbf{r}}{\partial \alpha} = \mathbf{r},_\alpha \tag{2.5a}$$

and

$$\frac{\partial \mathbf{r}}{\partial \beta} = \mathbf{r},_\beta \tag{2.5b}$$

are vectors that are tangent to the s_α and s_β coordinate lines, respectively. Because the coordinate lines are orthogonal, these tangent vectors are orthogonal as well, and their scalar product $\mathbf{r},_\alpha \cdot \mathbf{r},_\beta = 0$.

The vector joining the two points on the middle surface (α, β) and $(\alpha + d\alpha, \beta + d\beta)$ shown in Figure 2–2 is

$$\mathbf{ds} = \mathbf{r}_{,\alpha}d\alpha + \mathbf{r}_{,\beta}d\beta \tag{2.6}$$

If we form the scalar product of **ds** with itself, we have

$$\mathbf{ds} \cdot \mathbf{ds} = ds^2 = (\mathbf{r}_{,\alpha} \cdot \mathbf{r}_{,\alpha})d\alpha^2 + (\mathbf{r}_{,\beta} \cdot \mathbf{r}_{,\beta})d\beta^2 \tag{2.7}$$

Defining

$$A^2 = \mathbf{r}_{,\alpha} \cdot \mathbf{r}_{,\alpha} \tag{2.8a}$$

and

$$B^2 = \mathbf{r}_{,\beta} \cdot \mathbf{r}_{,\beta} \tag{2.8b}$$

then

$$ds^2 = A^2 d\alpha^2 + B^2 d\beta^2 \tag{2.9}$$

which is known as the *first quadratic form of the theory of surfaces.*

The quantities A and B are called *Lamé parameters or measure numbers* and are fundamental to the understanding of curvilinear coordinates. To interpret their meaning physically, consider two cases in which each of the coordinates α and β is varied individually and independently. For these cases, equation (2.9) becomes

$$ds_\alpha = A\,d\alpha \tag{2.10a}$$

and

$$ds_\beta = B\,d\beta \tag{2.10b}$$

Thus, ds_α is the change in arc length along coordinate line s_α when α is incremented by $d\alpha$, and ds_β is the change in arc length along coordinate lines s_β when β is incremented by $d\beta$. We see that the Lamé parameters are quantities which relate the *change in arc length* on the surface to the corresponding *change in the curvilinear coordinate;* hence the alternate name, *measure number.*

Example 2.2(a)

As a simple illustration, consider a circular arc of radius R in the Y-Z plane shown in Figure 2–3. If the curvilinear coordinate is chosen as the polar angle β,

$$ds = Rd\beta \tag{2.11}$$

and the Lamé parameter is R. Alternately, the curvilinear coordinate could be chosen as the Z coordinate, in which case we have (from Figure 2–3)

$$ds^2 = dY^2 + dZ^2 \tag{2.12}$$

From the equation of the circle $Y^2 + Z^2 = R^2$,

$$2Y\,dY = -2Z\,dZ$$

or

$$dY^2 = \frac{Z^2}{Y^2}\,dZ^2$$

so that equation (2.12) gives

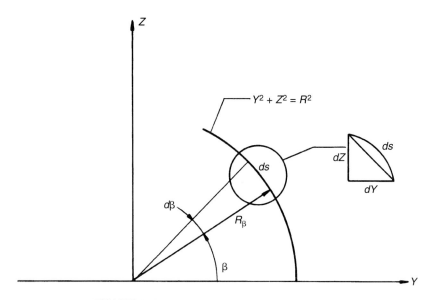

FIGURE 2–3   Lamé Parameter for a Circular Arc.

$$ds = \left(1 + \frac{Z^2}{Y^2}\right)^{1/2} dZ \qquad\qquad (2.13)$$

and the Lamé parameter is

$$\left(1 + \frac{Z^2}{Y^2}\right)^{1/2}$$

A third possibility is to choose the arc length itself as the curvilinear coordinate. Then

$$ds = 1\,(ds) \qquad\qquad (2.14)$$

and the Lamé parameter is the constant 1. It is apparent from this elementary example that the Lamé parameter may be a constant or a rather involved function. For a particular problem, the choice of curvilinear coordinates which correspond to the simplest possible expressions for the Lamé parameter can serve to expedite the mathematics of the solution greatly.

The preceding example illustrates that the Lamé parameters may sometimes be found by the geometric representation of equations (2.9) and (2.10). In more complicated cases, we may compute the Lamé parameters from equations (2.8) and (2.3) as

$$A^2 = (X_{,\alpha})^2 + (Y_{,\alpha})^2 + (Z_{,\alpha})^2 \qquad\qquad (2.15a)$$

$$B^2 = (X_{,\beta})^2 + (Y_{,\beta})^2 + (Z_{,\beta})^2 \qquad\qquad (2.15b)$$

Finally, with respect to the first quadratic form, note that it generally pertains to the *measurement of distances* on the surface, but does not involve the specific *shape* of the surface.

2.3   UNIT TANGENT VECTORS AND PRINCIPAL DIRECTIONS

2.3.1   Definition of Unit Tangent Vectors and Normal Section

It is convenient to refer all vector point functions to a triad of unit vectors composed of tangent vectors to the coordinate lines and a normal vector which define a right-hand system, as shown in Figure 2–4.

We have already defined tangent vectors to the coordinate lines in equation (2.5). Hence, in view of equation (2.8), the *unit* vectors are

$$\mathbf{t}_\alpha = \frac{\mathbf{r}_{,\alpha}}{\pm|\mathbf{r}_{,\alpha}|} = \frac{\mathbf{r}_{,\alpha}}{A} \tag{2.16a}$$

$$\mathbf{t}_\beta = \frac{\mathbf{r}_{,\beta}}{\pm|\mathbf{r}_{,\beta}|} = \frac{\mathbf{r}_{,\beta}}{B} \tag{2.16b}$$

and the normal vector is found by forming the vector product of $\mathbf{t}_\alpha$ and $\mathbf{t}_\beta$:

$$\mathbf{t}_n = \mathbf{t}_\alpha \times \mathbf{t}_\beta = \frac{1}{AB}(\mathbf{r}_{,\alpha} \times \mathbf{r}_{,\beta}) \tag{2.16c}$$

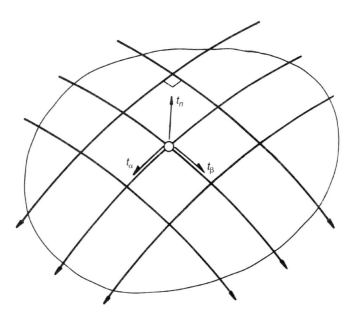

FIGURE 2–4   Unit Tangent Vectors.

A *normal section* of the surface is defined as a curve obtained by cutting the surface with a plane containing the normal to the surface, $\mathbf{t_n}$. In particular, normal sections containing the unit tangent vectors $\mathbf{t_\alpha}$ and $\mathbf{t_\beta}$ are of interest.

2.3.2 Principal Directions

Now consider the general point (α_i, β_j) on the middle surface as shown in Figure 2–1. At this point, each coordinate line s_β (α_i) and s_α (β_j), may be regarded as a normal section with a corresponding radius of curvature $R_\beta(\alpha_i, \beta_j)$ and R_α (α_i, β_j), respectively, which is directed from the center of curvature to the point (α_i, β_j) along $\mathbf{t_n}$. Obviously, there are an infinite number of possible R_α and R_β at any point, since an infinite number of orientations for the curvilinear coordinates exist. From the theory of surfaces, it may be shown that there is a system of orthogonal curvilinear coordinates (α^*, β^*) oriented such that one radius of curvature $R_\alpha^* = |\mathbf{R}_\alpha^*|$ is the maximum of all possible $|\mathbf{R}_\alpha|$, whereas the other radius of curvature $R_\beta^* = |\mathbf{R}_\beta^*|$ is the minimum of all possible $|\mathbf{R}_\beta|$.[1] We call this system the *principal orthogonal curvilinear coordinates* corresponding to the *principal directions,* α^* and β^*. The associated coordinate lines are known as the *lines of principal curvature,* and R_α^* and R_β^* as the *principal radii of curvature.* In the subsequent derivation of unit tangent vector derivatives, principal directions will be used exclusively so that α and β imply α^* and β^*.

2.3.3 Derivatives of Unit Tangent Vectors

To establish the relationships between the Lamé parameters and the principal radii of curvature for a surface, it is necessary to derive a set of relationships for the derivatives of the tangent vectors, $\mathbf{t_\alpha}$, $\mathbf{t_\beta}$, $\mathbf{t_n}$, with respect to a and β.[2] The derivatives of the unit tangent vectors are expressed in terms of the unit tangent vectors themselves. This is the major operational tool in the vector mechanics approach to shell theory which is developed in Chapter 3.

$$
\left\{\begin{array}{c} \mathbf{t_{\alpha,\alpha}} \\ \mathbf{t_{\alpha,\beta}} \\ \mathbf{t_{\beta,\alpha}} \\ \mathbf{t_{\beta,\beta}} \\ \mathbf{t_{n,\alpha}} \\ \mathbf{t_{n,\beta}} \end{array}\right\} =
\begin{bmatrix}
0 & \dfrac{-A,_\beta}{B} & \dfrac{-A}{R_\alpha} \\[2mm]
0 & \dfrac{B,_\alpha}{A} & 0 \\[2mm]
\dfrac{A,_\beta}{B} & 0 & 0 \\[2mm]
\dfrac{-B,_\alpha}{A} & 0 & \dfrac{-B}{R_\beta} \\[2mm]
\dfrac{A}{R_\alpha} & 0 & 0 \\[2mm]
0 & \dfrac{B}{R_\beta} & 0
\end{bmatrix}
\left\{\begin{array}{c} \mathbf{t_\alpha} \\ \mathbf{t_\beta} \\ \mathbf{t_n} \end{array}\right\} \tag{2.17}
$$

We now consider the verification of equation (2.17). It is convenient to write the equation once again with the elements grouped as shown:

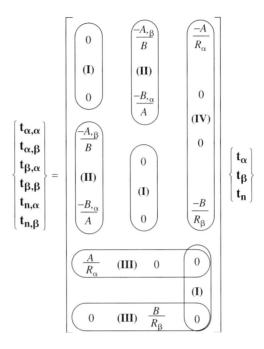

The grouping refers to the arguments in the following sections.

2.3.3.1 (I) Component in Direction of Differentiated Vector.

The derivative of any unit tangent vector is normal to the vector itself so that there are no components in the direction of the vector being differentiated; for example, there is no $\mathbf{t}_\alpha$ component for $\mathbf{t}_{\alpha,\alpha}$ or $\mathbf{t}_{\alpha,\beta}$.

This assertion is well known from elementary vector calculus. As a proof, consider the scalar product of two unit tangent vectors in the curvilinear coordinate system:

$$\mathbf{t_i} \cdot \mathbf{t_j} = \delta_{ij} \begin{cases} \delta_{ij} = 1, i = j; \delta_{ij} = 0, i \neq j \\ i, j = \alpha, \beta, n \end{cases}$$

Differentiating the product with respect to $k(k = \alpha$ or $\beta)$

$$(\mathbf{t_i} \cdot \mathbf{t_j})_{,k} = 0$$

from which

$$\mathbf{t_j} \cdot \mathbf{t}_{i,k} + \mathbf{t_i} \cdot \mathbf{t}_{j,k} = 0$$

Then set $j = i$ to get

$$2(\mathbf{t_i} \cdot \mathbf{t}_{i,k}) = 0$$

so that $\mathbf{t}_{i,k}$, has no component along $\mathbf{t_i}$. Taking $i = \alpha, \beta$ and n, and $k = \alpha$ and β, in turn, verifies the indicated zero terms in this group.

2.3.3.2 (II) Components of Derivatives of t_α and t_β in α and β Directions.

We multiply equations (2.16a) and (2.16b) by A and B to get $\mathbf{r}_{,\alpha}$ and $\mathbf{r}_{,\beta}$, respectively, and form the mixed second partial derivatives,

$$\mathbf{r}_{,\alpha\beta} = \mathbf{r}_{,\beta\alpha} \tag{2.18a}$$

With equations (2.16a,b) substituted into equation (2.18a), we obtain

$$(A\mathbf{t}_\alpha)_{,\beta} = \mathbf{t}_\alpha A_{,\beta} + A\mathbf{t}_{\alpha,\beta} = (B\mathbf{t}_\beta)_{,\alpha} = \mathbf{t}_\beta B_{,\alpha} + B\mathbf{t}_{\beta,\alpha} \tag{2.18b}$$

from which

$$\mathbf{t}_{\beta,\alpha} = \frac{1}{B}[-\mathbf{t}_\beta B_{,\alpha} + \mathbf{t}_\alpha A_{,\beta} + A\mathbf{t}_{\alpha,\beta}] \tag{2.18c}$$

Now consider the component of $\mathbf{t}_{\alpha,\alpha}$ in the $\mathbf{t}_\beta$ direction, given by $\mathbf{t}_\beta \cdot \mathbf{t}_{\alpha,\alpha}$. We take the derivative of the product $\mathbf{t}_\alpha \cdot \mathbf{t}_\beta$ with respect to α

$$(\mathbf{t}_\alpha \cdot \mathbf{t}_\beta)_{,\alpha} = \mathbf{t}_\beta \cdot \mathbf{t}_{\alpha,\alpha} + \mathbf{t}_\alpha \cdot \mathbf{t}_{\beta,\alpha} \tag{2.18d}$$

Since $\mathbf{t}_\alpha \cdot \mathbf{t}_\beta = 0$,

$$\mathbf{t}_\beta \cdot \mathbf{t}_{\alpha,\alpha} = -\mathbf{t}_\alpha \cdot \mathbf{t}_{\beta,\alpha} \tag{2.18e}$$

We replace $\mathbf{t}_{\beta,\alpha}$ by equation (2.18c) to get

$$\mathbf{t}_\beta \cdot \mathbf{t}_{\alpha,\alpha} = -\frac{1}{B}\mathbf{t}_\alpha \cdot [-\mathbf{t}_\beta B_{,\alpha} + \mathbf{t}_\alpha A_{,\beta} + A\mathbf{t}_{\alpha,\beta}] \tag{2.18f}$$

Since $\mathbf{t}_\alpha \cdot \mathbf{t}_\beta = \mathbf{t}_\alpha \cdot \mathbf{t}_{\alpha,\beta} = 0$, and $\mathbf{t}_\alpha \cdot \mathbf{t}_\alpha = 1$,

$$\mathbf{t}_\beta \cdot \mathbf{t}_{\alpha,\alpha} = \frac{-A_{,\beta}}{B} \tag{2.18g}$$

as given in the first row of equation (2.17). The other components of the derivatives of $\mathbf{t}_\alpha$ and $\mathbf{t}_\beta$ in the α and β directions may be verified in a similar manner.

2.3.3.3 (III) Derivatives of t_n.

Consider the normal section at the point p_1 on the s_α coordinate line, as shown in Figure 2–5. The vector $\mathbf{t}_n$ is shown at the point p_1 and also at point p_2, a small distance Δs_α away. The vector construction at p_1 shows that the change in $\mathbf{t}_n$, $\Delta\mathbf{t}_n$, is approximately parallel to the tangent to the curve at p_1 and the chord. Therefore,

$$\Delta\mathbf{t}_n = |\Delta\mathbf{t}_n|\mathbf{t}_\alpha \tag{2.19a}$$

By similar triangles, as $\Delta\alpha$ diminishes,

$$\frac{|\Delta\mathbf{t}_n|}{|\mathbf{t}_{n1}|} = \frac{\overline{p_1 p_2}}{R_\alpha} \tag{2.19b}$$

Considering Figure 2–5 and equation (2.10a),

$$\overline{p_1 p_2} \simeq \Delta s_\alpha = A\Delta\alpha \tag{2.19c}$$

Recognizing $|\mathbf{t}_{n1}| = 1$, and substituting equation (2.19a) and (2.19c) into (2.19b), we have

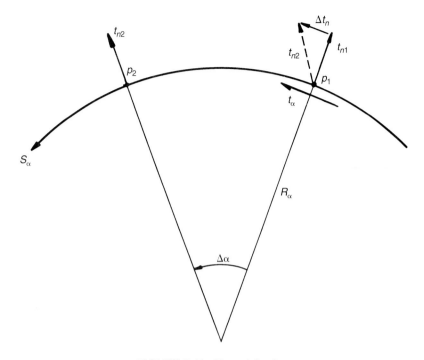

FIGURE 2–5   Normal Section.

$$\frac{\Delta \mathbf{t_n}}{\Delta \alpha} = \frac{A}{R_\alpha} \mathbf{t_\alpha} \tag{2.19d}$$

Taking the limit of both sides as $\Delta \alpha \to 0$,

$$\mathbf{t_{n,\alpha}} = \frac{A}{R_\alpha} \mathbf{t_\alpha} \tag{2.19e}$$

as given in the fifth row of equation (2.17). $\mathbf{t_{n,\beta}}$ is evaluated in a similar manner. Note that the argument employed in (III) is more general than that used in (II), because the entire derivative is computed instead of just one component.

2.3.3.4 (IV) Components of Derivatives of $\mathbf{t_\alpha}$ and $\mathbf{t_\beta}$ in Normal Direction.
Consider the normal component of $\mathbf{t_{\alpha,\alpha}}$ given by $\mathbf{t_n} \cdot \mathbf{t_{\alpha,\alpha}}$. Proceeding as before,

$$(\mathbf{t_n} \cdot \mathbf{t_\alpha})_{,\alpha} = \mathbf{t_\alpha} \cdot \mathbf{t_{n,\alpha}} + \mathbf{t_n} \cdot \mathbf{t_{\alpha,\alpha}} \tag{2.20a}$$

Since $\mathbf{t_n} \cdot \mathbf{t_\alpha} = 0$,

$$\mathbf{t_n} \cdot \mathbf{t_{\alpha,\alpha}} = -\mathbf{t_\alpha} \cdot \mathbf{t_{n,\alpha}} \tag{2.20b}$$

But, we have already evaluated $\mathbf{t_{n,\alpha}}$ in equation (2.19e); hence

$$\mathbf{t_n} \cdot \mathbf{t_{\alpha,\alpha}} = \frac{-A}{R_\alpha} \tag{2.20c}$$

as given in the first row, third column, of equation (2.17). The remaining normal components of the derivatives of $\mathbf{t}_\alpha$ and $\mathbf{t}_\beta$ may be verified in a similar manner.

2.4 SECOND QUADRATIC FORM OF THE THEORY OF SURFACES

Recall that in Section 2.2, we derived the first quadratic form of the theory of surfaces, which pertains to the measurement of distances on the surface but not specifically to the shape of the surface. In this section, we seek information with respect to the latter property, the shape.

Consider a normal section which traces a plane curve with arc length coordinate s_i and unit tangent and normal vectors $\mathbf{t_n}$ and $\mathbf{t_s}$, respectively. An example is the normal section along s_α shown in Figure 2–5. However, s_i is not necessarily restricted to only principal direction coordinate lines in the ensuing development. The curvature of such a section is known as the normal curvature κ_i. It is defined as a function of the unit tangent vector $\mathbf{t_s}$ and subsequently, as in equation (2.16), of the position vector $\mathbf{r}$ shown in Figure 2–2. The curvature is related to $\mathbf{r}$ by the Frenet-Serret formula[3] as

$$\kappa_i \mathbf{t_n} = -\frac{1}{R_i} \mathbf{t_n} = \mathbf{r}_{,s_i,s_i} \tag{2.21}$$

where the negative sign corresponds to the selection of $\mathbf{t_n}$ as the *outward* normal.

We want to express the normal curvature in terms of the curvilinear coordinates α and β. Starting with

$$\mathbf{r}_{,s_i s_i} = (\mathbf{r}_{,s_i})_{,s_i} = (\mathbf{r}_{,\alpha}\alpha_{,s_i} + \mathbf{r}_{,\beta}\beta_{,s_i})_{,s_i}$$

we first evaluate

$$(\mathbf{r}_{,\alpha}\alpha_{,s_i})_{,s_i} = \alpha_{,s_i}\mathbf{r}_{,\alpha s_i} + \mathbf{r}_{,\alpha}\alpha_{,s_i s_i}$$
$$= \alpha_{,s_i}(\mathbf{r}_{,\alpha\alpha}\alpha_{,s_i} + \mathbf{r}_{,\alpha\beta}\beta_{,s_i}) + \mathbf{r}_{,\alpha}\alpha_{,s_i s_i}$$

and

$$(\mathbf{r}_{,\beta}\beta_{,s_i})_{,s_i} = \beta_{,s_i}\mathbf{r}_{,\beta s_i} + \mathbf{r}_{,\beta}\beta_{,s_i s_i}$$
$$= \beta_{,s_i}(\mathbf{r}_{,\beta\alpha}\alpha_{,si} + \mathbf{r}_{,\beta\beta}\beta_{,s_i}) + \mathbf{r}_{,\beta}\beta_{,s_i s_i}$$

from which

$$\mathbf{r}_{,s_i s_i} = \mathbf{r}_{,\alpha\alpha}(\alpha_{,s_i})^2 + 2\mathbf{r}_{,\alpha\beta}\alpha_{,s_i}\beta_{,s_i} + \mathbf{r}_{,\beta\beta}(\beta_{,s_i})^2 + \mathbf{r}_{,\alpha}\alpha_{,s_i s_i} + \mathbf{r}_{,\beta}\beta_{,s_i s_i} \tag{2.22}$$

Next, we scalar multiply each side of equation (2.22) by $\mathbf{t_n}$. The last two terms on the right-hand side (rhs) vanish, since $\mathbf{t_n}$ is normal to $\mathbf{r}_{,\alpha}$ and $\mathbf{r}_{,\beta}$ by equation (2.16c). In order to simplify the first three terms of equation (2.22), we first consider

$$\mathbf{t_n} \cdot \mathbf{r}_{,\alpha} = 0 \tag{2.23a}$$
$$\mathbf{t_n} \cdot \mathbf{r}_{,\beta} = 0 \tag{2.23b}$$

and differentiate both equations with respect to α and then β, which gives

$$\mathbf{t_n} \cdot \mathbf{r,}_{\alpha\alpha} = -\mathbf{r,}_\alpha \cdot \mathbf{t}_{n,\alpha} \tag{2.23c}$$

$$\mathbf{t_n} \cdot \mathbf{r,}_{\alpha\beta} = -\mathbf{r,}_\alpha \cdot \mathbf{t}_{n,\beta} \tag{2.23d}$$

$$\mathbf{t_n} \cdot \mathbf{r,}_{\beta\alpha} = -\mathbf{r,}_\beta \cdot \mathbf{t}_{n,\alpha} \tag{2.23e}$$

$$\mathbf{t_n} \cdot \mathbf{r,}_{\beta\beta} = -\mathbf{r,}_\beta \cdot \mathbf{t}_{n,\beta} \tag{2.23f}$$

We now continue the scalar product of $\mathbf{t_n}$ with the first three terms of equation (2.22). In view of equations (2.21) and (2.23c–f), we have

$$\begin{aligned}
\kappa_i &= -\mathbf{r,}_\alpha \cdot \mathbf{t}_{n,\alpha}(\alpha,_{s_i})^2 - 2\mathbf{r,}_\alpha \cdot \mathbf{t}_{n,\beta}\alpha,_{s_i}\beta,_{s_i} - \mathbf{r,}_\beta \cdot \mathbf{t}_{n,\beta}(\beta,_{s_i})^2 \\
&= -A\mathbf{t}_\alpha \cdot \mathbf{t}_{n,\alpha}(\alpha,_{s_i})^2 - 2A\mathbf{t}_\alpha \cdot \mathbf{t}_{n,\beta}\alpha,_{s_i}\beta,_{s_i} - B\mathbf{t}_\beta \cdot \mathbf{t}_{n,\beta}(\beta,_{s_i})^2
\end{aligned} \tag{2.24}$$

We may evaluate the scalar products in equation (2.24) from equation (2.17). Therefore,

$$\kappa_i = \frac{-A^2}{R_\alpha}(\alpha,_{s_i})^2 + 0(\alpha,_{s_i}\beta,_{s_i}) - \frac{B^2}{R_\beta}(\beta,_{s_i})^2 \tag{2.25}$$

The second term on the rhs of equation (2.25), which vanishes in this case, would remain if nonorthogonal coordinates were used.

It is convenient to multiply equation (2.25) by

$$\frac{ds_i^2}{ds_i^2}$$

whereupon

$$\kappa_i = \frac{L(\alpha,_{s_i})^2 ds_i^2 + N(\beta,_{s_i})^2 ds_i^2}{ds_i^2} \tag{2.26a}$$

where

$$L = \frac{-A^2}{R_\alpha} \text{ and } N = \frac{-B^2}{R_\beta} \tag{2.26b}$$

Using equation (2.9), we may consolidate equation (2.26a) to get

$$\kappa_i = \frac{L\, d\alpha^2 + N\, d\beta^2}{A^2\, d\alpha^2 + B^2\, d\beta^2} \tag{2.27}$$

In equation (2.27), $d\alpha$ and $d\beta$ represent $(\partial\alpha/\partial s_i)ds_i$ and $(\partial\beta/\partial s_i)ds_i$, respectively, for a particular direction i on the surface.

Finally, we write equation (2.27) as

$$\kappa_i = \frac{\mathrm{II}}{\mathrm{I}} \tag{2.28a}$$

where

$$\mathrm{II} = L\, d\alpha^2 + N\, d\beta^2 \tag{2.28b}$$

and

$$I = A^2 \, d\alpha^2 + B^2 \, d\beta^2$$
$$= \text{First quadratic form (equation 2.9).} \tag{2.28c}$$

The second quadratic form, II, thus relates to the *shape* of the curve through the presence of radii of curvature R_α and R_β.

2.5 PRINCIPAL RADII OF CURVATURE

Thus far, we have assumed that the principal radii of curvature of the shell, R_α and R_β, are known or easily found. We now examine the details of this calculation.

Consider the case where the normal section corresponding to α is a curve $Z = f(X)$ in the X-Z plane.

$$R_\alpha = \frac{-[1 + (Z_{,X})^2]^{3/2}}{Z_{,XX}} \tag{2.29}$$

When the surface is specified parametrically in terms of the curvilinear coordinates as in equation (2.4), we may compute the radius of curvature using equation (2.21). If we take the s_i direction as one of the coordinate lines, for example, s_α, then

$$-\frac{1}{R_\alpha} = \kappa_\alpha = \mathbf{t_n} \cdot \mathbf{r}_{,s_\alpha s_\alpha} = \mathbf{t_n} \cdot [\mathbf{r}_{,\alpha\alpha}(\alpha_{,s_\alpha})^2 + \mathbf{r}_{,\alpha}\alpha_{,s_\alpha s_\alpha}] \tag{2.30}$$

Equation (2.30) was evaluated from equation (2.22), recognizing that $\beta_{,s_\alpha} = 0$. Since $\mathbf{t_n} \cdot \mathbf{r}_{,\alpha} = 0$, equation (2.30) reduces to

$$R_\alpha = \frac{-1}{\mathbf{t_n} \cdot \mathbf{r}_{,\alpha\alpha}(\alpha_{,s_\alpha})^2}$$

Noting that $\alpha_{,s_\alpha} = 1/A$ from equation (2.10a), and $\mathbf{t_n} = (1/AB)\,(\mathbf{r}_{,\alpha} \times \mathbf{r}_{,\beta})$ from equation (2.16c),

$$R_\alpha = \frac{-A^3 B}{(\mathbf{r}_{,\alpha} \times \mathbf{r}_{,\beta}) \cdot \mathbf{r}_{,\alpha\alpha}} \tag{2.31}$$

Similarly,

$$R_\beta = \frac{-B^3 A}{(\mathbf{r}_{,\alpha} \times \mathbf{r}_{,\beta}) \cdot \mathbf{r}_{,\beta\beta}} \tag{2.32}$$

Equations (2.31) and (2.32) may be evaluated from equation (2.4), (2.15), and (2.2) for a given geometry.

Another technique for computing the radii of curvature that is useful for shells of revolution is illustrated in Section 4.3.2.3.

A comprehensive example illustrating the evaluation of the terms defined in Sections 2.2 to 2.5 for a specific shell geometry is presented in Section 2.8.2.3.

2.6 GAUSS-CODAZZI RELATIONS

No connective relationships between the Lamé parameters, A and B, and the principal radii of curvature, R_α and R_β, have been set forth. To explore this further, consider the equality of the second mixed partials

$$t_{n,\alpha\beta} = t_{n,\beta\alpha} \tag{2.33}$$

or, from equation (2.17),

$$\left(\frac{A}{R_\alpha} t_\alpha\right)_{,\beta} = \left(\frac{B}{R_\beta} t_\beta\right)_{,\alpha} \tag{2.34a}$$

$$\left(\frac{A}{R_\alpha}\right)_{,\beta} t_\alpha + \frac{A}{R_\alpha} t_{\alpha,\beta} = \left(\frac{B}{R_\beta}\right)_{,\alpha} t_\beta + \frac{B}{R_\beta} t_{\beta,\alpha} \tag{2.34b}$$

Again, using the differential relationships in equation (2.17), we find

$$\left[\left(\frac{A}{R_\alpha}\right)_{,\beta} - \frac{A_{,\beta}}{R_\beta}\right] t_\alpha = \left[\left(\frac{B}{R_\beta}\right)_{,\alpha} - \frac{B_{,\alpha}}{R_\alpha}\right] t_\beta \tag{2.34c}$$

With t_α and t_β being mutually orthogonal, equation (2.34c) may only be satisfied if

$$\left(\frac{A}{R_\alpha}\right)_{,\beta} = \frac{A_{,\beta}}{R_\beta} \tag{2.35}$$

and

$$\left(\frac{B}{R_\beta}\right)_{,\alpha} = \frac{B_{,\alpha}}{R_\alpha} \tag{2.36}$$

If we now consider one of the other second mixed partial derivative identities

$$t_{\alpha,\alpha\beta} = t_{\alpha,\beta\alpha}$$

or

$$t_{\beta,\alpha\beta} = t_{\beta,\beta\alpha}$$

and manipulate these equations using equation (2.17), a third differential relationship

$$\left(\frac{B_{,\alpha}}{A}\right)_{,\alpha} + \left(\frac{A_{,\beta}}{B}\right)_{,\beta} = -\frac{AB}{R_\alpha R_\beta} \tag{2.37}$$

results. Equations (2.35), (2.36), and (2.37) are known as the Gauss-Codazzi relations and define the connectivity among A, B, R_α, and R_β, such that these parameters define a surface. Equation (2.37) is particularly useful in the derivation of the equations of equilibrium and is further discussed later.

Although we will not pursue the details now, it may be noted that a parallel set of equations may be derived for the deformed middle surface by starting with normal vector to the deformed surface in place of t_n in equation (2.33).[3] The resulting equations, properly termed *Gauss-Codazzi relations* for the *deformed* middle surface are the conditions for the *continuity* of the middle surface displacements and the *compatibility* between the strains and displacements. They serve the same role as the St. Venant equations in the theory of elasticity. This topic will be elaborated in Section 7.6.

2.7 GAUSSIAN CURVATURE

On the right side of equation (2.37), note the fraction $1/R_\alpha R_\beta$, which is the product of the principal curvatures. This is known as the *Gaussian curvature* and plays an important role in the characterization of shells. Although the Gaussian curvature may be readily computed by using the equations of Section 2.5, such a calculation is seldom required for purposes of classification; it is often sufficient to know only the algebraic sign.

    If we consider the normal sections corresponding to the principal directions, the Gaussian curvature is positive if both centers of curvature lie on the *same* side of the surface and is negative if the centers lie on *opposite* sides. If one of the radii of curvatures is equal to infinity, the Gaussian curvature is zero. Technically, the Gaussian curvature is a scalar point function, and a particular shell may have regions with positive, negative, and/or zero values. Nevertheless, a single sign predominates for most practical cases.

    Representative cases are shown in Figure 2–6. A plate is the degenerate case of a shell with zero Gaussian curvature, because both radii are infinite.

    As suggested by Calladine,[4] it is informative to recall the original concept of Gauss in order to derive a *geometrical* form of the Gaussian curvature. Considering first an arc of a plane curve such as Figure 2–5

$$\Delta s_\alpha = R_\alpha \, \Delta \alpha \qquad\qquad (2.38a)$$

or

$$\kappa_\alpha = \frac{1}{R_\alpha} = \frac{\Delta \alpha}{\Delta s_\alpha} \qquad\qquad (2.38b)$$

from equation (2.21) ignoring the negative sign.

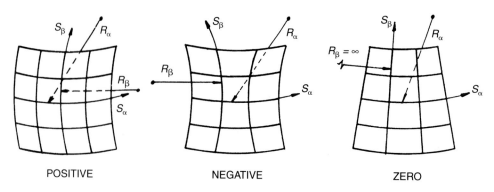

        POSITIVE   NEGATIVE   ZERO

FIGURE 2–6 Gaussian Curvature.

That is, the curvature, in other words, the reciprocal of the radius of curvature, is equal to the ratio of the angle subtended by an arc to the length of the arc. Then extending this notion to a surface, a measure of curvature is

$$\kappa = \frac{\Delta A}{\Delta S} \tag{2.39}$$

where ΔA is the solid angle subtended by the small area ΔS. Referring to Figures 2–1 and 2–2, we consider an element at (α,β) which has an area

$$\Delta S = R_\alpha \, d\alpha \, R_\beta \, d\beta \tag{2.40a}$$

subtending a solid angle

$$\Delta A = \Delta\alpha \, \Delta\beta \tag{2.40b}$$

so that

$$\kappa = \frac{\Delta\alpha\Delta\beta}{(R_\alpha \Delta\alpha R_\beta \Delta\beta)} = \frac{1}{R_\alpha R_\beta} \tag{2.41}$$

A classification of shells by Gaussian curvature is given in Figure 2–7. With respect to this classification, note that only the zero curvature shells are developable and hence may be

Classification \\ Gaussian Curvature	Positive	Negative	Zero
Surface	Doubly curved, synclastic	Doubly curved, anticlastic	Singly curved
Developability	Nondevelopable	Nondevelopable	Developable
Type of equation (discriminant)	Elliptic (positive)	Hyperbolic (negative)	Parabolic (zero)
Straight or ruled lines on surface	None	Two sets (A and B below)	One set (C below)
Examples			

Sphere

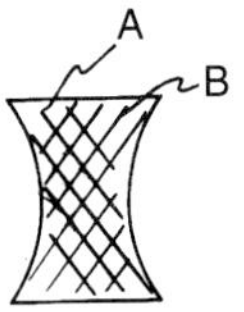

Paraboloid of Revolution

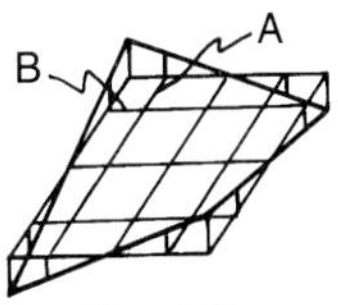

Hyperboloid of Revolution

Hyperbolic Paraboloid

Cylinder

Cone

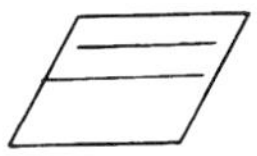

Flat Plate (Degenerate Case)

FIGURE 2–7   Classification of Shells by Gaussian Curvature.

formed from flat material. This property of zero curvature shells contributes to their wide usage. Also, note that the negative curvature shells have two sets of straight or ruled lines on the surface which correspond to the two sets of real characteristics associated with hyperbolic surfaces,[5] so that the formwork for such a shell can be fabricated from straight materials. This feature of negative curvature shells is largely responsible for the economical usage of reinforced concrete hyperbolic paraboloid (HP) shells in a variety of applications.[6]

The mathematical classification refers to the type of partial differential equation associated with quadratic surfaces having the indicated Gaussian curvature. If a shell is, or can be approximated logically as, a quadratic surface $Z = f(X, Y)$, then the algebraic sign of the *discriminant*

$$\delta = Z_{,XX} Z_{,YY} - (Z_{,XY})^2 \qquad (2.42)$$

determines the type of partial differential equation and, further, coincides with the sign of the Gaussian curvature,[5] as indicated in Figure 2–7.

A variety of thin shell and plate structures are shown in Figure 2–8 and will be referred to frequently throughout the book. Several of the shells shown in Figures 2-8 and many more important examples are shown and discussed on the following web sites:[7]

http://home.inet-access.com/mketchum   http://ketchum.org/shells.html

FIGURE 2–8a Open Hyperbolic Paraboloid Roof, Ponce Coliseum, Puerto Rico. (Courtesy Professor A.C. Scordelis)

(text continues on page 49)

FIGURE 2–8b   Hyperbolic Paraboloid Umbrellas, Memphis, TN.

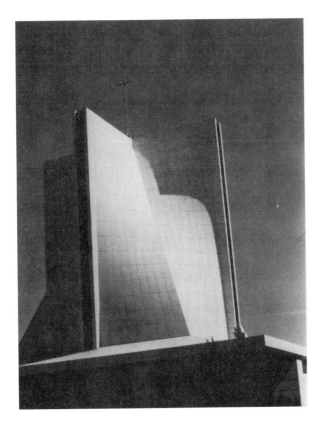

FIGURE 2–8c   Vertical Hyperbolic Paraboloids, Church, San Francisco, CA.

FIGURE 2–8d   Steel Water Storage Tank, Columbus, Ohio.

FIGURE 2–8e   Cylindrical Barrel Shell Supported on Beams, Union, Missouri.

FIGURE 2–8f   Parabolic Vaults,
Church, St. Louis County, MO.

FIGURE 2–8g   Hyperboloid of Revolution, Planetarium, St. Louis, MO.

FIGURE 2–8h Kingdome, Seattle, WA. (Courtesy Dudley, Hardin & Yang, Inc.)

FIGURE 2–8i Spherical Roof, Auditorium, Cambridge, MA.

FIGURE 2–8j   Intersecting Barrel Shells, Airport, St. Louis, MO.

FIGURE 2–8k   Parabolic Barrel Shells.

FIGURE 2–8l   Convention Center, Hong Kong.

FIGURE 2–8m   Reticulated Spherical Roof, Astrodome, Houston, TX.

FIGURE 2–8n   Saddledome, Calgary, Alberta, Canada. (Courtesy Dr. Jan Bobrowski, F.Eng.)

FIGURE 2–8o Cooling Tower, Callaway County, MO.

FIGURE 2–8p   Supporting Columns for Hyperbolic Cooling Tower.

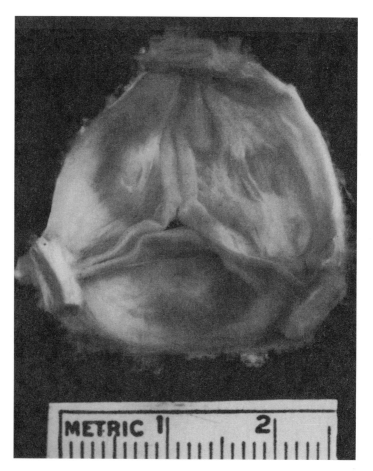

FIGURE 2–8q Human Aortic Heart Valve. Source: P.L. Gould et al., Stress analysis of the human aortic valve, *Journal of Computers nad Structures* 3 (1973): 379. (Courtesy Dr. R. Clark reprinted with permission of Pegamon Press.)

FIGURE 2–8r   Stiffened Cylindrical Shell. (Courtesy Chicago Bridge & Iron Co.)

FIGURE 2–8s   Ellipsoidal Water Tower. (Courtesy Chicago Bridge & Iron Co.)

FIGURE 2–8t Column-Supported Spherical Tanks. (Courtesy Chicago Bridge & Iron Co.)

FIGURE 2–8u Column-Supported, Stiffened Water Tower. (Courtesy Chicago Bridge & Iron Co.)

FIGURE 2–8v   Steel Hyperbolic Paraboloid Roof, Aircraft Hangar. (Courtesy Lev Zetlin Associates, Inc.)

FIGURE 2–8w   Waffle Slab, Library, St. Louis, MO.

FIGURE 2–8x   Folded Plate Roof, Law School, St. Louis, MO.

FIGURE 2–8y   Torospherical Head.

FIGURE 2–8z   Parliament Building, Solomon Islands. (Courtesy Baldridge & Associates Structural Engineering, Inc.)

2.8 SPECIALIZATION OF SHELL GEOMETRY

Because of the wide variety of plate and shell structures encountered in engineering practice, several geometrical classes are of particular interest.

2.8.1 Shallow Shell

The theory of shallow shells has wide application for roof shells that have a relatively small rise as compared to their spans. Considering the surface specified in Cartesian coordinates as in equation (2.1), the shell is said to be shallow if, in the subsequent mathematical analysis, $(Z_{,X})^2$ and $(Z_{,Y})^2$ may be neglected by virtue of smallness in comparison to unity. The effect of this simplification is seen by considering Figure 2–9.

Consider a differential element of the middle surface bound by the intersections with two planes parallel to the Y-Z plane and separated by a distance dX, and two planes parallel to the X-Z plane separated by a distance dY. As illustrated by the inset in Figure 2–9,

$$ds_X \simeq dX[1 + (Z_{,X})^2]^{1/2} \tag{2.43a}$$

$$ds_Y \simeq dY[1 + (Z_{,Y})^2]^{1/2} \tag{2.43b}$$

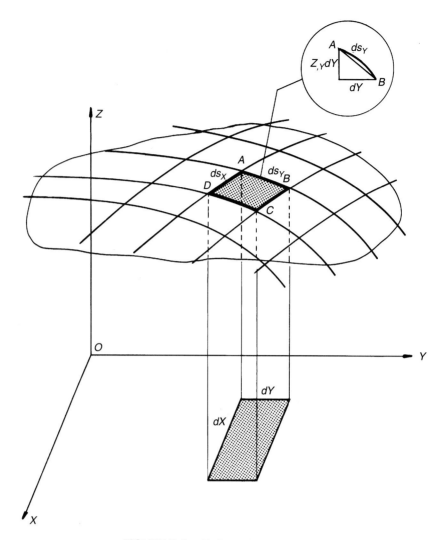

FIGURE 2–9   Shallow Shell Geometry.

which, when the geometric simplification is applicable, become

$$ds_X \simeq dX \tag{2.44a}$$

$$ds_Y \simeq dY \tag{2.44b}$$

The practical interpretation of equations (2.44) is that the curvilinear coordinates α and β may be selected as the Cartesian coordinates X and Y with the Lamé parameters $A = B = 1$. Additional approximations are introduced into shallow shell theory with respect to the equilibrium equations in Chapter 13.

The practical range of the theory of shallow shells is restricted to shells with a central rise of not more than one-fifth of the span. However, this theory has been applied to shells that may not fit the criteria in a global sense, by treating pieces or finite elements of the shell which can be considered shallow, and then assembling these elements to approximate the global geometry.[8]

2.8.2 Shells of Revolution

2.8.2.1 Definitions.
A surface of revolution is described by rotating a plane curve *generator* around an *axis of rotation* to form a closed surface, as illustrated in Figure 2–10. The lines of principal curvature are called the *meridians* (normal sections formed by planes containing the axis of rotation) and *parallel circles* (normal sections traced by planes perpendicular to the axis of rotation).

2.8.2.2 Curvilinear Coordinates.
In Figure 2–11, we show the meridian of a shell of revolution illustrating both positive and negative Gaussian curvature. The equation of the meridian is given by

$$Z = Z(R) \qquad\qquad (2.45a)$$

where

$$R = \sqrt{(X^2 + Y^2)} \qquad\qquad (2.45b)$$

Consider a reference point on the surface. The angle formed by the extended normal to the surface at this point and the axis of rotation is defined as the *meridional angle* ϕ; and the angle between the radius of the parallel circle at the point and the X axis is designated as the *circumferential angle* θ as shown on Figure 2–11. Correspondingly, the meridians are

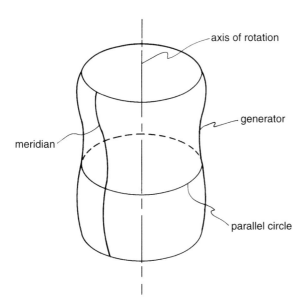

FIGURE 2–10   Shell of Revolution.

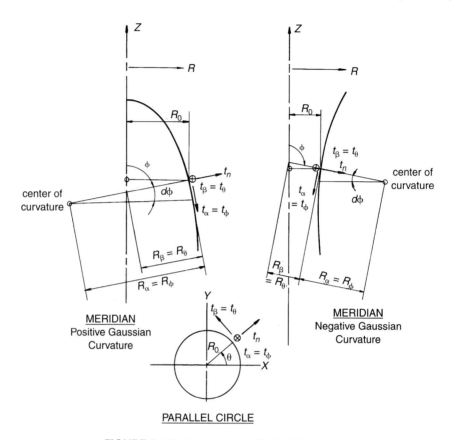

FIGURE 2–11   Geometry for Shells of Revolution.

taken as the s_α coordinate lines, and $R_\alpha = R_\phi$, the meridional radius of curvature. The parallel circles are the s_β coordinate lines, with $R_\beta = R_\theta$, the circumferential radius of curvature. The radius of the parallel circle, which is equal to R as defined in equation (2.45b), is termed the *horizontal radius* and is denoted by R_0. Note that R_0 is *not* a principal radius of curvature, because it is not normal to the surface. Rather, it is the projection of R_θ on the horizontal plane, i.e.,

$$R_0 = R_\theta \sin \phi \tag{2.46}$$

A closed shell of revolution is frequently called a *dome,* and the peak of such a shell is termed the *pole.* A pole introduces certain mathematical complications because, at this point, $R_0 \to 0$.

In most applications, the curvilinear coordinate in the β or θ direction is chosen as the circumferential angle θ. Therefore, from equation (2.10b),

$$ds_\beta = B \, d\beta = ds_\theta = R_0 d\theta \tag{2.47a}$$

and thus

$$B = R_0 = R_\theta \sin \phi \tag{2.47b}$$

In the α or ϕ direction, there are at least three useful choices for the curvilinear coordinate: (a) the meridional angle ϕ; (b) the axial coordinate Z; and (c) the arc length s_ϕ. The respective Lamé parameter A for each case is:

(a) Meridional angle, ϕ

$$ds_\phi = R_\phi \, d\phi \tag{2.48a}$$

and from equation (2.10)

$$A = R_\phi \tag{2.48b}$$

(b) Axial coordinate, Z:

Considering an element of arc length, similar to that shown on the inset of Figure 2–3, but with $dY = dR_0$

$$\begin{aligned}
ds_\phi^2 = ds_Z^2 &= dZ^2 + dR_0^2 \\
&= dZ^2 + (R_{0,Z})^2 dZ^2
\end{aligned} \tag{2.49a}$$

$$ds_Z = dZ[1 + (R_{0,Z})^2]^{1/2} \tag{2.49b}$$

and therefore

$$A = [1 + (R_{0,Z})^2]^{1/2} \tag{2.49c}$$

(c) Arc length s_ϕ:

$$A = 1 \tag{2.49d}$$

but because the meridian is defined by $Z = Z(R_0)$, the arc length coordinate must be computed by integrating equation (2.49b). Therefore,

$$s_Z = \int_{Z_0}^{Z} [1 + (R_{0,Z})^2]^{1/2} \, dZ \tag{2.50}$$

The limits of the integral indicate the coordinate Z at the origin for s_Z (perhaps the base or top of the shell), and at the section where the coordinate is being evaluated, respectively.

The choice of an appropriate curvilinear coordinate for the meridional direction is problem dependent and involves several considerations. In tower-type shells, which are essentially vertical structures, the axial coordinate Z has the greatest significance with respect to the physical construction of the shell. For relatively flat domes however, the axial coordinate approaches zero even for points relatively far from the pole. The meridional angle ϕ may behave in a similar manner, so the arc length will be the most stable coordinate for such cases. Also, if the meridian has an inflection point, the coordinate ϕ might cause difficulties because it may not provide a one-to-one correspondence with all points on the shell surface.

The most popular, but by no means universal, preference of theoreticians in the field of shells of revolution has been to choose the meridional angle ϕ as the basis for the development of the governing equations. This may be due somewhat to historical precedent, since the early work on shells of revolution focused on spherical shells[9] for which $A = R_\phi = $ constant, thereby greatly simplifying the ensuing treatment.

We noted in Section 2.6 that the parameters A, B, R_α, and R_β must satisfy the three Gauss-Codazzi conditions to define a surface. These conditions, as given by equations (2.35), (2.36), and (2.37), can be checked for the ϕ-θ curvilinear coordinate system:

Equation (2.35): Satisfied identically, since none of the parameters are functions of β or θ in this case.

Equation (2.36):

$$\left(\frac{R_\theta \sin \phi}{R_\theta}\right)_{,\phi} = \frac{(R_\theta \sin \phi)_{,\phi}}{R_\phi}$$

$$\cos \phi = \frac{(R_\theta \sin \phi)_{,\phi}}{R_\phi}$$

or

$$(R_\theta \sin \phi)_{,\phi} = R_{0,\phi} = R_\phi \cos \phi \tag{2.51}$$

Equation (2.37):

$$\left[\frac{(R_\theta \sin \phi)_{,\phi}}{R_\phi}\right]_{,\phi} = -\frac{R_\phi R_\theta \sin \phi}{R_\phi R_\theta} \tag{2.52}$$

Substituting equation (2.51) for the numerator on the left-hand side (lhs) of equation (2.52) gives an identity. Thus, the Gauss-Codazzi conditions for a shell of revolution described by the coordinates ϕ and θ are satisfied, provided equation (2.51) is valid. This equation is very useful in what follows, and it is instructive to derive it from a purely geometric argument. Consider the meridian shown in Figure 2–12:

Referring to points B and D

$$\text{At } B, \quad R_0(\phi) = AB$$
$$\text{At } D, \quad R_0(\phi + \Delta\phi) = CD$$
$$\Delta R_0 = CD - AB$$
$$= BD \sin\left(\frac{\pi}{2} - \phi\right)$$
$$= BD \cos \phi$$

Because
$$BD = R_\phi \Delta\phi$$
$$\Delta R_0 = R_\phi \cos \phi \Delta\phi$$
$$\Delta R_0 = \Delta(R_\theta \sin \phi) = R_\phi \cos \phi \Delta\phi$$

Or,

$$\frac{\Delta(R_\theta \sin \phi)}{\Delta\phi} = R_\phi \cos \phi$$

$\lim \Delta\phi \to 0$ gives equation (2.51).

One may desire to calculate the value of Z corresponding to a particular value of ϕ. With $R_\theta = R_\theta(\phi)$, $R_0(\phi)$ is found from equation (2.46) and substituted into equation (2.45a) to

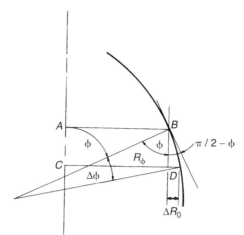

FIGURE 2–12 Geometrical Interpretation of Gauss-Codazzi Condition.

obtain the corresponding value of $Z(\phi)$. If the arc length is also required, equation (2.50) should be evaluated. Also, a direct transformation between s_ϕ and ϕ can be derived by integrating equation (2.48a).

It is expedient to be able to transform derivatives with respect to ϕ to derivatives with respect to Z, in other words, $(\ )_{,Z} = (\ )_{,\phi}\,(d\phi/dZ)$. If we consider equation (2.49c) with $R_{0,Z}$ given by $R_{0,\phi}\,(d\phi/dZ)$ and $R_{0,\phi}$ given by equation (2.51), we have

$$A = \left[1 + R_\phi^2 \cos^2\phi \left(\frac{d\phi}{dZ}\right)^2\right]^{1/2} \tag{2.53}$$

for the axial coordinate Z.

Next, we write ds_ϕ for both the ϕ and Z coordinates

$$ds_\phi = R_\phi d\phi = \left[1 + R_\phi^2 \cos^2\phi \left(\frac{d\phi}{dZ}\right)^2\right]^{1/2} dZ = ds_Z$$

or

$$R_\phi\left(\frac{d\phi}{dZ}\right) = \left[1 + R_\phi^2 \cos^2\phi \left(\frac{d\phi}{dZ}\right)^2\right]^{1/2}$$

Squaring both sides and solving for $(d\phi/dZ)$

$$\frac{d\phi}{dZ} = \pm\frac{1}{R_\phi(1 - \cos^2\phi)^{1/2}}$$

$$= \pm\frac{1}{R_\phi \sin\phi} \tag{2.54}$$

Substituting equation (2.54) into (2.53) gives

$$A = \pm\csc\phi \tag{2.55}$$

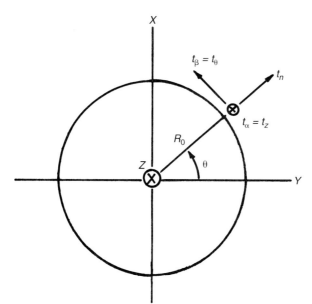

FIGURE 2–13   Alternate Cylindrical
Coordinates.

for the Z coordinate. Equation (2.54) is quite useful, because it is often preferable to differentiate with respect to ϕ first in order to utilize equation (2.51).

The $\pm$ signs in equations (2.54) and (2.55) indicate that an increment $d\phi$ may produce a change in arc length *equal* in magnitude but *opposite* in sense to that produced by dZ. For example, in Figure 2–11 where Z is directed *upward,* this would be the case. If it is desired to use Z as the primary meridional coordinate, it may be preferable to direct Z as positive *downward.*[10]

An alternate set of coordinates for which Z increases in direct proportion to ϕ is shown in Figure 2–13. The right-hand convention and the outward directed normal, in other words, $\mathbf{t_z} \times \mathbf{t_\theta} = \mathbf{t_n}$, are preserved, but θ is now measured from the Y axis.

2.8.2.3 Parametric Equations.
It is often convenient to express the position vector $\mathbf{r}$ in the form of equation (2.4). This specialization of equation (2.4) for a shell of revolution is called the parametric form by Kraus[11] and may be written as

$$\mathbf{r} = R_0 \cos \theta \mathbf{u} + R_0 \sin \theta \mathbf{v} + Z\mathbf{w} \qquad (2.56)$$

where $R_0 = R_0(Z)$ or $R_0(\phi)$ follows the notation of equation (2.4) and Figure 2–11.

Example 2.8

A toroidal shell has been considered by Raouf and Palazotto[12] as shown in Figure 2–14. Their equation for a point within the body of the torus is given by

$$\mathbf{r}(\theta_1, \theta_2, \zeta) = \cos \theta_1 [r_1 + (r_2 + \zeta)\cos \theta_2]\mathbf{u} + \sin \theta_1$$
$$[r_1 + (r_2 + \zeta) \cos \theta_2]\mathbf{v} + (r_2 + \zeta) \sin \theta_2\mathbf{w} \qquad (2.57)$$

using the notation of equation (2.3) and (2.4). In equation (2.57), the curvilinear coordinates of the middle surface are $\alpha = \theta_1$, $\beta = \theta_2$. In the notation of Figure 2–11, $\theta_1 = \theta$ and $\theta_2 = {}^\pi/_2 - \phi$. ζ is

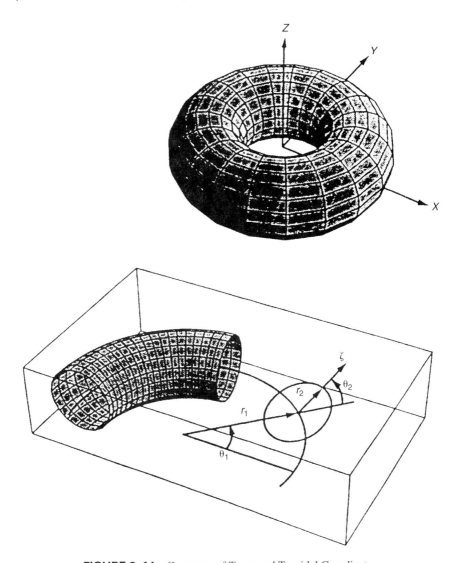

FIGURE 2–14   Geometry of Torus and Toroidal Coordinates.

a coordinate measured along the normal to the middle surface. For the present, we consider only the middle surface where $\zeta = 0$. Then equation (2.57) becomes

$$\mathbf{r}(\theta_1, \theta_2) = (r_1 + r_2 \cos \theta_2)\cos\theta_1 \ \mathbf{u}$$
$$+ (r_1 + r_2 \cos \theta_2)\sin \theta_1 \ \mathbf{v} \qquad (2.58)$$
$$+ (r_2 \sin \theta_2)\mathbf{w}$$

which is in the form of equation (2.56). Therefore,

$$R_0(\theta_2) = r_1 + r_2 \cos \theta_2 \qquad (2.59a)$$

$$Z(\theta_2) = r_2 \sin \theta_2 \qquad (2.59b)$$

and

$$\mathbf{r}(\theta_1,\theta_2) = R_0(\theta_2) \cos \theta_1 \mathbf{u} + R_0(\theta_2) \sin \theta_1 \, \mathbf{v} + Z(\theta_2)\mathbf{w} \tag{2.59c}$$

We may compute the Lamé parameters, A and B, as defined by equations (2.8a, b). Consider equations (2.59) and first evaluate

$$R_{0,\theta_2} = -r_2 \sin \theta_2 \tag{2.60a}$$
$$Z_{,\theta_2} = r_2 \cos \theta_2 \tag{2.60b}$$

so that

$$\mathbf{r}_{,\theta_1} = -(r_1 + r_2\cos\theta_2)\sin\theta_1 \, \mathbf{u} + (r_1 + r_2\cos\theta_2)\cos\theta_1 \, \mathbf{v} \tag{2.61a}$$

$$\mathbf{r}_{,\theta_2} = -r_2\sin\theta_2\cos\theta_1 \, \mathbf{u} - r_2\sin\theta_2\sin\theta_1\mathbf{v} + r_2\cos\theta_2\mathbf{w} \tag{2.61b}$$

Then, following equation (2.8a, b)

$$A^2 = \mathbf{r}_{,\theta_1} \cdot \mathbf{r}_{,\theta_1} = (r_1 + r_2\cos\theta_2)^2 = R_0^2 \tag{2.62a}$$

$$B^2 = \mathbf{r}_{,\theta_2} \cdot \mathbf{r}_{,\theta_2} = r_2^2 \tag{2.62b}$$

which checks with the results of reference 12 for the middle surface. The computation of the Lamé parameters for points off the middle surface but within the body of the shell is discussed in Section 7.3.2 where this example is continued.

Finally, we evaluate the radii of curvature. Corresponding to θ_1 and θ_2, we have R_1 and R_2, the circumferential and meridional radii, respectively. Since R_0 is given by equation (2.59a), and $\phi = \pi/2 - \theta_2$, equation (2.46) with $\theta = \theta_1$ and $\phi = \pi/2 - \theta_2$ give

$$R_1 = \frac{R_0}{\sin \phi} = \frac{R_0}{\cos \theta_2} = \frac{r_1}{\cos \theta_2} + r_2 \tag{2.63a}$$

Recalling the Gauss-Codazzi equation (2.37) specialized for shells of revolution as equation (2.51), we have from equation (2.59a)

$$R_0(\phi) = r_1 + r_2 \sin \phi$$

$$R_{0,\phi} = r_2 \cos \phi \tag{2.63b}$$

which checks since r_1 is a constant.

Obviously, the meridional radius

$$R_2 = r_2 \tag{2.63c}$$

The geometry of toroidal shells is discussed further in Section 4.3.2.4 where the curvilinear coordinates are taken as θ and ϕ, corresponding to θ_1 and θ_2 on Figure 2–14.

REFERENCES

1. H. Kraus, *Thin Elastic Shells* (New York: Wiley, 1967): 12–14.

2. V. V. Novozhilov, *Thin Shell Theory* [translated from 2nd Russian ed. by P. G. Lowe] (Groningen, The Netherlands: Noordhoff, 1964): 9–30.

3. A. A. Gol'denveizer. *Theory of Elastic Thin Shells* [translated from Russian by G. Herrmann] (New York: Pergamon Press, 1961): 13.

4. C. R. Calladine, *Gaussian Curvature and Shell Structures, The Mathematics of Surfaces* [J. A. Gregory Ed.] (Oxford: Claredon Press, 1986): 179–196.

5. A. M. Haas, *Design of Thin Concrete Shells,* vol. 2 (New York: Wiley, 1967): 5–11.

6. C. Faber, *Candela: The Shell Builder* (New York: Reinhold, 1963).

7. Ketchum, Milo, Private communication.

8. G. R. Cowper, G. M. Lindberg, M. D. Olson, A shallow shell finite element of triangular shape, *International Journal of Solids and Structures 6,* no. 8 (1970): 1133–1156.

9. H. Reissner, *Spannungen in Kugleschalen* (Leipzig: Muller-Breslau-Festschrift, 1912): 181–193.

10. V. Z. Vlasov, General Theory of Shells and Its Application In Engineering, NASA Technical Translation TTF-99 (Washington, DC: National Aeronautics and Space).

11. H. Kraus, *Thin Elastic Shells* (New York: Wiley, 1967): 8–9.

12. R. R. Raouf and A. N. Palazotto, Nonlinear dynamics of unidirectional, fiber-reinforced tori, *Journal of Engineering Mechanics, ASLE, 122,* no. 3 (March, 1996): 271–278.

EXERCISES

2.1. Verify the second, third, fourth, and sixth equations of equation (2.17).

2.2. Derive equation (2.37).

2.3. Verify that the Gauss-Codazzi relations are satisfied for a shell of revolution with A given by equation (2.49c).

2.4. The parametric equation for a shell of revolution was given by equation (2.56) as

$$\mathbf{r} = R_0 \cos \theta \mathbf{u} + R_0 \sin \theta \mathbf{v} + Z \mathbf{w}$$

where $R_0 = R_0 (Z)$ follows the notation of equation (2.4) and Figure 2–11.
   (a) Choosing Z and θ as the curvilinear coordinates, compute expressions for the first and second quadratic forms. Note that if the alternate Cartesian axes shown in Figure 2–13 are used, the parametric equation must be modified.
   (b) Repeat (a) using the meridional angle θ instead of Z.

2.5. Compute the principal ratio of curvature, and the first and second quadratic forms for the curvilinear coordinates specified in Exercise 2.4, considering the following geometries:
   (a) Right circular cylinder.
   (b) Ellipsoid of revolution. $\dfrac{R^2}{a^2} + \dfrac{Z^2}{b^2} = 1$; *a,b = lengths of major and minor axes*
   (c) Hyperboloid of one sheet. $\dfrac{R^2}{a^2} - \dfrac{Z^2}{b^2} = 1$; *a,b defined on Figure 4–6*

2.6. Investigate the relative merits of using $\mathbf{r}_{,\alpha}$ and $\mathbf{r}_{,\beta}$, nonunit tangent vectors, as the base vectors instead of $\mathbf{t}_{,\alpha}$ and $\mathbf{t}_{,\beta}$. See Gol'denveizer, *Theory of Elastic Thin Shells,*[3] for an example of such a formulation.

3
Equilibrium

3.1 STRESS RESULTANTS AND COUPLES

Consider the element of the shell shown in Figure 3–1(a), bounded by the normal sections α, $\alpha + d\alpha$, β, and $\beta + d\beta$. These correspond to the s_β and s_α coordinate lines, respectively, as defined in Figure 2–1. The geometry of the middle surface of such an element was considered in Chapter 2 (see Figures 2–2 and 2–4). Here, we show the entire thickness h with the coordinate ζ defined in the direction of $\mathbf{t_n}$ and depict a differential volume element $d\alpha\, d\beta\, d\zeta$ with thickness $d\zeta$, parallel to and displaced from the middle surface a distance ζ.

In Figure 3–1(b), the stresses acting on the volume element are indicated. The sign convention is that of the theory of elasticity, with the first subscript indicating the *surface* on which the stress acts, as identified by the normal to that surface, and the second subscript indicating the *direction* in which the stress acts. The solution for the various *stresses* shown in Figure 3–1(b) at a particular *point* of the continuum (α, β, ζ) is the fundamental problem of the theory of elasticity. Plate and shell theories deal with a simplification of the *stress at a point* problem. Instead of the stresses, it is considered sufficient to solve for the *total force and moment intensities* per length of middle surface, which are known as *stress resultants* and *stress couples,* respectively.[1] The stress at any point presumably can then be evaluated by back-substitution, although we will find that this can sometimes be done only approximately.

We have thus alluded to the major simplification present in the plate and shell problem as compared to the theory of elasticity problem: The plate or shell problem is formulated and

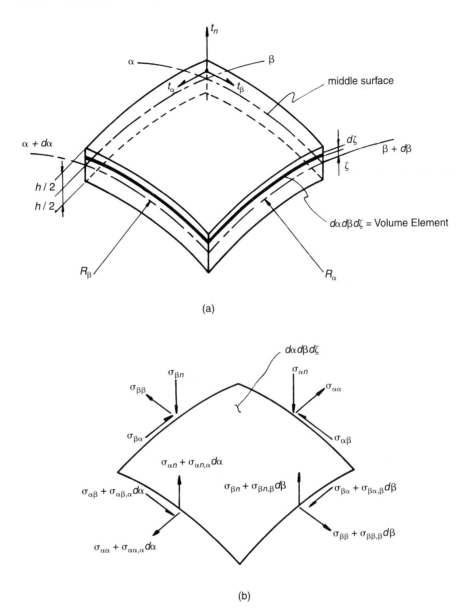

(a)

(b)

FIGURE 3–1   Shell Volume Element and Stresses.

solved for the forces and moments per unit length of the middle surface, rather than for the stresses at each point of the continuum. Toward this end, we now consider the definition of the stress resultants and couples in terms of the stresses.

In Figure 3–2, the stress resultants and couples are shown on an element of the middle surface. The extensional forces $\mathbf{N}_\alpha$ and $\mathbf{N}_\beta$, transverse shear forces $\mathbf{Q}_\alpha$ and $\mathbf{Q}_\beta$, and bending

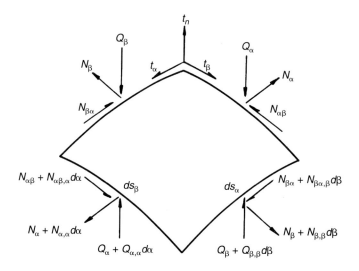

(a) Stress Resultants

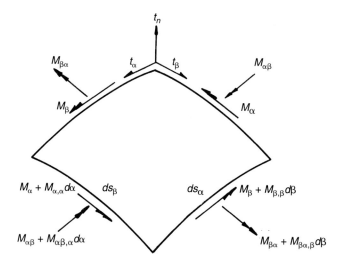

(b) Stress Couples

FIGURE 3–2   Stress Resultants and Stress Couples.

moments $\mathbf{M}_\alpha$ and $\mathbf{M}_\beta$ are represented by single subscripted variables indicating the surface on which the force or moment acts, whereas the in-plane shear forces $\mathbf{N}_{\alpha\beta}$ and $\mathbf{N}_{\beta\alpha}$, and the twisting moments $\mathbf{M}_{\alpha\beta}$ and $\mathbf{M}_{\beta\alpha}$, carry a double subscript. The first subscript corresponds to the surface *on which* the force or moment acts; the second indicates the direction in which the force or moment acts. Though this notation is not completely consistent with the theory of elasticity definitions shown in Figure 3–1(b), it has been used almost exclusively in classical texts and references, and a departure here is felt to be inappropriate.

We now focus on an arc of the middle surface

$$ds_\alpha = A\, d\alpha \tag{3.1}$$

along the s_α coordinate line and the corresponding arc of the volume element lying a distance ζ along the normal $\mathbf{t_n}$ from the middle surface as shown in Figure 3–3, which represents a view of Figure 3–1(a) normal to $\mathbf{t_\beta}$. The length of the volume element is

$$
\begin{aligned}
ds_\alpha(\zeta) &= \frac{R_\alpha + \zeta}{R_\alpha}\, ds_\alpha \\
&= ds_\alpha\left(1 + \frac{\zeta}{R_\alpha}\right)
\end{aligned}
\tag{3.2}
$$

and the projected area is

$$
\begin{aligned}
da_\beta(\zeta) &= ds_\alpha(\zeta)d\zeta \\
&= ds_\alpha\left(1 + \frac{\zeta}{R_\alpha}\right)d\zeta
\end{aligned}
\tag{3.3}
$$

A similar section normal to $\mathbf{t_\alpha}$ gives

$$ds_\beta(\zeta) = ds_\beta\left(1 + \frac{\zeta}{R_\beta}\right) \tag{3.4}$$

$$da_\alpha(\zeta) = ds_\beta\left(1 + \frac{\zeta}{R_\beta}\right)d\zeta \tag{3.5}$$

where $ds_\beta = B\, d\beta$.

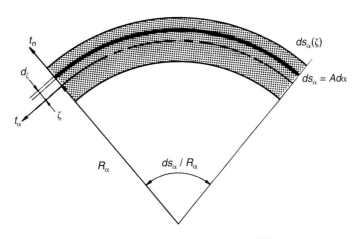

FIGURE 3–3   Normal Section on a Shell Element.

The magnitudes of the stress resultants acting on the section normal to $\mathbf{t}_\alpha$ are

$$\begin{Bmatrix} N_\alpha \\ N_{\alpha\beta} \\ Q_\alpha \end{Bmatrix} = \frac{1}{ds_\beta} \int_{-h/2}^{h/2} \begin{Bmatrix} \sigma_{\alpha\alpha} \\ \sigma_{\alpha\beta} \\ \sigma_{\alpha n} \end{Bmatrix} da_\alpha(\zeta)$$

$$= \int_{-h/2}^{h/2} \begin{Bmatrix} \sigma_{\alpha\alpha} \\ \sigma_{\alpha\beta} \\ \sigma_{\alpha n} \end{Bmatrix} \left(1 + \frac{\zeta}{R_\beta}\right) d\zeta \tag{3.6}$$

On the section normal to $\mathbf{t}_\beta$, we have

$$\begin{Bmatrix} N_\beta \\ N_{\beta\alpha} \\ Q_\beta \end{Bmatrix} = \frac{1}{ds_\alpha} \int_{-h/2}^{h/2} \begin{Bmatrix} \sigma_{\beta\beta} \\ \sigma_{\beta\alpha} \\ \sigma_{\beta n} \end{Bmatrix} da_\beta(\zeta) = \int_{-h/2}^{h/2} \begin{Bmatrix} \sigma_{\beta\beta} \\ \sigma_{\beta\alpha} \\ \sigma_{\beta n} \end{Bmatrix} \left(1 + \frac{\zeta}{R_\alpha}\right) d\zeta \tag{3.7}$$

The magnitudes of the stress couples are

$$\begin{Bmatrix} M_\alpha \\ M_{\alpha\beta} \end{Bmatrix} = \frac{1}{ds_\beta} \int_{-h/2}^{h/2} \begin{Bmatrix} \sigma_{\alpha\alpha} \\ \sigma_{\alpha\beta} \end{Bmatrix} \zeta\, da_\alpha(\zeta)$$

$$= \int_{-h/2}^{h/2} \begin{Bmatrix} \sigma_{\alpha\alpha} \\ \sigma_{\alpha\beta} \end{Bmatrix} \zeta \left(1 + \frac{\zeta}{R_\beta}\right) d\zeta \tag{3.8}$$

and

$$\begin{Bmatrix} M_\beta \\ M_{\beta\alpha} \end{Bmatrix} = \frac{1}{ds_\alpha} \int_{-h/2}^{h/2} \begin{Bmatrix} \sigma_{\beta\beta} \\ \sigma_{\beta\alpha} \end{Bmatrix} \zeta\, da_\beta(\zeta)$$

$$= \int_{-h/2}^{h/2} \begin{Bmatrix} \sigma_{\beta\beta} \\ \sigma_{\beta\alpha} \end{Bmatrix} \zeta \left(1 + \frac{\zeta}{R_\alpha}\right) d\zeta \tag{3.9}$$

At this point, we reiterate that once the plate or shell problem has been solved for the stress resultants and couples, the recovery of the elasticity problem would necessitate the inversion of equations (3.6) to (3.9), which might prove difficult in a mathematically exact sense. In a way, this is the price paid to realize a simplified theory—the sacrifice of mathematical precision to achieve a solution that is satisfactory from an engineering standpoint.

In practice, the stresses at a particular point in the continuum are usually recovered from the stress resultants and couples by (a) neglecting the terms $(1 + \zeta/R_\alpha)$ and $(1 + \zeta/R_\beta)$ in equations (3.6) to (3.9); and, (b) assuming that all stresses, except for the transverse shear stresses $\sigma_{\alpha n}$ and $\sigma_{\beta n}$, vary linearly across the cross-section. The preceding arguments imply that

$$\sigma_{ii}(\zeta) = \sigma_{ii}(0) + \left(\sigma_{ii}\left(\frac{h}{2}\right) - \sigma_{ii}(0)\right)\frac{\zeta}{h/2} \quad (i = \alpha, \beta)$$

$$\sigma_{ij}(\zeta) = \sigma_{ij}(0) + \left(\sigma_{ij}\left(\frac{h}{2}\right) - \sigma_{ij}(0)\right)\frac{\zeta}{h/2} \quad \begin{pmatrix} i = \alpha, \beta \\ j = \beta, \alpha \end{pmatrix} \tag{3.10a}$$

It is consistent with elementary theory to assume that the transverse shear stresses are distributed through the thickness as a second-degree parabola with a maximum at the middle surface. Thus,

$$\sigma_{in}(\zeta) \simeq \sigma_{in}(0) \left[1 - \left(\frac{\zeta}{h/2} \right)^2 \right] \quad (i = \alpha, \beta) \tag{3.10b}$$

When these relationships are introduced into equations (3.6) to (3.9), the maximum stresses on the cross-section are

$$\sigma_{ii}\left(\pm \frac{h}{2} \right) = \frac{N_i}{h} \pm \frac{6M_i}{h^2} \quad (i = \alpha, \beta) \tag{3.10c}$$

$$\sigma_{ij}\left(\pm \frac{h}{2} \right) = \frac{N_{ij}}{h} \pm \frac{6M_{ij}}{h^2} \quad \left(\begin{matrix} i = \alpha, \beta \\ j = \beta, \alpha \end{matrix} \right) \tag{3.10d}$$

$$\sigma_{in}(0) = \frac{3}{2} \frac{Q_i}{h} \quad (i = \alpha, \beta) \tag{3.10c}$$

Then, the stresses at any other level within the cross-section are found from equations (3.10a) and (3.10b).

3.2   EQUILIBRIUM OF THE SHELL ELEMENT

Now that all of the stresses acting on the shell are referred to quantities defined on the middle surface, we have reduced the problem to two dimensions and seek a set of relationships between the stress resultants and couples that reflects the equilibrium of the middle surface. In this derivation, we basically follow the approach of Novozhilov,[2] which is based on vector algebra. More mathematical approaches, based principally on tensor calculus,[3] as well as more physical approaches, relying mainly on the free-body diagram and trigonometry,[4,5] are available; in selecting the vector derivation, we hope to achieve an accommodation between mathematical elegance and sometimes cumbersome physical reasoning. The physical interpretation of the resulting equations will be explored later for selected geometrical forms.

The stress resultants and couples on each face of the middle surface element, shown in Figure 3–2, are combined into stress and stress couple vectors in Figure 3–4. Also shown is the load vector **q.** In terms of the stress resultants and couples, the resulting vectors are

$$\mathbf{F}_\alpha = (N_\alpha \mathbf{t}_\alpha + N_{\alpha\beta} \mathbf{t}_\beta + Q_\alpha \mathbf{t_n}) ds_\beta \tag{3.11a}$$

$$\mathbf{F}_\beta = (N_{\beta\alpha} \mathbf{t}_\alpha + N_\beta \mathbf{t}_\beta + Q_\beta \mathbf{t_n}) ds_\alpha \tag{3.11b}$$

$$\mathbf{C}_\alpha = (-M_{\alpha\beta} \mathbf{t}_\alpha + M_\alpha \mathbf{t}_\beta) ds_\beta \tag{3.11c}$$

$$\mathbf{C}_\beta = (-M_\beta \mathbf{t}_\alpha + M_{\beta\alpha} \mathbf{t}_\beta) ds_\alpha \tag{3.11d}$$

and the load vector is

$$\mathbf{q}\, ds_\alpha ds_\beta = [q_\alpha(\alpha,\beta) \mathbf{t}_\alpha + q_\beta(\alpha,\beta) \mathbf{t}_\beta + q_n(\alpha,\beta) \mathbf{t_n}] ds_\alpha ds_\beta \tag{3.11e}$$

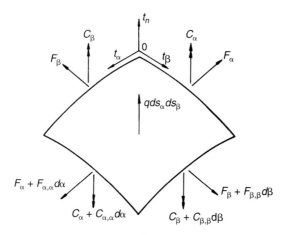

FIGURE 3–4 Stress and Stress-Couple Vectors.

where q_α, q_β, and q_n are the respective load intensities per unit area of the middle surface. Note with respect to equations (3.11c) and (3.11d) that the negative signs on the first terms arise because the stress couple vectors were chosen to correspond to the positive sense of the stresses (see Figures 3–1 and 3–2).

We now apply the equations of static equilibrium

$$\Sigma \mathbf{F} = \mathbf{0} \tag{3.12a}$$

$$\Sigma \mathbf{C} = \mathbf{0} \tag{3.12b}$$

Referring to Figure 3–4, the force equilibrium equation, (3.12a), becomes

$$(\mathbf{F}_\alpha + \mathbf{F}_{\alpha,\alpha}d\alpha - \mathbf{F}_\alpha) + (\mathbf{F}_\beta + \mathbf{F}_{\beta,\beta}d\beta - \mathbf{F}_\beta) + \mathbf{q}ds_\alpha ds_\beta = \mathbf{0}$$

or

$$\mathbf{F}_{\alpha,\alpha}d\alpha + \mathbf{F}_{\beta,\beta}d\beta + \mathbf{q}AB\, d\alpha\, d\beta = \mathbf{0} \tag{3.13}$$

Substituting equations (3.11a), (3.11b), and (3.11e) into equation (3.13), noting that $ds_\alpha = A d\alpha$ and $ds_\beta = B d\beta$, and dividing through by $d\alpha\, d\beta$, we have

$$[(N_\alpha \mathbf{t}_\alpha + N_{\alpha\beta} \mathbf{t}_\beta + Q_\alpha \mathbf{t_n})B]_{,\alpha} + [(N_{\beta\alpha}\mathbf{t}_\alpha + N_\beta \mathbf{t}_\beta + Q_\beta \mathbf{t_n})A]_{,\beta}$$
$$+ (q_\alpha \mathbf{t}_\alpha + q_\beta \mathbf{t}_\beta + q_n \mathbf{t_n})AB = \mathbf{0} \tag{3.14}$$

The differentiations indicated in equation (3.14) are carried out in a straightforward manner using equation (2.17) to evaluate the derivatives of the unit tangent vectors in terms of the vectors themselves. The resulting vector equations may be factored into the form

$$F_\alpha \mathbf{t}_\alpha + F_\beta \mathbf{t}_\beta + F_n \mathbf{t_n} = \mathbf{0} \tag{3.15}$$

Since the unit tangent vectors are independent, equation (3.15) can only be satisfied if

$$F_\alpha = 0; \quad F_\beta = 0; \quad F_n = 0 \tag{3.16}$$

which gives the three scalar equations of *force* equilibrium

$$F_\alpha = (BN_\alpha)_{,\alpha} + (AN_{\beta\alpha})_{,\beta} + A_{,\beta}N_{\alpha\beta} - B_{,\alpha}N_\beta + Q_\alpha \frac{AB}{R_\alpha} + q_\alpha AB = 0 \qquad (3.17a)$$

$$F_\beta = (BN_{\alpha\beta})_{,\alpha} + (AN_\beta)_{,\beta} + B_{,\alpha}N_{\beta\alpha} - A_{,\beta}N_\alpha + Q_\beta \frac{AB}{R_\beta} + q_\beta AB = 0 \qquad (3.17b)$$

$$F_n = (BQ_\alpha)_{,\alpha} + (AQ_\beta)_{,\beta} - N_\alpha \frac{AB}{R_\alpha} - N_\beta \frac{AB}{R_\beta} + q_n AB = 0 \qquad (3.17c)$$

The individual terms in equations (3.17a–c) will be interpreted in Section 3.3 for the shell of revolution geometry. However, note that there is considerable coupling of the equations; for example, N_α appears in F_β and F_n as well as in F_α as expected. This coupling is due to two general sources: (a) the curvature of the surface; and, (b) the possibility that the Lamé parameters A and B vary with both α and β.

If, for example, $A = A(\alpha)$ and $B = B(\beta)$ *only*, several of the coupling terms drop out. This emphasizes the importance of choosing the curvilinear coordinates so as to achieve the simplest mathematical formulation consistent with the problem. The moment equilibrium equation, (3.12b), is evaluated about axes through point O in Figure 3–4.

1. First, we have the contributions of the stress couple vectors

$$\mathbf{C}_\alpha + \mathbf{C}_{\alpha,\alpha}d\alpha - \mathbf{C}_\alpha + \mathbf{C}_\beta + \mathbf{C}_{\beta,\beta}d\beta - \mathbf{C}_\beta \qquad (3.18a)$$

or from equations (3.11c) and (3.11d),

$$[(-M_{\alpha\beta}\mathbf{t}_\alpha + M_\alpha\mathbf{t}_\beta)B]_{,\alpha}d\alpha\, d\beta + [(-M_\beta\mathbf{t}_\alpha + M_{\beta\alpha}\mathbf{t}_\beta)A]_{,\beta}d\alpha\, d\beta \qquad (3.18b)$$

2. Next, we investigate the contributions of the stress resultants. Observe in equation (3.18b) that the stress couple terms are of second differential order, $d\alpha\, d\beta$, so that terms of third or higher differential order will drop out in the limit. From Figure 3–2, we see that the extensional stress resultants N_α and N_β contribute moments about $\mathbf{t_n}$ of

$$-N_{\alpha,\alpha}d\alpha\, ds_\beta \frac{ds_\beta}{2}\, \mathbf{t_n} \rightarrow \text{third order}$$

and

$$N_{\beta,\beta}d\beta\, ds_\alpha \frac{ds_\alpha}{2}\, \mathbf{t_n} \rightarrow \text{third order}$$

and the in-plane shear stress resultants $N_{\alpha\beta}$ and $N_{\beta\alpha}$ give

$$N_{\alpha\beta}ds_\beta\, ds_\alpha\mathbf{t_n} \rightarrow \text{second order}$$

and

$$-N_{\beta\alpha}\, ds_\alpha ds_\beta\mathbf{t_n} \rightarrow \text{second order}$$

The transverse shear stress resultant Q_α contributes

$$-Q_\alpha ds_\beta ds_\alpha\mathbf{t_\beta} \rightarrow \text{second order}$$

about t_β whereas Q_β gives

$$Q_\beta ds_\alpha ds_\beta t_\alpha \rightarrow \text{second order}$$

about t_α. Collecting the second-order contributions of the stress vector to moment equilibrium, we have

$$[Q_\beta t_\alpha - Q_\alpha t_\beta + (N_{\alpha\beta} - N_{\beta\alpha})t_n]ds_\alpha ds_\beta \qquad (3.19)$$

3. Then, we note from Figure 3–4 that the contribution of q is of third differential order and therefore negligible.

4. Finally, we expand equation (3.18b) in accordance with equation (2.17), combine the resulting terms with equation (3.19), factor in terms of the unit vectors, and divide by $d\alpha\, d\beta$, giving

$$G_\alpha t_\alpha + G_\beta t_\beta + G_n t_n = 0 \qquad (3.20)$$

which is satisfied if and only if

$$G_\alpha = 0; \quad G_\beta = 0; \quad G_n = 0 \qquad (3.21)$$

Thus, we have the three scalar equations of moment equilibrium

$$G_\alpha = -(BM_{\alpha\beta}),_\alpha - (AM_\beta),_\beta - B,_\alpha M_{\beta\alpha} + A,_\beta M_\alpha + Q_\beta AB = 0 \qquad (3.22a)$$

$$G_\beta = (BM_\alpha),_\alpha + (AM_{\beta\alpha}),_\beta + A,_\beta M_{\alpha\beta} - B,_\alpha M_\beta - Q_\alpha AB = 0 \qquad (3.22b)$$

$$G_n = N_{\alpha\beta} - N_{\beta\alpha} + \frac{M_{\alpha\beta}}{R_\alpha} - \frac{M_{\beta\alpha}}{R_\beta} = 0 \qquad (3.22c)$$

The set of six equations of static equilibrium, equations (3.17a–c) and (3.22a–c), may immediately be reduced to five, as it is easily shown by the substitution of equations (3.6) to (3.9) into equation (3.22c) that the latter equation, commonly known as the *sixth* equation of equilibrium, is satisfied *identically* if the symmetry of the stress tensor $\sigma_{\beta\alpha} = \sigma_{\alpha\beta}$ is invoked.[2] Thus, equation (3.22c) may be disregarded, although we may mention it from time to time. There are several modified shell theories which attempt to redefine the stress resultants and couples so that the sixth equation can be satisfied in the form of equation (3.22c).[6] None of these alters the fact that the sixth equation is identically satisfied using a proposition from a higher theory, the theory of elasticity, and consequently cannot be germane in any further development. Therefore, five equilibrium equations in ten unknowns remain. Such a problem is classified as *statically indeterminate,* indicating that the solution awaits the introduction of additional relationships between the stress resultants and couples, and the deformations of the shell.

3.3   EQUILIBRIUM EQUATIONS FOR SHELLS OF REVOLUTION

3.3.1   Specialization of General Equations

The shell of revolution is a widely used geometry and affords a sufficiently general, yet easily visualized, model for the physical interpretation of the equilibrium equations.

For the shell of revolution shown in Figure 2–11, we set $\alpha = \phi$ and $\beta = \theta$. From equations (2.48b) and (2.47b), $A = R_\phi(\phi)$ and $B = R_0(\phi) = R_\theta(\phi) \sin \phi$. Also, recall the Gauss-Codazzi condition, equation (2.51), which gives

$$B_{,\alpha} = R_{0,\phi} = (R_\theta \sin \phi)_{,\phi} = R_\phi \cos \phi \qquad (3.23)$$

Writing the equilibrium equations including these relations, we have from equations (3.17a–c) and (3.22a–c),

$$(R_0 N_\phi)_{,\phi} + R_\phi N_{\theta\phi,\theta} - R_\phi \cos \phi \, N_\theta + R_0 Q_\phi + q_\phi R_\phi R_0 = 0 \qquad (3.24a)$$

$$(R_0 N_{\phi\theta})_{,\phi} + R_\phi N_{\theta,\theta} + R_\phi \cos \phi \, N_{\theta\phi} + R_\theta Q_\theta \sin \phi + q_\theta R_\phi R_0 = 0 \qquad (3.24b)$$

$$(R_0 Q_\phi)_{,\phi} + R_\phi Q_{\theta,\theta} - R_0 N_\phi - R_\phi \sin \phi \, N_\theta + q_n R_\phi R_0 = 0 \qquad (3.24c)$$

$$-(R_0 M_{\phi\theta})_{,\phi} - R_\phi M_{\theta,\theta} - R_\phi \cos \phi \, M_{\theta\phi} + R_\phi R_0 Q_\theta = 0 \qquad (3.24d)$$

$$(R_0 M_\phi)_{,\phi} + R_\phi M_{\theta\phi,\theta} - R_\phi \cos \phi \, M_\theta - R_\phi R_0 Q_\phi = 0 \qquad (3.24e)$$

$$N_{\phi\theta} - N_{\theta\phi} + \frac{M_{\phi\theta}}{R_\phi} - \frac{M_{\theta\phi}}{R_\theta} = 0 \qquad (3.24f)$$

The middle surface element for a shell of revolution geometry with positive Gaussian curvature is shown in Figure 3–5. Stress resultants, couples, and loads each are shown on an enlarged view of this element normal to $\mathbf{t_n}$ in Figure 3–6. Also shown in Figure 3–6 are horizontal and meridional sections, $HV1$, $MV1$, and $MV2$, with selected resultants and couples in-

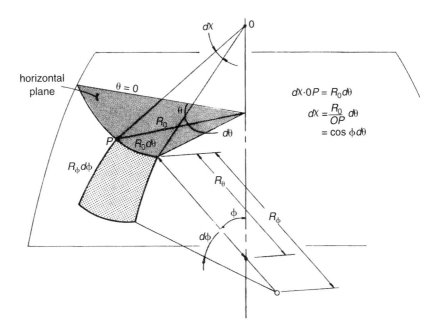

FIGURE 3–5   Middle Surface Element and Geometrical Relationships for a Shell of Revolution.

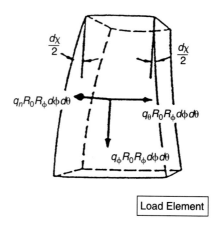

Load Element

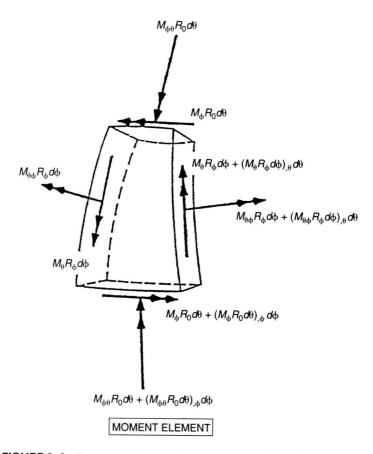

MOMENT ELEMENT

FIGURE 3–6   Forces and Moments Acting on a Shell of Revolution Element.

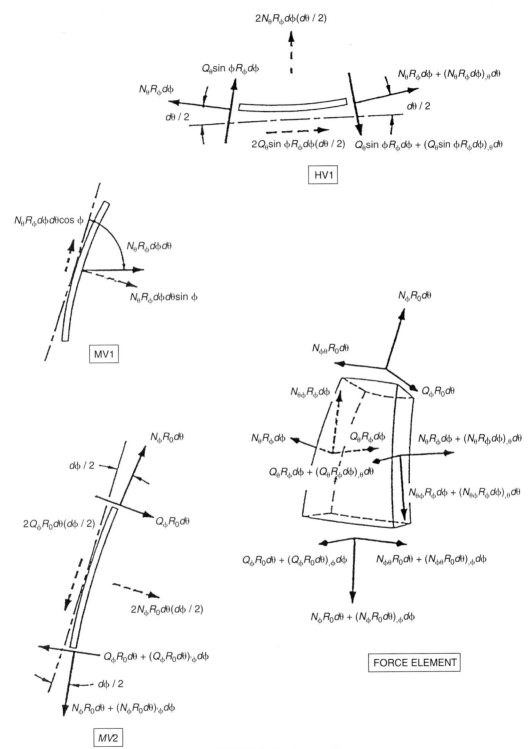

FIGURE 3–6 *Continued*

dicated. In comparing the terms of equation (3.24) with the forces and moments on Figure 3–6, recall that we divided through by $d\alpha d\beta$, or, in this case, $d\phi d\theta$, in the course of the derivation and that we have retained terms of second differential order only. Also, for the differential element, $\sin d\phi/2 = d\phi/2$, $\sin d\theta/2 = d\theta/2$, and $\cos d\phi/2 = \cos d\theta/2 = 1$.

3.3.2 Physical Interpretation of Equilibrium Equations

3.3.2.1 Force Equilibrium in the ϕ Direction: Equation (3.24a).
The first term represents the increment in the force, $N_\phi R_0 d\theta$ as ϕ changes, whereas the second gives the increment in the force $N_{\theta\phi} R_\phi d\phi$ with respect to θ, as seen on the FORCE element in Figure 3–6. Since R_ϕ is not a function of θ, it is treated as a constant in the second term. The third term indicates a contribution of the force $N_\theta R_\phi d\phi$ to the θ direction equilibrium and is explained by first referring to view $HV1$, where the radial component $2N_\theta R_\phi d\phi d\theta/2$, which is in the horizontal plane, is developed. This radial force is transferred to view $MV1$ and is further resolved in the ϕ and n directions, with the ϕ component being the third term in equation (3.24a). The negative sign indicates that the component acts in the negative ϕ direction as defined on Figure 2–11. The fourth term is the contribution of the projection of the radial force $Q_\phi R_0 d\theta$ and is substantiated by considering view $MV2$. The last term is, of course, the applied loading as shown on the LOAD element.

3.3.2.2 Force Equilibrium in the θ Direction: Equation (3.24b).
The first and second terms are analogous to the corresponding terms of equation (3.24a) and are easily verified on the FORCE element. The third term is the circumferential contribution of force $N_{\theta\phi} R_\phi d\phi$ and arises because forces $N_{\theta\phi} R_\phi d\phi$ and $N_{\theta\phi} R_\phi d\phi + (N_{\theta\phi} R_\phi d\phi)_{,\theta} d\theta$ are not parallel to one another, but rather are directed along tangents to the meridians at θ and $\theta + d\theta$, respectively. As shown in Figure 3–5 and the LOAD element, the extension of these tangents defines an intersection on the axis of rotation at an angle $d\chi$, which is expressed as $d\theta \cos \phi$. Therefore, from the FORCE element we have $2N_{\theta\phi} R_\phi d\phi d\chi/2 = N_{\theta\phi} R_\phi \cos \phi d\phi d\theta$ in the θ direction. The fourth term is the contribution of $Q_\theta R_\phi d\phi$ and is shown in view $HV1$, whereas the fifth term is the applied loading shown on the LOAD element.

3.3.2.3 Force Equilibrium in the n Direction: Equation (3.24c).
The first two terms are the increments of the transverse shear resultants Q_ϕ and Q_θ in the ϕ and θ directions, respectively, as shown on the FORCE element. The third term represents the contribution of force $N_\phi R_0 d\theta$ and is obtained by projection in view $MV2$. The fourth term, representing a similar contribution of $N_\theta R_\phi d\phi$, is the normal projection of the radial force $N_\theta R_\phi d\phi d\theta$ in view $MV1$. This force was introduced previously in connection with the third term in equation (3.24a). Both of these normal components are directed in the negative n direction. The fifth term again is the applied loading shown on the LOAD element.

3.3.2.4 Moment Equilibrium about the t_ϕ Axis: Equation (3.24d).
The first two terms of this equation are analogous to the corresponding terms of equation (3.24a) and may be verified accordingly on the MOMENT element, noting that the increments are directed in the negative ϕ direction. The third term is found analogously to the projection of

$N_\theta R_0 d\phi$ on views *HM*1 and *MV*1. The fourth term is the contribution of the force $Q_\theta R_\phi d\phi$ multiplied by $R_0 d\theta$ as seen on the FORCE element.

3.3.2.5 Moment Equilibrium about the t_θ Axis: Equation (3.24e).

Again, the first three terms are analogous to the force equilibrium, equation (3.24b), with the third term being opposite in sense to $N_{\theta\phi} R_\phi d\phi$ as seen from the MOMENT element. The last term is the couple $Q_\phi R_0 d\theta \, R_\phi d\phi$ which is directed along the negative θ direction.

3.3.2.6 Moment Equilibrium about the t_n Axis: Equation (3.24f).

The first two terms represent the moments of the in-plane shear resultants $N_{\phi\theta}$ and $N_{\theta\phi}$ about the normal t_n. The third term is opposite in sense but analogous to the normal projection of $N_\phi R_0 d\phi$ on view *MV*2; the fourth term follows the normal projection of $N_\theta R_\phi d\phi$ on views *HV*1 and *MV*1.

3.4 EQUILIBRIUM EQUATIONS FOR PLATES

Equations (3.17a–c) and (3.22a–c) may easily be reduced to sufficiently general equilibrium equations for medium-thin plates by allowing R_α and $R_\beta \to \infty$. After these simplifications, we have

$$(BN_\alpha)_{,\alpha} + (AN_{\beta\alpha})_{,\beta} + A_{,\beta} N_{\alpha\beta} - B_{,\alpha} N_\beta + q_\alpha AB = 0 \tag{3.25a}$$

$$(BN_{\alpha\beta})_{,\alpha} + (AN_\beta)_{,\beta} + B_{,\alpha} N_{\beta\alpha} - A_{,\beta} N_\alpha + q_\beta AB = 0 \tag{3.25b}$$

$$(BQ_\alpha)_{,\alpha} + (AQ_\beta)_{,\beta} + q_n AB = 0 \tag{3.25c}$$

$$-(BM_{\alpha\beta})_{,\alpha} - (AM_\beta)_{,\beta} - B_{,\alpha} M_{\beta\alpha} + A_{,\beta} M_\alpha + Q_\beta AB = 0 \tag{3.25d}$$

$$(BM_\alpha)_{,\alpha} + (AM_{\beta\alpha})_{,\beta} + A_{,\beta} M_{\alpha\beta} - B_{,\alpha} M_\beta - Q_\alpha AB = 0 \tag{3.25e}$$

$$N_{\alpha\beta} - N_{\beta\alpha} = 0 \tag{3.25f}$$

Further simplifications will be made for regular geometries and for bending in the absence of in-plane forces; for now, we note in equations (3.25a–c) that the in-plane stress resultants $(N_\alpha, N_\beta, N_{\alpha\beta}, N_{\beta\alpha})$ are uncoupled from the stress couples $(M_\alpha, M_\beta, M_{\alpha\beta}, M_{\beta\alpha})$ and the transverse shear stress resultants (Q_α, Q_β). This suggests that the plate resists in-plane loading (q_α, q_β) exclusively through extensional action, and transverse loading (q_n) only by flexure. This is an important contrast to shell action, as demonstrated by equation (3.17c) where the *in-plane* stress resultants provide resistance against transverse loading by virtue of the curvature of the shell.

3.5 NATURE OF THE APPLIED LOADING

Thus far, we have assumed that the applied loading is expressible as a distributed force per unit area of middle surface, with components in the directions of the unit vectors. This leaves open the cases of concentrated forces, and distributed and concentrated moments.

In general, shells are most efficient when the loading is distributed over the surface, because—provided certain geometrical and support conditions are met—such loading can be resisted primarily by the extensional and in-plane shear stress resultants, rather than by the transverse shear stress resultants and the bending and twisting stress couples. In addition to the distributed surface loading included in the formulation by $\mathbf{q}$, distributed moments about the $\mathbf{t}_\phi$ and $\mathbf{t}_\theta$ axes can easily be accommodated by including appropriate terms in equations (3.22a) and (3.22b). Moments about the $\mathbf{t}_n$ axis, however, are not admissible, since the corresponding equilibrium equation, (3.22c), was suppressed. A plate or shell generally offers very great stiffness to twisting about the normal, provided rigid body motion is restrained.

Concentrated forces and moments require special attention in plate and shell theory. Often, these loadings produce singular points, in which case admissible solutions may be found only away from the point of application of the load. In any case, these forces are usually resisted primarily by the transverse shears that are directly related to the bending and twisting stress couples. We may view concentrated forces and moments as limits of the corresponding distributed effect. Consider the area ΔA, subjected to a uniformly distributed loading of intensity p as shown in Figure 3–7(a). We define a concentrated force $\mathbf{P}_c$ as

$$\mathbf{P}_c = \lim_{\substack{\Delta A \to 0}} (\mathbf{p}\Delta A) \tag{3.26}$$

$\mathbf{p}\Delta A$ *remains constant*

Now consider two concentrated coplanar forces as defined in equation (3.26) and shown in Figure 3–7(b). The forces $\mathbf{P}_c$ are equal in magnitude and opposite in direction and a distance $\Delta\xi$ apart. We define a concentrated moment $\mathbf{M}_c$ about an axis normal to the plane of the forces as

$$\mathbf{M}_C = \lim_{\substack{\Delta\xi \to 0}} (\mathbf{P}_C\Delta\xi) \tag{3.27}$$

$\mathbf{P}_c\Delta\zeta$ *remains constant*

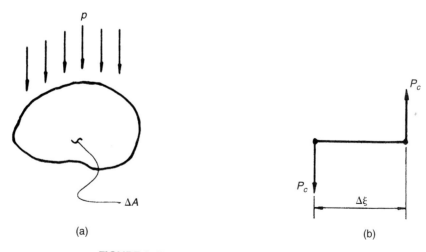

(a)

(b)

FIGURE 3–7   Concentrated Force and Moment.

Equations (3.26) and (3.27) are somewhat abstract in this form, but are useful in obtaining many solutions for plates and shells.

REFERENCES

1. H. Kraus, *Thin Elastic Shells* (New York: Wiley, 1967): 33.

2. V. V. Novozhilov, *Thin Shell Theory* [translated from 2nd Russian ed. by P. G. Lowe] (Groningen, The Netherlands: Noordhoff 1964): 34–39.

3. A. E. Green and W. Zerna, *Theoretical Elasticity* (Oxford; Oxford University Press, 1968).

4. S. Timoshenko and S. Woinowsky-Krieger, *Theory of Plates and Shells,* 2nd ed. (New York: McGraw-Hill, 1959).

5. W. Flügge, *Stresses in Shells,* 2nd ed. (Berlin: Springer-Verlag, 1973).

6. Kraus, *Thin Elastic Shells:* 58–65.

EXERCISES

3.1. Derive F_α, F_β, and F_n, as given in equation (3.17), from equation (3.14).

3.2. Derive G_α, G_β, and G_n, as given in equation (3.22), from equations (3.18b) and (3.19).

3.3. Verify that equation (3.22c) is identically satisfied if $\sigma_{\alpha\beta} = \sigma_{\beta\alpha}$. It is often incorrectly concluded that the identical satisfaction of equation (3.22c) implies that $N_{\alpha\beta} = N_{\beta\alpha}$ and $M_{\alpha\beta} = M_{\beta\alpha}$. For what geometries do these relations hold?

3.4. Redraw Figures 3–5 and 3–6 for a shell of revolution with negative Gaussian curvature and verify the physical interpretation shown in the views $HV1$, $MV1$ and $MV2$ for the negative curvature element.

3.5. Derive the equilibrium equations for a shell of revolution where the coordinate α is taken as the axial coordinate Z.

3.6. Consider Figure 3–5 for a shell with negative Gaussian curvature and re-derive the relationship between $d\chi$ and $d\theta$, considering $\phi > \pi/2$.

4

Membrane Theory:
Applications to Symmetrically Loaded Shells of Revolution

4.1 SIMPLIFICATION OF THE EQUILIBRIUM EQUATIONS

We examine the individual terms of the force equilibrium equations, (3.17a–c), and the moment equilibrium equations, (3.22a–c). We see that the equations are coupled only through the transverse shear stress resultants, Q_α and Q_β. If we suppose that for a certain class of shells, the stress couples are an order of magnitude smaller than the extensional and in-plane shear stress resultants, we may deduce from equations (3.22a–c) that the transverse shear stress resultants are similarly small and thus may be neglected in the force equilibrium equations, (3.17). This implies that the shell may achieve *force equilibrium* through the action of *in-plane forces* alone. From a physical viewpoint, this possibility is evident for the first two equilibrium equations which reflect in-plane resistance to in-plane loading, a natural and obvious mechanism. On the other hand, the third equilibrium equation refers to the normal direction, and the possibility of resisting *transverse* loading with *in-plane* forces *alone* is not as apparent. It is evident from equation (3.17c) that this mode of resistance is possible only if at least one radius of curvature is finite; in other words, R_α and/or $R_\beta \neq \infty$. Thus, flat plates are excluded from resisting transverse loading in this manner, within the limitations of small deformation theory (assumption [2], Table 1–1).

The applications and limitations of this idealized behavior, termed the *membrane theory of thin shells,* is examined in this chapter and in the following two chapters for a variety of shell forms. At present, it is constructive to consider some of the consequences of membrane behavior. First, it is quite desirable from a material efficiency standpoint. Recall from Chapter 1 that structural materials are generally far more efficient in extension rather than in flexure. Second, if we introduce an additional simplification with respect to the shear stress resultants,

$$N_{\alpha\beta} \simeq N_{\beta\alpha} = S \tag{4.1}$$

the force equilibrium equations become

$$\frac{1}{AB}\left[(BN_\alpha),_\alpha + (AS),_\beta + A,_\beta S - B,_\alpha N_\beta\right] + q_\alpha = 0 \tag{4.2a}$$

$$\frac{1}{AB}\left[(BS),_\alpha + (AN_\beta),_\beta + B,_\alpha S - A,_\beta N_\alpha\right] + q_\beta = 0 \tag{4.2b}$$

$$\frac{N_\alpha}{R_\alpha} + \frac{N_\beta}{R_\beta} = q_n \tag{4.2c}$$

In these three equations, there are but three unknowns, N_α, N_β, and S. Such a system is said to be *statically determinate* and thus independent of compatibility or constitutive considerations. Of course, this simplifies the subsequent mathematics considerably. The convenient assumption stated in equation (4.1) may be easily justified by considering the definition of $N_{\alpha\beta}$ and $N_{\beta\alpha}$ as given in equations (3.6) and (3.7). Since $\sigma_{\alpha\beta} = \sigma_{\beta\alpha}$ because the stress tensor is symmetric, $N_{\alpha\beta}$ and $N_{\beta\alpha}$ may differ only in the disparity of the terms $1 + \zeta/R_\beta$ and $1 + \zeta/R_\alpha$, respectively. Because ζ/R_β and ζ/R_α are themselves necessarily small in comparison to 1, the difference in $N_{\alpha\beta}$ and $N_{\beta\alpha}$ is obviously negligible. Thus, the simplification introduced in equation (4.1) is justified apart from other considerations.

Equations (4.2a–c), together with the requisite boundary conditions, constitute the essential components of the membrane theory. Strictly speaking, the entire solution includes the evaluation of the displacements as well. This requires the constitutive and compatibility relationships that have not yet been developed here. However, just as in the analysis of statically determinate beams and frames, the membrane theory displacements may be computed following the establishment of the force field with only a few exceptions.[1] Consequently, consideration of the membrane theory displacements is deferred until Chapter 8.

4.2 APPLICABILITY OF MEMBRANE THEORY

The shells for which the membrane theory is applicable fall into two general classes: (a) absolutely flexible shells or true membranes, which by virtue of their thinness have a negligible bending stiffness; and, (b) shells with finite bending stiffnesses which still develop relatively small bending stresses.[2]

Concerning the absolutely flexible shells, there is little question of the dominance of membrane action, provided that the principal stresses remained tensile. The presence of compressive stresses in this type of shell would likely cause buckling of the membrane. For flexible shells, the requirement of small displacements as stated in assumption [2] of Table 1–1 is often not attainable, and the equations as developed herein are not applicable without modification. This complication notwithstanding, membrane surfaces are widely used for long-span roofs, as either tentlike or air-supported structures. An unusual example of a thin, flexible, curved object to which shell theory has been applied for the purpose of estimating the stresses present is the human aortic heart valve [Figure 2–8(q)].[3]

A wide variety of shells with finite bending rigidities can be designed and constructed to resist external loading almost exclusively through membrane action. We should recognize that the establishment of the precise bounds of the membrane theory requires the consideration of the governing equations of the general theory of shells. Rigorously, a valid membrane theory solution must be a close approximation to the response that would be computed using the full set of shell equations. However, we may make some observations regarding the *geometry, boundary conditions,* and *loading* generally consistent with membrane behavior.

With respect to the *geometry,* a continuous curved surface is conducive to membrane action. Also recall that the second quadratic form of the theory of surfaces, equation (2.28b), contains the principal radii of curvature, and a smooth variation of these parameters is essential if equilibrium is to be achieved without transverse shears and bending and twisting moments.

The *boundary conditions* must be specified with respect to the membrane stress resultants, N_α, N_β, and S, or the corresponding displacements, and must indicate either a *constraint* that will fully develop the force on the boundary; a *release* that will enable the shell to displace freely in the corresponding direction; or, possibly, a linear combination of both, as in the case of an elastic support. Moreover, the boundary should *not* provide constraints that would develop moments and/or transverse shears. This implies that the ideal boundary must, in some cases, permit rotations and transverse displacements while providing complete restraint in the plane of the shell.

A *loading* that can be resisted by membrane action must be distributed over the surface without severe variations or concentrations. In the membrane equilibrium equations, (4.2a–c), concentrated loadings are not admissible. From a physical standpoint, it is apparent that a load normal to and concentrated over a small area of the shell surface would be resisted locally by transverse shears which, in turn, produce bending, as demonstrated by equations (3.22a–c).

From this brief discussion of the conditions that correspond to ideal membrane behavior, it is obvious that to design and construct an actual shell that satisfies all of these conditions may be quite difficult. However, even shells that do not meet all of the requirements entirely can be built so that membrane action predominates throughout most of the continuum, except for localized regions in which the required conditions are violated. In short, many shells exhibit basically membrane behavior overall, augmented by locally prominent bending action. This is not to say that all shells may be made to act primarily as membranes or that all bending effects are localized, but such behavior is a desirable and often attainable result of good design and careful construction. As such, the membrane theory stresses serve as the norm, even if the shell is analyzed by sophisticated numerical methods incorporating bending effects.

4.3   SHELLS OF REVOLUTION

4.3.1   Specialization of Equilibrium Equations

We now specialize the membrane theory equilibrium equations, (4.2), for the shell of revolution geometry, where the curvilinear coordinates are taken as the meridional angle, ϕ, and the circumferential angle θ.

With $\alpha = \phi$ and $\beta = \theta$, $A = R_\phi(\phi)$ and $B = R_0(\phi) = R_0 \sin \phi$ as previously established in equations (2.48b) and (2.47b). If we make these substitutions into equations (4.2a–c) and carry out the differentiations using the Gauss-Codazzi condition, equation (2.51), we find

$$\frac{1}{R_\phi} N_{\phi,\phi} + \frac{\cot \phi}{R_\theta}(N_\phi - N_\theta) + \frac{1}{R_0} S_{,\theta} + q_\phi = 0 \tag{4.3a}$$

$$\frac{1}{R_\phi} S_{,\phi} + \frac{2 \cot \phi}{R_\theta} S + \frac{1}{R_0} N_{\theta,\theta} + q_\theta = 0 \tag{4.3b}$$

$$\frac{N_\phi}{R_\phi} + \frac{N_\theta}{R_\theta} = q_n \tag{4.3c}$$

We immediately note that equations (4.3a–c) are identical to equations (3.24a–c) divided through by $R_\phi R_0$, with the bending terms suppressed and the in-plane shear stress resultants $N_{\phi\theta}$ and $N_{\theta\phi}$ set equal to S. As such, the physical interpretation of these equations is contained within Section 3.3.2. We also observe that the third equation, (4.3c), is an algebraic equation. Hence, either N_θ or N_ϕ may be directly eliminated from equations (4.3a) and (4.3b), leaving a second-order system of partial differential equations to be solved.

To simplify the solution of the equilibrium equations, it is convenient to introduce certain transformations at this stage. Timoshenko and Woinowsky-Krieger use integrating functions,[4] whereas Novozhilov suggests the auxiliary variables[5]

$$\psi = N_\phi R_\theta \sin^2 \phi \tag{4.4a}$$

and

$$\xi = S R_\theta^2 \sin^2 \phi \tag{4.4b}$$

We first solve equation (4.3c) for

$$N_\theta = R_\theta \left(q_n - \frac{N_\phi}{R_\phi} \right) \tag{4.5}$$

and then introduce equations (4.4) and (4.5) into equations (4.3a) and (4.3b) to get

$$\frac{R_\theta^2 \sin \phi}{R_\phi} \psi_{,\phi} + \xi_{,\theta} = (q_n \cos \phi - q_\phi \sin \phi) R_\theta^3 \sin^2 \phi \tag{4.6a}$$

$$\xi_{,\phi} - \frac{R_\theta}{\sin \phi} \psi_{,\theta} = -(q_\theta \sin \phi + q_{n,\theta}) R_\phi R_\theta^2 \sin \phi \tag{4.6b}$$

Equations (4.6a) and (4.6b) are the transformed membrane theory equilibrium equations. Once ψ and ζ are determined, N_ϕ and S follow from equations (4.4a) and (4.4b), and N_θ may then be calculated from equation (4.5). The subsequent combination and the eventual solution of the transformed equilibrium equations are dependent on the circumferential distribution of the applied surface loading. If the loading is distributed uniformly around the circumference, it is termed *axisymmetric,* whereas if the loading is a variable function of θ, it is obviously *nonsymmetric.* Each possibility is considered further in the subsequent sections.

4.3.2   Axisymmetrical Loading

4.3.2.1 Integration of Equilibrium Equations.   The case of axisymmetrically loaded shells of revolution is perhaps the most widely studied class of problems in the shell literature. From a standpoint of applications, this case is often the most important, because it corresponds to gravity loading and internal pressurization, as well as to many other practical loading situations.

We may drop the θ-dependent terms in equations (4.6a) and (4.6b) to derive the uncoupled set of ordinary differential equations

$$\psi_{,\phi} = (q_n \cos \phi - q_\phi \sin \phi)R_\phi R_\theta \sin \phi \tag{4.7a}$$

$$\xi_{,\phi} = -q_\theta R_\phi R_\theta^2 \sin^2 \phi \tag{4.7b}$$

Considering equation (4.7a) and integrating, we get

$$\psi(\phi) = \int (q_n \cos \phi - q_\phi \sin \phi)R_\phi R_\theta \sin \phi \, d\phi \tag{4.8}$$

In many cases, an alternate form of equation (4.8) is convenient. We recognize that evaluating the integral for a particular shell geometry will produce a single integration constant, and that this constant must be determined from a boundary condition on ψ, or on N_ϕ. We designate the boundary where the condition is specified as $\phi = \phi'$ and the corresponding boundary condition as $\psi(\phi')$. Then, equation (4.8) may be written as

$$\psi(\phi) = \psi(\phi') + \int_{\phi'}^{\phi} (q_n \cos \phi - q_\phi \sin \phi)R_\phi R_\theta \sin \phi \, d\phi \tag{4.9}$$

Of course, at $\phi = \phi'$, $\psi = \psi(\phi') = N_\phi(\phi')R_\theta(\phi')\sin^2 \phi'$. The solution for N_ϕ follows from equations (4.4a) and (4.9) as

$$
\begin{aligned}
N_\phi(\phi) &= \frac{N_\phi(\phi')R_\theta(\phi')\sin^2 \phi'}{R_\theta \sin^2 \phi} \\
&\quad + \frac{1}{R_\theta \sin^2 \phi} \int_{\phi'}^{\phi} (q_n \cos \phi - q_\phi \sin \phi)R_\phi R_\theta \sin \phi \, d\phi
\end{aligned}
\tag{4.10}
$$

Then, N_θ is computed from equation (4.5).

The in-plane shear stress resultant S is completely uncoupled from N_ϕ and N_θ as is apparent from equations (4.7a) and (4.7b). For common axisymmetric loading cases, $q_\theta = 0$ and the shear stress is also equal to 0; however, this is not necessarily so, since equation (4.7b) may be integrated in an identical fashion to equation (4.7a), yielding

$$\xi(\phi) = \xi(\phi'') - \int_{\phi''}^{\phi} q_\theta R_\phi R_\theta^2 \sin^2 \phi \, d\phi \tag{4.11}$$

or

$$S(\phi) = \frac{S(\phi'')R_\theta^2(\phi'') \sin^2 \phi''}{R_\theta^2 \sin^2 \phi} - \frac{1}{R_\theta^2 \sin^2 \phi} \int_{\phi''}^{\phi} q_\theta R_\phi R_\theta^2 \sin^2 \phi \, d\phi \tag{4.12}$$

In equations (4.11) and (4.12), $\xi(\phi'')$ and $S(\phi'')$ are the corresponding values of the functions specified at the boundary $\phi = \phi''$. This solution represents a state of pure shear produced by a torsional loading.

In summary, we observe that the membrane theory equilibrium equations for axisymmetrically loaded shells of revolution reduce to two uncoupled first-order ordinary differential equations. This system admits the prescription of two boundary conditions, one on the meridional stress resultant N_ϕ and the other on the in-plane shear stress resultant S.

One special case is worthy of consideration before we turn to the integration of equations (4.10) and (4.12) for specific shell geometries. If we consider a closed shell as first discussed in Section 2.8.2, we have the two possibilities illustrated in Figure 4–1. As mentioned in Chapter 1, the form of domed roofs evolved from the smooth to the pointed top in the Renaissance. In case (a), which we will call a *dome* [Figures 2–8(t) and (s)], the meridian remains continuous as ϕ and $R_0 \to 0$; whereas in case (b), which may be termed a *pointed* or *ogival* shell, $\phi = \phi_t$ at $R_0 = 0$. If we take the pole angle $\phi = \phi_p$ as the boundary ϕ' in equation (4.9), where $\phi_p = 0$ for case (a) and $\phi_p = \phi_t$ for case (b), and evaluate $\psi(\phi_p)$, we find

$$
\begin{aligned}
\psi(\phi_p) &= N_\phi(\phi_p)R_\theta(\phi_p)\sin^2\phi_p \\
&= N_\phi(\phi_p)R_0(\phi_p)\sin\phi_p \qquad\qquad (4.13) \\
&= 0
\end{aligned}
$$

because $R_0(\phi_p) = 0$. Similarly, considering equation (4.11) for ξ, $\xi(\phi_p) = 0$. Thus, the solutions for closed shells simplify to

$$
N_\phi = \frac{1}{R_\theta \sin^2\phi} \int_{\phi_p}^{\phi} (q_n \cos\phi - q_\phi \sin\phi)R_\phi R_\theta \sin\phi \, d\phi \qquad (4.14a)
$$

$$
S = -\frac{1}{R_\theta^2 \sin^2\phi} \int_{\phi_p}^{\phi} q_\theta R_\phi R_\theta^2 \sin^2\phi \, d\phi \qquad (4.14b)
$$

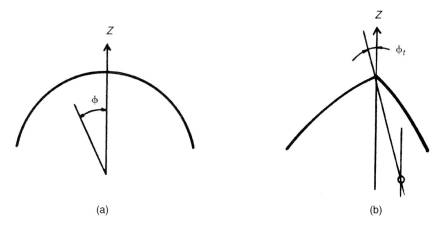

(a)   (b)

FIGURE 4–1   Closed Shells.

For the common case of a dome [Figure 4–1(a)], $\phi_p = 0$.

We may observe several interesting points about the preceding solutions for closed shells:

1. The pole condition does not necessarily imply that $N_\phi(\phi_p)$ or $S(\phi_p) = 0$.
2. For the case of a dome, $\phi_p = 0$, the integrated expressions may produce an indeterminate form which must be evaluated by some limiting operation.
3. Since the available boundary conditions on N_ϕ and S are *implied by* the pole, there is no opportunity to specify any further boundary conditions on the stress resultants within the membrane theory. This means that to maintain equilibrium, the remaining boundary of the shell must develop whatever values of the membrane stress resultants that are computed from equations (4.14). We recall that such an ideal boundary may not impose constraints which will develop bending forces or moments.
4. For a dome, all meridians meet at the pole, and any direction is parallel to one meridian and at right angles to another; that is, the curvilinear coordinates ϕ and θ are interchangeable and indistinguishable. Therefore, $R_\phi(0) = R_\theta(0) = R_p$ and, correspondingly, $N_\phi(0) = N_\theta(0)$. Then, from equation (4.3c),

$$N_\phi(0) = N_\theta(0) = q_n(0) \frac{R_p}{2} \tag{4.15}$$

4.3.2.2 Spherical Shells.   As the first case, consider the spherical shell illustrated in Figure 4–2. The shell has radius a, constant thickness h, mass density ρ, and is bound by angles ϕ_t and ϕ_b. Initially, the weight of the shell alone, known as the *self-weight,* is taken as the loading condition.

The principal radii of curvature are the definitive geometric parameters. The radius of curvature of the meridian is, of course, the radius of the generating circle a. This is easily verified by considering the equation of the meridian, $Z^2 + R_0^2 = a^2$ and substituting into equation (2.29) with $\alpha = \phi$ and $X = R_0$. With $Z_{,R_0} = -R_0/Z$ and $Z_{,R_0 R_0} = -(1/Z)(1 + R_0^2/Z^2)$, we find

$$R_\phi = \frac{-[1 + (R_0^2/Z^2)]^{3/2}}{-(1/Z)[1 + (R_0^2/Z^2)]}$$

$$= \sqrt{(Z^2 + R_0^2)} \tag{4.16a}$$

$$= a$$

The other principal radius may be computed from equation (2.46). We note from Figure 4–2 that the horizontal radius $R_0 = a \sin \phi$, so that

$$R_\theta = \frac{a \sin \phi}{\sin \phi} = a \tag{4.16b}$$

Equations (4.16a and b) obviously satisfy the Gauss-Codazzi condition, equation (2.51).

Next, consider the applied loading. The self-weight of the shell per unit area of the middle surface is given by

$$q = \rho g h \tag{4.17}$$

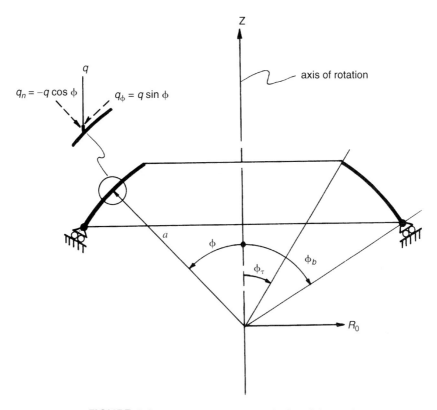

FIGURE 4–2   Spherical Shell Under Self-Weight Load.

in which g = the acceleration of gravity. The load q acts vertically, of course. The governing equilibrium equations require the loading to be resolved in the ϕ, θ, and n directions. Since the vertical load has no component in the circumferential direction, $q_\theta = 0$. On the inset in Figure 4–2, we have the resolution of q into q_ϕ and q_n. The signs are established by comparison with the unit vectors shown in Figure 2–11.

At present, the boundaries are assumed to satisfy ideal membrane theory restrictions. The upper boundary is presumed to be free of stresses and the lower boundary is taken to be unyielding in the ϕ or meridional direction, free to displace in the n or normal direction, and unrestrained against rotation about an axis along the θ or circumferential direction. The base conditions imply that the necessary N_ϕ to maintain equilibrium will be developed, but no Q_ϕ or M_ϕ can occur. The roller symbol used on the figure corresponds to a similar condition in plane structural analysis; however, here the boundary is a space curve, and for shells which have circumferential loading, a condition in the θ direction also must be specified. In the latter instance, the ideal membrane boundary must be extended beyond the simple conceptual model of a roller.

We now substitute the geometry, loading, and boundary conditions into equation (4.10). These calculated quantities are $R_\phi = R_\theta = a$; $q_\phi = q\frac{1}{2}\sin\phi$ and $q_n = -q\frac{1}{2}\cos\phi$. We take $\phi' = \phi_t$ so that $N_\phi(\phi_t) = 0$. Then, we have

$$N_\phi(\phi) = \frac{1}{a \sin^2 \phi} \int_{\phi_t}^{\phi} (-q \cos^2 \phi - q \sin^2 \phi) a^2 \sin \phi \, d\phi$$

$$= \frac{qa}{\sin^2 \phi} (\cos \phi - \cos \phi_t) \tag{4.18}$$

From equation (4.5), the circumferential or hoop stress resultant is given by

$$N_\theta = a\left(-q \cos \phi - \frac{N_\phi}{a}\right) \tag{4.19}$$

Example 4.3(a)

We now examine a spherical dome by letting $\phi_t = 0$. Using an elementary trigonometric identity, equation (4.18) becomes

$$N_\phi = \frac{qa}{(1 + \cos \phi)(1 - \cos \phi)} (\cos \phi - 1)$$

$$= \frac{-qa}{1 + \cos \phi} \tag{4.20}$$

and equation (4.19) follows as

$$N_\theta = a\left(-q \cos \phi + \frac{q}{1 + \cos \phi}\right)$$

$$= qa\left(\frac{1}{1 + \cos \phi} - \cos \phi\right) \tag{4.21}$$

At $\phi = 0$, we have

$$N_\phi = N_\theta = \frac{-qa}{2} \tag{4.22}$$

which is in accordance with equation (4.15). It is also notable that the indeterminate form mentioned as point (2) in Section 4.3.2.1 was resolved by evaluating the integral first, and then going to the limit as $\phi \to 0$.

It is quite instructive to consider equations (4.20) and (4.21) for increasing values of ϕ as shown in Figure 4–3, where nondimensional plots for N_ϕ and N_θ are provided. These graphs actually represent the self-weight stresses for *all* constant thickness spherical domes. For any particular dome for which the lower boundary is located by a given value of ϕ_b, the relevant parts are the regions $\phi \le \phi_b$. A shallow dome ($\phi_b \ll 90°$), a hemispherical dome ($\phi_b = 90°$), and a deep dome ($\phi_b > 90°$) are illustrated in the insets of the graphs. An interesting feature of the graphs is that N_ϕ, is always negative or compressive, whereas N_θ changes from negative to positive. This exact transition point may be computed from equation (4.21) as the value of ϕ satisfying

$$\frac{1}{1 + \cos \phi} - \cos \phi = 0$$

which has the solution $\phi = 51°49'$. The transition angle has some general practical ramifications, because a shell with $\phi_b < 51°$ will be entirely in compression under gravity loading, which is especially desirable for shells constructed of concrete.

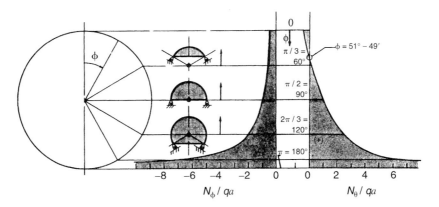

FIGURE 4-3 Self-Weight Stress Resultants for Spherical Domes of Constant Thickness.

With reference to the ancient masonry domes described in Chapter 1, the necessity for maintaining the entire dome in compression is obvious and is apparently reflected in the change from the spherical to the more parabolic profile used in the Renaissance.

The apparent desireability of providing an entirely compressive state of stress is countered by another practical consideration. Recall that the support for the spherical shell must develop the calculated value of N_ϕ at $\phi = \phi_b$. An idealized typical support is shown in Figure 4-4(a), where a circumferential ring beam is employed to resist the thrust. The shell is as-

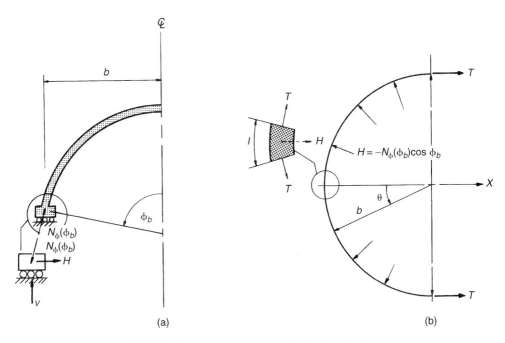

FIGURE 4-4 Ring Beam for a Shell of Revolution.

sumed to extend to the centroid of the ring beam to eliminate the introduction of eccentricity.[6] A section through the ring beam is shown in the inset. The vertical component of the thrust, $V = N_\phi(\phi_b)\sin\phi_b$, is transmitted to the foundation, whereas the horizontal component must be developed by the beam. A half-plan of the ring is shown in Figure 4–4(b), with a unit length segment in the inset; there, the horizontal reaction $H = -N_\phi(\phi_b)\cos\phi_b$ is countered by the radial component of the hoop tension **T**.

We may evaluate **T** by summing forces in the X direction on Figure 4–4(b).

$$2\mathbf{T} = \int_{-\pi/2}^{\pi/2} -N_\phi(\phi_b)\cos\phi_b\cos\theta(b\,d\theta)\mathbf{t_x} \tag{4.23a}$$

with the magnitude

$$T = -N_\phi(\phi_b)b\cos\phi_b \tag{4.23b}$$

Example 4.3(b)

For a spherical shell, the radius of the ring beam is

$$b = a\sin\phi_b \tag{4.24}$$

so that, from equation (4.23b),

$$T = -N_\phi(\phi_b)a\cos\phi_b\sin\phi_b \tag{4.25}$$

For the self-weight case, we find by substituting equation (4.20) into equation (4.25)

$$T_{DL} = qa^2\frac{\cos\phi_b\sin\phi_b}{1+\cos\phi_b} \tag{4.26}$$

It is clear from comparing equation (4.26) and the graph for N_θ in Figure 4–3 that there will be a strain incompatibility at the junction between the shell and the ring beam. For $\phi_b \le 51°49'$, the hoop stress N_θ is compressive; the ring force T_{DL} is tensile for all $\phi_b < 90°$. Thus a relatively shallow dome, which has been shown to be in a state of compression through membrane theory analysis, must somehow accommodate the circumferential expansion of the base ring accompanying the tensile force T_{DL}. Clearly, this cannot be accomplished with membrane action alone, and bending forces must be considered to satisfy deformational continuity. In the terminology of Section 4.2, we have violated the ideal *boundary conditions*. In some instances, prestressing the base ring effectively reduces the circumferential strain incompatibility. In general, the bending effects introduced by this type of support are usually confined to a relatively small portion of the shell adjacent to the base, and membrane action will still predominate throughout most of the shell. It should also be noted that equation (4.23) is not restricted to spherical shells, but is valid for all shells of revolution supported in this fashion.

It is apparent from Figure 4–3 that the stresses increase rapidly for deep spherical shells with $\phi_b > 120°$. For such shells, it is logical to support the shell somewhere above the base. Such a situation is depicted in Figure 2–8(t), although the description of discrete col-

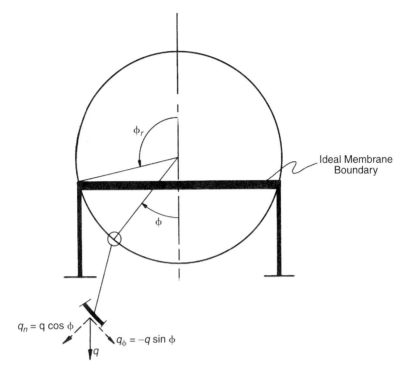

FIGURE 4–5 Complete Spherical Shell with an Intermediate Continuous Support.

umn supports requires an extension of the ideal membrane case, as we will see later. If we have, for example, a complete sphere supported by an ideal membrane boundary at an angle $\phi = \phi_r$ as shown in Figure 4–5, we can easily obtain the stress pattern from Figure 4–3. For $\phi \le \phi_r$, we compute the stresses as before. For $\phi > \phi_r$, the lower portion of the shell behaves just like a spherical cap with a boundary angle $\phi_b = \pi - \phi_r$ and with a loading which is opposite in sense to that shown on Figure 4–2, as shown in the inset of Figure 4–5. Thus, if we enter $\overline{\phi} = \pi - \phi$ for ϕ on Figure 4–3, the dead load stress resultants for the lower portion of the shell are identical in magnitude but opposite in sign to those read from the graph.

Closely related to the dead load case just considered is a loading which is considered as uniform over the *horizontal projection* of the shell, rather than the *middle surface*. This is termed *live load* and may simulate an elevated or suspended platform supported on a closely spaced grid of columns or hangers.

Example 4.3(c)

Another common loading condition for spherical domes is that of uniform normal pressure. Consider a position internal pressure p on the spherical dome. We substitute $R_\phi = R_\theta = a$; $q_\phi = 0$; $q_n = p$ and $\phi_p = 0$ into equation (4.14a), whereupon

$$N_\phi(\phi) = \frac{1}{a \sin^2 \phi} \int_0^\phi p \cos \phi \, a^2 \sin \phi \, d\phi$$

$$= \frac{pa^2}{a \sin^2\phi} \left[\frac{1}{2} \sin^2 \phi \right]_0^\phi \qquad (4.27)$$

$$= \frac{pa}{2}$$

From equation (4.5)

$$N_\theta(\phi) = a \left(p - \frac{N_\phi}{a} \right)$$

$$= \frac{pa}{2} \qquad (4.28)$$

so that we have a uniform state of stress, $N_\phi = N_\theta = pa/2$, throughout the shell.

A variation on the spherical water tank is shown in Figure 2–8(s), which is an ellipsoidal shell of revolution. An extension of the basic spherical geometry is the spheroid which is discussed in Exercise 4.15 and shown on Figure 4–28.[7] At the pole, the spheroid is an example of a pointed or ogival shell, Figure 4–1(b).

4.3.2.3 Hyperboloidal Shells. As a second illustration of a membrane theory solution for an axisymmetrically loaded shell of revolution, we consider the form illustrated in Figures 4–6, 2–8(d), 2–8(g) and 2–8(o). This surface is generated by the rotation of a hyperbola of one sheet and is a shell with *negative* Gaussian curvature, because the centers of curvature corresponding to R_ϕ and R_θ lie on opposite sides of the meridian (see Section 2.7 and Figures 2–6 and 2–11). This form has considerable practical importance, because such shells are widely used for reinforced concrete *hyperbolic cooling towers* [Figure 2–8(o)]. These massive structures may approach a height of 600 ft (183 m), span 300 ft (92 m) in diameter, and yet have an average thickness of less than 9 in. (23 cm), a striking testimony to the efficiency of thin shell structures. The same shape can be built in steel as shown in Figure 2–8(d).

The equation of the generating curve is

$$\frac{R_0^2}{a^2} - \frac{Z^2}{b^2} = 1 \qquad (4.29)$$

in which b is a characteristic dimension of the shell that may be evaluated by substituting the base coordinates (s, S) or the top coordinates (t, T) into equation (4.29) as

$$b = \frac{aT}{\sqrt{(t^2 - a^2)}} = \frac{aS}{\sqrt{(s^2 - a^2)}} \qquad (4.30)$$

The ratio a/b is the slope of the asymptote to the generating hyperbola shown on Figure 4–6, and the parameter

$$k = \sqrt{\left(1 + \frac{a^2}{b^2} \right)} \qquad (4.31)$$

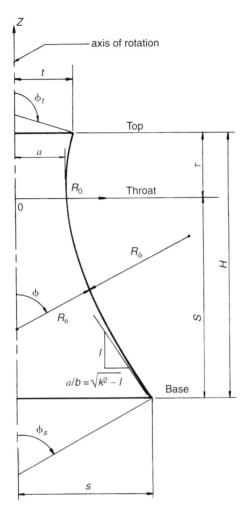

FIGURE 4–6   Hyperboloidal Shell Geometry.

may be viewed as an indicator of the deviation of the profile from the degenerate case of the cylinder, $k = 1$, with a larger k corresponding to a more pronounced curvature of the meridian. As may be seen from equation (4.30), dimension b and hence the geometric profile can be set independently using specified top or base dimensions. For cooling tower applications, it is sometimes desirable to use different values of b for the top and bottom, thus creating a compound shell with a smooth transition at $Z = 0$.

With the parameter k defined, we may rewrite equation (4.29) as

$$R_0^2 - (k^2 - 1)Z^2 = a^2 \qquad (4.32)$$

The next step is to derive expressions for the principal radii of curvature in terms of the curvilinear coordinate ϕ. Direct substitution of the equation of the meridian into equation (2.29), as previously done for the spherical shell, leads to cumbersome expressions since Z

and R_0 are not given as explicit functions of ϕ. Instead we consider equation (4.32) along with the expression for the differential arc length, equation (2.49a), which becomes for this case,

$$ds_\phi^2 = dZ^2 + dR_0^2$$

or

$$R_\phi^2 d\phi^2 = Z_{,\phi}^2 d\phi^2 + R_{0,\phi}^2 d\phi^2 \tag{4.33}$$

We then solve equation (4.32) for Z

$$Z = \left(\frac{R_0^2 - a^2}{k^2 - 1}\right)^{1/2} \tag{4.34}$$

and compute

$$Z_{,\phi} = \frac{R_0 R_{0,\phi}}{Z(k^2 - 1)}$$

Because

$$R_{0,\phi} = R_\phi \cos \phi \tag{4.35a}$$

$$Z_{,\phi} = \frac{R_0 R_\phi \cos \phi}{[(R_0^2 - a^2)(k^2 - 1)]^{1/2}} \tag{4.35b}$$

After substituting equations (4.35a) and (4.35b) into equation (4.33) and clearing fractions, we may cancel $R_\phi^2 d\phi^2$ in each term, so that

$$(R_0^2 - a^2)(k^2 - 1) = R_0^2 \cos^2 \phi + (R_0^2 - a^2)(k^2 - 1)\cos^2 \phi \tag{4.36}$$

from which we find, after some manipulation,

$$R_\theta = \frac{R_0}{\sin \phi} = \frac{a\sqrt{(k^2 - 1)}}{[k^2 \sin^2 \phi - 1]^{1/2}} \tag{4.37}$$

Finally, from equations (4.37) and (4.35a),

$$R_\phi = \frac{-a\sqrt{(k^2 - 1)}}{[k^2 \sin^2 \phi - 1]^{3/2}} \tag{4.38}$$

We now consider the elementary case of the self-weight. The principal radii of curvature are given by equations (4.37) and (4.38), and the loading components are $q_\phi = q \sin \phi$ and $q_n = -q \cos \phi$ as derived in Figure 4–2. Also, we take $\phi' = \phi_t$ in equation (4.10), which becomes

$$N_\phi(\phi) = N_\phi(\phi_t) \frac{\sin^2 \phi_t}{\sin^2 \phi} \frac{[k^2 \sin^2 \phi - 1]^{1/2}}{[k^2 \sin^2 \phi_t - 1]^{1/2}} - \frac{[k^2 \sin^2 \phi - 1]^{1/2}}{a \sin^2 \phi \sqrt{(k^2 - 1)}}$$

$$\cdot \int_{\phi_t}^{\phi} (-q \cos^2 \phi - q \sin^2 \phi) \frac{a^2(k^2 - 1)\sin \phi \, d\phi}{[k^2 \sin^2 \phi - 1]^2} \tag{4.39}$$

Assuming a stress-free top edge, $N_\phi(\phi_t) = 0$, and q = constant (uniform thickness), equation (4.39) integrates to [8]

$$N_\phi(\phi) = \frac{qa[k^2 \sin^2 \phi - 1]^{1/2}}{\sin^2 \phi \sqrt{(k^2 - 1)}} [\zeta_1(\phi) - \zeta_1(\phi_t)] \tag{4.40a}$$

in which the integrated function

$$\zeta_1(\phi) = \frac{-\cos \phi}{2[k^2 \sin^2 \phi - 1]} + \frac{1}{4k\sqrt{(k^2 - 1)}} \ln \left(\frac{\sqrt{(k^2 - 1)} - k \cos \phi}{\sqrt{(k^2 - 1)} + k \cos \phi} \right) \tag{4.40b}$$

The circumferential stress resultant N_θ is computed from equation (4.5) as

$$N_\theta(\phi) = \frac{a\sqrt{(k^2 - 1)}}{[k^2 \sin^2 \phi - 1]^{1/2}} \left[-q \cos \phi + \frac{N_\phi(\phi)[k^2 \sin^2 \phi - 1]^{3/2}}{a\sqrt{(k^2 - 1)}} \right] \tag{4.41}$$

Some further comments are in order regarding the preceding solution for the self-weight stress resultants. If the shell thickness changes with the height of the shell, the solution as derived must be generalized slightly. Assuming that the shell thickness is or may be approximated as *piecewise constant* within a defined subregion of the shell, the basic solution, equation (4.39), may be applied in a stepwise fashion. For any subregion r bounded by ϕ_t^r and ϕ_s^r as shown in Figure 4–7,

$$\phi_s^{r-1} = \phi_t^r \geq \phi^r \geq \phi_s^r = \phi_s^{r+1} \tag{4.42a}$$
$$r = 1, \ldots, n$$

and, from equation (4.17),

$$q^r = pgh^r \tag{4.42b}$$

where h^r is the local constant thickness. We start at the top of the shell, $r = 1$, for which the solution is given by equation (4.40a). Then, we evaluate equation (4.40a) at $\phi = \phi_s^1 = \phi_t^2$ to compute $N_\phi(\phi_s^1) = N_\phi(\phi_t^2)$, which is substituted into the first term of equation (4.39). Thus, the solution for region $r = 2$ is given by equation (4.40a), plus the first term of equation (4.39), with ϕ_t taken as ϕ_t^2. For the general region r, we have

$$N_\phi(\phi^r) = N_\phi(\phi_t^r) \frac{\sin^2 \phi_t^r [k^2 \sin^2 \phi^r - 1]^{1/2}}{\sin^2 \phi^r [k^2 \sin^2 \phi_t^r - 1]^{1/2}}$$

$$+ q^r a \frac{[k^2 \sin^2 \phi^r - 1]^{1/2}}{\sin^2 \phi^r \sqrt{(k^2 - 1)}} [\xi_1(\phi^r) - \xi_1(\phi_t^r)] \tag{4.43}$$

$N_\theta(\phi^r)$ is calculated as before from equation (4.41).

Although the coordinate ϕ has proven convenient for integrating the membrane theory equations, it is somewhat awkward for physically locating a particular position on the shell. It is obvious that the axial coordinate Z would be more meaningful from the standpoint of practical construction. Corresponding to any value of ϕ, R_0 may be computed from equation (4.37), and then Z can be found from equation (4.34). Conversely, for a specified Z, R_0 is computed from equation (4.32), and ϕ is conveniently obtained by solving equation (4.37) for $\sin \phi$. Then,

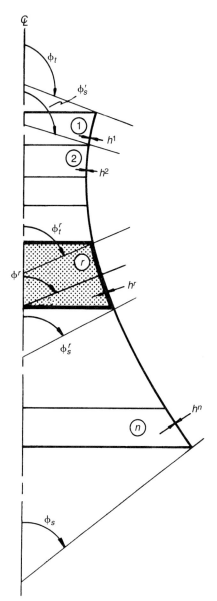

FIGURE 4–7 Hyperboloidal Shell Divided into Constant Thickness Regions.

$$\phi = \sin^{-1}\left\{\frac{R_0}{[a^2 + k^2(R_0^2 - a^2)]^{1/2}}\right\} \tag{4.44}$$

Values of ϕ in the first and second quadrant correspond to the lower and upper portions of the shell, respectively.

To compute the limits of integration, ϕ_t and ϕ_s, when the corresponding value of R_0, t, and s, are known, equation (4.44) can be used directly.

Example 4.3(d)

For spherical shells, we were able to deduce some general characteristics of the response to self-load by making a nondimensional plot of the membrane theory stress resultants. It is also useful to attempt such a study for the hyperboloidal shell. If we define a nondimensional meridional coordinate

$$\Phi = \frac{\phi_t - \phi}{\phi_t - \phi_s} \tag{4.45}$$

and assume the thickness to be constant, we may plot nondimensional membrane stress resultants for various values of k^2, a/s, and a/t. Such a plot is shown in Figure 4–8 for a shell of typical proportions, $a/t = 0.90$ and $a/s = 0.55$. To give some idea of the actual size of a corresponding cooling tower with $k^2 = 1.18$ and $a = 70$ ft (23 m), the dimensions would be $s = 127$ ft (39 m), $t = 78$ ft (24 m), $S = 258$ ft (79 m), and $T = 80$ ft (25 m) so that the total height is 338 ft (104 m).

From Figure 4–8, we see that under self-weight load, the entire shell is in a state of compression, except for a portion above the throat which has a relatively small tensile hoop stress. From the standpoint of reinforced concrete, this stress pattern is quite favorable. With respect to the base region, $\Phi = 1.0$, there are several points to be noted. First, the state of circumferential compression does not match the anticipated tension in the ring support, as previously illustrated in the discussion of spherical shells. Second, hyperboloidal shells which are used in cooling tower applications must have openings at the base to allow air to enter. This is accomplished by supporting the shell on an annular ring of closely spaced columns, as shown in Figure 2–8(p),

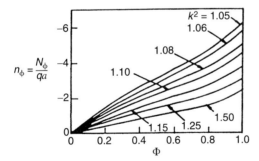

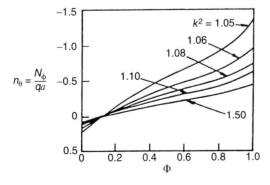

FIGURE 4–8 Nondimensional Self-Load Stress Resultants for a Hyperboloidal Shell with $a/t = 0.90$ and $a/s = 0.55$. Source: P. L. Gould and S. L. Lee, Hyperbolic cooling towers under seismic load, *Journal of the Structural Division*, ASCE, 93, no. ST3, (June 1967): 95.

producing a so-called mixed boundary, where the shell may be presumed to have zero meridional displacement within the column width and zero meridional stress between the columns. The combination of the discrete support and the ring beam clearly indicates a boundary far more complex than the idealized membrane theory case. This is considered further in subsequent sections. (Complete tabulations of membrane theory stress resultants for hyperboloidal shells of typical dimensions are provided in several available articles.)[8]

4.3.2.4 Toroidal Shells.

The toroidal shell is quite useful and efficient for pressure vessel applications. Also, segments of toroidal shells are frequently used as transitions between cylindrical tanks and shallow spherical or flat caps [Figure 2–8(y)]. This type of *compound* shell will be treated in some depth later.

Examining the toroidal geometry, Figure 4–9, we see that the surface is generated by the rotation of a closed curve, usually a circle, about an axis lying inside the curve. We have introduced the geometry of a torus in Section 2.8.2.3, example 2.8. Continuing on from that development with the curvilinear coordinates now taken as θ and ϕ, the definitive geometry is conveniently established from Figure 4–9. The meridian is the circle *ABCD*, with radius *a*. If we examine a normal to the surface defined by a meridional angle ϕ, we observe that it pierces the surface at two points; that is, there is not a one-to-one correspondence between curvilinear coordinate ϕ and a unique point on the surface. The consequence of this anomaly

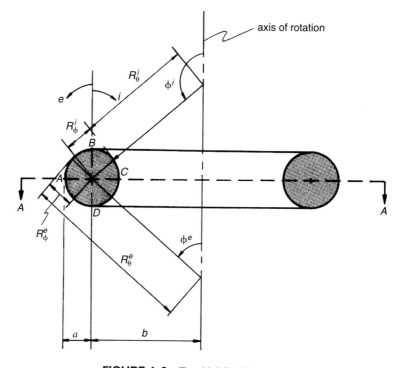

FIGURE 4–9   Toroidal Shell Geometry.

is that the segment of the shell within the semicircle *BAD* must be considered separately from that within *BCD*.

Considering the exterior segment *BAD* denoted by the superscript *e*,

$$R_\theta^e = b \csc \phi^e + a \qquad (4.46)$$

where *b* = the constant distance from the center of the generating circle to the axis of rotation.

The center of curvature of the meridian *BAD* is on the same side of the meridian as the center of curvature of R_θ^e signifying a shell of *positive Gaussian curvature*, so that

$$R_\phi^e = + a \qquad (4.47)$$

For the interior segment *BCD* denoted by the superscript *i*

$$R_\theta^i = b \csc \phi^i - a \qquad (4.48)$$

Observe that, for this segment, the center of curvature of the meridian lies on the *opposite* side of the meridian from the center of curvature of R_θ^i, indicating that this segment has *negative Gaussian curvature*. Thus,

$$R_\phi^i = - a \qquad (4.49)$$

At this point, it is appropriate to comment further on the choice of the correct algebraic sign of the principal radii of curvature when the magnitudes are obtained from a geometric rather than from a mathematical argument, as we have done here. Specifically, the condition of negative Gaussian curvature implies only that the *product* $1/R_\phi R_\theta$ is negative and does not indicate which of the radii has the negative sign. Also, a shell with positive Gaussian curvature may seemingly have both principal radii with negative signs. From the fundamental geometrical definition of the shell of revolution, Figure 2–11, the horizontal radius $R_0 = R_\theta \sin \phi$ is always positive. With $\sin \phi$ remaining positive for $0 \le \phi \le \pi$, R_0 and hence R_θ will always be positive, and thus the signs on equation (4.47) and (4.49) are necessary to give the correct sign for the Gaussian curvature. It is easily verified that equations (4.46) and (4.48) satisfy the Gauss-Codazzi condition. In the toroid, we have encountered a shell which has positive Gaussian curvature in one region and negative curvature in the other, with transition points at *B* and *D*. At *B* and *D*, $R_\theta \to \infty$, indicating a zero Gaussian curvature condition.

Example 4.3(e)

The most common loading case for toroids is a uniform internal pressure *p*. Referring to Figure 2–11, this loading corresponds to the positive sense of $\mathbf{t_n}$ for the positive curvature segment (*e*) and the negative sense of $\mathbf{t_n}$, for the negative curvature segment (*i*). Therefore, for segment *BAD* (*e*), $q_n = q_n^e = +p$, and for segment *BCD* (*i*), $q_n = q_n^i = -p$. Further, because of symmetry, we may restrict our consideration to the quarter-circles *AB* for $e(\pi/2 \ge \phi^e \ge 0)$ and *BC* for $i(\pi \ge \phi \ge \pi/2)$. We select the indefinite integral form, equation (4.8), for ψ, since this shell does not have an external boundary where ψ or N_ϕ is known or can be specified *a priori*. For *e*, $q_n = q_n^e = p$, $R_\phi = R_\phi^e = +a$ and $R_\theta = R_\theta^e = b \csc \phi^e + a$, so that

$$\psi_1 = \int p \cos \phi^e (a)(b \csc \phi^e + a)\sin \phi^e \, d\phi^e$$

$$= pa \left(a \sin \phi^e + \frac{a}{2} \sin^2 \phi^e + C_1 \right)$$

and

$$N_\phi^e = \frac{\psi_1}{R_\theta^e \sin^2 \phi^e}$$

$$= \frac{pa}{(b \csc \phi^e + a) \sin^2 \phi^e} \left(b \sin \phi^e + \frac{a}{2} \sin^2 \phi^e + C_1 \right)$$

$$= \frac{pa}{b + a \sin \phi^e} \left(b + \frac{a}{2} \sin \phi^e + \frac{C_1}{\sin \phi^e} \right)$$

For a finite meridional stress at $\phi^e = 0$ or π (points B and D), $C_1 = 0$, so that

$$N_\phi^e = \frac{pa(2b + a \sin \phi^e)}{2(b + a \sin \phi^e)} \tag{4.50}$$

To evaluate the circumferential stress resultant N_θ, we use equation (4.5)

$$N_\theta^e = (b \csc \phi^e + a)\left[p - \frac{pa(2b + a \sin \phi^e)}{2(b + a \sin \phi^e)} \right]$$

Multiplying the numerator and denominator of the rhs by $\sin \phi^e$ and simplifying, we find

$$N_\theta^e = \frac{p}{\sin \phi^e} \left(b + a \sin \phi^e - b - \frac{a \sin \phi^e}{2} \right)$$

$$= \frac{pa}{2} \tag{4.51}$$

which, of course, does not vary with ϕ.

   Using a parallel argument for segment BCD with $q_n = q_n^i = -p$, $R_\phi = R_\phi^i = -a$ and $R_\theta = R_\theta^i = b \csc \phi^i - a$,

$$N_\phi^i = \frac{pa(2b - a \sin \phi^i)}{2(b - a \sin \phi^i)} \tag{4.52}$$

and

$$N_\theta^i = \frac{pa}{2} \tag{4.53}$$

At the common points B and D, $\phi^e = 0$ and π, and $\phi^i = \pi$ and 0, respectively, and

$$N_{\phi B}^e = N_{\phi B}^i = N_{\phi D}^e = N_{\phi D}^i = pa \tag{4.54}$$

___

Interestingly, we see that the circumferential stress resultant N_θ throughout the entire toroidal shell, as well as the meridional stress resultant N_ϕ at the top and bottom circles, are independent of the relative plan size of the torous as represented by b, and are only dependent on the radius of the generating circle a. Also, as the mean plan radius b becomes large as compared to a, N_ϕ^e and $N_\phi^i \rightarrow pa$ throughout the shell. These observations have rather sim-

ple physical interpretations which give some insight into the general load-resisting characteristics of shells.

First, to investigate the circumferential stress resultant N_θ, we consider the half-plan
free body as shown in Figure 4–10. We have a resultant force of magnitude $p \times$ [projected
area of outer circle – projected area of inner circle], acting in the Y direction. From Figure
4–9, the difference in the projected areas is $[4ab + 2(1/2\pi a^2)] - [4ab - 2(1/2\pi a^2)] = 2\pi a^2$, or
simply twice the cross-sectional area, so that the resultant force is $2\pi p a^2$. This force must be
balanced by the force in the shell wall arising from N_θ. Since the pressure p is constant
throughout the cross-section, it is reasonable to regard N_θ as constant. Therefore,

$$2N_\theta(2\pi a) = 2p\pi a^2$$

or

$$N_\theta = \frac{pa}{2} \qquad\qquad (4.55)$$

which verifies equations (4.51) and (4.53). In the next section, we further illustrate that equations of *overall* (as opposed to *differential*) equilibrium on strategically chosen sections of
shells can frequently lead to simple solutions for the membrane theory stress resultants.

Next, we seek to interpret the values of N_ϕ at the top and bottom circles. From equations (4.46) and (4.48), the circumferential radius $R_\theta \to \infty$ at B and D. Considering the so-
called third equation of membrane equilibrium, equation (4.3c),

$$\frac{N_\phi}{R_\phi} + \frac{N_\theta}{R_\theta} = q_n$$

we have

$$\frac{N_\phi}{a} + \frac{N_\theta}{\infty} = p$$

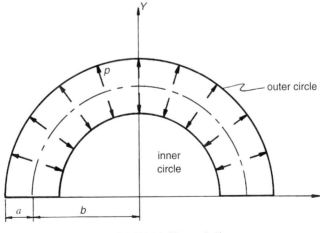

SECTION AA (Figure 4–9)

FIGURE 4–10   Toroidal Shell under Uniform Pressure Load.

or

$$N_\phi = pa \tag{4.56}$$

which verifies equation (4.54). Once N_θ has been derived from the consideration of overall equilibrium [i.e., Figure 4–10 and equation (4.55)], N_ϕ may be computed at any point on the meridian using equation (4.3c).

In general, we make frequent recourse to equation (4.3c) to interpret the results of mathematical analyses from a physical standpoint. This equation, which describes the resistance of the shell to transverse loading, is perhaps the most incisive single equation in the theory of shells. Moreover, it clearly illustrates the basic difference between plate and shell action in terms of the influence of the initial curvature of the surface.

4.4 ALTERNATE FORMULATION

4.4.1 Overall Equilibrium

In the previous section, the circumferential stress computation for a symmetrically loaded toroidal shell was verified using an equation expressing the overall equilibrium across an internal section. This procedure is quite efficient in a variety of cases. Consider the general axisymmetrically loaded shell of revolution shown in Figure 4–11 and pass a reference section normal to the axis of rotation. The trace of this section on the shell is the horizontal circle defined by the meridional angle ϕ.

To maintain equilibrium in the Z direction, the axial component of the meridional stress resultant N_ϕ multiplied by the circumference of the section must balance the resultant

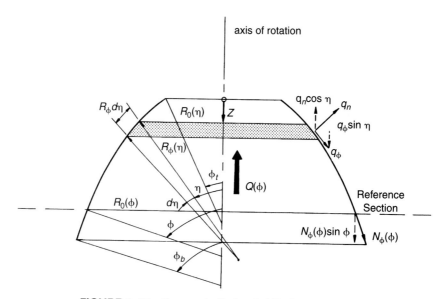

FIGURE 4–11   Symmetrically Loaded Shell of Revolution.

axial load above the section $\mathbf{Q}(\phi)$, taken positive in the *negative Z* direction to ensure that a compressive force produces a negative meridional stress.

The axial component of $N_\phi(\phi)$ is given by $N_\phi(\phi)\sin\phi$ so that

$$\mathbf{Q}(\phi) = N_\phi(\phi)\sin\phi[2\pi R_0(\phi)]\mathbf{t}_z \tag{4.57}$$

In terms of the magnitude $Q(\phi)$, we find

$$N_\phi(\phi) = \frac{Q(\phi)}{2\pi R_0(\phi)\sin\phi} \tag{4.58}$$

Then, $N_\theta(\phi)$ is found from equation (4.3c).

To compute the resultant vertical load $Q(\phi)$ when the shell is subject to a distributed surface loading, we examine a differential ring element above the section. This ring element is defined by the auxiliary meridional angle η and has a meridional length equal to $R_\phi(\eta)d\eta$. The distributed surface loading, represented by q_n and q_ϕ, contributes axial components $+q_n(\eta)\cos\eta$ and $-q_\phi(\eta)\sin\eta$ per unit area of middle surface, respectively, as shown in Figure 4–11. Then

$$Q(\phi) = \int_{\phi_t}^{\phi} [q_n(\eta)\cos\eta - q_\phi(\eta)\sin\eta][2\pi R_0(\eta)][R_\phi(\eta)d\eta] \tag{4.59}$$

which may be directly substituted into equation (4.58) to evaluate $N_\phi(\phi)$.

Example 4.4(a)

As an elementary example, we will again solve the spherical dome under self-weight load q previously considered in Section 4.3.2.2 and illustrated in Figure 4–2. From Figure 4–2, we have $q_n = -q\cos\eta$, $q_\phi = q\sin\eta$, $R_0 = a\sin\eta$, and $R_\phi = a$, so that

$$Q(\phi) = \int_{\phi_t}^{\phi} (-q\cos^2\eta - q\sin^2\eta)(2\pi a\sin\eta)a\,d\eta$$

$$= 2\pi qa^2(\cos\phi - \cos\phi_t) \tag{4.60a}$$

From equation (4.58),

$$N_\phi = \frac{Q(\phi)}{2\pi a\sin^2\phi}$$

$$= \frac{qa}{\sin^2\phi}(\cos\phi - \cos\phi_t) \tag{4.60b}$$

which is identical to equation (4.18).

Alternatively we can directly evaluate $Q(\phi)$ from the dead load q as

$$Q_\phi = \int_{\phi_t}^{\phi} (-q)(2\pi a\sin\eta)a\,d\eta$$

$$= 2\pi qa^2(\cos\phi - \cos\phi_t) \tag{4.60c}$$

We may also consider a closed rotational shell of arbitrary shape under uniform normal pressure, p. From equation (4.58)

$$N_\phi(\phi) = \frac{p\pi[R_0(\phi)]^2}{2\pi R_0(\phi)\sin\phi} = \frac{1}{2}pR_\theta \tag{4.61a}$$

whereupon

$$N_\theta(\phi) = pR_\theta\left[1 - \frac{R_\theta}{2R_\phi}\right] \tag{4.61b}$$

from equation (4.3c). These simple equations hold for *any* shell of revolution.

4.4.2 Circular Cylindrical Shells

A similar approach may be used for a circular cylindrical shell subject to an internal pressure $p(Z)$. Consider a unit length axial slice of such a cylinder with radius a as shown in Figure 4–12(a). In this case, the magnitude of the total resultant force along the $\theta = 0$ (vertical) axis is

$$Q(\theta) = 2\int_0^\theta p(Z)\cos\eta(1)(a\,d\eta) \tag{4.62a}$$
$$= 2p(Z)a\sin\theta$$

The corresponding component of the circumferential force, which must balance $Q(\theta)$, is

$$2N_\theta(\theta)\sin\theta(1) \tag{4.62b}$$

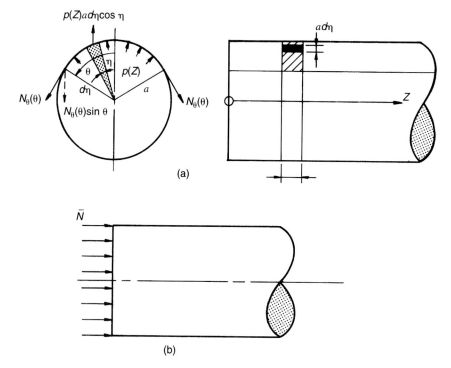

(a)

(b)

FIGURE 4–12   Symmetrically Loaded Cylindrical Shell.

Equating forces, we have

$$2N_\theta(\theta) \sin\theta = 2p(Z)a \sin\theta$$

or

$$N_\theta = p(Z)a \qquad (4.62c)$$

For a uniform pressure distribution, $p(Z) = p$ and

$$N_\theta = pa \qquad (4.62d)$$

which is, of course, well known from elementary strength of materials. Obviously equations (4.62c) and (4.62d) follow directly from equation (4.3c), since $R_\phi = \infty$. This solution also holds for incomplete or open cylindrical shells, bounded at a maximum $\theta = \pm\theta''$. This type of shell, illustrated in Figure 2–8(e), is examined in detail in Chapters 6 and 12.

We now turn to the computation of the meridional stress resultant. Here, ϕ is not suitable for the meridional coordinate, since $\phi = \pi/2$ all along the axis. Instead, the axial coordinate Z is appropriate. We return to equations (4.2) with $\alpha = Z$, $\beta = \theta$, $A = 1$, and $B = a$. For axisymmetrical loading, equation (4.2b) vanishes and equation (4.2c) confirms the result given in equation (4.62c). This leaves equation (4.2a), which becomes

$$\frac{1}{a}(aN_Z)_{,Z} + q_Z = 0$$

or

$$N_{Z,Z} = -q_Z$$

so that

$$N_Z = -\int q_Z dZ \qquad (4.63a)$$

We may also write the integral in the alternate form introduced in Section 4.3.2.1.

$$N_Z(Z) = N_Z(0) - \int_0^Z q_Z dZ \qquad (4.63b)$$

Equations (4.63a) and (4.63b) indicate that for a symmetrically loaded cylindrical shell, the meridional stress resultant is uncoupled from both the normal loading and the circumferential stress resultant and is a function only of the axial loading and the boundary conditions. For example, if the shell shown in Figure 4–12(a) is subjected to a uniform axial edge load N per unit length of circumference, as shown in Figure 4–12(b), and the pressure $q_Z = 0$, then equation (4.63b) gives

$$N_Z(Z) = \overline{N} = \text{constant} \qquad (4.64)$$

so that the effect of the edge load penetrates along the axis of the cylinder and must be developed on the opposite boundary. This unattenuated propagation of edge loads is quite plausible if one recalls that the cylindrical shells have straight-line meridians parallel to such axial loads. A similar situation arises in the membrane behavior of shells with negative Gaussian curvature, where tangential edge loads tend to propagate along the straight lines on the surface.[9]

It is interesting to compare the solutions for the toroidal shell and the cylindrical shell under uniform normal pressure loading. For this purpose, the coordinates ϕ and θ on the toroid should be regarded as equivalent to θ and Z, respectively, for the cylinder. Now, if $b \gg a$, we find [from equation (4.50) or (4.52) for the toroid and equation (4.62d) for the cylinder] that $(N_\phi)_{\text{toroid}} = (N_\theta)_{\text{cylinder}} = pa$. Furthermore, if the cylindrical shell is closed by a flat circular plate, $N_Z = p \times (\text{area/circumference}) = pa/2$, which would match N_θ as computed for the toroid in equations (4.51) and (4.53). Thus, the membrane resistance mechanism of the two geometries is remarkably similar, with the torus representing, in effect, a self-closing cylinder.

4.4.3 Conical Shells

We now investigate conical shells and initially examine the frustum, as shown in Figure 4–13(a). The shell is conveniently described in terms of the axial coordinate Z, because the meridional angle $\phi = \pi/2 - \alpha = $ constant. At any section,

$$R_0 = t + Z \tan \alpha \tag{4.65}$$

and

$$R_\theta = \frac{R_0}{\sin \phi} = \frac{t + Z \tan \alpha}{\cos \alpha} \tag{4.66}$$

Since the meridian is straight, $R_Z = \infty$, and from equation (4.3c) we immediately obtain

$$N_\theta = q_n \frac{t + Z \tan \alpha}{\cos \alpha} \tag{4.67}$$

Then we may concentrate on the meridional stress resultant N_Z.

Assume that the shell is subjected to an internal suction q, and that the top is closed by a flat plate which is free to move radially and to rotate freely at the junction with the conical shell. The idealized junction detail is shown in the inset. Under these conditions, the plate transmits a total force of only $q \pi t^2$ in the $+Z$ direction to the shell. Therefore, the resultant axial force at the top is of magnitude

$$Q_1 = -q\pi t^2 \tag{4.68}$$

This force is distributed over the circumference of the top circle, $2\pi t$, providing an edge load of intensity $qt/2$ on the shell, as shown in Figure 4–13(b).

We now want to derive the resultant force $Q_2(Z)$ due to the uniform suction. We define an auxiliary axial coordinate η and resolve the load accordingly. Then we evaluate the magnitude

$$
\begin{aligned}
Q_2(Z) &= \int_0^Z [-q \sin \alpha][2\pi(t + \eta \tan \alpha)] \frac{d\eta}{\cos \alpha} \\
&= -2\pi q \tan \alpha \int_0^Z (t + \eta \tan \alpha) d\eta \\
&= -2\pi q Z \tan \alpha \left(t + \frac{Z}{2} \tan \alpha \right)
\end{aligned}
\tag{4.69}
$$

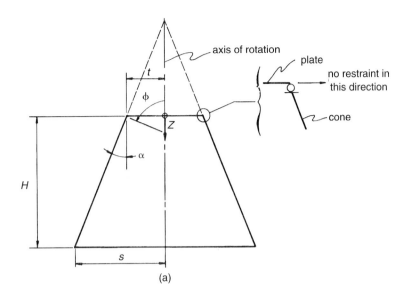

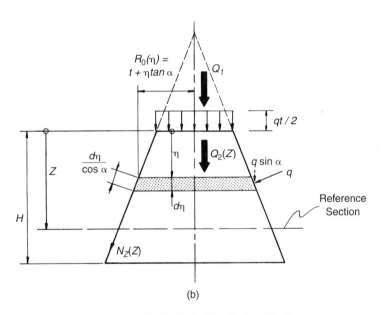

FIGURE 4–13   Conical Shell.

Because

$$N_Z(Z) \sin \phi = \frac{Q(Z)}{2\pi R_0(Z)} \tag{4.70a}$$

and

$$Q(Z) = Q_1 + Q_2(Z) \tag{4.70b}$$

then,

$$N_Z(Z) = \frac{Q_1 + Q_2(Z)}{2\pi \cos \alpha (t + Z \tan \alpha)}$$
$$= \frac{-q\{t^2 + 2Z \tan \alpha [t + (Z/2) \tan \alpha]\}}{2 \cos \alpha (t + Z \tan \alpha)} \tag{4.71a}$$

From equation (4.67), with $q_n = -q$,

$$N_\theta(Z) = -q \frac{t + Z \tan \alpha}{\cos \alpha} \tag{4.71b}$$

We note in passing that if α is set equal to 0, we have a cylinder with radius t. Equations (4.71) (a) and (b) would reduce to

$$N_Z = -\frac{qt}{2} \tag{4.72a}$$

and

$$N_\theta = -qt \tag{4.72b}$$

Since the N_Z comes entirely from the top plate reaction in the axial direction, these equations agree with the solution for cylindrical shells found in the previous section.

We may now investigate a complete cone by letting $t \to 0$, whereby

$$N_Z = -\frac{q}{2} Z \frac{\tan \alpha}{\cos \alpha} \tag{4.73a}$$

and

$$N_\theta = -qZ \frac{\tan \alpha}{\cos \alpha} \tag{4.73b}$$

The complete cone is an illustration of a pointed closed shell, shown in Figure 4–1(b). In contrast to the dome, which was examined in detail in Sections 4.3.2.1 and 4.3.2.2, N_Z and $N_\theta \to 0$ at the pole. Also, recall that for a closed spherical shell under uniform pressure, we found $N_\phi = N_\theta$ throughout; whereas, for the closed conical shell, the circumferential stress is double the meridional stress.

4.5 COMPOUND SHELLS

4.5.1 Transitional Segments

We now examine shells which have a meridian defined by more than one geometric curve; such combinations may be called *compound shells.* A common usage of this form is for pressure vessels, in which a basic cylindrical tube may be capped at each end by a so-called head and bottom, which are often shallow spherical shells. Examples of such shells are shown in Figures 2–8(r and u), and an idealized case is depicted in Figure 4–14. One's first inclination might be to make the ends hemispherical in order to provide a smooth transition; however, this form is neither developable nor readily formed by bending flat sheets. Rather, it is considerably easier to produce a relatively shallow spherical cap by rolling and dishing operations guided by suitable templates.

   Turning to the shell shown in Figure 4–14 under a positive internal pressure p, we find from equations (4.27) and (4.28) that for the spherical segments

$$N_\phi(\phi) = N_\theta(\phi) = \frac{pa_1}{2} \tag{4.74a}$$

For the cylindrical shell, from equation (4.62c),

$$N_\theta = pa_2 \tag{4.74b}$$

As far as the meridional stress resultant in the cylindrical shell is concerned, we have shown in equation (4.64) that N_Z is a constant dependent on the boundary value and the axial loading, if present. As shown in the inset of Figure 4–14, the meridional stress in the sphere imparts an edge force per unit length of $(pa_1/2)\sin\phi_1$ to the cylinder so that

$$N_Z(Z) = \left(\frac{pa_1}{2}\right)\sin\phi_1 \tag{4.75}$$

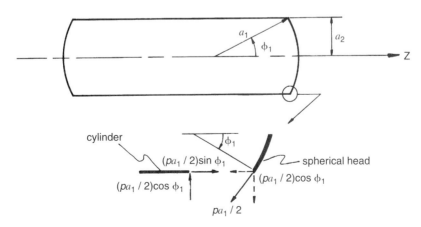

FIGURE 4–14 Cylindrical Shell with a Spherical Head.

We attempt to ascertain the admissibility of the membrane theory solution for the compound shell. First, note that the sudden transition between the cylindrical and spherical shell is a violation of the *geometrical* guideline for the existence of a membrane state of stress, as discussed in Section 4.2. It is rather obvious that the radial component of $N_\phi(\phi_1)$, $(pa_1/2)$ $\cos \phi_1$, cannot be balanced by an extensional force in the cylinder, but will introduce transverse shears and, consequently, bending which will affect both shells. Another result apparent from equation (4.74) is that the circumferential or hoop stress resultant N_θ will likely be different for both shells at the junction. This leads to discontinuities in the circumferential strain and the radial deformation, which we cannot quantify at this stage of the development but which serve as further evidence of the inadequacy of the membrane theory in the junction region. Note that even if a complete hemisphere is used for a cap making $a_1 = a_2 = a$, then

$$N_\theta \text{ (hemisphere)} = \frac{pa}{2} \tag{4.76a}$$

and

$$N_\theta \text{ (cylinder)} = pa \tag{4.76b}$$

so that the circumferential strain discrepancy will still remain, even though the unbalanced radial force $(pa/2) \cos \phi_1$ vanishes.

A common method of balancing the radial component of the thrust in the cap is to provide a ring stiffener or collar at the junction. An example is provided by Figure 2–8(r), and an idealization is shown in Figure 4–15. The ring tension **T** in the stiffener can balance the radial thrust in the same manner as a cylindrical shell resists normal pressure. Evaluating *T*, the magnitude of the radial force on the spherical head in Figure 4–14, $2T = (pa_1/2)$ $\cos \phi_1(2a_2)$, with the assumption that $w \ll a_2$. The ring, however, will restrain the radial expansion of the cylinder, and thus violate the ideal membrane *boundary* requirements.

We consider two additional possibilities of providing a smooth transition. In Figure 4–16(a), we show an ellipsoidal cap; in Figure 4–16(b), we insert a segment of a toroidal shell between the cylinder and the spherical cap. The latter form is sometimes called a *torospherical head*.

Considering the ellipsoidal head first, we note that a smooth transition is provided between the cylinder and the cap along the meridian, so that transverse shear in the cylinder is

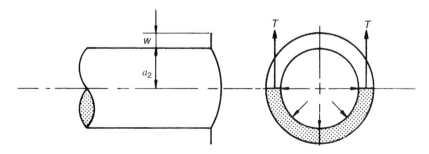

FIGURE 4–15   Ring Stiffener at a Shell Junction.

not required for radial equilibrium. However, the hoop stresses computed for an ellipsoid again do not match those for a cylinder as calculated from equation (4.76b)[10]; thus, an incompatibility in the circumferential strain and subsequent radial deformation occurs, which, as before, must be corrected by transverse shears on the cylinder and on the ellipsoid. Also, the problem of forming a deep curved cap remains.

The torospherical head shown in Figure 4–16(b) circumvents the deep cap problem but it turns out that the hoop stress and the corresponding radial deformations, as computed from membrane theory, result in incompatibilities at both the cylindrical-toroidal and torodial-spherical junctions. The torospherical shell will be treated in detail in the following section.

Meanwhile, we may illustrate an interesting property of compound pressure vessels by considering the normal equilibrium equation

$$\frac{N_\phi}{R_\phi} + \frac{N_\theta}{R_\theta} = q_n \tag{4.76c}$$

which was earlier termed the single most incisive equation in thin shell theory. We first apply the overall equilibrium approach developed in Section 4.4.1 to the shells shown in Figure 4–16, and particularly consider equation (4.61a). At any section defined by the coordi-

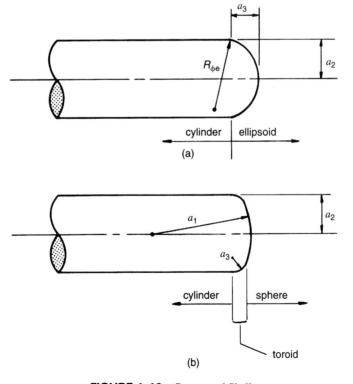

FIGURE 4–16   Compound Shells.

nate ϕ, $N_\phi(\phi)$ cannot change abruptly as long as R_θ does not change. With this in mind, we rewrite equation (4.3c) with $q_n = p$ as

$$N_\theta = R_\theta \left(p - \frac{N_\phi}{R_\phi} \right) \tag{4.77}$$

With the foregoing arguments in mind, at a smooth transition where R_θ is constant,

$$N_\theta \text{ is proportional to } \left[\frac{-N_\phi}{R_\phi} + p \right]$$

Thus, even if the transition is smooth along the meridian and the loading is also smooth, a sudden change in R_ϕ can *radically* change the magnitude and even the algebraic sign of N_θ. Recall that the cylindrical shell in Figure 4–16 resists the pressure p by N_θ/R_θ, because R_ϕ is ∞. This occurs regardless of the magnitude of N_ϕ. Suddenly, at the transition, we encounter a finite value of $R_\phi, R_{\phi e}$ for the ellipsoid and a_3 for the toroid. Since N_ϕ is positive, equation (4.77) reveals that N_θ decreases in proportion to the now-finite value of R_ϕ, even perhaps becoming *negative*. Both cases in Figure 4-16 may produce a large negative value of the hoop stress resultant N_θ,[11] and such shells may fail because of circumferential wrinkling (buckling) if proper strengthening and stiffening is not provided. An actual case where this occurred is described in Fino and Schneider.[12] The shell segment shown in Figure 2–8(y) was fabricated to quantify this failure mechanism further through an experimental study.

4.5.2 Torospherical Head

In the preceding section, the possibility of inserting a torodial segment between a spherical cap and a cylindrical shell was discussed. The geometry of this shell is shown in more detail in Figure 4–17. In terms of the radii of curvature of the sphere, a_1, the toroid, a_3, and the cylinder, a_2, the meridional angle at the sphere-toroid junction is given by

$$\sin \phi_1 = \frac{a_2 - a_3}{a_1 - a_3} \tag{4.78}$$

In practice, the radius of the toroid is selected so as to provide smooth transitions at both junctions.

To derive the membrane theory stress resultants, we again use the overall equilibrium method. At a general section within the toroid, $\phi_1 < \phi < \pi/2$, the resultant axial load consists of two parts: $\mathbf{Q_1}$ = the load on the spherical cap, and $\mathbf{Q_2}(\phi)$ = the load between ϕ_1 and ϕ on the toroid. For a uniform positive pressure p, the magnitude of the first force is the pressure multiplied by the projected base area of the cap,

$$Q_1 = p\pi(a_1 \sin \phi_1)^2 \tag{4.79a}$$

For the second force, we first compute the horizontal radius of the toroid

$$R_0(\phi) = (a_1 - a_3)\sin \phi_1 + a_3 \sin \phi \tag{4.79b}$$

whereupon the magnitude equals the pressure multiplied by the area of the projected annulus,

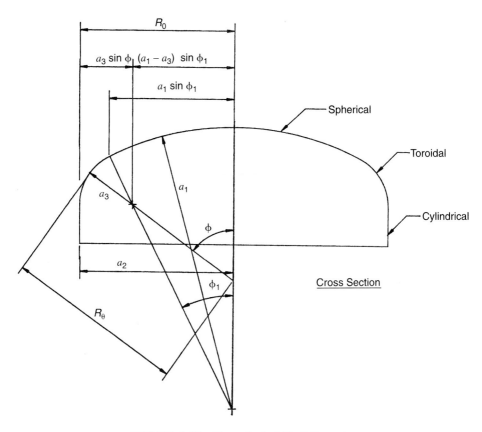

FIGURE 4–17   Torospherical Shell Geometry.

$$Q_2(\phi) = p\pi[R_0^2 - (a_1 \sin \phi_1)^2]$$
$$= p\pi[(a_1 - a_3)^2 \sin^2 \phi_1 + a_3^2 \sin^2 \phi$$
$$+ 2(a_1 - a_3)a_3 \sin \phi_1 \sin \phi - a_1^2 \sin^2 \phi_1]$$

which reduces to

$$Q_2(\phi) = p\pi a_3(\sin \phi - \sin \phi_1)[a_3(\sin \phi - \sin \phi_1) + 2a_1 \sin \phi_1] \qquad (4.79c)$$

Summing equations (4.79a) and (4.79c), we have

$$Q(\phi) = p\pi \overline{Q}(\phi) \qquad (4.80a)$$

where

$$\overline{Q}(\phi) = (a_1 \sin \phi_1)^2 + a_3(\sin \phi - \sin \phi_1)[a_3(\sin \phi - \sin \phi_1) + 2a_1 \sin \phi_1] \qquad (4.80b)$$

From equation (4.58),

$$N_\phi = \frac{1}{2\pi R_0 \sin\phi} Q(\phi)$$

$$= \frac{p\overline{Q}(\phi)}{2[(a_1 - a_3)\sin\phi_1 + a_3 \sin\phi] \sin\phi} \tag{4.81a}$$

To compute the circumferential stress resultant, we use equation (4.3c), which gives

$$N_\theta = R_\theta \left(p - \frac{N_\phi}{R_\phi} \right) \tag{4.81b}$$

From equation (4.79b) since $R_0 = R_\theta \sin\phi$,

$$R_\theta = (a_1 - a_3)\frac{\sin\phi_1}{\sin\phi} + a_3 \tag{4.82a}$$

From Figure 4–17,

$$R_\phi = a_3 \tag{4.82b}$$

Substituting equations (4.81a) and (4.82a and b) into equation (4.81b) gives

$$N_\theta = p \left[\frac{(a_1 - a_3) \sin\phi_1 + a_3 \sin\phi}{\sin\phi} \right]$$
$$\cdot \left\{ 1 - \frac{\overline{Q}(\phi)}{2a_3 \sin\phi[(a_1 - a_3)\sin\phi_1 + a_3 \sin\phi]} \right\} \tag{4.83}$$

Equations (4.81a) and (4.83) constitute the membrane theory solution for the so-called toroidal knuckle portion of the torospherical head. As implied in the previous section, the form of equations (4.81b) and (4.83) suggests that N_θ will probably be negative (compressive) in the toroidal segment, whereas the circumferential stress in both adjacent segments will be positive (tensile).

Example 4.5(a)

The dimensions of the test specimen shown in Figure 2–8(y) are given in Figure 4–18(a). This shell has been subjected to extensive analytical and experimental investigation and will be used here to compute the linear stress pattern caused by internal pressure.

From Figure 4–18(a), allowing for the wall thickness $h = 0.2$ in., the geometrical properties are

$$a_1 = 172.9 \text{ in.}$$
$$a_2 = 96.1 \text{ in.}$$
$$a_3 = 32.74 \text{ in.}$$

as shown in Figure 4–18(b), so that from equation (4.78)

$$\sin\phi_1 = \frac{96.1 - 32.74}{172.9 - 32.74}$$
$$= 0.45205$$

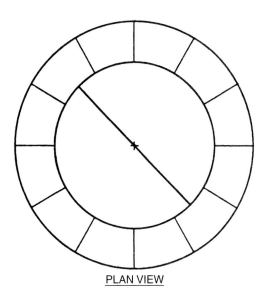

PLAN VIEW

HEAD NO.1 $h = 0.20''$
HEAD NO.2 $h = 0.25''$

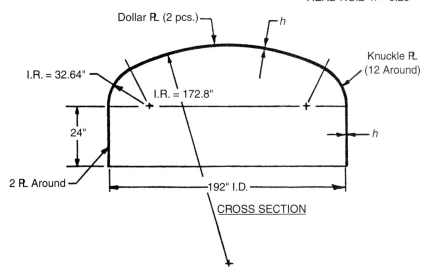

CROSS SECTION

FIGURE 4–18a   Torospherical Test Specimen.

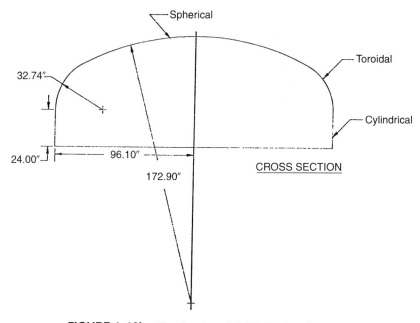

FIGURE 4–18b   Test Specimen Middle Surface Geometry.

and

$$\phi_1 = 26.88°$$

The equation of the meridian was established with the origin set at the pole, and the resulting equations are as follows:

Spherical cap:

$$Z^2 + R^2 - 345.8Z = 0 \tag{4.84a}$$

Toroidal knuckle:

$$Z^2 + R^2 - 95.7574Z - 126.72R + 5234.9519 = 0 \tag{4.84b}$$

Cylindrical segment:

$$R - 96.1 = 0 \tag{4.84c}$$

Pertinent material properties are Young's modulus E = 30,000 psi and Poisson's ratio $\mu = 0.3$. Equations (4.84) may be used to verify the continuity of slopes at the junctions.

In Figure 4–19, the circumferential stress resultant N_θ is shown as a function of the arc length s_ϕ. The change from tension in the spherical cap to compression in the torus and again to tension in the cylinder is clearly illustrated. Along with N_θ, a scale for the extensional component of the circumferential stress, $\sigma_{\theta\theta} = N_\theta/h$, is also given to facilitate comparisons with bending theory solutions, which are somewhat thickness-dependent. The abrupt jumps are mitigated by bending effects, which will be discussed in Chapter 13. The extension of this analysis is shown on Figure 13–4.

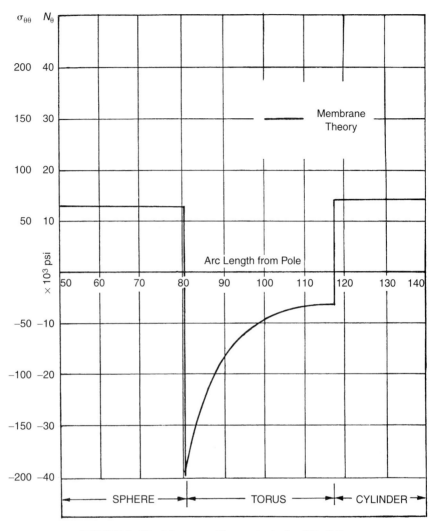

FIGURE 4–19   Membrane Theory Analysis of Shell Cap.

REFERENCES

1. W. Flügge, *Stresses in Shells,* 2nd ed. (Berlin: Springer-Verlag, 1973): 100–102.

2. V. V. Novozhilov, *Thin Shell Theory* [translated from 2nd Russian ed. by P. G. Lowe] (Groningen, The Netherlands: Noordhoff, 1964): 105–107.

3. P. L. Gould, A. Cataloglu, G. Dhatt, A. Chattopadhyay, and R. E. Clark, Stress analysis of the human heart valve, *Journal of Computers and Structures, 3,* 1973: 377–384.

4. S. Timoshenko and S. Woinowsky-Krieger, *Theory of Plates and Shells,* 2nd ed. (New York: McGraw-Hill, 1959): 449–450.

5. Novozhilov, *Thin Shell Theory:* 117–119.

6. D. P. Billington, *Thin Shell Concrete Structures,* 2nd ed. (New York: McGraw-Hill, 1982): 114.

7. E. H. Baker, L. Kovalevsky, and F. L. Rish, *Structural Analysis of Shells* (New York: McGraw-Hill, 1972): 256.

8. P. L. Gould and S. L. Lee, Hyperbolic cooling towers under seismic design load, *Journal of the Structural Division,* ASCE 93, no. ST3, (June 1967): 87–109. Closure, *Journal of the Structural Division,* ASCE 94, no. ST10 (October 1968): 2487–2493; S. L. Lee and P. L. Gould, Hyperbolic cooling towers under wind load, *Journal of the Structural Division,* ASCE 93, no. ST5 (October 1967): 487–514.

9. Flügge, *Stresses in Shells,* 2nd ed.: 171–179.

10. Novozhilov, *Thin Shell Theory:* 130–138.

11. Ibid.

12. A. Fino and R. W. Schneider, Wrinkling of a large thin code head under internal pressure, *Welding Research Council Bulletin* no. 69 (New York: Welding Research Council, June 1961): 11–13.

13. V. S. Kelkar and R. T. Sewell, *Fundamentals of the Analysis and Design of Shell Structures* (Englewood Cliffs, N. J.: Prentice-Hall, Inc., 1987): 79, 193–194, 307–309.

EXERCISES

4.1. Consider the spherical shell shown in Figure 4–2, with a snow load of p, constant per unit area of plan projection.
   (a) Complete a membrane theory analysis for this loading, including a graphical study similar to Figure 4–3.
   (b) At what angle ϕ does the circumferential stress change from compression to tension?

4.2. For the spherical shell under snow load described in the preceding exercise and bound by a base angle ϕ_b, compute the tensile force that would be developed in the ring beam at $\phi = \phi_b$.

4.3. Consider the spherical lantern shell as shown in Figure 4–20.
   (a) Compute the membrane theory stress resultants caused by:
       (1) A uniform downward vertical line load of 500 (force/unit length of circumference) on the upper ring.

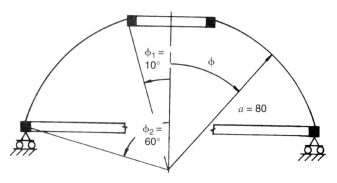

FIGURE 4–20 Exercise 4.3

(2) A live load of 30 (force/unit area of horizontal projection).

(3) A dead load of 75 (force/unit area of middle surface).

Compute each case separately, and then combine the results.

(b) Compute the forces that must be developed by horizontal ring beams at the upper and lower edges, so that the shell will behave approximately as a membrane.

4.4. In Section 4.3.2.2, we saw that the self-weight meridional stress resultant in a uniform thickness spherical dome increases as ϕ increases. The corresponding stress $\sigma_{\phi\phi} = N_\phi/h$ will, of course, increase in the same fashion. To try to equalize the self-weight meridional stress in a hemispherical dome, it is proposed to increase the thickness from a basic value $h = h_0$ at $\phi = 0$ to $h = h_b$ at $\phi = 90°$ by using the transition $h(\phi) = h_0 + (h_b - h_0)\sin \phi$.

(a) Compute the N_ϕ and N_θ for the resulting self-weight load, where the shell material has mass density p.

(b) Show the variation of the meridional stress $\sigma_{\phi\phi}$ from $\phi = 0$ to $\phi = 90°$.

(c) Select a value for h_b as a function of h_0, so that the condition at the base will govern.

4.5. Consider the ellipsoid of revolution shown in Figure 4–21.

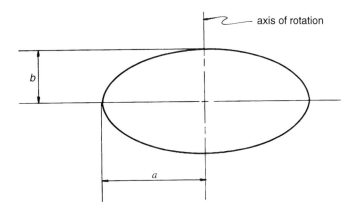

FIGURE 4–21   Exercise 4.5

(a) Compute the expressions for the principal radii of curvature.

(b) Derive expressions for the membrane theory stress resultants caused by a uniform internal suction q.

(c) Is it possible to have tensile stresses in this shell under the suction q? For what dimensional combination will this occur?

4.6. An open ogival shell is shown in Figure 4–22.

(a) Determine the membrane theory stress resultants caused by

(1) Dead load, q (force/unit area of middle surface).

(2) Live load, p (force/unit area of plan projection).

(3) Ring load, w (force/unit length of circumference).

Evaluate the effects of each loading separately. For cases (1) and (2), derive the expressions by using the general formulation for axisymmetrically loaded shells of revolution. Check cases (1) and (2) as well as (3) by using the alternative formulation based on overall vertical equilibrium.

(b) Investigate the singularity as $R_1 \to 0$.

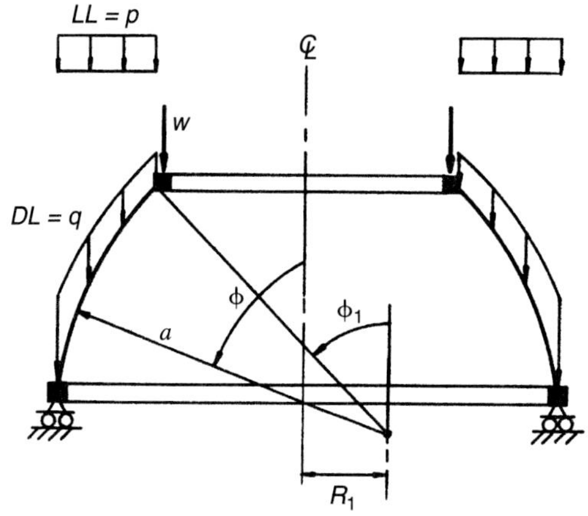

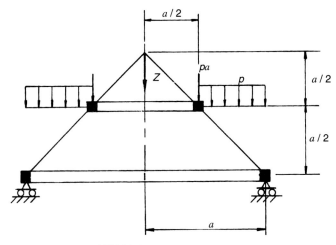

FIGURE 4–22   Exercise 4.6

4.7. Consider a hyperboloid of revolution as shown in Figure 4–6.
  (a) Derive R_ϕ, equation (4.38), directly from equation (2.29).
  (b) Show that a transformation between ϕ and R_0 equivalent to equation (4.44) can be derived from equation (4.36).
  (c) In terms of the parameter k and the meridional angle ϕ or the axial coordinate Z, determine the membrane theory stress resultants caused by a hydrostatic loading $q_h = \gamma\,(T - Z)$, where γ is the unit weight of the fluid.

4.8. For a constant thickness vertical cylindrical tank of radius a and height H, compute the membrane theory stress resultant distribution attributed to hydrostatic pressure, using a fluid with unit weight γ, for the following cases:
  (a) The tank is full.
  (b) The tank is half-full.

4.9. Consider the conical shell shown in Figure 4–23. The shell has a line load, pa, applied to the collar at midheight, and a uniform live load, p, applied only on the area outside of the collar.

FIGURE 4–23   Exercise 4.9

   (a) Compute the membrane theory stress resultants in the shell.
   (b) Compute the force in the ring beam at the base.
4.10. The double conical pressure vessel shown in Figure 4–24 is supported at midheight by a circular
       ring beam in the horizontal plane.

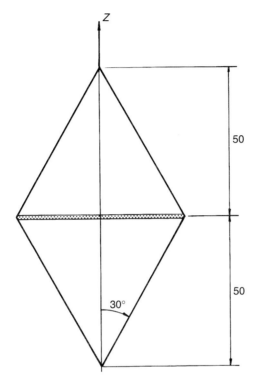

FIGURE 4–24   Exercise 4.10

   (a) Determine the membrane theory stress resultants caused by a uniform internal pressure
       $p = 30$ (force/unit area) and a self-weight of $q = 3$ (force/unit area).
   (b) For each loading case, show the variations of N_ϕ and N_θ with Z and compute the reactions
       on the ring beam. In particular, note the maxima for N_ϕ and N_θ.
4.11. Analyze the paraboloidal shell of revolution with dimensions as shown in Figure 4–25 for

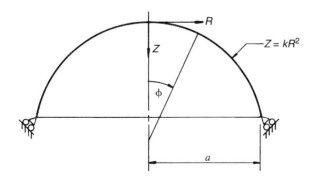

FIGURE 4–25   Exercise 4.11

(a) Self-weight, q.

(b) Uniform internal pressure, p.

4.12. Re-solve the compound shell shown in Figure 4–14 for a hydrostatic loading γZ, where Z is measured from the pole of the spherical cap at the left end. Determine the incompatibility in the circumferential stress resultant at the junctions of the two shells.

4.13. A cylindrical tank with an ellipsoidal bottom is shown in Figure 4–26. The shell is subject to a hydrostatic load from a fluid with unit weight γ.

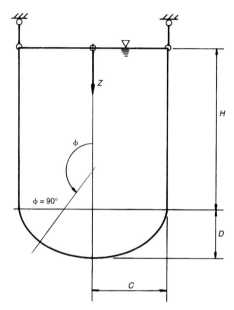

FIGURE 4–26   Exercise 4.13

(a) Compute the membrane theory stress resultants in the shell.

(b) Plot the stress resultants along the meridian.

(c) For what range of C and D will the circumferential stress in the ellipsoidal bottom become compressive?

4.14. Analyze the cylindrical pressure vessel with conical dish ends shown in Figure 4–27 using the membrane theory. Consider an internal pressure of 100. Discuss the shortcomings of the solution and suggest at least two remedies.

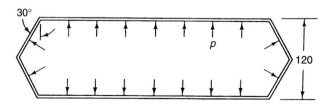

FIGURE 4–27   Exercise 4.14. (Adapted from Kelkar and Sewell, Ref. 13.)

4.15. The meridian of the shell roof covering a mosque in Islamic-style architecture (Figure 1–5) is shown in Figure 4–28. The pointed or ogival cap (Figure 4–1b) is formed by a constant meridional radius with the center of curvature offset from the axis of rotation. This is termed a *spheroidal* shell. The lower segment is a toroidal shell tangent to the spheroidal shell at the horizontal circle B.

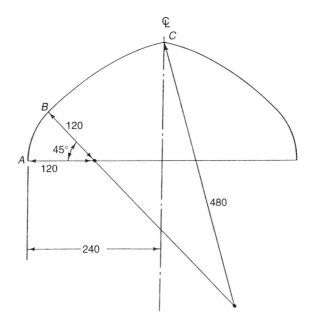

FIGURE 4–28   Exercise 4.15. (Adapted from Kelkar and Swell, Ref. 13.)

   (a) For a uniform live load of 100, compute the stress resultants in the dome
   (b) Discuss the shortcomings of the solution at junction B

4.16. Consider a hyperbolodial shell of revolution as shown in Figure 4–6. The shell is subject to a vertical static seismic loading $q = fq_{DL}$, where f is a known fraction. Determine the membrane theory stress resultants.

4.17. To optimize the efficiency of the shell material, the meridional and circumferential stress resultants N_ϕ and N_θ should be equal at all points on the shell. This can be accomplished by specifying that $N_\phi(\phi,\theta) = N_\theta(\phi,\theta) = N^*$ at every point of a dome of constant thickness for a given axisymmetric loading condition and then finding the requisite shell of revolution geometry, i.e., $R_\phi(\phi)$, $R_\theta(\phi)$, $R_0(\phi)$.

Investigate this possibility for the following loading cases:
   (a) Uniform normal pressure, q = constant.
   (b) Hydrostatic pressure $q = \gamma Z$, where Z is measured from the pole. Part (b) is analogous to the problem of determining the shape of a drop of liquid resting on a plane surface, where N^* is the surface tension. (See A. Pfluger, *Elementary Statics of Shells,* 2nd ed., trans. E. Galantay, New York: F. W. Dodge, 1961, pp. 42–44.)

5

Membrane Theory:
Applications to Nonsymmetrically Loaded Shells of Revolution

5.1 NONSYMMETRICAL LOADING

5.1.1 Transformation of Equilibrium Equations

Thus far, we have considered the membrane theory analysis of shells of revolution subject to loading that is nonvarying with respect to the circumferential coordinate θ, so-called symmetric loading. The solutions obtained in the previous chapter were based principally on equation (4.7a), which is written in terms of the auxiliary variable ψ. This equation was obtained by suppressing the θ dependence. We now return to equation (4.6) and eliminate ξ by taking $\partial/\partial\phi$ of equation (4.6a); $\partial/\partial\theta$ of equation (4.6b); then taking the difference of the two equations; and finally by dividing both sides by the leading term $R_\phi R_\theta \sin\phi$, which gives[1]

$$
\frac{1}{R_\phi R_\theta \sin\phi} \left(\frac{R_\theta^2 \sin\phi}{R_\phi} \psi,_\phi \right),_\phi + \frac{1}{R_\phi \sin^2\phi} \psi,_{\theta\theta}
$$

$$
= \frac{1}{R_\phi R_\theta \sin\phi} [q_n \cos\phi - q_\theta \sin\phi) R_\theta^3 \sin^2\phi],_\phi + R_\theta(q_{n,\theta\theta} + q_{\theta,\theta} \sin\phi)
$$

(5.1)

Equation (5.1) is a second-order partial differential equation in ψ. Once ψ is determined, ξ may be found by solving equation (4.6a) or (4.6b).

5.1.2 Fourier Series Representation

Since we have a partial differential equation to solve, we try the standard method of separation of variables. Specifically, we apply the Fourier series technique, whereby all loadings and dependent variables are taken in the form of Fourier series:

Loading:

$$\{q\} = \begin{Bmatrix} q_\phi \\ q_\theta \\ q_n \end{Bmatrix} = \sum_{j=0}^{\infty} \begin{Bmatrix} q_\phi^j \cos j\theta \\ q_\theta^j \sin j\theta \\ q_n^j \cos j\theta \end{Bmatrix} \tag{5.2a}$$

Stress Resultants:

$$\begin{Bmatrix} N_\phi \\ N_\theta \\ S \end{Bmatrix} = \sum_{j=0}^{\infty} \begin{Bmatrix} N_\phi^j \cos j\theta \\ N_\theta^j \cos j\theta \\ S^j \sin j\theta \end{Bmatrix} \tag{5.2b}$$

Auxiliary Variables:

$$\begin{Bmatrix} \psi \\ \xi \end{Bmatrix} = \sum_{j=0}^{\infty} \begin{Bmatrix} \psi^j \cos j\theta \\ \xi^j \sin j\theta \end{Bmatrix} \tag{5.2c}$$

Substituting equations (5.2) into equations (4.6a), (4.6b), and (5.1), we get

$$\sum_{j=0}^{\infty} \left\{ \left[\frac{R_\theta^2 \sin\phi}{R_\phi} \psi,_\phi^j + j\xi^j \right] \cos j\theta = [(q_n^j \cos\phi - q_\phi^j \sin\phi)R_\theta^3 \sin^2\phi] \cos j\theta \right\} \tag{5.3a}$$

$$\sum_{j=0}^{\infty} \left\{ \left[\xi,_\phi^j + j\frac{R_\theta}{\sin\phi} \psi^j \right] \sin j\theta = -[(q_\theta^j \sin\phi - jq_n^j)R_\phi R_\theta^2 \sin\phi] \sin j\theta \right\} \tag{5.3b}$$

and

$$\sum_{j=0}^{\infty} \left\{ \left[\frac{1}{R_\phi R_\theta \sin\phi} \left(\frac{R_\theta^2 \sin\phi}{R_\phi} \psi,_\phi^j \right),_\phi - \frac{j^2}{R_\phi \sin^2\phi} \psi^j \right] \cos j\theta = \right.$$
$$\left. \left[\frac{1}{R_\phi R_\theta \sin\phi} [(q_n^j \cos\phi - q_\phi^j \sin\phi)R_\theta^3 \sin^2\phi],_\phi - jR_\theta(jq_n^j - q_\theta^j \sin\phi) \right] \cos j\theta \right\} \tag{5.3c}$$

We may treat each equation of the series separately, and then obtain the total solution by summing to a suitable truncation limit on j. We recognize that the individual equations for each harmonic are ordinary, rather than partial, differential equations, because the θ-dependent terms may be cancelled. This is, of course, a formidable mathematical simplification. Also, we see that equations (5.3a) and (5.3b) with $j = 0$, reduce to equations (4.7a) and (4.7b), respectively, which have previously been investigated with respect to axisymmetric loading. Thus, the previous solution for axisymmetric loading, as given by equation (4.8) and (4.9), also serves as the solution of the $j = 0$ harmonic for the nonsymmetric loading situation, with the loading taken as q_ϕ^0 and q_n^0 which are the $j = 0$ components of the surface loading. These solutions do not include the case of a $j = 0$ component of q_θ which, as mentioned in Section 4.3.2.1, is quite rare. This case can be accommodated, however, by simply interchanging the $\sin j\theta$ and $\cos j\theta$ terms in equations (5.2a–c), obtaining an analogous set of equations to equations (5.3a–c), and combining the results of the two solutions.

Since the $j = 0$ case has already been solved, we now seek a solution to equation (5.3c) for a general harmonic $j > 0$. After cancelling the $\cos j\theta$ terms, we have for harmonic j

$$\frac{1}{R_\phi R_\theta \sin \phi} \left(\frac{R_\theta^2 \sin \phi}{\phi} \psi,_\phi^j \right),_\phi - \frac{j^2}{R_\phi \sin^2 \phi} \psi^j$$
$$= \frac{1}{R_\phi R_\theta \sin \phi} [(q_n^j \cos \phi - q_\phi^j \sin \phi) R_\theta^3 \sin^2 \phi],_\phi - jR_\theta(jq_n^j - q_\theta^j \sin \phi) \tag{5.4}$$

We postpone consideration of the general solution of equation (5.4) until further specializations are introduced. For now, note that once ψ^j has been determined, we may solve for ξ^j from either equation (5.3a) or (5.3b). Because the variable ξ^j itself occurs in equation (5.3a), this equation is usually selected. For each harmonic, the stress resultants at any circumferential angle may then be computed by evaluating N_ϕ^j and S^j using equations (4.4), and N_θ^j using equation (4.3c), and then substituting the appropriate value of θ into equations (5.2b). In passing, note that the "price paid" for the mathematical simplification of reducing the partial differential equation to an uncoupled series of ordinary differential equations is reflected in the total number of harmonic solutions that must be summed to adequately represent the effect of the particular surface loading in a given case. Each harmonic solution entails a complete analysis of the shell, whereas the partial differential equation, (5.1), need be solved only once. In spite of this, the Fourier series approach has been widely used to solve shells of revolution subject to unsymmetric loading because it is simpler to carry out.

The eventual solution of equation (5.4) is strongly dependent on the harmonic number j. The case $j = 1$ is called *antisymmetric* loading and is somewhat more tractable than the case $j > 1$, which will be called *asymmetric* loading.

5.2 ANTISYMMETRICAL LOADING

5.2.1 Integration of Equilibrium Equations and Boundary Requirements

Consider equation (5.4) with $j = 1$ and introduce a further auxiliary variable[1]

$$Y(\phi) = \psi^1(\phi) R_\theta \sin \phi \tag{5.5a}$$

which gives

$$\left[\frac{1}{R_\phi \sin \phi} Y,_\phi \right],_\phi = \frac{1}{R_\theta \sin \phi} [(q_n^1 \cos \phi - q_\phi^1 \sin \phi) R_\theta^3 \sin^2 \phi],_\phi$$
$$- R_\theta R_\phi (q_n^1 - q_\theta^1 \sin \phi) \tag{5.5b}$$

The transformation between equations (5.4) and (5.5b) is most easily verified by substituting equation (5.5a) into equation (5.5b) and performing the indicated differentiation with the aid of the Gauss-Codazzi relation, equation (2.51).

Equation (5.5b) is in a form which permits Y to be evaluated by successive integrations with respect to ϕ. Then, ψ^1 is found from equation (5.5a), and the remaining auxiliary variables and stress resultants are determined from equations (5.3a), (5.2a), (4.4a), (4.4b), and (4.5). This procedure is straightforward and, since it is discussed in detail in Novozhilov,[2] it is not repeated here. Rather, we present the results which may be easily verified as a solution by direct substitution into the membrane theory equilibrium equations, (4.3a–c). The stress resultants corresponding to $j = 1$ are

$$N_\phi = N_\phi^1 \cos\theta; \quad N_\theta = N_\theta^1 \cos\theta; \quad S = S^1 \sin\theta \qquad (5.6a, b, c)$$

where the Fourier coefficients are given by

$$N_\phi^1 = \frac{1}{R_0^2 \sin\phi} \left\{ C_2 + C_1 \int_{\phi'}^{\phi} R_\phi \sin\phi \, d\phi + \int_{\phi'}^{\phi} \Phi(\phi) R_\phi \sin\phi \, d\phi \right\} \qquad (5.6d)$$

$$N_\theta^1 = q_n^1 R_\theta - \frac{R_\theta}{R_\phi} N_\phi^1 \qquad (5.6e)$$

$$S^1 = N_\phi^1 \cos\phi - \frac{C_1}{R_0} + \chi(\phi) \qquad (5.6f)$$

and the functions $\Phi(\phi)$ and $\chi(\phi)$ are

$$\Phi(\phi) = (q_n^1 \cos\phi - q_\phi^1 \sin\phi) R_0 R_\theta - \int_{\phi'}^{\phi} (q_n^1 \sin\phi + q_\phi^1 \cos\phi - q_\theta^1) R_\phi R_0 \, d\phi \qquad (5.7a)$$

and

$$\chi(\phi) = \frac{1}{R_0} \int_{\phi'}^{\phi} (q_n^1 \sin\phi + q_\phi^1 \cos\phi - q_\theta^1) R_\phi R_0 \, d\phi \qquad (5.7b)$$

Here, we assume that both N_ϕ and S are specified on the same boundary, $\phi = \phi'$. C_1 and C_2 are integration constants evaluated from equations (5.6d) and (5.6f) as

$$C_2 = N_\phi^1(\phi') R_0^2(\phi') \sin\phi' \qquad (5.8a)$$

and

$$C_1 = R_0(\phi') [N_\phi^1(\phi') \cos\phi' - S^1(\phi')] \qquad (5.8b)$$

so that a stress-free top edge gives $C_1 = C_2 = 0$. Also, for a dome we take $\phi' = 0$. Then, $R_0 = 0$ and $C_1 = C_2 = 0$. The more general case, where N_ϕ and S are specified at different boundaries, may be treated in an analogous manner.

To evaluate the stress resultants at the pole of a dome for $j = 1$, we first consider equation (5.6d) with $\phi' = 0$ and $C_1 = C_2 = 0$, and recognize that we have a 0/0 indeterminate form. We apply L'Hospital's rule by differentiating the numerator and denominator of equation (5.6d). Noting that

$$\begin{aligned} (R_0^2 \sin\phi),_\phi &= (\sin\phi) 2 R_0 R_\phi \cos\phi + R_0^2 \cos\phi \\ &= R_0 \cos\phi (2 R_\phi \sin\phi + R_0) \end{aligned} \qquad (5.9)$$

we get

$$\lim_{\phi \to 0} N_\phi^1(\phi) = \lim_{\phi \to 0} \frac{\Phi R_\phi \sin \phi}{R_0 \cos \phi (2R_\phi \sin \phi + R_0)} \tag{5.10}$$

which remains an indeterminate form. Therefore, we must apply L'Hospital's rule once more to equation (5.10). Considering the numerator, after canceling $\sin \phi$ with $R_0 = R_\theta \sin \phi$,

$$(\Phi R_\phi)_{,\phi} = R_\phi \Phi_{,\phi} + \Phi R_{\phi,\phi}$$

Because $\Phi(0) = 0$, the second term vanishes. Evaluating $\Phi_{,\phi}(0)$, we find that this term also vanishes provided $q_n^1(0) = 0$, which is a reasonable smoothness requirement as will be apparent later. Thus,

$$\lim_{\phi \to 0} (\Phi R_\phi \sin \phi)_{,\phi} = 0 + 0 = 0 \tag{5.11}$$

Differentiating the denominator as modified by replacing R_0 with $R_\theta \sin \phi$, and then cancelling $\sin \phi$ gives

$$2(R_\theta \sin \phi \, R_\phi \cos \phi)_{,\phi} + (R_0 R_\theta \cos \phi)_{,\phi} \tag{5.12}$$

At $\phi = 0$, the second term in equation (5.12) goes to zero. Considering the first term, we have

$$2(R_\theta \sin \phi \, R_\phi \cos \phi)_{,\phi} = 2[R_\phi \cos \phi (R_\phi \cos \phi) + R_\theta \sin \phi (R_\phi \cos \phi)_{,\phi}] \tag{5.13}$$

As $\phi \to 0$, the second term of equation (5.13) also goes to 0, and the first term becomes $2R_\phi^2(0)$. Thus, we finally conclude that

$$\lim_{\phi \to 0} N_\phi^1(\phi) = \frac{0}{R_\phi(0)[2R_\phi(0) + R_\theta(0)]} = 0 \tag{5.14a}$$

or

$$N_\phi^1(0) = 0 \tag{5.14b}$$

We may then establish from equation (5.6f) that

$$S^1(0) = 0 \tag{5.15}$$

because $C_1 = 0$ and $\chi(0) = 0$ by L'Hospital's rule applied to equation (5.7b).

Examining equation (5.6e) for N_θ^1 as $\phi \to 0$, the second term goes to 0, and the first term remains. Thus, we find

$$N_\theta^1(0) = q_n^1(0) R_\theta(0) \tag{5.16}$$

In Section 4.3.2.1, we proved that $N_\phi(0) = N_\theta(0)$ for a dome. Equation (5.14b) implies that also $N_\theta^1(0) = 0$ which, in view of equation (5.16), requires that $q_n^1(0)$ must be equal to 0 for a properly defined antisymmetric loading. An example is presented in the following section.

Finally with respect to the general analysis of antisymmetrically loaded shells of revolution, note that we have evaluated the constants of integration through the consideration of (a) specified values of N_ϕ^1 and S^1 at given boundaries for open shells; or, (b) pole conditions for domes. It is required that the remaining boundary develops the calculated values of N_ϕ and S to ensure membrane action. As illustrated in Figure 5–1(a), the displacements corresponding to these stress resultants, in the meridional direction for N_ϕ and in the tangential direction for S, must be 0.

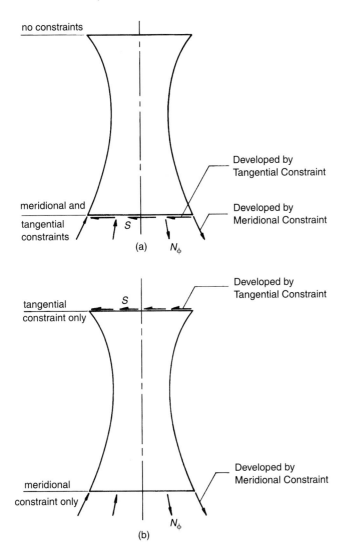

no constraints

Developed by
Tangential Constraint

meridional and

Developed by
Meridional Constraint

tangential
constraints

S

(a) N_ϕ

tangential
constraint only

S

Developed by
Tangential Constraint

meridional
constraint only

Developed by
Meridional Constraint

N_ϕ

(b)

FIGURE 5–1 Boundary Requirements for Shells of Negative Gaussian Curvature.

Such a set of boundary conditions (meridional and tangential displacements equal to 0 on one boundary) are apparently sufficient to ensure membrane behavior, provided that the corresponding geometrical and loading restrictions are observed. The question arises as to the possibility of specifying an alternate set of conditions for an open shell, e.g., meridional displacement equal to 0 on one boundary, and circumferential displacement equal to 0 on the other, as shown in Figure 5–1(b). The study of this possibility requires the consideration of *pure bending deformations,* which are deformations that produce neither extension nor in-plane shearing of the middle surface.

Although a detailed study of pure bending deformations is beyond the scope of this text, one can show that shells for which this behavior is possible are inherently unstable. The requisite circumstances are not primarily a function of *applied loading,* but are related to the *geometrical properties* and the *boundary conditions.* Mathematically, one can show that such deformations constitute a nontrivial solution to the equations describing the middle surface displacement of the shell when the loading terms q_ϕ, q_θ, and q_n are equal to 0, for example, a nontrivial homogeneous solution to the equations for the displacements of the middle surface. A closely related configuration, *inextensional deformations,* originally conceived by Lord Rayleigh in his study of the vibrations of cylindrical shells,[3] has been used to facilitate instability analysis of shells. The relationship between such deformations and the Gaussian curvature of the shell is discussed in Section 7.6.2, and an illustrative example is presented for a cylindrical shell.

Obviously the possibility of pure bending deformations in actual structures is to be avoided. One can show that a support for which both the meridional and circumferential displacements are constrained on one boundary, Figure 5–1(a), is *sufficient* to prevent pure bending deformations for all shells of revolution and is also *necessary* to remove this possibility for shells with *negative* Gaussian curvature.[4] But, there are some geometries for which a shell with negative Gaussian curvature might exhibit pure bending deformations if both the meridional and tangential displacements are not prevented at one boundary, as in Figure 5–1(b). The sufficiency condition is shown for a circular cylindrical (zero Gaussian curvature) shell in Section 7.6.2.

An interesting parable based on an awareness of the possibility of these deformations drastically reducing the stable configuration of lattice structures is found in the treatise of V. Z. Vlasov.[5] He wrote:

> The lattice designs of Engineer Shukov which are well known in construction practice, outlined by the surface of a hyperboloid of one sheet and consisting of rods arranged in straight lines, forming and tied together in horizontal rings, may constitute instantaneously varying unstable systems under certain combinations of geometric sizes and types of asymmetrical loads (wind loads, for example). Unless such structures are properly reinforced by additional structural elements (for example, tension members located in the planes of the rings), they possess very little load-bearing capacity.[5,a]

Shukov designed the hyperboloidal water tower shown in Figure 5–2 for the 16th Russian Industrial Exhibition in 1894. It was later reconstructed in Polibino where it was still standing in 1989.[6] From the photos, it appears that he recognized the importance of the ring elements and also the necessity of providing the boundary constraints as shown in Figure 5–3.[b]

5.2.2 Spherical Shell under Antisymmetrical Wind Loading

It is common practice to represent dynamic loading such as wind and earthquake by statically equivalent or pseudostatic loading. Although this procedure is not universally applica-

---

[a]Reprinted with permission of National Aeronautics and Space Administration.

[b]Although he is considered the leading engineer in Russia and the Soviet Union in the late-nineteenth and early-twentieth century and a contemporary of such giants as Eiffel, Maillart, and Freyssinet in engineering and the art of building, Shukov and his works are little known in the West. Outside of some references to the great Ninji-Novogorod Exhibitions in The Engineer, London, 1896 to 1897, little has apparently been written about him in the English language, and only recently has he been noted in the German literature.[6]

FIGURE 5–2   Hyperboloidal Water Tower.

(a) Ring Elements

FIGURE 5–3   Details of Water Tower.

ble and can often lead to erroneous conclusions, it is expedient. If due care is exercised, it can provide an approximation that is adequate for design, particularly when the pseudostatic loading is developed from consideration of the dynamic forces and the response characteristics of the structure.

With this preface, we show a static wind loading on a spherical dome in Figure 5–4.[7] The wind pressure is considered to be normal to the surface, and the components are

(b) Lower Boundary

FIGURE 5–3a   *Continued*

$$q_\phi = 0 \tag{5.17a}$$
$$q_\theta = 0 \tag{5.17b}$$
$$q_n = -p \sin \phi \cos \theta \tag{5.17c}$$

where p = the static wind pressure intensity.

Therefore, from equation (5.2a),

$$q_n^1 = -p \sin \phi \tag{5.18}$$

At this point, a word of warning is in order. It is fairly obvious that the pseudostatic wind loading given on Figure 5–4 and in equation (5.17c) has been conceived as a reasonable facsimile of a possible pressure distribution to fit the antisymmetical load form, $j = 1$. It is not based on measured dynamic wind pressures. *Extreme caution* should be exercised in using this loading for actual design. In Section 5.3.5, output from the analysis of a spherical shell for an experimentally determined pressure distribution are presented, and quite different results from those using the antisymmetrical form are obtained. Nevertheless, the antisymmetical pressure case is interesting.

Proceeding with the results for the spherical dome, we have $R_\phi = R_\theta = a$; $C_1 = C_2 = 0$; $\phi^1 = 0$; and $q_\phi^1 = q_\theta^1 = 0$, $q_n^1 = -p \sin \phi$ substituted into equations (5.6) and (5.7) After integration, the solution is[7]

$$N_\phi^1 = -pa \frac{\cos \phi}{3 \sin^3 \phi} (2 + \cos \phi)(1 - \cos \phi)^2 \tag{5.19a}$$

$$N_\theta^1 = -(pa \sin \phi + N_\phi^1) \tag{5.19b}$$

$$S^1 = -pa \frac{(2 + \cos \phi)(1 - \cos \phi)^2}{3 \sin^3 \phi} \tag{5.19c}$$

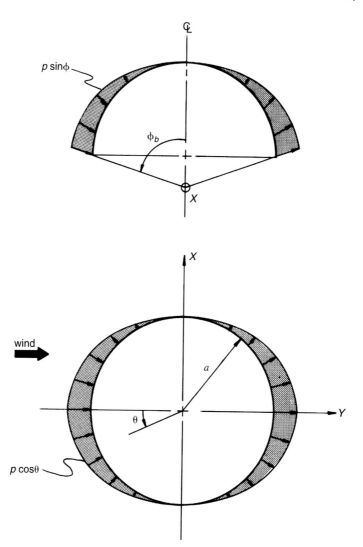

FIGURE 5–4   Static Design Wind Load on a Spherical Shell.

This solution is readily verified by direct substitution in the equilibrium equations, (4.3a–c).

Equations (5.18) and (5.19) are graphed in nondimensional form on Figure 5–5 for a hemisphere. As discussed earlier with respect to Figure 4–3, the graph holds as well for spherical domes with $\phi_b < \pi/2$ by considering the portion above that ordinate. Two points of special interest on this graph are (a) S^1 is maximum at $\phi = \pi/2$; and, (b) $N_\phi^1 = 0$ at $\phi = \pi/2$. To provide a physical explanation of these results, we again find the overall equilibrium approach to be helpful. The appropriate equilibrium equations, written with respect to the X-Y coordinates shown in Figure 5–6, are $\Sigma M_X = 0$ and $\Sigma F_Y = 0$. Here, the X axis is drawn

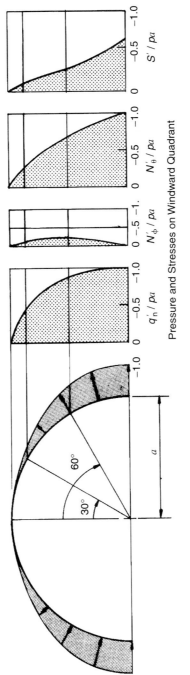

FIGURE 5-5 Stress Resultants Caused by Antisymmetrical Wind Load on a Spherical Dome.

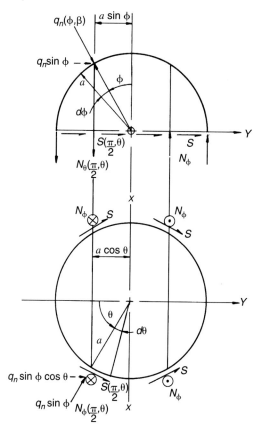

FIGURE 5–6   Equilibrium of a Spherical Shell Under Antisymmetrical Wind Load.

through the center of curvature and Y is parallel to the assumed direction of the wind. We show the appropriate free-body diagrams in Figure 5–6, where the positive signs of N_ϕ and S are established from Figure 3–6.

First, considering $\Sigma \mathbf{M_X} = \mathbf{0}$, we have

$$\int_{-\pi}^{\pi} N_\phi\left(\frac{\pi}{2}, \theta\right)(a\, d\theta)(a\cos\theta) + \int_{0}^{\pi/2}\int_{-\pi}^{\pi} q_n(\phi, \theta)(a\, d\theta)(a\, d\phi)(0)$$

$$+ \int_{-\pi}^{\pi} S\left(\frac{\pi}{2}, \theta\right)(a\, d\theta)(0) = 0 \tag{5.20}$$

The last two integrals vanish because $q_n(\phi, \theta)$ and $S(\pi/2, \theta)$ pass through the X axis, so that we have remaining

$$\int_{-\pi}^{\pi} N_\phi\left(\frac{\pi}{2}, \theta\right)a^2\cos\theta\, d\theta = N_\phi^1\left(\frac{\pi}{2}\right)a^2\int_{-\pi}^{\pi}\cos^2\theta\, d\theta$$

$$= \pi a^2 N_\phi^1\left(\frac{\pi}{2}\right)$$

$$= 0$$

We thus confirm that

$$N_\phi^1\left(\frac{\pi}{2}\right) = 0 \tag{5.21}$$

as shown on Figure 5–5.

Next, we evaluate $\Sigma \mathbf{F_Y} = \mathbf{0}$

$$\int_{\theta=-\pi}^{\pi} \int_{\phi=0}^{\pi/2} -q_n(\phi,\theta) \sin\phi \cos\theta(a\, d\phi)(a \sin\phi\, d\theta)$$

$$+ \int_{-\pi}^{\pi} S\left(\frac{\pi}{2},\theta\right) \sin\theta(a\, d\theta)$$

$$= pa^2 \int_{-\pi}^{\pi} \int_{0}^{\pi/2} \sin^3\phi \cos^2\theta\, d\phi\, d\theta$$

$$+ aS^1\left(\frac{\pi}{2}\right)\int_{-\pi}^{\pi} \sin^2\theta\, d\theta \tag{5.22}$$

$$= pa^2\pi\left[-\frac{1}{3}\cos\phi(\sin^2\phi + 2)\right]_0^{\pi/2} + aS^1\left(\frac{\pi}{2}\right)\pi$$

$$= \pi a\left[\frac{2}{3}pa + S^1\left(\frac{\pi}{2}\right)\right]$$

$$= 0$$

so that

$$\frac{S^1(\pi/2)}{pa} = -\frac{2}{3} \tag{5.23}$$

as shown in Figure 5–5.

The verification of the stress resultants using the overall equilibrium equations suggests that these two equations, written with reference to a general coordinate ϕ, should provide an independent derivation for N_ϕ^1 and S^1, whereupon N_θ^1 would be computed, as usual, from equation (4.3c). Note that in the general case $\phi_b \neq \pi/2$, so that N_ϕ and S would participate in $\Sigma \mathbf{M_X} = \mathbf{0}$, and N_ϕ in $\Sigma \mathbf{F_Y} = \mathbf{0}$. The procedure of summing forces and moments across a section to compute bending and shear stresses is strikingly similar to beam theory. In fact, the $j = 1$ case is indeed equivalent to elementary flexure except, of course, that theory alone does not admit the computation of the circumferential stress resultant N_θ. The use of beam theory to compute the stress in antisymmetrically loaded shells of revolution was developed in detail by Popov.[8]

5.2.3 Rotational Shells under Seismic Loading

Another dynamic force frequently simulated by a static representation is an earthquake or seismic load. Records of actual or artificial earthquakes are generally available in the form of two horizontal components (N-S) and (E-W), and a vertical component. The vertical component is an axisymmetric loading which fits the solution developed in Section 4.3.2.1. Regarding the horizontal components, it is frequently assumed that they are not in phase; i.e., the peak effects from the two directions do not occur at the same instant on the structure. Because shells of revolution are axisymmetric in construction, here it is regarded as sufficient to consider only the stronger of the two components. The X axis in Figure 5–7 is presumed to be oriented in this direction.

From the results of a linear dynamic analysis or from some approximations thereto, a total design base shear, **V**, may be computed corresponding to the horizontal motion earthquake component. The base shear is frequently written as a percentage, C, of the dead weight of the structure (e.g., 15% g). Details of the computation of **V** are beyond the scope of this book because the dynamic characteristics of the structure must enter into any rational treatment. An excellent comprehensive discussion of this subject is given in Clough,[9] and an application to shells of revolution is contained in Gould, Sen, and Suryoutomo.[10] The remain-

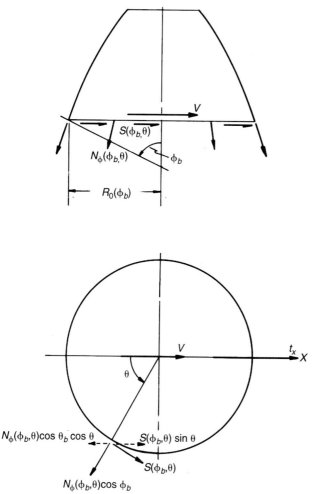

FIGURE 5–7   Base Shear Caused by Horizontal Seismic Loading.

der of this section is concerned with the determination of the distribution of surface loading that produces a specified value of the base shear, $\mathbf{V}$, and the evaluation of the corresponding membrane theory stress resultants.

As depicted in Figure 5–7, the magnitude of the base shear, V, is equal to the total horizontal resultant of N_ϕ and S, or

$$V = \int_{-\pi}^{\pi} [S(\phi_b, \theta) \sin\theta - N_\phi(\phi_b, \theta) \cos\phi_b \cos\theta] R_0(\phi_b)\, d\theta \qquad (5.24\text{a})$$

We assume that the base moves uniformly without circumferential distortion, so that only the $j = 1$ harmonic, as defined in Section 5.1.2 and equation 5.2(b), participates. Then, we have

$$V = R_0(\phi_b)\left[S^1(\phi_b) \int_{-\pi}^{\pi} \sin^2\theta\, d\theta - N_\phi^1(\phi_b) \cos\phi_b \int_{-\pi}^{\pi} \cos^2\theta\, d\theta \right]$$
$$= \pi R_0(\phi_b)[S^1(\phi_b) - N_\phi^1(\phi_b) \cos\phi_b] \qquad (5.24\text{b})$$

The subsequent analysis for the membrane theory stress resultants is dependent on the surface loading which produces $\mathbf{V}$. We know that the equivalent surface loading should have a resultant in the direction of $\mathbf{V}$; that is,

$$\iint_{\text{Surface}} \mathbf{q}\, ds_\alpha\, ds_\beta = V = V\mathbf{t_x} \qquad (5.25)$$

where $\mathbf{q} = q\mathbf{t_x}$ is defined in equation (3.11e). Because $\mathbf{V}$ is assumed to act in the horizontal plane, at any level $\mathbf{q}$ acts in the horizontal plane and is parallel to the assumed direction of the earthquake, indicated here by the X axis. Referring to Figure 5–8(b), we see that $q(\phi)$ has a circumferential component of $q(\phi)\sin\theta$, and a radial component of $q(\phi)\cos\theta$. Considering the meridional view, Figure 5–8(a), the radial component, $q(\phi)\cos\theta$, is projected from the lower figure and resolved into a meridional component, $q_\phi\cos\theta\cos\phi$, and a normal component, $q(\phi)\cos\theta\sin\phi$. Thus, referring to Figure 3–6 for the correct signs we have

$$q_\phi = -q(\phi) \cos\phi \cos\theta \qquad (5.26\text{a})$$

$$q_\theta = q(\phi) \sin\theta \qquad (5.26\text{b})$$

$$q_n = -q(\phi) \sin\phi \cos\theta \qquad (5.26\text{c})$$

We compare equations (5.26a–c) with equation (5.2a) and conclude that

$$q_\phi^1 = -q(\phi) \cos\phi \qquad (5.26\text{d})$$

$$q_\theta^1 = q(\phi) \qquad (5.26\text{e})$$

$$q_n^1 = -q(\phi) \sin\phi \qquad (5.26\text{f})$$

It remains to determine the meridional variation of q, $q(\phi)$ so that equation (5.25) is satisfied.

Consider a cooling tower structure in the form of a hyperboloid of revolution as illustrated in Figure 5–9. An actual cooling tower is shown in Figure 2–8(o). First, we relate the total base shear magnitude V to the total shell weight W. Following the overall equilibrium method,

$$W = N_\phi(\phi_b) \sin\phi_b[2\pi R_0(\phi_b)] \qquad (5.27\text{a})$$

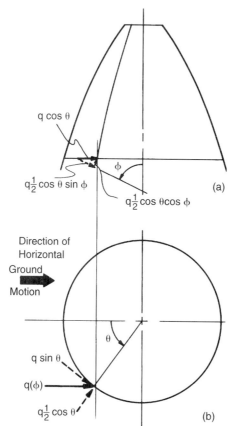

q cos θ

$q\frac{1}{2}\cos\theta\sin\phi$

ϕ

(a)

$q\frac{1}{2}\cos\theta\cos\phi$

Direction of
Horizontal
Ground
Motion

θ

q sin θ

q(φ)

$q\frac{1}{2}\cos\theta$

(b)

FIGURE 5–8 Static Horizontal Seismic Loading on a Shell of Revolution.

where $N_\phi(\phi_b)$ is caused by self-weight load. Substituting equation (4.40a) and (4.37) into equation (5.27a),

$$W = 2\pi q_{DL}a^2[\zeta_1(\phi_b) - \zeta_1(\phi_t)] \qquad (5.27b)$$

where ζ_1 is defined in equation (4.40b). Then,

$$C = \frac{V}{W} \qquad (5.28a)$$

or

$$V = CW \qquad (5.28b)$$

Either C or V is assumed to be specified by consideration of the dynamic characteristics of the system, as described, for example, in Clough.[9] Note also that equation (5.27b) may be used for calculating the total weight of a tower of given dimensions.

Now, we refer to some characteristics of the linear dynamic response of the shell in harmonic $j = 1$. We can show that this response can be represented as a linear combination of

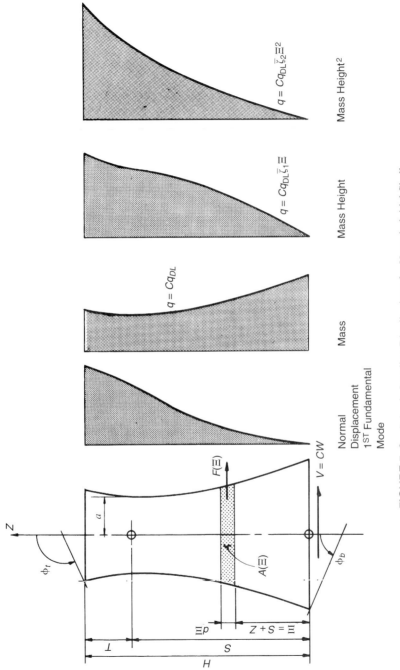

FIGURE 5-9 Seismic Loading Distributions for Hyperboloidal Shells.

the longitudinal modes of free vibration,[11] and that the first mode is the most important for the seismic analysis. The first-mode normal displacement for a tower of typical dimensions is shown in Figure 5–9. Keeping in mind that the total base shear V is presumed to be known, the actual magnitude of the normal displacement is not needed for this discussion; what is important here is the *shape*. The inertial forces produced by a harmonic response are, in fact, proportional to the mode shape. This follows from D'Alembert's principle. Thus, it is reasonable to try to arrive at a distribution of the base shear V which resembles the dynamic deflected shape. Of course, the actual longitudinal response is a combination of the participating modes and is an ever-changing function of time; the best we can attempt with a static approach is a one-term approximation at a given time, and this is reasonably selected as the first-mode shape. Thus, we shall try to arrive at a distribution $q(\phi)$ which has a shape similar to the first mode of the normal displacement.

The most common distribution used for shell design has been to assume that the seismic load is everywhere proportional to the gravity load, or

$$q(\phi) = Cq_{DL} \tag{5.29}$$

This is referred to as the *mass* (M) distribution of Figure 5–9 and the components are computed by substituting equation (5.29) into equation (5.26). It is obvious from the figure that this distribution does *not* realistically represent the first-mode shape.

Another possibility, frequently adopted in the analysis of tall building structures, is to assume that the load is distributed in proportion to the weight and the height of each element of the structure above the base. We refer to this as the *mass-height* (MH) distribution on the figure. This distribution has been adapted to cooling towers by the author.[12] To derive this shape, we refer to a differential element $d\Xi$, which is located at a distance $\Xi = S + Z$ above the base, as shown on Figure 5–9. Ξ is thus an axial coordinate measured from the base. The surface loading acting on the differential element $d\Xi$, which represents the appropriate portion of the base shear V computed by equation (5.25), is given by

$$F(\Xi) = V \frac{W_{,\Xi}\Xi d\Xi}{\int_0^H W_{,\Xi}\Xi d\Xi} \tag{5.30}$$

where W is given by equation (5.27b). Then, $F(\Xi)$ is divided by the differential area

$$A(\Xi) = 2\pi R_0(\Xi)d\Xi \tag{5.31}$$

to get $q(\Xi) = F(\Xi)/A(\Xi)$. The resulting expression is easily transformed into a function of ϕ

$$q(\phi) = Cq_{DL}\bar{\zeta}_1\Xi \tag{5.32}$$

through equations (4.34) and (4.37). ζ_1 is a constant defined in Appendix 5A. From the graph of the MH distribution, it is obvious that this appears to be far more realistic than the M distribution when compared to the mode shape.

A third possibility may be included, whereby the shear is distributed according to the weight and the *square* of the height of each element above the base. This is termed the *mass-height²* (MH²) distribution. Following the same procedure, we get[13]

$$q(\phi) = Cq_{DL}\bar{\zeta}_2\Xi^2 \tag{5.33}$$

where ζ_2 is a constant defined in the Appendix 5A. From the graph, we see that perhaps the MH^2 distribution is even a better approximation than the MH.

The components of surface load for each of the assumed distributions may be substituted into equations (5.6), (5.7), and (5.8) to evaluate the stress resultants, as has been done in various publications. Since the results are not readily available in collected form, the pertinent formulae are summarized in Appendix 5A. All quantities are nondimensionalized and are tersely presented in a form suitable for computer input preparation. Tabulated results for a wide range of tower dimensions are also available for the M distribution.[14]

Although the detailed derivation of the various solutions are probably of interest only to the specialist, it is instructive to look at a comparative study of the stress resultants. In Figure 5–10, we show the nondimensional meridional stress resultant $n_\phi = N_\phi^1/(q_{DL}a)$ for a shell of typical proportions using the three distributions. Also shown is the negative of the dead-load stress resultant, equation (4.40a), suitably nondimensionalized. When the seismic load graph falls to the right of the dead load graph, this indicates a net tension at $\theta = 0$ because of the combined loading. It is for this net tension that a concrete shell should be reinforced.

Also, it is appropriate to note here that the designer should be very careful when considering combined lateral and gravity loadings which produces a net tension situation.

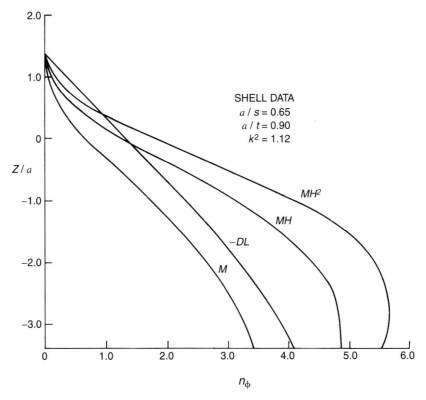

FIGURE 5–10   Effect of Base Shear Distribution on the Meridional Stress Resultant for a Hyperboloidal Shell.

Each loading should be appropriately factored by an appropriate amount prior to combining effects with opposite signs. For example, consider Figure 5–10 at $Z/a = -1.0$ where $N_\phi^{(DL)} = -2.5$ and $N_\phi^{1(MH)} = 3.3$. Using load factors from a standard code,

$$N_\phi^{(NET)} = 1.3(3.3) - 0.9(2.5) = 2.04$$

An alternative (not recommended) would be to apply a factor of safety to the net stress. For example, using a nominal factor of safety of 1.5,

$$N_\phi^{(NET)} = 1.5(3.3 - 2.5) = 1.20$$

It is apparent that the latter procedure can lead to unconservative results when the lateral and dead-load stresses are close in magnitude but opposite in sign.

In evaluating the effects of seismic loading on rotational shells, it is often necessary to consider the influence of the contents, for example, oil or grain. It is convenient to take the so-called hydrodynamic pressure as the sum of an *impulsive* component, which represents the effect of the portion of the contents that moves with the shell as a rigid mass; and a *convective* component, which represents the portion of the contents undergoing a sloshing motion. The impulsive component may be represented by an antisymmetrical normal pressure in the form of equation (5.2b). This is applied to liquid-filled cylindrical tanks by Veletsos and Tang, and by Malhetra and Veletsos.[15] Such models are discussed briefly in Section 7.6.2.

5.3 ASYMMETRICAL LOADING

5.3.1 Integration of Equations

The analysis of shells of revolution for harmonics $j > 1$ is quite different from the $j = 0$ and $j = 1$ cases. Whereas a relative abundance of quadrature solutions are available for the symmetric and antisymmetric loading cases, this is not the case for asymmetric loading; the governing equation, equation (5.4), is not solvable in general terms but, rather, the few solutions which exist are dependent on some special form of the variable coefficients.

At this point, recall that the $j = 0$ and $j = 1$ solutions were directly related to and derivable from static equations of overall equilibrium. The $j = 0$ solution represents force equilibrium parallel to the axis of rotation, whereas the $j = 1$ case reflects equilibrium of forces normal to the axis and equilibrium of moments about the axis of overturning. In contrast, $j > 1$ does not correspond to an equation of overall equilibrium. All loading cases for $j > 1$ are said to be *self-equilibrated* with respect to the overall equilibrium of the shell. Thus, the differential equations offer the most promising approach. One should note at the outset (and it is demonstrated in subsequent sections) that the membrane theory stress resultants in shells of revolution may be *very sensitive* to the influence of the higher harmonic components of the surface loading, and that shells that are subject to such loading must be investigated very thoroughly.

5.3.2 Spherical Shell with Edge Loading

One specific geometry that can be solved for $j > 1$ is the spherical shell. We limit our investigation to the case of no surface loading, which produces a homogeneous set of equations. This form is sufficient to describe an important practical situation, the edge-loaded shell of revolution. Once a homogeneous solution is available, a particular solution for a given surface loading is easily obtained using the standard method of variation of parameters.

It is expedient to return to the original untransformed membrane equations, equation (4.3). We introduce the separated form of the stress resultants, equation (5.2b), for harmonic j into these equations. Noting that equation (4.3c) gives $N_\theta = -N_\phi$ since $q_n = 0$, we have, after multiplying through by $R_\phi = R_0$, two equations:

$$N^j_{\phi,\phi} + 2N^j_\phi \cot \phi + \frac{j}{\sin \phi} S^j = 0 \tag{5.34a}$$

and

$$S^j_{,\phi} + 2S^j \cot \phi + \frac{j}{\sin \phi} N^j_\phi = 0 \tag{5.34b}$$

These equations may be separated by introducing the auxiliary variables[16]

$$\Omega^j_1 = N^j_\phi + S^j \tag{5.35a}$$

$$\Omega^j_2 = N^j_\phi - S^j \tag{5.35b}$$

giving

$$\Omega^j_{1,\phi} + \left(2 \cot \phi + \frac{j}{\sin \phi}\right) \Omega^j_1 = 0 \tag{5.36a}$$

and

$$\Omega^j_{2,\phi} + \left(2 \cot \phi - \frac{j}{\sin \phi}\right) \Omega^j_2 = 0 \tag{5.36b}$$

which have solutions[16]

$$\Omega^j_1 = C^j_1 \frac{\cot^j(\phi/2)}{\sin^2 \phi} \tag{5.37a}$$

and

$$\Omega^j_2 = C^j_2 \frac{\tan^j(\phi/2)}{\sin^2 \phi} \tag{5.37b}$$

From equation (5.35)

$$N^j_\phi = \frac{1}{2 \sin^2 \phi} \left[C^j_1 \cot^j \left(\frac{\phi}{2}\right) + C^j_2 \tan^j \left(\frac{\phi}{2}\right) \right] \tag{5.38a}$$

and

$$S^j = \frac{1}{2 \sin^2 \phi} \left[C^j_1 \cot^j \left(\frac{\phi}{2}\right) - C^j_2 \tan^j \left(\frac{\phi}{2}\right) \right] \tag{5.38b}$$

The complete homogeneous solution is given in the Fourier series form of equation (5.2b) by

$$N_\phi(\phi,\theta) = -N_\theta(\phi,\theta)$$

$$= \frac{1}{2 \sin^2 \phi} \sum_{j=0}^\infty \left[C^j_1 \cot^j \left(\frac{\phi}{2}\right) + C^j_2 \tan^j \left(\frac{\phi}{2}\right) \right] \cos j\theta \tag{5.39a}$$

and

$$S(\phi,\theta) = \frac{1}{2 \sin^2 \phi} \sum_{j=0}^\infty \left[C^j_1 \cot^j \left(\frac{\phi}{2}\right) - C^j_2 \tan^j \left(\frac{\phi}{2}\right) \right] \sin j\theta \tag{5.39b}$$

The constants for each harmonic, C^j_1 and C^j_2, must be evaluated from a set of nonzero force boundary conditions which are also expanded in Fourier series. One possibility is

$$N_\phi(\phi',\theta) = \sum_{j-0}^{\infty} N_\phi^j(\phi') \cos j\theta \qquad (5.40a)$$

$$S(\phi'',\theta) = \sum_{j=0}^{\infty} S^j(\phi'') \sin j\theta \qquad (5.40b)$$

In these equations, ϕ' and ϕ'' denote the respective boundaries of the spherical shell where N_ϕ and S are specified. Also, the form of equations (5.39) indicates that N_ϕ and S may both be designated at the same boundary, for example, $\phi' = \phi''$ in equation (5.40), or that N_ϕ or S alone, may be chosen at both boundaries. In any case, only two boundary conditions are available, and the computed values of N_ϕ and/or S at the other boundaries must be developed by the support in order for the solution to remain valid.

For a spherical dome, $\cot(\phi/2) \to \infty$ at the pole, so that $C_1^j = 0$ and only one condition, N_ϕ or S, may be specified on the lower boundary.

5.3.3   Shell of Revolution Supported on Columns

It is frequently necessary to provide interruptions at the base of a shell above the ground for functional reasons. One obvious possibility is to support the shell on a series of concentrically placed columns, Figures 2–8(p, t, u). This obviously introduces an interruption in the continuous, nondeforming boundary required to maintain membrane action and suggests that the stress pattern, particularly in the vicinity of the support, may be severely altered.

A fairly common approach is to analyze the shell by a superposition technique as shown in Figure 5–11. In what follows, the discussion is restricted to axisymmetric loading and membrane theory to introduce the concept expeditiously, but this type of model has been generalized for nonsymmetric loading and bending effects as well.[17] Also, here the columns are assumed to be equispaced around the circumference, and the axis of the column is taken to be tangent to the meridian at the boundary $\phi = \phi_b$. For columns oriented vertically, the same theory can be applied with some minor modifications.

We now consider the superposition in Figure 5–11 in detail. We assume that the axisymmetrical surface loading q is carried equally by the n_c equispaced columns. As shown in Figure 5–11(a), the column reaction R_c is taken to be uniformly distributed over the column width w_c, so that the intensity of the reaction per unit length of circumference R_{cl} is R_c/w_c. Loading case (a) is represented by the superposition of cases (b) and (c). Case (b) is the familiar axisymmetric loading case previously studied in great detail in Sections 4.3.2 and 4.4. The continuous boundary meridional stress resultant at $\phi = \phi_b$ is denoted by $N_{\phi(b)}(\phi_b)$. Case (c) has an edge loading applied along $\phi = \phi_b$. It consists of (i) the negative of the continuous boundary reaction from (b) in the region *between* the columns; and, (ii) the difference between the column reaction and the continuous boundary reaction *within* the column width. Since case (b) is readily solvable, the remaining problem is to address case (c).

First, we evaluate the column reaction in terms of the known meridional stress resultant from the continuous boundary analysis, case (b), as

$$R_c = N_{\phi(b)}(\phi_b)\frac{2\pi R_0(\phi_b)}{n_c} \qquad (5.41)$$

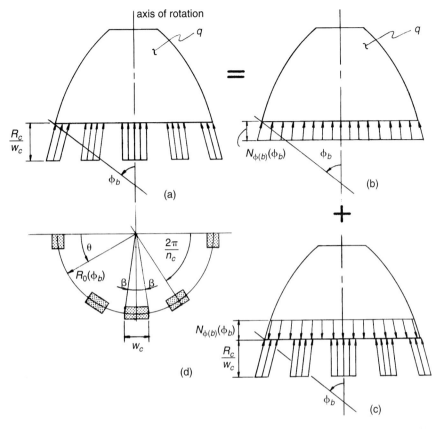

FIGURE 5-11   Superposition Representation of a Column-Supported Shell of Revolution.

The intensity is given by

$$R_{cl} = \frac{R_c}{w_c} \tag{5.42}$$

Referring to Figure 5-11(d), we write w_c in terms of the subtended angle 2β measured in the horizontal plane

$$w_c = 2\beta R_0(\phi_b) \tag{5.43}$$

Then, substituting equations (5.43) and (5.41) into equation (5.42), we have

$$R_{cl} = \frac{\pi N_{\phi(b)}(\phi_b)}{\beta n_c} \tag{5.44}$$

R_{cl} has units of force/length.

Now, we wish to express the edge loading shown in Figure 5-11(c) as a function of the circumferential coordinate θ. Referring to Figure 5-11(d), we have, formally,

$$N_{\phi(c)}(\phi_b, \theta) = [-N_{\phi(b)}(\phi_b) + \delta R_{cl}]$$

$$\text{where } \delta = 1 \text{ if } m(2\pi/n_c) - \beta \le \theta \le m(2\pi/n_c) + \beta$$

$$(m = -n_c/2, \ldots, -1, 0, 1, \ldots, n_c/2) \tag{5.45}$$

$$\delta = 0 \text{ for all other } \theta.$$

The limits on δ are set presuming an even number of columns and can be adjusted accordingly for an odd number, and $N_{\phi(c)}$ is taken as positive in the positive sense of $N_{\phi(b)}$.

We expand $N_{\phi(c)}$ in a Fourier series similar to equation (5.40a),

$$N_{\phi(c)}(\phi_b, \theta) = \sum_{j=0}^{\infty} N_{\phi(c)}^j(\phi_b) \cos j\theta \tag{5.46}$$

First, the Fourier coefficient $N_{\phi(c)}^0$, is given by

$$N_{\phi(c)}^0(\phi_b) = \frac{1}{2\pi} \int_{-\pi}^{\pi} N_{\phi(c)}(\phi_b, \theta) \, d\theta \tag{5.47a}$$

From equations (5.45) and (5.44),

$$N_{\phi(c)}^0(\phi_b) = \frac{1}{2\pi} \int_{-\pi}^{\pi} \left[-N_{\phi(b)}(\phi_b) + \delta\pi \frac{N_{\phi(b)}(\phi_b)}{\beta n_c} \right] d\theta$$

$$= \frac{N_{\phi(b)}(\phi_b)}{2\pi} \left[-2\pi + 2\beta \frac{n_c\pi}{\beta n_c} \right] \tag{5.47b}$$

$$= 0$$

This confirms that the *net* meridional resultant in Figure 5–11(c) is 0, since the total of the column reactions is equal and opposite to the uniform boundary reaction. This edge loading is then said to be *self-equilibrated*.

Next for $j > 0$, we have

$$N_{\phi(c)}^j(\phi_b) = \frac{1}{\pi} \int_{-\pi}^{\pi} N_{\phi(c)}(\phi_b, \theta) \cos j\theta \, d\theta \tag{5.48}$$

Referring to Figures 5–11(d), where the circumferential spacing of the columns is given as $2\pi/n_c$, we have

$$N_{\phi(c)}^j(\phi_b) = \frac{1}{\pi} \left\{ \int_{-\pi}^{\pi} - N_{\phi(b)}(\phi_b) \cos j\theta \, d\theta + R_{cl} \left[\int_0^{\beta} \cos j\theta \, d\theta \right. \right.$$

$$+ \int_{(2\pi/n_c)-\beta}^{(2\pi/n_c)+\beta} \cos j\theta \, d\theta + \int_{2(2\pi/n_c)-\beta}^{2(2\pi/n_c)+\beta} \cos j\theta \, d\theta + \tag{5.49a}$$

$$\cdots + \int_{(n_c-1)(2\pi/n_c)-\beta}^{(n_c-1)(2\pi/n_c)+\beta} \cos j\theta \, d\theta + \int_{2\pi-\beta}^{2\pi} \cos j\theta \, d\theta \left. \right] \right\}$$

$$N_{\phi(c)}^{j}(\phi_b) = \frac{N_{\phi(b)}(\phi_b)}{\pi} \left[-\frac{2}{j} \sin j\pi + \frac{\pi}{j\beta n_c} \left\{ \sin j\beta \right. \right.$$

$$\left. + \sum_{m=1}^{n_c - 1} \left[\sin j \left(m\frac{2\pi}{n_c} + \beta \right) - \sin j \left(m\frac{2\pi}{n_c} - \beta \right) \right] \right. \tag{5.49b}$$

$$\left. \left. + \sin j2\pi - \sin j(2\pi - \beta) \right\} \right]$$

For an integer value of j, the first and next-to-last terms drop out. Furthermore, it can be shown that for $j \neq in_c$, where i is an integer, the remaining terms sum to zero, so that $N_{\phi(c)}^{j}(\phi_b) = 0$ if $j \neq in_c$. When $j = in_c$, the complete symmetry results in the total integral being equal to n_c times the integral over one column, or

$$\begin{aligned} N_{\phi(c)}^{j}(\phi_b) &= n_c \frac{N_{\phi(b)}(\phi_b)}{\pi} \frac{2\pi \sin j\beta}{j\beta n_c} \\ &= \frac{2}{j\beta} N_{\phi(b)}(\phi_b) \sin j\beta \quad (j = in_c) \end{aligned} \tag{5.50}$$

The preceding assertions are easily verified for a specified numerical value of n_c.

We thus conclude that only those harmonics j which are integer multiples of the total number of columns n_c participate in the solution, and we may rewrite equation (5.46) as

$$N_{\phi(c)}(\phi_b, \theta) = \sum_{j=n_c, 2n_c \dots}^{\infty} N_{\phi(c)}^{j}(\phi_b) \cos j\theta \tag{5.51}$$

This is an example of *cyclic symmetry* for rotational shells.

We now turn to the spherical shell treated in Section 5.3.2 as an example of a column-supported shell. An actual case is shown in Figure 2–8(t). If the shell is not a dome, there are two constants of integration to be evaluated. We have already specified $N_{\phi}(\phi_b)$, and since we cannot provide an ideal condition because of the columns, the lower boundary must develop the corresponding value of $S(\phi_b)$. This leaves us the option of specifying $N_{\phi}(\phi_t)$ or $S(\phi_t)$; $N_{\phi}(\phi_t) = 0$ is probably the more realistic choice. Thus, we must provide suitable resistance to S at both boundaries in the case of the open shell, or at the base in the case of a dome. This can be accomplished with a circumferentially stiff edge member.

Example 5.3(a)

For our example, we choose a hemispherical dome supported on four columns as shown in Figure 5–12, subject to dead load q. From equation (4.20), the solution for case (b) of the superposition is

$$N_{\phi(b)} = \frac{-qa}{1 + \cos \phi} \tag{5.52}$$

At $\phi = \phi_b = \pi/2$, $N_{\phi(b)} = N_{\phi(b)}(\phi_b) = -qa$ and equation (5.50) becomes

$$N_{\phi(c)}^{j}(\phi_b) = -\frac{2qa}{j\beta} \sin j\beta \tag{5.53a}$$

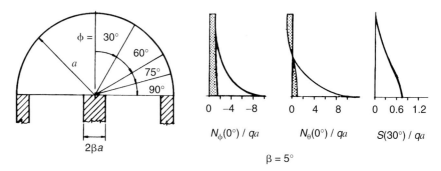

$$N_\phi(0°)\,/\,qa \qquad N_\theta(0°)\,/\,qa \qquad S(30°)\,/\,qa$$

$$\beta = 5°$$

FIGURE 5–12   Stress Resultants for a Column-Supported Hemispherical Dome under Dead Load.

From equation (5.51) with $n_c = 4$ and $\phi_b = \pi/2$,

$$N_{\phi(c)}\left(\frac{\pi}{2}, \theta\right) = \sum_{j=4,8,12,\ldots}^{\infty} -\frac{2qa}{j\beta} \sin j\beta \cos j\theta \tag{5.53b}$$

Recalling the Fourier series expansion for the stresses in the spherical shell, equations (5.39a) and (5.39b); taking $C_1^j = 0$ because we are considering a dome; and evaluating equation (5.39a) at $\phi_b = \pi/2$,

$$N_\phi\left(\frac{\pi}{2}, \theta\right) = \frac{1}{2}\sum_{j=0}^{\infty} C_2^j \tan^j\left(\frac{\pi}{4}\right) \cos j\theta \tag{5.54}$$

Equating the coefficients of $\cos j\theta$ for each term in the series in equations (5.53) and (5.54), we find

$$C_2^j = \frac{-4qa}{j\beta} \sin j\beta \quad (j = 4,8,12,\ldots) \tag{5.55}$$

because $\tan^j(\pi/4) = 1$, and $C_2^j = 0$ for $j \neq 4, 8, 12$. Now, we may write the solution to case (c) of the superposition by substituting C_2^j into equations (5.39)

$$N_{\phi(c)}(\phi, \theta) = -N_{\theta(c)}(\phi, \theta)$$

$$= \frac{-2qa}{\beta \sin^2 \phi} \sum_{j=4,8,12,\ldots}^{\infty} \frac{\sin j\beta}{j} \tan^j\left(\frac{\phi}{2}\right) \cos j\theta \tag{5.56a}$$

$$S_c(\phi, \theta) = \frac{2qa}{\beta \sin^2 \phi} \sum_{j=4,8,12,\ldots}^{\infty} \frac{\sin j\beta}{j} \tan^j\left(\frac{\phi}{2}\right) \sin j\theta \tag{5.56b}$$

The complete solution consists of case (b) plus case (c), which gives

$$N_\phi(\phi, \theta) = \frac{-qa}{1 + \cos \phi} + N_{\phi(c)}(\phi, \theta) \tag{5.57a}$$

$$N_\theta(\phi, \theta) = qa\left(\frac{1}{1 + \cos \phi} - \cos \phi\right) - N_{\phi(c)}(\phi, \theta) \tag{5.57b}$$

$$S(\phi,\theta) = S_c(\phi,\theta) \tag{5.57c}$$

in view of equations (4.20) and (4.21). Again, we stress that the value of $S = S_c$ computed from equation (5.57) must be developed at the lower boundary.

To study the characteristics of this solution, we show the nondimensionalized stress resultants for the hemisphere in Figure 5–12. The half-angle β is taken as $5°$; N_ϕ and N_θ are graphed at $\theta = 0°$, the center line of the column; and S is shown at $\theta = 30°$. The series converges rather slowly with about fourteen terms ($n = 56$) required to compute, within 1%, the known value of N_ϕ at the base, $[360°/4 \times 2 \times 5°)]$ $N_{\phi b}$ (ϕ_b), or 9 times the continuous boundary reaction. The continuous boundary stress resultants, which correspond to Figure 5–11(b) and are available from Figure 4–3, are superimposed and shaded to indicate the penetration of the discontinuous boundary reaction amplification into the shell. While such problems as solved above are routinely treated today using finite element analysis, introduced in Section 13.1.4, the membrane theory solution provides considerable insight into the primary load-resisting mechanism of the shell.

Although we have obtained a convergent membrane theory solution for the spherical dome supported on columns, all shells with concentrated edge effects are not able to sustain such loading in this way. Only a limited number of specific geometries have been investigated, and no general proof is offered; however, it appears that the attenuation of concentrated meridional edge effects through principally membrane action occurs only in positive curvature shells. Physically, it can be surmised that the straight characteristic lines present in zero- and negative-curvature shells offer a path for the unimpeded propagation of the edge effects to the opposite boundaries, unless bending distortions occur.

While the quantification of bending effects is beyond the scope of this chapter, we may expect that the shell may have to be strengthened to provide adequate resistance. In concrete shells, bending stresses are conveniently resisted by local thickening. The wall thicknesses of the shells shown in Figure 2–8(g) and (p) are approximately equal to the column depth near the base and are gradually reduced away from this region. For metal shells, thickening alone may not be sufficient to provide an economical solution, because buckling in the region of compressive stresses must be considered. Auxiliary stiffeners, such as the ring beam shown in Figure 2–8(u), are one possibility. Another configuration used for bin structures is to provide full-height columns, along with eave and transition ring beams, as shown in Figure 5–13.[18]

Rotter has proposed a membrane theory solution for the case of axisymmetrical loaded bins, based on the cyclic symmetry of the structure.[18] As shown in Figure 5–14, the shell is loaded locally by a normal pressure $q_n(Z) = p_n(Z)$ and a wall-friction loading $q_Z(Z) = \mu p_n(Z)$, where the normal pressure $p_n(Z)$ and the friction coefficient μ are determined from the properties of the material in the bin. Containments for solids differ markedly from liquid storage tanks with the additional need to provide resistance to the wall-friction loading imparted by the solids. These forces may be amplified during drawdown and should be considered carefully by the designer.[19]

Also acting are the self-load of the cylindrical shell $q_Z = q = W_1/(2\pi aL)$, where W_1 is the total weight of the cylindrical portion of the bin with length L; and a circumferential line load $\bar{q}_Z(0) = W_2/(2\pi a)$, where $W_2 =$ the superimposed vertical load at the top of the cylinder, $Z = 0$. Note that $N_Z(0) = -\bar{q}_Z(0)$.

The vertical load in the shell wall is transferred into the adjacent column by the in-plane shear stress resultant, S. By symmetry, $S = 0$ along the meridian equidistant between

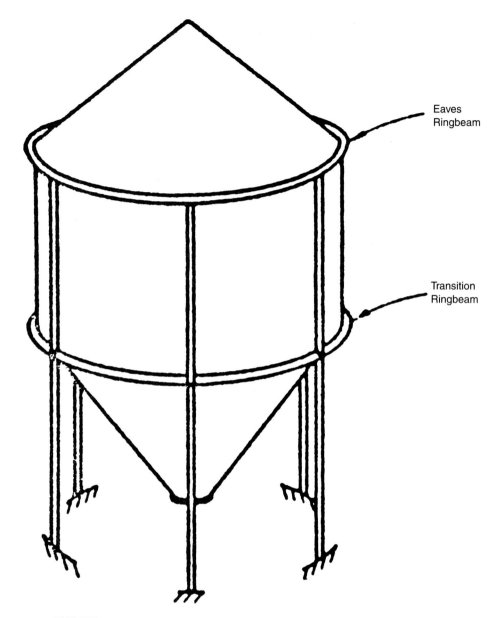

Eaves
Ringbeam

Transition
Ringbeam

FIGURE 5–13 Column-Supported Bin with Columns Extending to the Eaves
Ring.

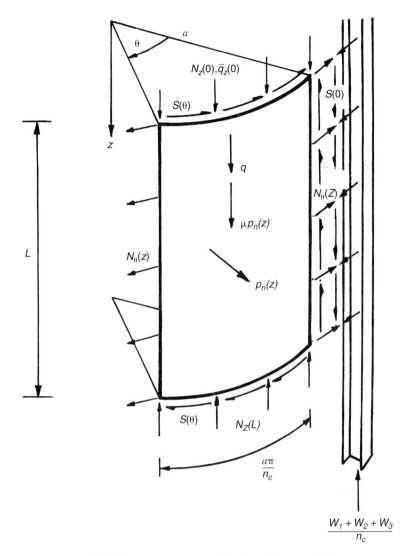

FIGURE 5–14   Half-Panel of Column-Supported Bin.

two columns. Assuming S to be *uniform* along the column and to vary *linearly* in the circumferential direction, we have

$$S(\theta) = \frac{W}{2n_c L}\left(1 - \frac{n_c\theta}{\pi}\right) \tag{5.58a}$$

where

$$W = \text{total load carried by the } n_c \text{ columns}$$
$$= W_1 + W_2 + W_3$$

and

$$W_3 = \text{total weight of the stored material in the bin}$$

This implies that the total load is transferred *uniformly* into the columns along the length L. The circumferential stress resultant is easily found from equation (4.3c) as

$$N_\theta(Z) = ap_n(Z) \tag{5.58b}$$

For the meridional stress resultant, $N_Z(Z)$, we follow the overall equilibrium approach, Section 4.4.1. We have the resultant force from the weight of the shell above the section Z, $W_1(Z/L)$; the resultant force from the superimposed vertical load, W_2; the resultant force from the frictional loading, $2\pi a\mu\int_0^Z p_n(Z)dZ$; and the total force in the columns, $W(Z/L)$. Substituting into equation (4.58) and recalling that a resultant axial load in the *negative* Z direction is taken as *positive*, we have

$$N_Z(Z) = -\mu \int_0^Z p_n(Z)dZ + \frac{1}{2\pi a}\left[(W-W_1)\frac{Z}{L} - W_2\right] \tag{5.58c}$$

At the top of the panel, $Z = 0$, N_Z is attributed to only the superimposed load W_2; at the bottom of the panel, $Z = L$, the stress is proportional to the *difference* of the weight of the contents and the frictional load on the wall.

Just as in the case of a spherical shell supported on columns, the ring beams must resist the shear force $S(\theta)$ applied around the circumference, probably eccentric to the axis of the shell. This is examined in greater detail by Rotter.[18] The shell thickness must be adequate to resist buckling caused by the maximum principal compressive stresses produced by $S(0)$, $N_\theta(L)$, and $N_Z(L)$, which probably occur near the intersection of the shell, the column, and the transition ring beam.

5.3.4 Cylindrical Shell Under Circumferentially Varying Normal Pressure

Consider a circumferentially varying normal pressure $p(\theta)$ acting on a cylindrical shell. This could simulate a pseudostatic wind loading acting on a tower, as shown in Figure 5–15, or the convective component of the hydrodynamic seismic pressure as mentioned in Section 5.2.3. The circumferential variation of the pressure would generally be established through field measurements or wind tunnel tests. For this example, p is taken as constant with respect to Z, but a variation with Z may be treated by a simple extension of the following development.

We follow a procedure suggested in Rish and Steel.[20] It is convenient to return to the membrane theory equations in terms of curvilinear coordinates α and β, equations (4.2). We take $\alpha = Z$, $\beta = \theta$, $A = 1$, $B = a$, $R_\alpha = \infty$, $R_\beta = a$; $q_\alpha = q_\beta = 0$, and $q_n = -p(\theta)$. Then, we have

$$N_{Z,Z} + \frac{1}{a}S_{,\theta} = 0 \tag{5.59a}$$

$$S_{,Z} + \frac{1}{a}N_{\theta,\theta} = 0 \tag{5.59b}$$

$$\frac{N_\theta}{a} = -p(\theta) \tag{5.59c}$$

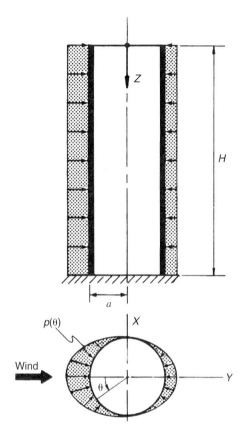

FIGURE 5–15   Wind Loading on a
Cylindrical Shell.

Because these equations are relatively simple, we can treat the partial differential equation form directly rather than introducing separated variables. However, we later find it convenient to represent the integrated solution in terms of the separated variables.

Proceeding from equation (5.59c), we have

$$N_\theta(Z,\theta) = -p(\theta)a \tag{5.60}$$

Next, we take $\partial/\partial\theta$ of equation (5.60), substitute into equation (5.59b), and integrate to get

$$S(Z,\theta) = p(\theta),_\theta Z + f_1(\theta)$$

With a stress-free top edge, $S(0,\theta) = 0$; therefore, $f_1(\theta) = 0$ and

$$S(Z,\theta) = p(\theta),_\theta Z \tag{5.61}$$

Finally, we take $\partial/\partial\theta$ of equation (5.61), substitute into equation (5.59a), and integrate once more to obtain

$$N_Z(Z,\theta) = -p(\theta),_{\theta\theta} \frac{Z^2}{2a} + f_2(\theta)$$

Again invoking the stress-free condition at $Z = 0$, $N_Z(0,\theta) = 0$ gives $f_2(\theta) = 0$. Then,

$$N_Z(Z,\theta) = -p(\theta),_{\theta\theta} \frac{Z^2}{2a} \tag{5.62}$$

We now represent $p(\theta)$ by the Fourier series

$$p(\theta) = \sum_{j=0}^{\infty} p^j \cos j\theta \tag{5.63}$$

and write the stress resultants in the separated form, similar to equation (5.2b), with Z replacing ϕ:

$$\begin{Bmatrix} N_Z \\ N_\theta \\ S \end{Bmatrix} = \sum_{j=0}^{\infty} \begin{Bmatrix} N_Z^j(Z)\cos j\theta \\ N_\theta^j(Z)\cos j\theta \\ S^j(Z)\sin j\theta \end{Bmatrix} \tag{5.64}$$

where

$$N_Z^j(Z) = j^2 p^j \frac{Z^2}{2a} \tag{5.65a}$$

$$N_\theta^j(Z) = -p^j a \tag{5.65b}$$

$$S^j(Z) = -jp^j Z \tag{5.65c}$$

These very simple expressions for the stress resultants for a general harmonic j offer a rare insight into the interesting properties of the higher harmonic components. Consider, for example, the solution expressed by equations (5.64) and (5.65) contrasted with a solution for the resultant lateral force and overturning moment found by assuming that the shell is a cantilever beam with a cross-section which remains circular.

1. The circumferential stress resultant N_θ is not computable by beam theory.
2. The shear stress resultant is linearly proportional to the axial coordinate Z, just as in beam theory. However, the shell solution shows that S is also proportional to the harmonic number j. This discrepancy, of course, increases in proportion to the contribution of the higher harmonics.
3. The meridional stress resultant is proportional to the square of the axial coordinate by either theory; however, the shell solution reveals that N_Z is also proportional to j^2, indicating that the meridional stress resultant is highly sensitive to the higher harmonic components.

To help put these observations in perspective, note two characteristics of the circumferential normal pressure distribution $p(\theta)$:

1. High harmonic number components of $p(\theta)$ result from distributions of the pressure that vary rapidly around the circumference. To a point, the more rapid the fluctuations, the greater relative participation of the higher harmonics. However, the case of ex-

tremely rapid fluctuations associated with large values of j would be tempered by the physical limitations of accompanying significant amplitudes. That is, the p^j values would usually tend to diminish as j becomes very large.

2. Harmonic components $j = 0$ and $j > 1$ do not contribute to the resultant lateral force or overturning moment on the shell. Only $j = 1$ affects these resultants. If the meridian were curved, the $j = 0$ component would influence the axial force equilibrium, but the $j > 1$ components are always self-equilibrated with respect to the overall equilibrium of the shell.

Example 5.3(b)

As an illustration, consider the cylinder shown in Figure 5–15 subject to the pressure distribution shown in Figure 5–16. This loading was derived from measurements taken from a wind tunnel experiment[21] and is representative of the circumferential pressure distribution on large cylindrical and hyperboloidal towers with a roughened surface. The Fourier coefficients for this distribution are also shown on the figure. The loading is given in terms of pressure coefficients multiplied by a reference pressure P_r, which is generally a function of geographical location, terrain, exposure, and

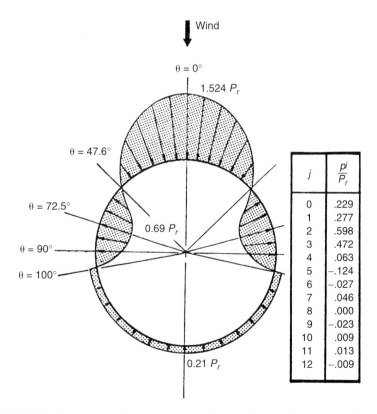

FIGURE 5–16   Design Wind Pressure for Circular Towers. (Reprinted with permission of American Society of Civil Engineers)

height; it is usually taken as the stagnation pressure of the wind ($\frac{1}{2}$) ρV^2, where ρ is the mass density and V is the velocity of the wind. The 0.524 increase shown on Figure 5–16 is caused by internal suction, generally fairly constant around the circumference. The maximum external or side suction, however, has been found to be a function of the angle where flow separation occurs, which, in turn, is dependent on the surface roughness. The value of the side suction, here at $\theta = 72.5°$, $0.69 + 0.52$ or about $1.2\ P_r$ without the internal suction, is characteristic of rough towers, whereas a peak approaching $3.0\ P_r$ is theoretically possible for smooth towers. Roughness in the form of vertical ribs is often built into cooling tower shells to reduce the side suction[21] [see Figure 2–8(o)].

For simplicity, P_r is taken as unity here, and also as constant with height. A stepwise variation in P_r is easily handled by applying the solution, equations (5.59) through (5.65), in a piecewise fashion as detailed in Section 4.3.2.3.

We focus the computation of the meridional stress N_z at $\theta = 0$ at the base, $Z = H$. With $P_r = 1.0$, equations (5.64) and (5.65) give

$$N_Z(H,0) = \frac{H^2}{2a} \sum_{j=0}^{\infty} j^2 p^j \tag{5.66}$$

for the Fourier coefficients shown on Figure 5–16. Summing for harmonics 0 to 12, we get

$$N_Z(H,0) = 5.42 \frac{H^2}{2a} \tag{5.67}$$

Now, for comparison, we evaluate $N_Z(H,0)$ from beam theory. Because only the first harmonic contributes to the resultant force per unit height R_Y, we have

$$R_Y = \int_{-\pi}^{\pi} p^1 \cos\theta(a,\,d\theta)\cos\theta$$
$$= p^1 a\pi$$

and the bending moment is

$$M_X = R_Y \frac{H^2}{2} = \frac{p^1 a\pi H^2}{2}$$

Since we are considering the stress resultant that acts on the circular middle surface line, we compute N_Z using the familiar linear stress formula $N_Z = M_Z a/I_X$, where I_X is the moment of inertia of a circle of radius a about a diameter, which is equal to πa^3. Thus, we have

$$N_Z^1(H,0) = p^1 \frac{H^2}{2a} \tag{5.68}$$

For the tabulated value of p^1 found in Figure 5–16, equation (5.68) becomes

$$N_Z^1(H,0) = 0.277 \frac{H^2}{2a} \tag{5.69}$$

Comparing equations (5.69) and (5.67) clearly illustrates the sensitivity of N_Z to the higher harmonic components. This emphasizes the practical importance of obtaining realistic circumferential pressure distributions on tower-type structures.

The perhaps unexpected sensitivity of N_Z^j and S^j to the harmonic number j and the resulting large departure from the values that might be anticipated from elementary beam theory are clearly demonstrated in the preceding solution. From a physical standpoint, it is instructive to consider a "slice-beam" model proposed by Brissoulis and Pecknold.[22] As shown on Figure 5–17(a), the slice-beam has an arc cross-section which is defined by the value of θ for which N_Z and $N_\theta = 0$ in a particular harmonic j, that is, $\cos j\theta = 0$ or $\theta = \pm\pi/2j$.

On Figure 5–17(b), a longitudinal slice bounded by the meridians $\theta = \pm(\pi/2j)$ and loaded by a normal pressure is shown. The beam is a cantilever supported at $Z = L$. At the section $Z = L$, the loading $p(Z,\theta)$ produces a bending moment about an axis parallel to the chord of the circular sector cross-section of the slice beam. The applied moment is resisted by a couple consisting of (a) the resultant of the meridional stresses $N_Z^j(L)\cos j\theta$ acting within the depth of the beam cross-section, $a[1 - \cos(\pi/2j)]$; and, (b) the accumulated shearing force, $\int_0^L S^j(Z)\,dZ$, acting along both edges of the slice. As the harmonic number j increases, the depth of the slice beam and, hence, the available internal lever arm *decreases*. Likewise, the resultant transverse (vertical) component of $p(Z,\theta)$ from $Z = 0$ to $Z = L$ must be resisted only by a similar component of the resultant of $S^j(L,\theta) = S^j(L)\sin j\theta$ acting on the sector cross-section, which is *shortening* as j increases. N_θ^j does not participate since $\cos \pi/2j$ vanishes at the boundary. Therefore, as j increases, both N_Z^j and S^j must increase proportionally to balance both the applied bending moment and transverse force. This model has been quantified by Gould to verify equations (5.64) and (5.65).[23]

Although some additional resistance may be provided by meridional bending and transverse shear, the cantilevered slice-beam appears to reveal the essentials of the resisting mechanism of cylindrical-type shells subject to normal pressures with substantial circumferential fluctuation (i.e., significant $j > 1$ content). In addition, Brissoulis and Pecknold suggested that the basic beam could be generalized into a "propped-cantilever," with the reaction at the free end representing the effect of a ring stiffener or a roof.[22]

5.3.5 Spherical Shell Under Measured Pseudostatic Wind Pressure

The separated flow pattern for wind acting on a cylindrical shell, Figure 5–16, can also be expected to occur on spherical shells, but it is more complex because the flow moves over as well as around the body. A set of pressure coefficients from a wind-tunnel test of a truncated sphere is shown in Figure 5–18(a). It is apparent that separation occurs both meridionally and circumferentially, with external suction acting over most of the surface.

Each curve in Figure 5–18(a) would require a set of Fourier coefficients to be properly represented. This would complicate the harmonic-by-harmonic solution of equation (5.4), re-

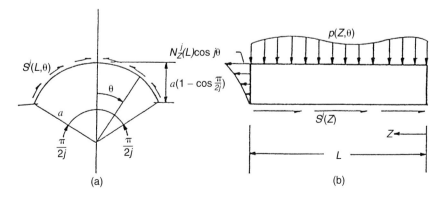

(a)   (b)

FIGURE 5–17   Slice-Beam Model.

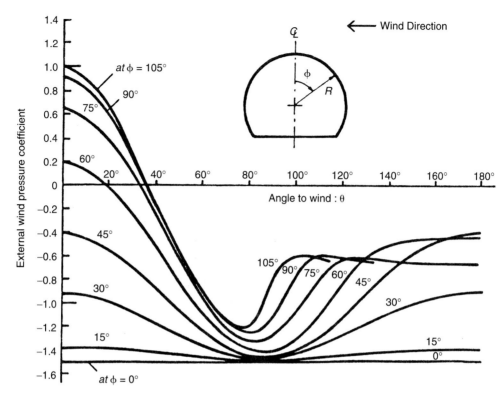

FIGURE 5–18a   External Wind Pressure Distribution for Truncated Spheres.
[From Bicknell, J. and Davis, P., Wind tunnel studies of spherical tower radomes,
MIT, Lincoln Laboratories, *Group Report 76-7*(1958).]. Reprinted with permission
of Lincoln Laboratories, Massachusetts Institute of Technology, Lexington, Massachusetts.

quiring the shell to be divided into meridional segments within which the Fourier coefficients for the loading could be taken as constant. Ultimately, this produces a step-by-step solution to an initial value problem starting at the free edge, which is similar to the procedure
for variable thickness shells discussed in Section 4.3.2.3 and illustrated in Figure 4–7.

In such cases, much of the efficiency of the harmonic superposition method is lost, and
it may become attractive to consider the partial differential equations (4.3) in the variables
ϕ and θ. This approach requires only one solution over the surface, but must usually be accomplished by a numerical method such as finite differences. The curves for N_ϕ and N_θ
shown in Figure 5–18(b) were obtained by Bellworthy and Croll[24] in that way.

Example 5.3(c)

It is of interest to compare these results with those presented on Figure 5–5 for the antisymmetric loading. The latter values are scaled by 0.9 to correspond to the pressure coefficient at

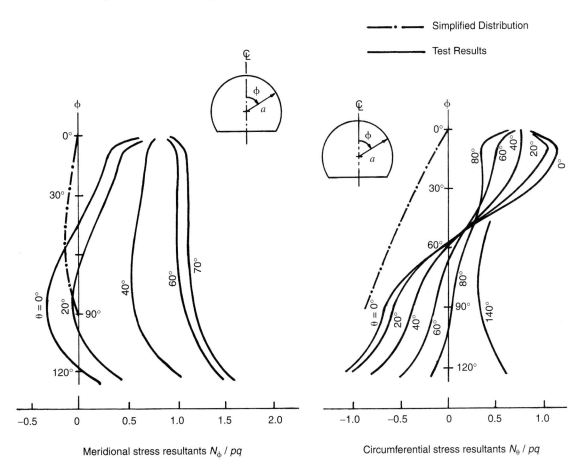

FIGURE 5–18b   Membrane Stress Resultant for Spherical Shell under External Wind Pressure.

$\phi = 90°$, $\theta = 0°$ on Figure 5–18(a), and are plotted on Figure 5–18(b) and labeled "simplified distribution." Any similarity between the curves is probably coincidental. However, it might be fortuitous for thin-metal shells, which are susceptible to buckling and may have been designed for an antisymmetric pressure distribution, that the more realistic loading produces conservative values for the maximum negative hoop stress.

As in the previous section, the importance of establishing a realistic distribution of wind pressure on the shell surface is apparent.

5.3.6   Hyperboloidal Shells Under Wind Loading

As mentioned previously, analytical solutions to the membrane theory equations for $j > 1$ are only available for isolated cases. As a means of demonstrating a possible approach to the general solution of equation (5.4) and also of describing a problem with considerable practical utility, consider the hyperboloidal shell of revolution, Figure 4–6, subject to the normal wind pressure shown in Figure 5–16.

In this case, the loading takes the form

$$q_\phi = 0 \tag{5.70a}$$

$$q_\theta = 0 \tag{5.70b}$$

$$q_n = \sum_{j=0}^{\infty} p^j \cos j\theta \tag{5.70c}$$

where the Fourier coefficients are tabulated in Figure 5–16.

The solutions for harmonics $j = 0$ and $j = 1$ are obtained in identical fashion to the dead-load case, Section 4.3.2.3, and the seismic-load case, Section 5.2.3, respectively. They are available in Lee and Gould.[25]

For the case $j > 1$, a different approach is required. We begin with equation (5.4) and differentiate the first term on each side using the product rule. The resulting equation is

$$\psi_{,\phi\phi} + \chi_1(\phi)\psi_{,\phi} + \chi_2(\phi)\psi = \chi_3(\phi) \tag{5.71}$$

after dividing through by the leading coefficient. In equation (5.71),

$$\chi_1(\phi) = \left(2\frac{R_\phi}{R_\theta} - 1\right)\cot\phi - \frac{1}{R_\phi}R_{\phi,\phi} \tag{5.72a}$$

$$\chi_2(\phi) = -\frac{j^2 R_\phi}{R_0 \sin\phi} \tag{5.72b}$$

$$\chi_3(\phi) = \frac{R_\phi}{R_0 R_\theta}[(q_n^j \cos\phi - q_\phi^j \sin\phi) R_0^2 R_\theta]_{,\phi} - jR_\phi^2 (jq_n^j - q_\theta^j \sin\phi) \tag{5.72c}$$

The expansion in equations (5.71) and (5.72) has been retained in a general form to point out that we are confronted with a second-order linear differential equation with *variable* coefficients. It is obvious from equations (5.72a) and (5.72b) that the variable coefficients may be very complicated expressions for many surfaces.

A strategy for dealing with such an equation is suggested in Novozhilov.[26] The technique involves a straightforward elimination of the first order term $\psi_{,\phi}$ through the introduction of an additional auxiliary variable, and also another less obvious transformation on the independent variable ϕ. The resulting equation has the form

$$\overline{\psi}_{,\overline{\beta\beta}} + [\rho(\overline{\beta})]^2\overline{\psi} = \chi_4(\phi) \tag{5.73}$$

in which $\overline{\psi}$ and $\overline{\beta}$ are the transformed variables, and $\chi_4(\phi)$ is the appropriately modified rhs. The details of the transformation are left to the cited references; here, we examine the remaining variable coefficient ρ in more detail. The coefficient is a very involved function of the geometry and harmonic number j and is given in Lee and Gould[25] in terms of a new function, χ_5, as

$$[\rho(\overline{\beta})]^2 = [\rho(\phi)]^2 = 1 + \chi_5(\phi) \tag{5.74}$$

The remaining solution procedure is outlined as follows:

1. Expand $[\rho\,(\bar{\beta})]^2$ in a Fourier cosine series in $\bar{\beta}$

$$[\rho(\bar{\beta})]^2 = \sum_{k=1}^{\infty} \alpha^k \cos 2k\bar{\beta} \qquad (5.75)$$

   and substitute into equation (5.73).

2. The equation is now in the form of Hill's equation, which is solved using techniques described in Hill, and also in Whittaker and Watson.[27] The solution is the form of a rapidly converging infinite series. From a historical viewpoint, it is interesting to note that this equation was first studied by the noted astronomer, G. W. Hill, with reference to the motion of the lunar perigee, which is the point of the orbit of the moon that is nearest to the earth.

3. Using the solution derived in (2), the original stress resultant variables are recovered through the various transformations.

4. Through some numerical studies using the rather involved series solutions obtained from Hill's technique, a very close approximation was found by considering only one term $k = 0$, of the Fourier series expansion for $[\rho\,(\bar{\beta}^2]$, in other words, α^o.[28] In effect, this reduces equation (5.73) to a second-order linear differential equation with a *constant* coefficient. With the coefficient $[\rho(\bar{\beta})]^2$ replaced by the constant α^o in equation (5.73), the equation is easily solvable in terms of elementary trigonometric functions.

A comprehensive set of tables for hyperboloidal shells with the general properties of hyperbolic cooling towers that are subject to the wind loading shown in Figure 5–16 is given in Lee and Gould.[25]

The solution for a hyperbolic cooling tower exhibits similar characteristics to those previously observed for the cylindrical shell under wind loading. In particular, two points should be reemphasized:

1. The solution is very sensitive to the harmonic number j and may be interpreted physically using the slice-beam model of Section 5.3.4.

2. The wind-load stresses are opposite in sign to the gravity-load stresses, and the net stresses, critical for design, should be calculated using proper load factors. The sensitivity of the cooling tower to the *net* tension is similar to the combined seismic-gravity loadings discussed in Section 5.2.3 and illustrated in Figure 5–10.

5.3.7 Commentary

The preceding example indicates that the solution to the $j > 1$ case may be quite complicated for other than cylindrical, spherical, and, perhaps, conical geometries. The transformation to Hill's equation form has apparently only been investigated for the hyperboloidal shell and may be useful for other geometries as well. Also, the hyperboloidal shell under wind load has been solved in the partial differential equation form, similar to that used for the cylindrical shell in Section 5.3.4, as opposed to the harmonic decomposition approach stressed here.[29]

The investigation of a doubly curved shell for asymmetric loading probably approaches the limit to which analytical solutions are efficiently employed. Computer-based numerical techniques have been used extensively for rotational shell analysis,[30] but the independent check of selected cases, with such analytical solutions that are available, serves to establish confidence in the numerical procedures. Also, note that a comprehensive collection of analytical solutions to rotational shell problems has been compiled in a semitabular form by Baker et al.[31] and provides a valuable resource for the design engineer.

In addition to illuminating some of the difficulties encountered in solving a nonsymmetrically loaded shell of revolution, equation (5.71) also provides a good basis for discussing a widely used and apparently logical approximate solution technique for such shells. That is, replace the meridian with a combination of cylindrical, conical, and spherical segments, all of which can be solved for the general harmonic. Although this approach seems reasonable when comparing the corresponding coordinates of the shell profile along the meridian, it should be noted that the definitive geometric parameters present in equations (5.72), most notably R_ϕ are *not* necessarily approximated closely by these simpler geometries. With mathematically based, computer-implemented solution techniques now available for such problems, it is recommended that *ad hoc* substitute curve procedures not be relied upon.

REFERENCES

1. V. V. Novozhilov, *Thin Shell Theory*, (Translated from 2nd Russian ed. by P. G. Lowe (Groningen, The Netherlands: Noordhoff, 1964): 117–119.

2. Ibid., 147–151.

3. H. Kraus, *Thin Elastic Shells*, (New York: Wiley, 1967): 307–314.

4. Novozhilov, *Thin Shell Theory:* 151–163.

5. V. Z. Vlasov, *General Theory of Shells and Its Application in Engineering*, NASA Technical Translation TTF-99 (Washington, D.C.: National Aeronautics and Space Administration, 1964): 200–201.

6. Shukov, *Die Kunst der Sparsamen Konstruktion*, Deutsch Verlags-Anstalt, Stuttgart, Germany, 1990.

7. S. Timoshenko and S. Woinowsky-Krieger, *Theory of Plates and Shells*, 2nd ed. (New York: McGraw-Hill, 1959): 449–456.

8. E. P. Popov, Earthquake stresses in spherical domes and in cones, *Journal of the Structure Division*, ASCE 82, no. ST3 (May 1956): 974-1–974-14.

9. R. W. Clough, Earthquake response of structures, in R. L. Wiegel, ed., *Earthquake Engineering*, (Englewood Cliffs, N.J.: Prentice-Hall, 1970): 307–334.

10. P. L. Gould, S. K. Sen, and H. Suryoutomo, Dynamic analysis of column-supported hyperboloidal shells, *Earthquake Engineering and Structural Dynamics 2* (1974): 269–280.

11. Ibid.

12. P. L. Gould, *Hyperbolic Cooling Towers under Seismic Design Loading*, Proc. of the Fourth Symposium, on Earthquake Engineering I., University of Roorkee, Roorkee, India, November 1970.

13. P. L. Gould, Quasistatic seismic loading distributions for hyperbolic cooling towers, *Bulletin of the Indian Society of Earthquake Technology 8*, no. 4 (December 1971): 163–168.

14. P. L. Gould and S. L. Lee, Hyperbolic cooling towers under seismic design load. *Journal of the Structural Division,* ASCE, no. ST10 (October 1968): 2487–2493.

15. A. S. Veletsos and Y. Tang, Soil-structure interaction of structures for laterally excited liquid-storage tanks, *Earthquake Engineering and Structure Dynamics,* 19:473–496; P. K. Malhotra and A. S. Veletsos, Uplifting response of unanchored liquid-storage tanks, *Journal of Structural Engineering,* ASCE, 120, no. 12 (Dec. 1994): 3525–3547.

16. S. Timoshenko and S. Woinowsky-Krieger, *Theory of Plates and Shells:* 453–456.

17. P. L. Gould and S. L. Lee, Column-supported hyperboloids under wind load, *Publications of the International Association for Bridge and Structural Engineering,* Zurich, Switzerland 31–11, 1971: 47–64; S. K. Sen and P. L. Gould, Hyperboloidal shells on discrete supports, technical note, *Journal of the Structural Division,* ASCE 99, no. ST3 (March 1973): 595–603.

18. J. M. Rotter, Membrane Theory of Shells for Bins and Silos, *Dept. of Civil and Mining Engineering,* Univ. of Sydney, NSW, Australia.

19. M. E. Killion, Design pressures in circular bins, *Journal of the Structural Division,* ASCE 111, no. 8 (August 1985): 1760–1774.

20. R. F. Rish and T. F. Steel, Design and selection of hyperbolic cooling towers, *Journal of the Power Division,* ASCE 85, no. PO5 (October 1959): 89–117.

21. *Reinforced Concrete Cooling Tower Shells—Practice and Commentary* (ACI 334.2R-84) (Detroit: American Concrete Institute, 1984): 6–7.

22. D. Brissoulis and D. A. Pecknold, Behavior of empty steel grain silos under wind loading part 1: The stiffened cylindrical shell, *Engineering Structures* (October 1986): 260–274.

23. P. L. Gould, The cylindrical shell slice beam, *Journal of Engineering Mechanics,* ASCE 114, no. 7 (July 1988): 905–911.

24. A. J. Bellworthy and J. G. A. Croll, Dielictric space frame domes, *Space Structures 1,* no. 1 (1985): 41–50.

25. S. L. Lee and P. L. Gould, Hyperbolic cooling towers under wind load. *Journal of the Structural Division,* ASCE 93, no. ST5 (October 1967): 487–514.

26. V. V. Novozhilov, *Thin Shell Theory* [translated from 2nd Russian ed. by P. G. Lowe] (Groningen, The Netherlands: Noordhoff, 1964): 315–319.

27. G. W. Hill, *Collected Mathematical Works,* vol. 1 (Washington, D.C.: Carnegie Institute of Washington, 1905–1907): 243–270; E. T. Whittaker and G. N. Watson, *A Course of Modern Analysis,* 4th ed. (Cambridge: Cambridge University Press, 1935): 412–417.

28. P. L. Gould, Unsymmetrically loaded hyperboloids of revolution, *Journal of the Engineering Mechanics Division,* ASCE 94, no. EMS (October 1968): 1029–1043.

29. E. Ingerslev, Design of cooling tower shells, *Proceedings of the June 1966 Bratislava, Czechoslovakia, Symposium on Tower Shaped Steel and Reinforced Concrete Structures,* IASS. Madrid, Spain, 1968.

30. P. L. Gould, *Finite Element Analysis of Shells of Revolution* (London: Pitman, 1985).

31. E. H. Baker, L. Kovalevsky, and F. L. Rish, *Structural Analysis of Shells* (New York: McGraw-Hill, 1972).

32. T. B. Quimby, Seismic forces on spherical domes, *Journal of Engineering Mechanics,* ASCE 121, no. 11 (November, 1995): 1272–1275.

EXERCISES

5.1. (a) Verify the first harmonic shell of revolution solution presented in equations (5.6) and (5.7).
  (b) Verify that for a dome, $S^1(0) = 0$, by considering equations (5.6) and (5.7).

5.2. For the spherical shell shown in Figure 4–2:
  (a) Determine the membrane theory stress resultants caused by a horizontal quasistatic seismic loading.
  (b) For a hemisphere, determine the total base shear and overturning moment.
  (c) Verify the values of N_ϕ and S at $\phi = 90°$ computed by (a), using the results of (b).
  (d) Consider a vertical snow load which varies as $\cos \phi$ on the surface.[32] Modify the expressions for the loading components given in equation (5.26 a–c) accordingly and repeat (a) through (c).

5.3. Consider the column-supported spherical shell shown in Figure 5–11:
  (a) Reanalyze this shell for a uniform live load p.
  (b) For eight equispaced columns and a value of $\beta = 2°$, compute the membrane-theory stress-resultant distribution and show the results in a graph similar to Figure 5–12.

5.4. Consider the wind-loaded cylindrical shell shown in Figure 5–15. Assume that the vertical pressure distribution is $q_n(Z) = -[(Z - H)^2/(H^2)]\, p(\theta)$ and derive the corresponding expressions to equations (5.65).

5.5. Consider a conical shell as shown in Figure 4–13 and derive expressions for the membrane-theory stress resultants for circumferentially varying normal pressure, similar to the cylindrical shell analysis in Section 5.3.4.

APPENDIX 5A

Summary of Surface Loading
and Stress Resultants
for Quasistatic Seismic Loading
on Hyperboloidal Shells of Revolution[12]

See Figure 5–9 for Notation.

NONDIMENSIONAL GEOMETRY

$$a = \text{throat radius}$$

$$s = \frac{S}{a}$$

$$\xi = \frac{\Xi}{a}$$

$$r_\theta = \frac{R_\theta}{a} = \frac{\sqrt{(k^2 - 1)}}{[k^2 \sin^2 \phi - 1]^{1/2}}$$

$$r_\phi = \frac{R_\phi}{a} = \frac{-\sqrt{(k^2 - 1)}}{[k^2 \sin^2 \phi - 1]^{3/2}}$$

$$r_0 = r_\theta \sin \phi$$

SURFACE LOADING

$$\bar{q}(\phi) = \frac{q(\phi)}{q_{DL}}$$

$$\bar{q}_\phi^1 = \bar{q}(\phi) \cos \phi$$

$$\bar{q}_\theta^1 = \bar{q}(\phi)$$

$$\bar{q}_n^1 = -\bar{q}(\phi) \sin \phi$$

STRESS RESULTANTS

$$n_\phi^1 = \frac{N_\phi^1}{q_{DL}a} = \frac{1}{r_0^2 \sin \phi} \Phi_1(\phi)$$

$$n_\theta^1 = \frac{N_\theta^1}{q_{DL}a} = r_\theta \left(\bar{q}_n^1 - \frac{n_\phi^1}{r_\phi} \right)$$

$$s^1 = \frac{S^1}{q_{DL}a} = n_\phi^1 \cos \phi - \frac{\Phi_2(\phi)}{r_0}$$

FUNCTIONS

M Distribution

$$q(\phi) = C q_{DL}$$
$$\Phi_1(\phi) = C\{2\zeta_1(\phi_t)[\zeta_3(\phi) - \zeta_3(\phi_t)] - [\zeta_4(\phi) - \zeta_4(\phi_t)]\}$$
$$\Phi_2(\phi) = 2C[\zeta_1(\phi) - \zeta_1(\phi_t)]$$

MH Distribution

$$q(\phi) = C\bar{\zeta}_1\xi$$
$$\Phi_1 = \frac{2C\bar{\zeta}_1}{k^2 - 1} \left\{ \frac{1}{12k^2} (3\zeta_1 - \zeta_3 r_\theta^3) - [\zeta_2(\phi_t) + s\zeta_1(\phi)]r_\theta \cos \phi + s \frac{k^2 - 1}{3k^2} r_\phi \right\}$$
$$\Phi_2 = 2C\bar{\zeta}_1[\zeta_2(\phi) - \zeta_2(\phi_t)]$$

MH² Distribution

$$q(\phi) = C\bar{\zeta}_2\xi^2$$
$$\Phi_1 = 2C\bar{\zeta}_2 \left[\zeta_1\zeta_8 + \zeta_9 + \xi\zeta_5(\phi_t) \right]_{\phi_t}^{\phi}$$
$$\Phi_2 = 2C\bar{\zeta}_2[\zeta_5(\phi) - \zeta_5(\phi_t)]$$

General

$$\zeta_1(\phi) = \frac{1}{4(k^2 - 1)} \left[\frac{\sqrt{(k^2 - 1)}}{k} \ln K - 2r_\theta^2 \cos \phi \right]$$

$$K = \frac{\sqrt{(k^2 - 1)} - k \cos \phi}{\sqrt{(k^2 - 1)} + k \cos \phi}$$

$$\zeta_2(\phi) = \frac{r_\phi}{3k^2} - \zeta_1 s$$

$$\zeta_3(\phi) = \frac{\cos \phi \, r_\theta}{k^2 - 1}$$

$$\zeta_4(\phi) = \frac{1}{2k^2} \, \zeta_3 \left[\frac{1}{\sqrt{(k^2 - 1)}} \ln K + \frac{2}{k \cos \phi} + \frac{r_\phi}{3k^2} \right]$$

$$\zeta_5(\phi) = \zeta_1 \zeta_6 + \xi \zeta_2 + \zeta_7$$

$$\zeta_6(\phi) = \frac{1}{4k^2(k^2 - 1)} - \zeta_3 s$$

$$\zeta_7(\phi) = \frac{r_\phi}{12k^2} \left[\zeta_3 + 4s \right]$$

$$\zeta_8(\phi) = \frac{\zeta + s}{4k^2(k^2 - 1)} + \zeta_3 \left[\frac{1}{2k^2(k^2 - 1)} - s^2 \right]$$

$$\zeta_9(\phi) = \frac{r_\phi}{3k^2} \left[\frac{\zeta_3(\xi + s)}{4} + s^2 - \frac{1}{2k^2(k^2 - 1)} + \frac{r_\theta^2}{10k^2(k^2 - 1)} \right]$$

$$\overline{\zeta}_1 = \frac{\zeta_1(\phi_b) - \zeta_1(\phi_t)}{\zeta_2(\phi_b) - \zeta_2(\phi_t)}$$

$$\overline{\zeta}_2 = \frac{\zeta_1(\phi_b) - \zeta_1(\phi_t)}{\zeta_5(\phi_b) - \zeta_5(\phi_t)}$$

6

Membrane Theory:
Applications to Shell
of Translation

6.1 GENERAL

A surface of translation may be generated by passing one plane curve over another. The translational shell is often treated separately from the rotational shell, especially in a theroretical context; however, there may be considerable overlap. For example, zero-curvature shells of revolution, such as cylinders and cones, may be formed by the translation of a straight line over a circle, or, in the case of the cylinder, by the translation of the circle along the straight line. Even the negative curvature hyperboloid of revolution may be produced by the trace of a skewed line around a circle. In the context of shell analysis, it is generally simpler to regard shells as rotational rather than translational surfaces when there is a choice. Therefore, in our subsequent treatment, we are only concerned with shells that do not possess rotational symmetry.

A popular use of the translational shell is to provide a roof without interior supports over what can be a quite large plan area. Some notable examples are shown in Figures 2–8. As can be seen from Figures 2–8(a) and (i), such shells are often supported at only a few points to provide an uninterrupted open space. While the popularity of monolithic formed concrete roof shells has diminished, space frames following singly or doubly curved surfaces and resisting loading in the same general manner as monolithic shells are currently prominent, for example, Figure 2–8(*l*). Because of the wide variety of boundary conditions encountered, we may find that the membrane theory solution, by itself, is a poor approximation to the stress distribution in such a shell, and that bending effects will play an important or even dominant role in the design. The bending analysis of translational shells is discussed in Chapter 13.

Further, we find it convenient to separate the membrane analysis of translational shells into two parts: (a) circular cylindrical shells; and, (b) shells of double curvature.

6.2 CIRCULAR CYLINDRICAL SHELLS

Single or multiple circular cylindrical shells have been widely employed in *barrel-shell* roofs, such as that illustrated in Figure 2–8(e). Extensive discussions of this application are available.[1] Here, we concentrate on the basic features of such shells.

The geometry is defined in Figure 6–1. We have attempted to remain consistent with Section 4.4.2 and Figure 4–12. An exception is the adoption of X rather than Z as the axial coordinate to conform to the prevalent practice in the literature. Thus, we have $\alpha = X$, $\beta = \theta$, $A = 1$, $B = \alpha$ and $R_\alpha = \infty$, $R_\beta = \alpha$. The shell is bounded by the two arcs at $X = 0$ and $X = L$, and the two straight edges at $\theta = \pm\theta_k$.

We start from the membrane theory equilibrium equations, (4.2), with the appropriate geometric parameters inserted:

$$N_{X,X} + \frac{1}{a} S_{,\theta} + q_X(X,\theta) = 0 \tag{6.1a}$$

$$S_{,X} + \frac{1}{a} N_{\theta,\theta} + q_\theta(X,\theta) = 0 \tag{6.1b}$$

$$\frac{N_\theta}{a} = q_n(X,\theta) \tag{6.1c}$$

If we compare equations (6.1a–c) with equations (5.59), we find that they are identical except for the loading terms.

As such, with reference to Section 5.3.4, the equations are readily solved to give

$$N_\theta(X,\theta) = q_n a \tag{6.2a}$$

$$S(X,\theta) = -\int \frac{1}{a} N_{\theta,\theta} \, dX - \int q_\theta \, dX + f_1(\theta) \tag{6.2b}$$

$$N_X(X,\theta) = -\frac{1}{a} \int S_{,\theta} \, dX - \int q_X \, dX + f_2(\theta) \tag{6.2c}$$

As an elementary example, we consider a circumferentially uniform but axially varying gravity load $q_d(X)$, as shown in Figure 6–1:

$$q_X(X,\theta) = 0 \tag{6.3a}$$

$$q_\theta(X,\theta) = q_d(X)\sin\theta \tag{6.3b}$$

$$q_n(X,\theta) = -q_d(X)\cos\theta \tag{6.3c}$$

Proceeding, we find

$$N_\theta = -aq_d(X)\cos\theta$$
$$N_{\theta,\theta} = aq_d(X)\sin\theta \tag{6.4a}$$

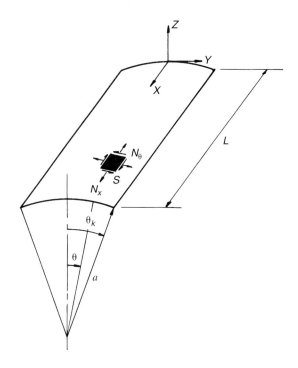

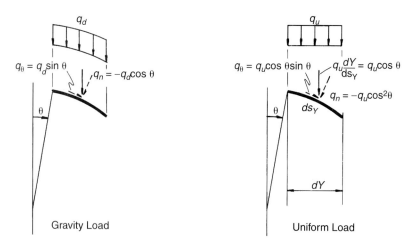

FIGURE 6-1   Open Cylindrical Shell.

$$S = -2 \sin \theta \int q_d(X) \, dX + f_1(\theta) \tag{6.4b}$$

$$S_{,\theta} = -2 \cos \theta \int q_d(X) \, dX + f_1(\theta)_{,\theta}$$

$$N_X = -\frac{1}{a} \left\{ -2 \cos \theta \int \left[\int q_d(X) \, dX \right] dX + f_1(\theta)_{,\theta} \int dX \right\} + f_2(\theta) \tag{6.4c}$$

We must now inquire further into the longitudinal distribution of the dead load $q_d(X)$. There are two cases of practical interest (a) uniform distribution:

$$q_d(X) = q = \text{constant}$$

and (b) periodic distribution:

$$q_d(X) = \sum_{j=0}^{\infty} q_d^j \sin \frac{j\pi X}{L}$$

For the latter case, the function $q_d(X)$ may be considered as an odd function on the interval $-L \leq X \leq L$, as shown in Figure 6–2, whereupon

$$q_d(X) = \sum_{j=1,3,\text{odd}}^{\infty} q_d^j \sin \frac{j\pi X}{L} \tag{6.5a}$$

where the Fourier coefficients are

$$q_d^j = \frac{2}{L} \int_0^L q_d(X) \sin \frac{j\pi X}{L} \, dX \quad (j = 1, 3, \ldots \text{odd}) \tag{6.5b}$$

As an example we may approximate the uniform distribution q_d, by the first term $(j = 1)$ of the Fourier series, for which

$$q_d^1 = \frac{2q_d}{L} \int_0^L \sin \frac{\pi X}{L} \, dX$$

$$= \frac{4}{\pi} q_d \tag{6.6a}$$

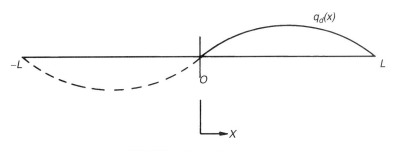

FIGURE 6–2   Odd Function.

and

$$q_d(X) \simeq \frac{4}{\pi} q_d \sin \frac{\pi X}{L} \tag{6.6b}$$

Finally, we evaluate the stress resultants, equations (6.2a–c), for both a uniform dead load q_d, = constant, and for the first term approximation of q_d as given by equation (6.6b). We proceed with the integration from the boundary $X = 0$, and assume that the shell is simply supported; e.g., $N_X(0) = N_X(L) = 0$.

For $q_d(X) = q_d$ = constant, we have from equations (6.4b) and (6.4c)

$$S = -2q_d X \sin \theta + f_1(\theta) \tag{6.7a}$$

$$N_X = -\frac{1}{a}[-q_d X^2 \cos \theta + Xf_1(\theta),_\theta] + f_2(\theta) \tag{6.7b}$$

Applying the boundary conditions to equation (6.7b),

$$f_2(\theta) = 0$$

$$f_1(\theta),_\theta = q_d L \cos \theta$$

from which

$$f_1(\theta) = q_d L \sin \theta$$

and

$$S(X,\theta) = q_d L\left(1 - \frac{2X}{L}\right) \sin \theta \tag{6.8a}$$

$$N_X(X,\theta) = -q_d \frac{L}{a} X\left(1 - \frac{X}{L}\right) \cos \theta \tag{6.8b}$$

Also, equation (6.4a) becomes

$$N_\theta(X,\theta) = -q_d a \cos \theta \tag{6.8c}$$

For the single-term Fourier series approximation, we substitute equation (6.6b) into equations (6.4b) and (6.4c) and carry out an analogous integration, which gives

$$S = \frac{8q_d}{\pi}\left(\frac{L}{\pi}\right) \cos \frac{\pi X}{L} \sin \theta + f_1(\theta) \tag{6.9a}$$

$$N_X = -\frac{1}{a}\left[\frac{8q_d}{\pi}\left(\frac{L}{\pi}\right)^2 \sin \frac{\pi X}{L} \cos \theta + Xf_1(\theta),_\theta\right] + f_2(\theta) \tag{6.9b}$$

We first apply the boundary condition $N_X(0) = 0$, from which $f_2(\theta) = 0$. Also, we note that the loading $q_d(X)$ is symmetrical about $X = L/2$, so that $S(L/2) = 0$. Then from equation (6.9a), $f_1(\theta) = 0$, and the stress resultants S and N_X are

$$S(X,\theta) = \frac{8q_d L}{\pi^2} \cos \frac{\pi X}{L} \sin \theta \tag{6.10a}$$

$$N_X(X,\theta) = -\frac{8q_d L^2}{\pi^3 a} \sin \frac{\pi X}{L} \cos \theta \tag{6.10b}$$

Also, equation (6.4a) becomes

$$N_\theta(X,\theta) = -\frac{4q_d a}{\pi} \sin \frac{\pi X}{L} \cos \theta \qquad (6.10c)$$

It is interesting to compare the stress resultants computed from the actual loading q_d with those evaluated from the single-term Fourier approximation. Note the maximum stress resultants $S(0)$, N_X, $(L/2)$, and N_θ, $(L/2)$, as shown in Table 6–1. From this comparison, we see that except for $S(0)$, the maximum stress resultants are conservatively computed from the one-term solution. If further refinement is desired, additional terms of the series may be included.[2]

An interesting aspect of this solution was the enforcement of the symmetry boundary condition, $S(L/2) = 0$. In general, the appropriate boundary conditions, in addition to $N_X(0) = N_X(L) = 0$, would be kinematic constraints which can develop the shear stress resultant S at the supports.

For this problem, the end-point values of S are easily found from either equation (6.8a) or equation (6.10a) as

$$S(0,\theta) = -S(L,\theta) = q_d L \sin \theta \qquad (6.11a)$$

or

$$S(0,\theta) = -S(L,\theta) = \frac{8q_d L}{\pi^2} \sin \theta \qquad (6.11b)$$

respectively.

If the symmetry situation was not recognized or enforced, the appropriate kinematic conditions would have been

$$D_\theta(0) = D_\theta(L) = 0 \qquad (6.11c)$$

The direct application of the later constraints creates a *statically indeterminate membrane theory* problem, because the stress resultants cannot be evaluated without consideration of the displacements. This is discussed further in Section 8.3.4.1.

Another case of practical interest is that of a load acting in the Z direction and uniformly distributed in the Y direction, q_u. The resolution of this load is shown in Figure 6–1.

$$q_X(X,\theta) = 0 \qquad (6.12a)$$

$$q_\theta(X,\theta) = q_u \cos \theta \sin \theta \qquad (6.12b)$$

$$q_n(X,\theta) = -q_u \cos^2 \theta \qquad (6.12c)$$

TABLE 6–1   Comparison of Maximum Stress Resultants

Maximum Stress Resultant	Common Factor	Coefficient	
		Exact	Approximate
$S(0)$	$q_d L \sin \theta$	1	$8/\pi^2 = 0.81$
$N_x(L/2)$	$-q_d (L/a) \cos \theta$	$L/4 = 0.25 L$	$8L/\pi^3 = 0.258L$
$N_\theta(L/2)$	$-q_d a \cos \theta$	1	$4/\pi = 1.27$

A parallel set of solutions for the stress resultants, using both the actual loading and the Fourier series approximation, can be derived in a straightforward manner.

When the load, q_u, is uniformly distributed in the X direction as well, it is then constant over the X-Y plane. This form is commonly used to represent snow loading on a shell roof.

We now inquire into the accuracy of the membrane theory stress analysis for a barrel shell. Obviously, the cases we have discussed thus far satisfy the requirements of smooth *loading* and regular *geometry*. When it comes to the adequacy of the physical *boundaries* to meet the requirements of the membrane theory, however, we need a closer examination.

The transverse boundaries at $X = 0$ and $X = L$ have been assumed to be simply supported, implying that N_X should vanish. Moreover, our analysis indicates that the maximum value of S must be developed there, as shown by equations (6.11a, b). These conditions suggest the use of a diaphragm that is perfectly flexible in the X, or longitudinal direction, but infinitely rigid in the θ, or circumferential direction. A typical practical solution is an arch following the shell form in the Y-Z plane. In general, the membrane theory condition is considered to be fairly well approximated by such a support.

Now we examine the longitudinal boundaries at $\theta = \pm\theta_k$. The solution requires $N_\theta(\theta_k)$ and $S(\theta_k)$, as evaluated from equations (6.8) or (6.10), to be developed on the boundary. Furthermore, the boundary must be free to displace in the normal direction so that no transverse shear will develop. In practice, several situations are found.[3] In multiple barrel shells, the outer edge is often free and the interior edges are continuous with the adjacent barrel, but unrestrained in the vertical direction (see Figure 12–9a). Also, supplementary stiffening beams have been employed in some structures, as shown in Figure 2–8(e). However, unless a continuous support, such as a bearing wall or a stiff frame, is provided beneath all longitudinal edges, there will be serious violation of the membrane theory conditions and a resulting drastic alteration of the load-resisting mechanism.

This situation is easily understood by reference to Figure 6–3, where we can see that the membrane theory solution implies that the shell resists transverse loading primarily as a series of circular arches. Also, referring to Figure 2–8(e), the overhanging edge beams are evidently designed to resist the circumferential forces imparted by the "arch action," with the beam cross-section stepped to balance the longitudinal bending moments. Although this is a statically admissible solution, a question could be raised as to whether the beams are actually supporting the shell, or whether the shell is, at least in part, carrying the beams. This cannot be resolved using the membrane theory.

In contrast, where supports are not provided along the edge, the shell must resist transverse loading primarily by flexural action in the longitudinal direction. Because the stress resultant N_X must develop the required longitudinal resisting couple at any section, the value computed from membrane theory will be considerably altered by bending, as we will see in Chapter 12. Because the contribution of $N_\theta(\pm\theta_k)$ in resisting the vertical load will be lost, the vertical equilibrium must be provided by $S(0)$ and $S(L)$. From these heuristic observations, we may anticipate a significant readjustment of the extensional and in-plane shear stress resultants, as well as the possible addition of significant bending moments and transverse shear forces from the solution of the general equations for open cylindrical shells. This is discussed further in Section 12.4.3.

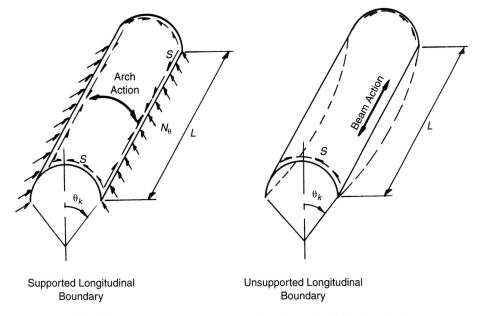

FIGURE 6–3   Longitudinal Boundary Conditions for Cylindrical Shells.

6.3 SHELLS OF DOUBLE CURVATURE

If we refer to the shell element shown in the center of Figure 6–4, the membrane theory equilibrium equations may be written directly from equations (4.2a–c) by selecting the arc lengths as the curvilinear coordinates. Thus, we would have $\alpha = s_x$ and $\beta = s_y$. However, a somewhat different approach is widely used because it simplifies the subsequent mathematics.[4,5] First, we choose the Cartesian coordinate system X-Y-Z as shown on the figure. Then, the differential area element $ds_x \, ds_y$ located at $Z(X,Y)$ is mapped onto the lower element $dXdY$ in the X-Y plane. Also, a shorthand notation for the incremental change of stress resultants is introduced on the figure: $N^+_x = N_x + N_{x,x} \, dx$, etc.

Projected stress resultants N_X, N_Y, and S_{XY} are defined, such that the X-Y components of the *resulting forces* on the edges ds_x or ds_y are equal to the *total forces* on the corresponding mapped edges dX or dY. On the shell element, we have noted the local tangents

$$\left. \begin{aligned} \tan \gamma_x(X,Y) &= Z_{,X} \\ \tan \gamma_y(X,Y) &= Z_{,Y} \end{aligned} \right\} \tag{6.13}$$

which are useful in what follows. Proceeding, we set

$$N_X dY = N_x \, ds_y \cos \gamma_x$$
$$S_{XY} \, dY = S \, ds_y \cos \gamma_y$$
$$N_Y \, dX = N_y ds_x \cos \gamma_y$$
$$S_{XY} \, dX = S \, ds_x \cos \gamma_x$$

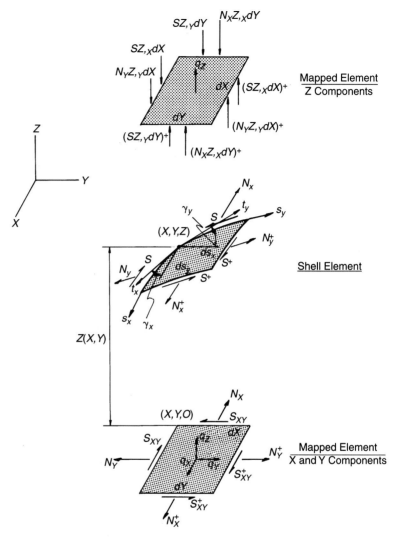

FIGURE 6–4   Shell Element and Mapping into Cartesian Coordinates.

Because

$$\left. \begin{aligned} \frac{dY}{ds_y} &= \cos \gamma_y \\[2mm] \frac{dX}{ds_x} &= \cos \gamma_x \end{aligned} \right\} \tag{6.14}$$

then

$$N_x = N_X \frac{\cos \gamma_y}{\cos \gamma_x} \tag{6.15a}$$

$$N_y = N_Y \frac{\cos \gamma_x}{\cos \gamma_y} \tag{6.15b}$$

$$S = S_{XY} \tag{6.15c}$$

Now, we derive the equilibrium equations for the mapped element. After solving for the projected stress resultants, the actual membrane stress resultants are recovered from equations (6.15a–c). Because $S_{XY} = S$ from equation (6.15c), we drop the subscript on this term.

Summing the forces in the X and Y directions, we have

$$(N_X + N_{X,X}\, dX)dY - N_X\, dY + (S + S_{,Y}\, dY)dX - S\, dX + q_X\, dX\, dY = 0$$

or

$$N_{X,X} + S_{,Y} + q_X = 0 \tag{6.16a}$$

and

$$N_{Y,Y} + S_{,X} + q_Y = 0 \tag{6.16b}$$

To form the third equilibrium equation, we project onto the plane the Z components of the forces corresponding to the actual stress resultants, as shown on the top element in Figure 6–4.

We have from N_x

$$N_x ds_y \sin \gamma_x = N_X \frac{\cos \gamma_y}{\cos \gamma_x} \frac{dY}{\cos \gamma_y} \sin \gamma_x$$

$$= N_X \tan \gamma_x\, dY \tag{6.17a}$$

$$= N_X Z_{,X}\, dY$$

and from N_y

$$N_y\, ds_x \sin \gamma_y = N_Y \frac{\cos \gamma_x}{\cos \gamma_y} \frac{dX}{\cos \gamma_x} \sin \gamma_y$$

$$= N_Y \tan \gamma_y\, dX \tag{6.17b}$$

$$= N_Y Z_{,Y}\, dX$$

From the shear forces, we find

$$S\, ds_y \sin \gamma_y = S \frac{dY}{\cos \gamma_y} \sin \gamma_y$$

$$= S \tan \gamma_y\, dY \tag{6.17c}$$

$$= S Z_{,Y}\, dY$$

$$S \, ds_x \sin \gamma_x = S \, \frac{dX}{\cos \gamma_x} \sin \gamma_x$$

$$= S \tan \gamma_x \, dX \tag{6.17d}$$

$$= SZ_{,X} \, dX$$

Summing the projected forces in the Z direction,

$$(N_X Z_{,X})_{,X} + (N_Y Z_{,Y})_{,Y} + (SZ_{,Y})_{,X} + (SZ_{,X})_{,Y} + q_Z = 0 \tag{6.18}$$

It is convenient to expand and rearrange equation (6.18) as

$$Z_{,XX} N_X + 2Z_{,XY} S + Z_{,YY} N_Y + Z_{,X}(N_{X,X} + S_{,Y}) + Z_{,Y}(N_{Y,Y} + S_{,X}) + q_Z = 0 \tag{6.19}$$

We then replace the terms in parentheses in equation (6.19) by q_X and q_Y from equations (6.16a) and (6.16b), leaving

$$Z_{,XX} N_X + 2Z_{,XY} S + Z_{,YY} N_Y - Z_{,X} q_X - Z_{,Y} q_Y + q_Z = 0 \tag{6.20}$$

as the third equilibrium equation.

We now introduce the Pucher stress function, $\mathscr{F}$,[6] which is defined such that

$$\mathscr{F}_{,YY} = N_X + \int q_X \, dX \tag{6.21a}$$

$$\mathscr{F}_{,XX} = N_Y + \int q_Y \, dY \tag{6.21b}$$

$$-\mathscr{F}_{,XY} = S \tag{6.21c}$$

It is easily verified by substitution that if $\mathscr{F}$ is a continuous function with continuous partial derivatives, equations (6.16a) and (6.16b) are identically satisfied. Therefore, the membrane theory stress analysis for shells of double curvature reduces to the single equation

$$Z_{,XX} \mathscr{F}_{,YY} - 2Z_{,XY} \mathscr{F}_{,XY} + Z_{,YY} \mathscr{F}_{,XX} = -q_Z + Z_{,X} q_X$$
$$+ Z_{,Y} q_Y + Z_{,XX} \int q_X \, dX + Z_{,YY} \int q_Y \, dY \tag{6.22}$$

The mapping onto the X-Y plane, and the subsequent introduction of the Pucher stress function, has produced a single equation which encompasses a large class of shells. Once the stress function $\mathscr{F}$ is evaluated, the projected stress resultants are computed from equation (6.21), and the actual resultants from equation (6.15). The mathematical complications in solving equation (6.22) for a given geometry, and the resulting properties of the solution, are dependent on the type of the surface equation $Z(X,Y)$—for example, elliptic, hyperbolic, or parabolic—and the complexity of the applied loading q_X, q_Y, and q_Z, specified in the Cartesian coordinate system.

For the purposes of illustrating the Pucher stress function technique, we consider three representative surfaces encountered quite often in shell roof structures: (a) hyperbolic paraboloid, (b) elliptic paraboloid, and (c) conoid.

6.4 HYPERBOLIC PARABOLOIDS

6.4.1 Surface Properties

The *hyperbolic paraboloid* (HP) is perhaps the most fascinating of the translational shells. The basic form is generated by passing a convex parabola over a concave parabola, producing an anticlastic surface, as shown in Figure 6–5. For obvious reasons, such a hyperbolic paraboloid is commonly referred to as a "saddle" shell.

The equation of the surface shown in Figure 6–5 is

$$Z = \left(\frac{c_X}{a^2}\right) X^2 - \left(\frac{c_Y}{b^2}\right) Y^2 \tag{6.23}$$

We observe that the hyperbolic paraboloid is a surface of negative Gaussian curvature. This is easily verified from equation (2.42) with $Z_{,XX} = 2(c_X/a^2)$, $Z_{,YY} = -2\ (c_Y/b^2)$ and $Z_{,XY} = 0$, from which the discriminant $\delta = Z_{,XX}Z_{,YY} - (Z_{,XY})^2 < 0$.

The stress analysis of the basic form of the hyperbolic paraboloid shown in Figure 6–5 is postponed so that we can first consider a more common configuration. We observe from Figure 6–5 that the surface has negative Gaussian curvature. Referring to the classification in Figure 2–7, we expect that there should be two sets of straight lines on the surface. Mathematically, these families of straight lines are called the *real characteristics*. In order to arrive at the equations for the characteristics, we use a construction suggested by Billington.[7]

First, we pass an arbitrary vertical plane through the shell. This plane intersects the *X-Y* plane along the straight line

$$X = a_1 - \left(\frac{a_1}{b_1}\right) Y \tag{6.24}$$

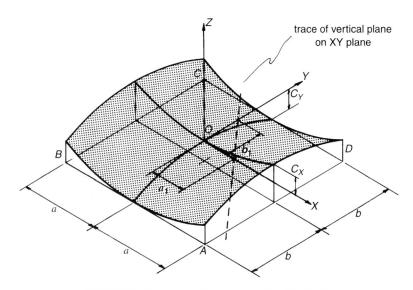

FIGURE 6–5   Hyperbolic Paraboloid Saddle Shell.

shown as a dashed line in Figure 6–5. In equation (6.24), a_1 and b_1 are the intercepts of the line on the X and Y axes, respectively, and $-a_1/b_1$ is the slope of the line. Next, we locate the intersection of the vertical plane with the shell surface. Since the X-Y projection of the intersection is the straight line given by equation (6.24), we substitute the latter equation into equation (6.23) to get the Z coordinate of the intersection of the plane and the surface as a function of Y:

$$Z = \left[\left(\frac{c_X}{a^2}\right)\left(\frac{a_1}{b_1}\right)^2 - \left(\frac{c_Y}{b^2}\right)\right] Y^2 - 2\left(\frac{c_X}{a^2}\right)\left(\frac{a_1^2}{b_1}\right) Y + \left(\frac{c_X}{a^2}\right) a_1^2 \tag{6.25}$$

Equation (6.25) describes a straight line if the coefficient of Y^2 vanishes, which occurs when

$$\frac{a_1}{b_1} = \pm \left[\frac{c_Y/b^2}{c_X/a^2}\right]^{1/2} \tag{6.26}$$

Recalling that $(-a_1/b_1)$ is the slope of the projection of the intersection onto the X-Y plane, the families of straight lines on the surface are defined by those vertical planes intersecting the X-Y plane which produce traces with slopes given by the rhs of equation (6.26). Finally, we substitute the computed (a_1/b_1), equation (6.26), into the equation of the intersection with the surface, equation (6.25), to get the equations of the families of straight lines, or characteristics

$$Z = \pm 2\frac{\sqrt{(c_X c_Y)}}{ab} a_1 Y + \frac{c_X}{a^2} a_1^2 \tag{6.27}$$

Through any point on the surface pass two characteristics. To identify the straight lines passing through a given point $(\overline{X}, \overline{Y})$, we substitute, $\overline{X}, \overline{Y}$, and equation (6.26) into equation (6.24) and find

$$\overline{a}_1 = \overline{X} \mp \left[\frac{(c_Y/b^2)}{(c_X/a^2)}\right]^{1/2} \overline{Y} \tag{6.28}$$

Then, each of the computed values of $\overline{a}_1$ is substituted into equation (6.27) to obtain the equations of the intersecting straight lines.

Now that we have identified two families of intersecting straight lines on the surface, a new possibility arises. Instead of visualizing the shell as produced by the translation of one parabola over another, why not generate the shell by the translation of one or the other of the sets of straight lines? Correspondingly, we could choose our reference coordinates to coincide with a pair of the straight lines. This is especially attractive if the boundaries of the shell coincide with these straight lines.

To explore this possibility in terms of our previous equations, we restrict our discussion to a hyperbolic paraboloid with a square plan-form, for which $b = a$ and $c_X = c_Y = c$. Then, the slopes of the straight lines on the surface in the X-Y plane are computed from equation (6.26) as

$$\left(\frac{a_1}{b_1}\right) = \pm 1 \tag{6.29}$$

Therefore, the families of straight lines are orthogonal, and we can define a new coordinate system $\widetilde{X}$-$\widetilde{Y}$ at 45° from the X-Y system, as shown in Figure 6–6. The more general transformation is treated in Billington.[8]

The relationship between the original X-Y coordinates and a new pair rotated 45° is easily derived from Figure 6–6:

$$X = \widetilde{X} \cos 45° - \widetilde{Y} \sin 45°$$
$$Y = \widetilde{X} \sin 45° + \widetilde{Y} \cos 45°$$
(6.30)

which, upon substitution into equation (6.23), gives

$$Z = \frac{c}{a^2} [\widetilde{X}^2 \cos^2 45° + \widetilde{Y}^2 \sin^2 45° - 2 \widetilde{X} \widetilde{Y} \cos 45° \sin 45°$$
$$- \widetilde{X}^2 \sin^2 45° - \widetilde{Y}^2 \cos^2 45° - 2\widetilde{X} \widetilde{Y} \sin 45° \cos 45°]$$
(6.31)
$$= -\frac{2c}{a^2} \widetilde{X} \widetilde{Y}$$

Now, we choose new boundaries for the surface that are parallel to the straight lines, instead of to the parabolas. These are indicated by dashed lines on Figure 6–6 as EF, FG, GH, and HE. The plan dimension referred to in the rotated coordinate system is $d = (\sqrt{2}/2)a$, so that we may rewrite equation (6.31) as

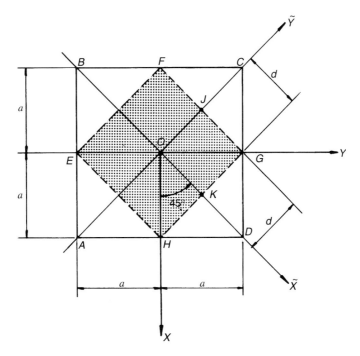

FIGURE 6–6   Coordinate Rotation for Hyperbolic Paraboloid.

$$Z = \frac{-2c}{(2d/\sqrt{2})^2} \widetilde{X} \widetilde{Y}$$

$$= \frac{-c}{d^2} \widetilde{X} \widetilde{Y} \tag{6.32}$$

$$= k \widetilde{X} \widetilde{Y}$$

where $k = -c/d^2$.

We show a quadrant of the hyperbolic paraboloid defined by equation (6.32) in Figure 6–7. W may visualize this surface as being formed by a straight line parallel to the $\widetilde{X}$ axis, called a *generator,* translated between the two straight lines *OJ* and *KG,* known as *directrices,* where point *G* has been displaced vertically a distance c from *K* in the negative *Z* direction. This construction produces a warped hyperbolic paraboloidal surface with straight boundaries.

Hyperbolic paraboloids of this general form have become a symbol of grace and elegance in architectural applications of thin-shell roofs. Popularized by F. Candela in Mexico,[9] several examples of actual shells are shown in Figures 2–8(a–c). A noteworthy construction feature of this surface is that the formwork for reinforced concrete shells may be fabricated entirely from straight boards.

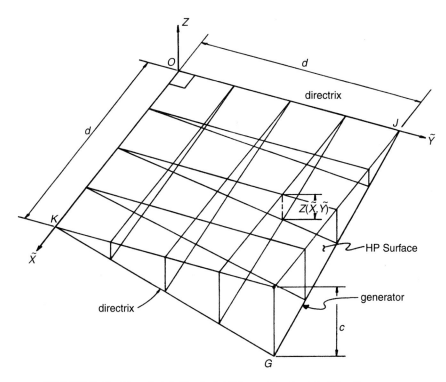

FIGURE 6–7   Hyperboloic Paraboloid Formed by Straight-Line Generators.

Various combinations of four quadrants, similar to those shown in Figure 6–7, are depicted in Figure 6–8. These basic forms are referred to freely in the ensuing discussion. In turn, these units may be combined into larger surfaces, such as the multiple saddle roof in Figure 2–8(a) and the multiple umbrella shells in Figure 2–8(b).

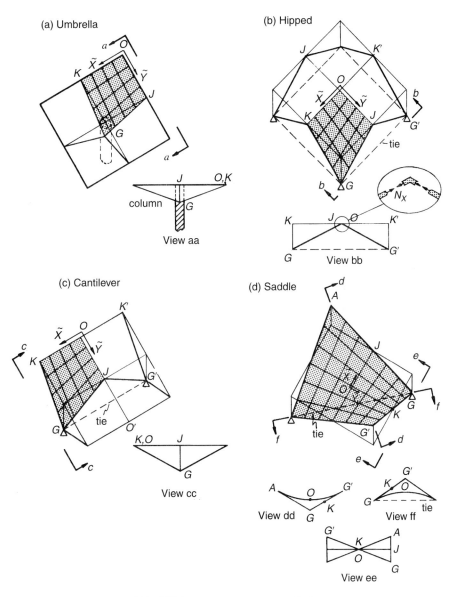

FIGURE 6–8   Variations of HP Shells.

6.4.2   Stresses in HP Shell with Straight Boundaries

Now we proceed to the stress analysis. We rewrite the governing equation for the Pucher stress function, equation (6.22), to match the $\tilde{X}$-$\tilde{Y}$ coordinate system:

$$Z_{,\tilde{X}\tilde{X}} \, \mathcal{F}_{,\tilde{Y}\tilde{Y}} - 2Z_{,\tilde{X}\tilde{Y}} \, \mathcal{F}_{,\tilde{X}\tilde{Y}} + Z_{,\tilde{Y}\tilde{Y}} \, \mathcal{F}_{,\tilde{X}\tilde{X}}$$

$$= -q_Z + Z_{,\tilde{Y}} \, q_{\tilde{Y}} + Z_{,\tilde{X}} \, q_{\tilde{X}} + Z_{,\tilde{X}\tilde{X}} \int q_{\tilde{X}} \, d\tilde{X} + Z_{,\tilde{Y}\tilde{Y}} \int q_{\tilde{Y}} \, d\tilde{Y} \tag{6.33}$$

with equations (6.21) correspondingly redefined. Then, from equation (6.32),

$$Z_{,\tilde{X}} = k\tilde{Y}$$
$$Z_{,\tilde{Y}} = k\tilde{X}$$
$$Z_{,\tilde{X}\tilde{X}} = 0$$
$$Z_{,\tilde{Y}\tilde{Y}} = 0$$
$$Z_{,\tilde{X}\tilde{Y}} = k$$

whereupon equation (6.33) becomes

$$-\mathcal{F}_{,\tilde{X}\tilde{Y}} = \frac{1}{2k} \left(-q_Z + k\tilde{X} \, q_{\tilde{Y}} + k\tilde{Y} \, q_{\tilde{X}} \right) \tag{6.34}$$

Now, referring to equation (6.21), we see that

$$-\mathcal{F}_{,\tilde{X}\tilde{Y}} = S = \frac{1}{2k} \left(-q_Z + k\tilde{X} \, q_{\tilde{Y}} + k\tilde{Y} \, q_{\tilde{X}} \right) \tag{6.35}$$

Note that there are no terms in S that can be affected by the boundary conditions. Rather, the boundaries of the shell are required to develop the computed values at $\tilde{X} = d$ or $\tilde{Y} = d$.

We now proceed with the solution of equation (6.34) for $\mathcal{F}$. First, we take $\partial/\partial\tilde{X}$ and integrate with respect to $\tilde{Y}$, giving

$$\mathcal{F}_{,\tilde{X}\tilde{X}} = -\frac{1}{2k} \int [-q_{Z,\tilde{X}} + k(\tilde{X} \, q_{\tilde{Y}} + \tilde{Y} \, q_{\tilde{X}})_{,\tilde{X}}] \, d\tilde{Y} + f_1(\tilde{X})$$

$$= N_{\tilde{Y}} + \int q_{\tilde{Y}} \, d\tilde{Y} \tag{6.36}$$

by equation (6.21). Similarly,

$$\mathcal{F}_{,\tilde{Y}\tilde{Y}} = -\frac{1}{2k} \int [-q_{Z,\tilde{Y}} + k(\tilde{X} \, q_{\tilde{Y}} + \tilde{Y} \, q_{\tilde{X}})_{,\tilde{Y}}] \, d\tilde{X} + f_2(\tilde{Y})$$

$$= N_{\tilde{X}} + \int q_{\tilde{X}} \, d\tilde{X} \tag{6.37}$$

From equations (6.35), (6.36), and (6.37), we may write explicit expressions for the projected stress resultants and then evaluate the actual stress resultants from equation (6.15), with X and Y replaced by $\tilde{X}$ and $\tilde{Y}$. The functions of integration, $f_1(\tilde{X})$ and $f_2(\tilde{Y})$, can then be determined from the specified boundary conditions along $\tilde{X} = \pm d$ and $\tilde{Y} = \pm d$.

Frequently when the hyperbolic paraboloid is used as a roof and subjected principally to a distributed vertical loading, further assumptions are justified. A first step is to consider only the Z component of loading, which greatly simplifies the integrals in equations (6.36) and (6.37). A further step is to assume that the remaining load q_Z is constant and uniformly distributed over the projected surface. If this is the case, equations (6.35), (6.36), and (6.37) reduce to

$$S = \frac{-q_Z}{2k} \tag{6.38}$$

and

$$N_{\widetilde{Y}} = f_1(\widetilde{X}) \tag{6.39a}$$

$$N_{\widetilde{X}} = f_2(\widetilde{Y}) \tag{6.39b}$$

Referring to Figure 6–7, if the boundary $\widetilde{X} = 0$ or $\widetilde{X} = d$ is assumed to be stress-free in the $\widetilde{X}$ direction $[N_{\widetilde{X}}(0,\widetilde{Y}) = 0$ or $N_{\widetilde{X}}(d,\widetilde{Y}) = 0]$, then as $\widetilde{Y}$ varies along that boundary, $N_{\widetilde{X}} = f(\widetilde{Y}) = 0$. Similarly, a free boundary at $\widetilde{Y} = 0$ or $\widetilde{Y} = d$ gives $f_1(\widetilde{X}) = 0$, so that the expressions for the stresses on the hyperbolic paraboloid simplify to

$$S = \frac{-q_Z}{2k} \tag{6.40}$$

and

$$N_{\widetilde{x}} = N_{\widetilde{y}} = 0 \tag{6.41}$$

where $\widetilde{x}$ and $\widetilde{y}$ are curvilinear coordinates on the shell surface corresponding to $\widetilde{X}$ and $\widetilde{Y}$ on the projected surface.

A loading of the form q_Z = constant is frequently termed *live load* and is often taken to represent snow loading, as discussed for cylindrical shells in Section 6.2. For shells that are relatively shallow, a uniform distribution for q_Z may also be used to approximate the dead load closely. If a uniform load approximation is used for the dead load, an equivalent value q_Z should be computed by dividing the total shell weight by the plan area. This will ensure that overall statics will not be violated.

We now reflect on the remarkably simple stress distribution found for a uniform load on a hyperbolic paraboloid. Equations (6.40) and (6.41) indicate a state of pure shear in the shell. To compute S, it is necessary only to evaluate $k = -c/a^2$ and divide q_Z by $2k$. In designers' jargon, this is termed a "back of the envelope" calculation. But, one should not be deceived by the apparent simplicity of the mathematics. We must inquire further into the physical boundary conditions which have to be provided to sustain the shell in this simple-to-calculate state of stress.

Example 6.4 (a)

As an example, consider a uniform load $q_Z = p_u$, constant per unit area of projected surface, acting on the umbrella shell shown in Figure 6–8(a). The shell is shown in detail in Figure 6–9, where the four segments are numbered. From equation (6.40) and Figure 6–9, $k = -c/d^2$ and

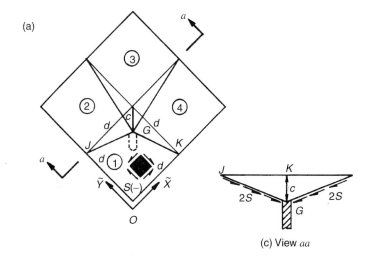

(c) View aa

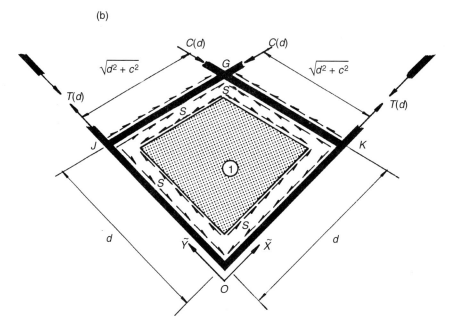

FIGURE 6–9   Umbrella Shell.

$$S = \frac{-(-p_u)}{-2(p/d^2)} = \frac{-p_u d^2}{2c} \tag{6.42}$$

The sense of S corresponding to the negative value in equation (6.42) is obtained by referring to Figure 6–4, and is shown in Figure 6–9(a). An enlarged view of quadrant 1 is shown in Figure 6–9(b), with the shell proper separated from the edges OJ, JG, GK, and KO. Clearly, the uniform shear S must be developed by axially loaded edge members along each of these lines. Edge OJ is subject to a tensile force $T(\tilde{Y})$, which increases from 0 at point O to a maximum value of $T(d) = Sd$ at J, where it is balanced by an equal tension coming from the edge member of adjacent quadrant 2. An identical tensile force is present along edge OK. The situation along the interior edge members JG and KG is somewhat different. Considering JG, the compression $C(X)$ contributed from quadrant 1 builds from 0 to J to a maximum of $S\sqrt{(d^2 + c^2)}$ at G. Adjacent quadrant 2 contributes an identical force, so the maximum compression $C(d) = 2S\sqrt{(d^2 + c^2)}$. An identical situation occurs along KG and the other two interior edge members. Although the horizontal component of the maximum compression at G for each interior edge member, $C(d)[d/\sqrt{(d^2 + c^2)}]$, is balanced by the similar component from the opposite interior edge member, the vertical components of these four forces, $4C(d)[c/\sqrt{(d^2 + c^2)}]$, is transmitted through the column to the foundation. Of course, the total vertical force $4[2S\sqrt{(d^2 + c^2)}]$ $[c/\sqrt{(d^2 + c^2)}] = 8Sc$ should balance the total applied load $4p_u d^2$. From equation (6.42),

$$8Sc = \frac{8p_u d^2 c}{2c} = 4p_u d^2 \tag{6.43}$$

We observe from Figure 6–9(c) that the compressive forces flow naturally to the support. Also note that the internal self-balancing of the maximum tensile forces and the horizontal components of the maximum compressive forces occurs only if the dimensions and loading are identical for the four segments. Any unbalanced loading will produce a resultant shear and overturning moment, which would have to be resisted by the single column. For this reason, the single free-standing umbrella is usually restricted to rather modest dimensions. The umbrella has also been used in multiple repetitions to cover large plan areas where occasional columns are acceptable[10] as shown in Figure 2–8(b). In such applications, greater lateral stability can be provided through frame action of the interconnected shells.

The stress pattern for any of the other cases shown in Figure 6–8 is easily obtained by locating the corresponding quadrant $OJGK$ on each figure. Then, the forces in the edge members are computed by starting from a point of assumed zero axial force, usually the junction of the edge member with an external boundary.

Next, consider a dead load p_d, constant per unit area of *middle* surface. Referring to Figure 6–4,

$$-p_d ds_x ds_y = q_z dXdY$$

From equation (6.14), $dX/ds_x = \cos\gamma_x$ and $dY/ds_y = \cos\gamma_y$, so that

$$q_Z = \frac{-p_d}{\cos\gamma_X \cos\gamma_Y} \tag{6.44}$$

In terms of the $\tilde{X}$-$\tilde{Y}$ coordinates

$$\tan\gamma_x = Z_{,\tilde{x}} = k\tilde{Y}$$
$$\tan\gamma_y = Z_{,\tilde{y}} = k\tilde{X}$$

and

$$
\left.
\begin{aligned}
\cos \gamma_x &= \frac{1}{\sqrt{(1 + \tan^2 \gamma_x)}} = \frac{1}{\sqrt{(1 + k^2 \widetilde{Y}^2)}} \\
\cos \gamma_y &= \frac{1}{\sqrt{(1 + \tan^2 \gamma_y)}} = \frac{1}{\sqrt{(1 + k^2 \widetilde{X}^2)}}
\end{aligned}
\right\}
\tag{6.45}
$$

so that

$$
\begin{aligned}
q_Z &= -p_d[(1 + k^2\widetilde{Y}^2)(1 + k^2 \widetilde{X}^2)]^{1/2} \\
&= -p_d[1 + k^2(\widetilde{X}^2 + \widetilde{Y}^2) + k^4 \widetilde{X}^2 \widetilde{Y}^2]^{1/2}
\end{aligned}
\tag{6.46a}
$$

For most practical cases, $k^4 \ll k^2$, and the last term in equation (6.46a) can be neglected. We may also approximate the remaining square root term with the binomial expansion

$$
q_Z \simeq -p_d \left[1 + \frac{1}{2} k^2 (\widetilde{X}^2 + \widetilde{Y}^2) + \cdots \right]
\tag{6.46b}
$$

Substituting equation (6.46b) into equation (6.40) gives

$$
S = \frac{-q_Z}{2k} = \frac{p_d}{2k} \left[1 + \frac{1}{2} k^2 (\widetilde{X}^2 + \widetilde{Y}^2) \right]
\tag{6.47}
$$

Noting that $\mathscr{F},_{\widetilde{X}\widetilde{Y}} = -S$, we integrate equation (6.47) to get

$$
\mathscr{F} = \frac{-p_d}{2k} \left[\widetilde{X} \widetilde{Y} + \frac{k^2}{6} (\widetilde{X}^3 \widetilde{Y} + \widetilde{Y}^3 \widetilde{X}) + f_1(\widetilde{Y}) + f_2(\widetilde{X}) \right]
\tag{6.48}
$$

Then, we compute the extensional stress resultants from equations (6.21a) and (6.21b) as

$$
\begin{aligned}
N_{\widetilde{X}} &= \mathscr{F},_{\widetilde{Y}\widetilde{Y}} \\
&= \frac{-p_d}{2k} [k^2 \widetilde{X} \widetilde{Y} + f_1(\widetilde{Y}),_{\widetilde{Y}\widetilde{Y}}]
\end{aligned}
\tag{6.49}
$$

and

$$
\begin{aligned}
N_{\widetilde{Y}} &= \mathscr{F},_{\widetilde{X}\widetilde{X}} \\
&= \frac{-p_d}{2k} [k^2 \widetilde{X} \widetilde{Y} + f_2(\widetilde{Y}),_{\widetilde{X}\widetilde{X}}]
\end{aligned}
\tag{6.50}
$$

Whereas S is independent of any boundary conditions, just as in the case of the uniform loading, the extensional stress resultants are dependent on the boundary conditions. Also, because we may specify $N_{\widetilde{X}}$ and $N_{\widetilde{Y}}$ on only one of the two parallel boundaries, the opposite boundaries must develop the computed stress resultants. If, for example, we consider the umbrella shell shown in Figure 6–9, it is reasonable to presume that these stress resultants vanish on the exterior boundaries, so that

$$
N_{\widetilde{X}} (0,\widetilde{Y}) = 0
\tag{6.51a}
$$

$$
N_{\widetilde{Y}} (\widetilde{X},0) = 0
\tag{6.51b}
$$

giving

$$f_1(\widetilde{Y}),_{\widetilde{Y}\widetilde{Y}} = f_2(\widetilde{X}),_{\widetilde{X}\widetilde{X}} = 0 \tag{6.52}$$

The expressions for the dead-load stress resultants in this shell are evaluated from equations (6.47), (6.52), (6.49), (6.50), (6.15), and (6.45):

$$S = \frac{-p_d d^2}{2c} \left[1 + \frac{c^2}{2d^4} (\widetilde{X}^2 + \widetilde{Y}^2) \right] \tag{6.53a}$$

$$N_{\widetilde{x}} = \frac{p_d c}{2d^2} \widetilde{X}\widetilde{Y} \left[\frac{1 + (c^2/d^4)\widetilde{Y}^2}{1 + (c^2/d^4)\widetilde{X}^2} \right]^{1/2} \tag{6.53b}$$

$$N_{\widetilde{y}} = \frac{p_d c}{2d^2} \widetilde{X}\widetilde{Y} \left[\frac{1 + (c^2/d^4)\widetilde{X}^2}{1 + (c^2/d^4)\widetilde{Y}^2} \right]^{1/2} \tag{6.53c}$$

The forces in the edge members are found in a similar manner as for the uniform load case. However, because $S = S(X,Y)$ we must proceed more formally. Referring to Figure 6–9,

$$T(\widetilde{Y}) = \int_0^{\widetilde{Y}} S(0,\widetilde{Y}) \, d\widetilde{Y} \tag{6.54a}$$

$$C(\widetilde{X}) = 2\int_0^{\widetilde{X}} S(\widetilde{X},d) \, \frac{d\widetilde{X}}{\cos \gamma_x} \tag{6.54b}$$

Note that the compressive force $C(\widetilde{X})$ is evaluated in equation (6.54b) by integrating along the inclined edge member. The integrands for equations (6.54a) and (6.54b) are computed by substituting $\widetilde{X} = 0$ and $\widetilde{Y} = d$, respectively, in equation (6.53a); then, the maximum values of T and C are found by inserting $\widetilde{Y} = d$ and $\widetilde{X} = d$ as the upper limits of integration.

Earlier, we alluded to the possibility of replacing the dead load, which is constant over the shell surface, with a uniform load which is constant over the projected area. A comparison of the solutions for the loadings p_u and p_d, as given by equations (6.42) and (6.53) respectively, provides an obvious motivation in terms of algebraic simplicity if the errors introduced are not significant. As mentioned before, if we replace p_d by an equivalent p_u, we should use an adjusted load q_z, which is the total weight of the shell divided by the plan area, to satisfy overall statics. We find the total weight of the shell using equation (6.46b) as

$$Q = \int_0^d \int_0^d p_d \left[1 + \frac{1}{2} k^2(\widetilde{X}^2 + \widetilde{Y}^2) \right] d\widetilde{X} \, d\widetilde{Y} \tag{6.55}$$

and then compute the equivalent uniform load

$$q_x = \overline{p}_u = -\frac{Q}{d^2} \tag{6.56}$$

for use in equation (6.40). The effect of this simplification on the accuracy of the ensuing solution is not great for many practical geometries.

An interesting alternative to the free-edge boundary conditions stated in equations (6.51) is implied in a design example based on an actual shell having the basic hipped or gable form shown in Figure 6–8(b).[11] If the exterior edges KG and GJ were taken as stress-free, the integration functions could be readily evaluated by substituting

$$N_{\widetilde{X}}(d, \widetilde{Y}) = 0$$

and

$$N_{\widetilde{Y}}(\widetilde{X}, d) = 0$$

into equations (6.49) and (6.50), respectively. However, the subsequent expressions for $N_{\widetilde{x}}$ and $N_{\widetilde{y}}$, analogous to equations (6.53b) and (6.53c), would give nonzero (tensile) values along the ridge lines *OJ* and *OK;* that is,

$$N_{\widetilde{x}}(0, \widetilde{Y}) > 0$$

and

$$N_{\widetilde{y}}(\widetilde{X}, 0) > 0$$

The horizontal components of these stress resultants would be balanced by identical forces from the adjacent segments, but the vertical (*Z*) components would add to produce an unbalanced force that would have to be sustained by the edge member. This is illustrated in the inset of view *bb* in Figure 6–8(b). To avoid this situation, the designer apparently decided it was preferable to assume that $N_{\widetilde{x}}$ and $N_{\widetilde{y}}$ vanish along the ridge lines, and then to develop the calculated values of the stress resultants on the perimeter. For this case, $N_{\widetilde{x}}$ and $N_{\widetilde{y}}$ become compressive as well in the interior of the shell. We surmise that occasionally there is some latitude in specifying boundary conditions in a membrane theory solution, and that the subsequent problem of providing physical constraints consistent with the assumed conditions should be kept in mind. This shell is discussed further in Sections 6.4.3 and 6.4.4.

6.4.3 Arch Action

The state of pure shear found for uniformly loaded hyperbolic paraboloid shells can be interpreted from another standpoint. We study the same quadrant *OJGK* shown in Figure 6–9, but, for variety, we consider the saddle shell illustrated in Figure 6–8(d). The membrane theory stress analysis is the same, and the uniform negative shear *S* is shown in Figure 6–10. From elementary strength of materials, it is easily shown that a state of pure shear with respect to the $\widetilde{X}$-$\widetilde{Y}$ axes corresponds to principal tensile and compressive stresses of the same magnitude acting in the *X* and *Y* directions as shown. If we then take sections through the element parallel to the *X* and *Y* axes, we reveal segments of the parabolic generators of the basic hyperbolic paraboloid shown in Figure 6–5. The convex parabola *MN* parallel to the *Y* axis acts as a *compression arch* subject to a uniformly distributed load p_u, producing a constant axial compression of *S* per unit width. This is the elementary problem of the second-degree parabola subject to a uniform load per unit projected length. Now, examining the parabola parallel to the *X* axis *PQ,* we find a similar situation, except that we have a concave parabola forming a *tension* arch. Each parabola imparts a constant reaction *S* per unit width of shell on the edge members, e.g., at points *N* and *P* on *JG*. Of course, the other ends of the parabolas, indicated by dashed extensions, will intersect edge members in adjacent quadrants.

To complete the arch representation, we investigate the edge forces. Considering a typical point on the boundary (e.g., *N*), observe that the unit width compression arch imparts a force of *S* in the +*Y* direction, whereas the tension arch acting on the same segment of the

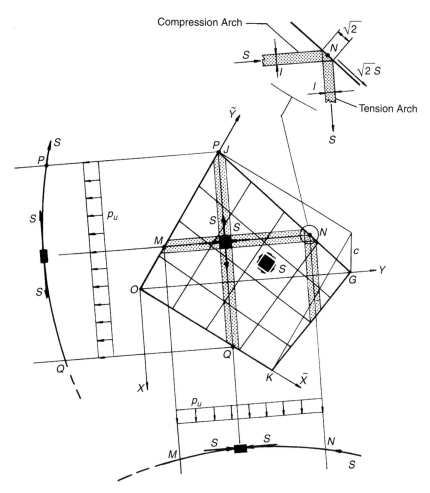

FIGURE 6–10   Arch Action for a Uniformly Loaded Hyperbolic Paraboloic.

edge beam also produces a force of S directed along the $+X$ axis. The intersecting arches are coplanar at N, and we may resolve the intersecting forces into components along and normal to the edge member. The normal components cancel, while the axial components produce a resultant force of $\sqrt{2}S$. This force acts over a length of edge beam = $\sqrt{2}$, so that we find a constant force, S, per unit length of edge beam as before. Thus, the arch representation of the hyperbolic paraboloid is certainly equivalent to the previous shell solution, and often provides valuable insight—especially if irregular shapes or boundaries are encountered.

The gable or hipped HP, Figure 6–8(b), which was discussed in Section 6.4.2, should carry load primarily by compression arches parallel to OG and tension arches parallel to KJ. Subsequent studies by Shaaban and Ketchum[12] have shown that the tension arches are ineffective, and that practically the entire load is transmitted to the corners by the intersecting compression arches passing through the crown. Of course, the stresses are essentially doubled.

This is illustrated in Figure 6–11(a), which corresponds to quadrant $OK'\ G'\ J'$ in Figure 6–8(b). The dominant wide diagonal band of compressive arch action spanning between the opposite supports, G and G', and the very small tensile stresses are indicated. This behavior is attributed to the inability of the edge members to resist the tensile stresses in the shell; rather, they displace inward, forcing the load to flow to the corners[11,12] as shown in Figure 6–11(b).

6.4.4 Edge Members

We have seen that the membrane theory stress analysis for hyperbolic paraboloids with straight-line boundaries is fairly straightforward. We have also established the necessity of

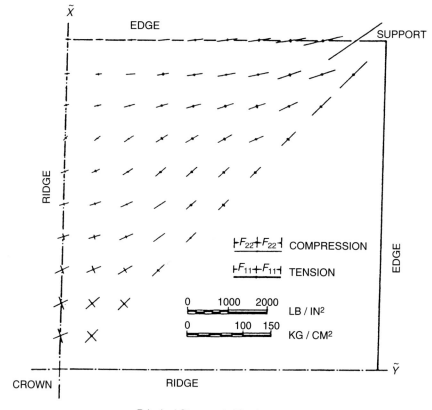

Principal Stresses in Membrane

FIGURE 6–11a Principal Stresses in Membrane. [From Shaaban and Ketchum, Design of hipped hypershells, ASCE, Journal of the Structural Division, vol. 102, no. ST11 (1976). Reprinted by permssion of the American Society of Civil Engineers.]

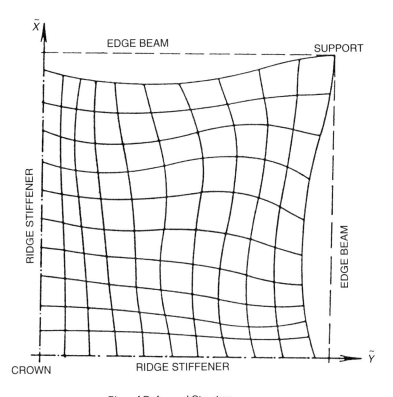

Plan of Deformed Structure

FIGURE 6–11b   Plan of Deformed Structure.

developing the in-plane shear forces on the boundary. We now look at the requirements for *ideal* edge members:

1. In order to develop the shear fully, the edge members should be inextensible.
2. Because the entire weight of the shell is ultimately carried to the foundation through only four to eight compression members for the cases illustrated in Figure 6–8, the edge members should be capable of sustaining sizable axial forces. These forces will produce corresponding strains that will conflict with the first requirement.
3. The edge members should be self-supporting and should not be carried by the shell. Because these members may be rather sizable, they would impart a sizable additional load on the shell that might be unacceptable.

From this list, it is obvious that in practice it may be difficult to attain any of these ideal conditions.

Example 6.4(b)

As a specific illustration, consider the cantilever shell shown in Figure 6–8(c). For this shell, we have four different edge members, as seen in Table 6–2. Typical sections for such members in reinforced concrete shells are shown in Figure 6–12. The dashed lines indicate possible smooth transitions between the shell and the edge member that may be desirable for architectural and/or constructional purposes. The edge members may be proportioned as struts under axial tension or compression.

Next, consider the self-weight of the edge members. Referring to Figure 6–8(c), the ridge member *OJ* could be considered as a beam spanning between *O* and *J*. The valley member *GJ* forms a triangular frame *GJG'* with the corresponding member on the other half of the shell and resists the beam reactions at *O* from the two ridge members, *OJ* and *O'J*, in addition to their self-weight and the shear forces imparted by the shell. Exterior member *KK'* is regarded as a beam spanning between *K* and *K'* and carrying the beam reaction at *O* from *OJ*, their self-weight, and the shell shear forces. This brings us to the exterior member *KG* and *K'G'*, which are required to take the beam reactions from *KK'* along with their self-weight and the shear force, seemingly as cantilevers.

Other alternatives are possible. In some cases, it is feasible to utilize exterior walls or window wall framing to support the perimeter edge members. Also, some designers might convert the self-weight of some of the edge members—particularly those that cannot sustain themselves—to equivalent distributed loads, and include these loads in the uniform load q_z. This essentially requires the shell to support the edge beams.

Perhaps the most challenging cases from the design standpoint are cantilevered compression edge members, such as *KG,* or the even longer members, such as *G'G*, that occur in saddle shells as shown in Figure 6–8(d). A spectacular example is shown in Figure 2–8(a).

Referring again to Figures 6–8(c) and 6–9, if the neutral axis of the edge beam for member *KG* does not coincide with the middle surface of the shell, the shear force acts through an eccentricity that produces a uniformly distributed moment. For the downturned beam (Figure 6–12), the moment is opposite to that produced by the cantilevered self-weight and thus counters the self-weight somewhat; whereas for the upturned beam, the moment would add to that arising from the self-weight. Depending on whether tension or compression is introduced into the edge member by the shell and the particular exterior support condition, a countering prestressing moment can generally be produced by appropriate upturning or downturning of the member to give the required eccentricity.

TABLE 6–2   Maximum Forces in Edge Members

Member	Location	Maximum Force (+ = tension)
OK	Exterior	$-Sd$ @ 0
KG	Exterior	$-S\sqrt{(c^2 + d^2)}$ @ *G*
OJ	Ridge	$+2Sd$ @ 0
JG	Valley	$-2S\sqrt{(c^2 + d^2)}$ @ *G*

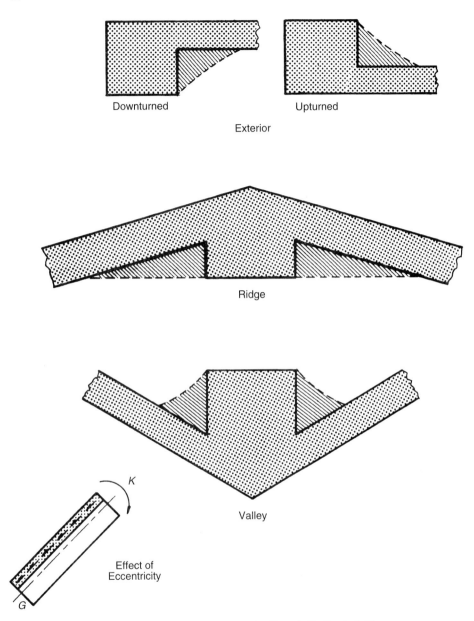

Downturned   Upturned

Exterior

Ridge

Valley

K

Effect of
Eccentricity

G

FIGURE 6–12   Edge Members for Hyperbolic Paraboloids.

Another possibility—particularly effective for tension edge members—is to produce mechanical prestressing with rods or cables. Also, as shown in the examples of Figure 6–8, it is desirable to use ties between the supported ends of the inclined compression members to counter the horizontal component of the thrust. Insofar as the reduction of the dominance of the membrane type action and the subsequent introduction of bending into the shell and the edge member are concerned, a potentially detrimental effect is a relative horizontal displacement of the abutments.[13] A comprehensive discussion of the forces in the edge members of a hipped hyperbolic paraboloid and some pertinent design recommendations for such shells are presented in Schnobrich and in Shaaban and Ketchum.[14]

Again referring to the hipped shell discussed in the preceding two sections, it has been suggested that if the edge members do not effectively transmit the in-plane shear stresses to the supports, then they may be reduced in size to avoid loading the shell excessively.[14] In fact a recent study by Jadik and Billington,[15] using a popular commercial finite element code, concluded that an improved design could be realized by reducing the size of the crown beams (ridge stiffeners), thickening the supported corners, and eliminating the edge beams. Also for a continuous group of hipped hypars, the effect of prescribing a relative displacement of the abutments consistent with the strain in a steel tie was investigated.[16] The state of stress in the shell closely resembled the membrane theory pattern, but the edge members received a nonsymmetrical loading. Limited field observations supported this characterization, as opposed to the unrestrained or undeforming idealizations.

Finally with respect to hyperbolic paraboloids with straight boundaries, note that although the presentation has been directed primarily at reinforced concrete shells, there are many interesting applications of this very efficient structural form in other materials: for example, in steel[17] and wood.[18] A cantilevered, cable-supported hyperbolic paraboloid forms the roof for the jumbo jet hangar shown in Figure 2–8(v), and metal hyperbolic paraboloid units were placed between long-span arched trusses to cover a large arena in Mexico City constructed for the 1968 Olympic games.

6.4.5 Hyperbolic Paraboloid with Curved Boundaries

We now resume our investigation of the basic hyperbolic paraboloid described by equation (6.23) and shown in Figure 6–5. We restrict ourselves to a q_Z loading, for which equation (6.22) becomes

$$Z_{,XX}\mathscr{F}_{,YY} - 2Z_{,XY}\mathscr{F}_{,XY} + Z_{,YY}\mathscr{F}_{,XX} = -q_Z \tag{6.57}$$

From equation (6.22)

$$Z_{,XX} = \frac{2c_X}{a^2}$$

$$Z_{,YY} = -\frac{2c_Y}{b^2}$$

and

$$Z_{,XY} = 0$$

so that equation (6.57) reduces to

$$\frac{2c_X}{a^2}\mathscr{F}_{,YY} - \frac{2c_Y}{b^2}\mathscr{F}_{,XX} = -q_Z(X, Y) \tag{6.58}$$

As a specific case, we take the uniform load $q_Z = -p_u$. There are obviously multiple particular solutions, since the derivatives on $\mathscr{F}$ are uncoupled. Two possibilities are

$$\mathcal{F}_1(X) = -\frac{1}{4} p_u \frac{b^2}{c_Y} X^2 \tag{6.59}$$

and

$$\mathcal{F}_2(Y) = \frac{1}{4} p_u \frac{a^2}{c_X} Y^2 \tag{6.60}$$

Also, we may have any linear combination of $\mathcal{F}_1$ and $\mathcal{F}_2$

$$\mathcal{F}(X, Y) = \lambda_1 \mathcal{F}_1(X) + \lambda_2 \mathcal{F}_2(Y)$$

where

$$\lambda_1 + \lambda_2 = 1 \tag{6.61}$$

This presents us with a multiplicity of particular solutions for the projected stress resultants, which can be written using equations (6.21) as

$$N_X(X, Y) = \frac{\lambda_2}{2} p_u \frac{a^2}{c_X} \tag{6.62a}$$

$$N_Y(X, Y) = -\frac{\lambda_1}{2} p_u \frac{b^2}{c_Y} \tag{6.62b}$$

$$S(X, Y) = 0 \tag{6.62c}$$

The actual stress resultants, N_x and N_y, may then be found from equations (6.15).

For a given shell, the solution which will prevail is a function of the relative stiffnesses of the boundaries. Referring to Figure 6–5, if the boundaries at $X = \pm a$ are fully developed in the x direction, and the shell is unrestrained at $Y = \pm b$, we would choose $\lambda_1 = 0$ and $\lambda_2 = 1$. This, of course, corresponds to a tension arch in the x direction, such as we encountered in a previous section. Similarly, $\lambda_1 = 1$ and $\lambda_2 = 0$ indicate full support in the y direction at $Y = \pm b$, and no restraint at $X = \pm a$, or a compression arch in the y direction. Obviously, with finite in-plane resistance along all boundaries, we would have nonzero values of both λ_1 and λ_2, but the necessary requirement is that at least one pair of the boundaries must develop the ensuing forces.

The possibility of "directing" the extensional stresses by adjusting the support conditions makes this version of the hyperbolic paraboloid quite versatile. Although a designer may prefer the state of uniform compression, $\lambda_1 = 1$, for a reinforced concrete shell, a tensile state, $\lambda_2 = 1$, might be more desirable for a steel structure. Numerous structures in this shape have been built using high-strength steel cables as the principal members. In such applications, *sagging* cables resist the principal tension and *hogging* cables span in the normal direction. A roofing system consisting of precast panels or a fabric membrane may be applied to such a structure.

Another variation on the hyperbolic paraboloid with curved boundaries is to leave the edges free of extensional stresses and to resist the surface loading by shear forces along only two parallel boundaries. In turn, the resulting reactions can be provided by a pair of axially loaded boundary arches. This would correspond to the hyperbolic paraboloid shown in Figure 6–5 with boundary conditions

$$N_X(\pm a, Y) = 0 \qquad\qquad (6.63a)$$
$$N_Y(X, \pm b) = 0 \qquad\qquad (6.63b)$$
$$S(X, \pm b) = 0 \qquad\qquad (6.63c)$$

Moreover, the boundaries AD and BC can develop the in-plane shear stresses

$$S(\pm a, Y) \neq 0 \qquad\qquad (6.63d)$$

This case has been studied in detail by Flügge.[19] In his analysis, stress discontinuities are found between various regions of the shell, making the membrane theory state of stress only an idealized possibility for this case. Careful detailing and construction are required to approach this stress distribution in an actual structure.

The Saddledome, Figure 2–8(n), is an elegant anticlastic shell with the boundary defined by a space curve. The form is developed from a sphere of about 136 m in diameter, as shown in Figure 6–13(a). The floor is circular, formed by the intersection of a horizontal plane and the sphere (b) and the roof perimeter is the intersection of an HP with the sphere (c). This produces the definitive form of the appropriately named Saddledome (d) bounded by an undulating ring beam.[20]

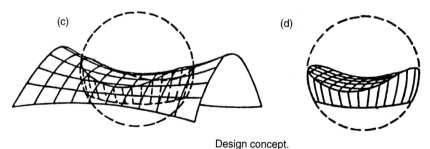

Calgary's Olympic Saddledome

(a)   (b)

(c)   (d)

Design concept.

FIGURE 6–13   Calgary's Olympic Saddledome Design Concept.

The roof is a stressed cable network with precast lightweight concrete panels placed on the cables and the gaps filled with concrete. The result is a solid ribbed thin shell.

The ring beam is subjected to a complex system of loading that is basically counterbalanced by prestressing. A member of this sophistication could only be analyzed accurately by a numerical procedure, the finite element method, which is well within the state of the art of modern technology.

The loading on the roof is a function of the stiffness of the ring and the restraining A-frames, located at the two low points at the ends of the center hogging cables and visible on Figure 2–8(n). A very flexible ring would minimize the forces in the net but allow large horizontal deflections, while the restraint produced by a rigid A-frame would minimize the deflections at the expense of a larger horizontal reaction transferred from the ring. The optimized final design produced an elastic support which controlled the deflections while resisting the reactions caused by the prestressing, dead and live loads, creep and thermal effects.[20]

Comparing this structure with the straight boundary HPs, Figure 6–8, the evolution of this form, facilitated by high-strength cables, prestressing, and computer technology, is striking.

An interesting extension of the basic hyperbolic paraboloidal geometry can be achieved by intersecting two shells, as shown in Figure 6–14. Each quadrant is known as a *groined* vault. The basic resistance is essentially in the directions parallel to the exterior boundaries. Along the edges, the loading is transferred by the shell, which acts as an assemblage of compression arches, to the groin arches, which pick up reactions from two adjacent vaults and transmit the forces to the foundation. Again, it is desirable to provide horizontal restraint at the base of the groin arches through a system of ties. Many variations on the groined vault concept are found. An elegant example with a parabolic cross-section is shown in Figure 2–8(f). This shell form is treated in detail by Billington, who provides a simplified model to compute the stresses in the groin arches as well as a detailed analysis procedure for the shell.[21]

The very similar shells illustrated in Figures 2–8(j) and (k) show that the geometry of the shell surface may have a great effect on the size and shape of the groin arch. The designer of the former shell, with a cylindrical cross-section, used heavy groin arches while the latter shell, with a parabolic cross-section, treated the intersection in a more graceful manner.

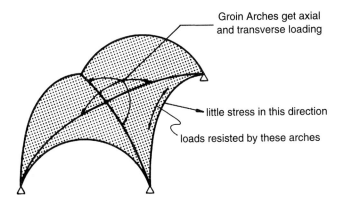

Groin Arches get axial and transverse loading

little stress in this direction

loads resisted by these arches

FIGURE 6–14   Groined Vault.

6.5 ELLIPTIC PARABOLOIDS

The elliptic paraboloid is generated by the translation of one convex paraboloid over another, as shown in Figure 6–15. The equation of the shell is given by

$$Z = -\left(\frac{c_X}{a^2}\right) X^2 - \left(\frac{c_Y}{b^2}\right) Y^2 \tag{6.64}$$

It is easily observed and may be demonstrated numerically using equation (2.42) that this shell has positive Gaussian curvature, and, as such, we do not find straight lines on the surface.

The name *elliptic paraboloid* comes from the fact that if a horizontal plane $Z = \tilde{Z} =$ constant ($\tilde{Z} < 0$) is passed through the shell, the equation of the intersection

$$\left(\frac{c_X}{a^2}\right) X^2 + \left(\frac{c_Y}{b^2}\right) Y^2 = -\tilde{Z} \tag{6.65}$$

is an ellipse. Also, if $c_X/a^2 = c_Y/b^2 = c/d^2$ we have a paraboloid of revolution

$$\left(\frac{c_X}{a^2}\right) X^2 + \left(\frac{c_Y}{b^2}\right) Y^2 = -\tilde{Z} \tag{6.66}$$

with the intersection at $Z = \tilde{Z}$ ($\tilde{Z} < 0$) given by

$$X^2 + Y^2 = -\left(\frac{d^2}{c}\right) \tilde{Z} \tag{6.67}$$

which is, of course, a circle because Z is negative.

Evaluating the required partial derivatives from equation (6.64) and substituting into equation (6.57), we obtain

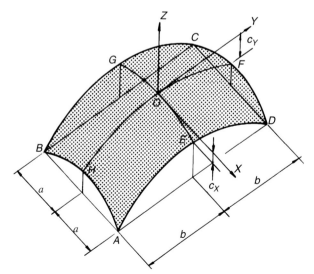

FIGURE 6–15   Elliptic Paraboloid Shell.

$$\left(\frac{2c_X}{a^2}\right)\mathscr{F}_{,YY} + \left(\frac{2c_Y}{b^2}\right)\mathscr{F}_{,XX} = q_Z(X,\,Y) \tag{6.68}$$

which is very similar to equation (6.58), and suggests that this shell would resist a uniform load $q_Z = -p_u$ as an intersecting system of compression arches parallel to the X and Y axis. It is easily shown by comparison with equations (6.58), (6.59), and (6.62) that particular solutions of the form

$$\mathscr{F}_3 = -\frac{1}{4}p_u\frac{b^2}{c_Y}X^2 \tag{6.69}$$

and

$$\mathscr{F}_4 = -\frac{1}{4}p_u\frac{a^2}{c_X}Y^2 \tag{6.70}$$

will produce projected stress resultants $N_X = 0$ and $N_Y = 0$, respectively. However, these simple unidirectional patterns are often not compatible with the physical situation, because it may be somewhat difficult to provide the reactions required to develop N_Y along $Y = \pm b$ and/or N_X along $X = \pm a$. Rather, we wish to explore the possibility of this system transmitting the loading directly to the four corners, without the necessity of providing in-plane restraint along the exterior boundaries.

The specific requirement is to find a solution to equation (6.68), such that $N_X\,(\pm a,\,Y) = N_Y\,(X,\,\pm b) = 0$ except at the corners, $(\pm a,\,\pm b)$. Obviously, it is only possible to satisfy one or the other condition using equation (6.69) or (6.70), so that, at best, $\mathscr{F}_3$ or $\mathscr{F}_4$ serve as a particular solution. We may add a homogeneous solution in the form of an infinite series that *identically* satisfies the same boundary conditions as the selected particular solution, $\mathscr{F}_3$ or $\mathscr{F}_4$. Then, the amplitudes of the terms of the infinite series are adjusted so that the total solution, homogeneous plus particular, will satisfy the remaining boundary conditions.

As an illustration, we select

$$\mathscr{F} = \mathscr{F}_4 + \mathscr{F}_5 \tag{6.71}$$

where

$$\mathscr{F}_5 = \sum_{j=1,3,5\,\ldots}^{\infty} A_j\cosh mX\cos nY$$

$$m = \frac{j\pi}{2a}\sqrt{\frac{c_X}{c_Y}}$$

$$n = \frac{j\pi}{2b}$$

$\mathscr{F}_5$ is easily shown to be the homogeneous solution of equation (6.68), and we have seen that $\mathscr{F}_4$ produces $N_Y = 0$ throughout. From equation (6.21b), we observe that because of $\mathscr{F}_5$, N_Y is proportional to $\cos\,(j\pi/2b)Y$, which vanishes at $Y = +b$. Thus, the boundary conditions on $N_Y(X,\,\pm b)$ are identically satisfied.

To proceed with the solution for the stress resultants, $\mathscr{F}_4$ must be expanded in a Fourier series compatible with $\mathscr{F}_5$. $N_X\,(\pm a,\,Y)$ is then written from equation (6.21a), and the A_j coeffi-

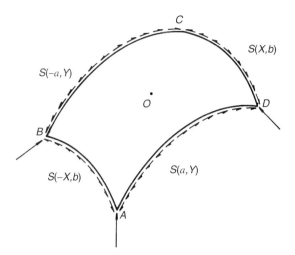

FIGURE 6–16   Boundary Forces in Elliptic Paraboloid.

cients are chosen to satisfy the boundary condition $N_X\,(\pm a,\ Y) = 0$. Then, the stress function and stress resultants are found in series form from equations (6.71), (6.21), and (6.15). The details are rather involved algebraically and are not given here. Complete solutions, including parametric tabulations, are available.[22]

It is apparent that the resulting membrane theory solution requires that the boundary arches develop the in-plane shear and thereby transmit a portion of the applied load to the corners, as shown in Figure 6–16. However, the enforced boundary conditions prohibit direct stresses along the exterior boundaries. The shear S builds up to a theoretically infinite value at the corners, where a concentrated reaction acts, and the remaining load is carried directly to the supports by arch strips AOC and BOD in a fashion similar to that of the hipped HP, as discussed in Section 6.4.3. It is apparent that the actual stress state cannot be fully described by membrane theory, and, therefore, bending effects should be considered as well. Also, the details of construction might well include some stiffening and rounding of the corner to spread the high shear forces dictated by membrane theory equilibrium.

6.6 CONOIDS

A surface produced by the translation of a straight-line generatrix between two directrices— one of which is itself a straight line and the other is a plane curve—is called a *conoid* (Figure 6–17). If the generator is perpendicular to the directrices, we have a square conoid; if not, the conoid is said to be skewed.

Shells in the form of conoids are used for roof applications when it is desired to maintain a relatively flat profile and still provide openings to admit natural sunlight.

The coordinate system shown in Figure 6–17 has the Y-Z plane normal to the symmetry line of the curved directrix, whereas the X axis is taken along the straight directrix. With respect to this coordinate system, the equation of the curved directrix is of the form $Z = f(X)$ and the square conoid is described by

$$Z(X,\ Y) = \frac{Y}{L}f(X) \tag{6.72}$$

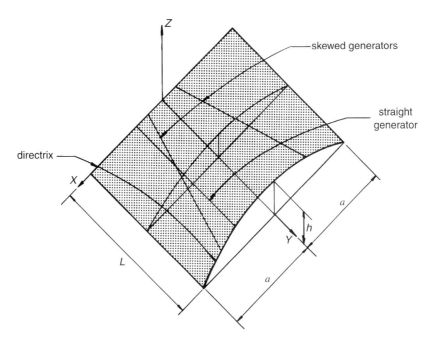

FIGURE 6–17   Conoidal Shell.

For a skewed conoid, the surface equation is more complicated. In either case, the conoid is not a second-order surface, and the discriminant test for Gaussian curvature, equation (2.42), is inapplicable. However, it is readily observed from Figure 6–17 that the skewed conoid has negative Gaussian curvature because there are two sets of straight-line generators, whereas the square conoid has zero Gaussian curvature, because the two sets of generators become coincident.

Consider a square parabolic conoid,

$$f(X) = h \left(1 - \frac{X^2}{a^2} \right) \tag{6.73a}$$

and

$$Z = \frac{h}{L} Y \left(1 - \frac{X^2}{a^2} \right) \tag{6.73b}$$

We now write equation (6.22) for a uniform load $q_z = -p_u$:

$$-\left(\frac{2h}{a^2 L} Y \right) \mathscr{F}_{,YY} + \left(\frac{4h}{a^2 L} X \right) \mathscr{F}_{,XY} = p_u \tag{6.74}$$

Although, as we have observed with respect to the hyperbolic and elliptic paraboloids that multiple solutions for stress functions can frequently be generated, a solution of practical utility must be consistent with the physical boundary conditions. Because of symmetry, we should have S vanish along $X = 0$, so that $-\mathscr{F}_{,XY}(0,Y) = 0$. Also, if the curved directrix cannot sustain forces normal to the plane of the curve, $N_X(X,L) = 0$, implying that $\mathcal{F}_{,YY}(X, L) = 0$.

Again, comprehensive results are readily available in the literature,[23] so that we will not dwell on the details of the various solutions for conoidal shells. However, it is of general interest to consider a simplified model of the conoid that is suggested by our earlier study of hyperbolic paraboloids. From Figure 6–17, it is apparent that the conoid could possibly transmit the uniform load as a series of arches in the X-Z plane, with the attendant thrusts developed along the boundaries $X = \pm a$. The provision of the appropriate boundary conditions is plausible, because the horizontal components of the thrust would balance along the interior boundaries, and the vertical components could be resisted by walls or beams in the Y direction. Further, the shell would seemingly be entirely in compression.

With respect to the equation (6.74), we may obviously select a solution of the form

$$\mathcal{F}_{,YY} = N_X = -\frac{a^2 L}{2hY} p_u \tag{6.75}$$

which would yield the one-way arch action with N_Y and $S = 0$. Then if we return to equation (6.15a) for the actual stress resultant N_X and compute

$$\tan \gamma_x = Z_{,X} = -\frac{2h}{a^2 L} X Y$$

$$\tan \gamma_y = Z_{,Y} = \frac{h}{a^2 L} X^2$$

from equation (6.73b), we find

$$N_x = -\frac{a^2 L}{2hY} \left\{ \frac{1 + [(-2h/a^2 L) XY]^2}{1 + [(h/a^2 L) X^2]^2} \right\}^{1/2} p_u \tag{6.76}$$

from equations (6.45) and (6.75). For small values of Y, N_x grows very large, with a singularity at $Y = 0$. Essentially, this reflects that the shell becomes very shallow and the arches correspondingly flatter, until they degenerate into a straight line along the direction $Y = 0$. Obviously, this region of the conoid cannot be sustained in equilibrium by extensional forces alone, and bending must be considered. The use of a plate approximation for this region would seem appropriate.

REFERENCES

1. Design of cylindrical concrete roofs, *ASCE Manual of Engineering Practice* no. 31 (New York: American Society of Civil Engineers, 1952); Billington, *Thin Shell Concrete Structures,* 2nd ed. (New York: McGraw Hill, 1982) chaps. 5 and 6.

2. Billington, *Thin Shell Concrete Structures:* 186–191.

3. Billington, *Thin Shell Concrete Structures:* 210–211.

4. Billington, *Thin Shell Concrete Structures:* 258–262.

5. A. L. Parme, Shells of double curvature, *Trans. ASCE,* vol. 123, 1958: 990–1025.

6. Billington, *Thin Shell Concrete Structures:* 261–262.

7. Billington, *Thin Shell Concrete Structures:* 272–274.

8. Ibid.

9. C. Faber, *Candela: The Shell Builder* (New York: Reinhold, 1963).

10. The New Newark Airport, *Civil Engineering 44,* no. 9 (September 1974): 74–76.

11. Billington, *Thin Shell Concrete Structures* (New York: McGraw-Hill, 1965): 250–254; 2nd ed., 1982: 277–283.

12. A. Shaaban and M. Ketchum, Design of hipped hyper shells, *Journal of the Structural Division,* ASCE, 102, no. ST11 (November 1976): 2151–2161.

13. W. C. Schnobrich, Analysis of hyperbolic paraboloid shells, *Symposium on Concrete Thin Shells,* ACI Special Publication SP-28, paper SP-28-13 (Detroit: American Concrete Institute, 1971): 275–311.

14. W. C. Schnobrich, Analysis of hipped roof hyperbolic paraboloid structures, *Journal of the Structural Division,* ASCE 98, no. ST7 (July 1972): 1575–1583; discussion by M. S. Ketchum, vol. 99, no. ST4 (April 1973): 796–797; closure, vol. 100, no. ST2 (February 1974): 467–469; A. Shaaban and M. S. Ketchum, Design of hipped hyper shells, *Journal of the Structural Division,* ASCE 102, no. ST11 (November 1976): 2151–2161.

15. T. Jadik and D. P. Billington, Gabled hyperbolic paraboloid roofs without edge beams, *Journal of Structural Engineering,* ASCE, 121, No. 2 (February 1985): 328–335.

16. S. H. Simmonds, Continuous Hyper Roofs for Water Treatment Plant, ACI Fall Convention, Baltimore, November 1986.

17. S. S. Tezcan, K. M. Agrawal, and G. Kostro, Finite element analysis of hyperbolic paraboloid shells, *Journal of the Structural Division,* ASCE 97, no. ST1 (January 1971): 407–424; P. V. Banavalkar and P. Gergely, Thin-steel hyperbolic paraboloid shells, *Journal of the Structural Division,* ASCE 98, no. ST11 (November 1973): 2605–2621.

18. R. C. Liu and N. C. Teter, Hyperbolic paraboloid shells built with wood products, *Bulletin of IASS 30* (June 1967): 3–8.

19. W. Flügge, *Stresses in Shells* (Berlin: Springer-Verlag, 1962): 184–191.

20. J. Bobrowski, Calgary's olympic saddledome, *Space Structures,* v. 1, no. 1, 1985, pp. 13–26; and The Saddledome: The Olympic ice stadium in Calgary, *Proc. Canadian Society for Civil Engineering Annual Conference,* Saskatoon, SK, May 27–31, 1985.

21. Billington, *Thin Shell Concrete Structures* (1982): 29–32, 265–272.

22. Billington, *Thin Shell Concrete Structures* (1982): 283–290.

23. A. M. Haas, *Thin Concrete Shells,* vol. 2 (New York: Wiley, 1967): 103–133.

24. V. S. Kelkar and R. T. Sewell, *Fundamentals of the Analysis and Design of Shell Structures* (Englewood Cliffs, N.J.: Prentice-Hall, Inc., 1987): 79, 193–194, 307–309.

EXERCISES

6.1. For an open cylindrical shell as shown in Figure 6–1, develop membrane theory solutions for the case where the load q_u is uniformly distributed in the Y direction and:

(a) The load is uniformly distributed in X direction.

(b) The load is harmonically distributed in X direction.

6.2. Consider the open cylindrical shell shown in Figure 6–1. In some respects, it may be visualized as a beam of span L. With this in mind, it is proposed to increase the thickness progressively from the ends to the center according to the relation

$$h(X) = h(0) + \left[h\left(\frac{L}{2}\right) - h(0) \right] \sin \frac{\pi X}{L}$$

Compute the membrane theory stress resultants due to the self-weight of this shell.

6.3. For the hyperbolic paraboloid defined by equation (6.23), verify that the shell has negative Gaussian curvature.

6.4. For the hipped hyperbolic paraboloid shell shown in Figure 6–18, compute

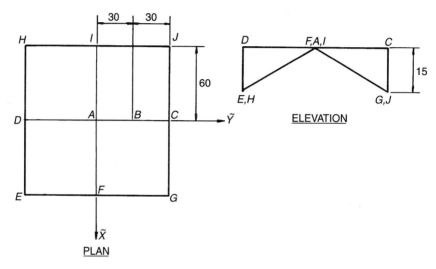

FIGURE 6–18 Exercise 6.4

(a) $N_{\tilde{x}}$ and $N_{\tilde{y}}$ caused by a dead load q and a live load p.
(b) The axial force in the edge member AC at points A and B.
(c) Evaluate (a) and (b) for the following numerical values:
  (1) Dead load of 40 (force per unit area of middle surface).
  (2) Snow load of 20 (force per unit area of horizontal projection).

6.5. Consider the hyperbolic paraboloid shown in Figure 6–19. Compute the membrane theory stress resultants caused by the fill load shown. Take the boundary conditions as $N_{\tilde{x}}(0, Y) = N_{\tilde{y}}(X, 0) = 0$.

6.6. Verify the positive Gaussian curvatures of the elliptic paraboloid defined by equation (6.64).

6.7. Consider the shell shown in Figure 6–20. The shell is a paraboloid of revolution on an equilateral triangular plan. It was shown in Section 6.5 that this geometry is a specialized case of the elliptic paraboloid. Assume that extensional stress resultants cannot be developed normal to the boundaries.

(a) Solve for the membrane-theory stress resultants in the shell caused by a dead load q.
(b) Determine the stress distribution along the boundaries and the stresses at the corners.

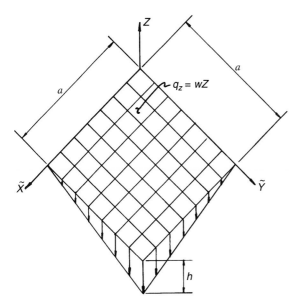

FIGURE 6–19   Exercise 6.5

6.8. Consider the slice-beam shown in Figure 5–17 with $p(Z, \theta) = -p^j \cos j\theta$, and verify equation (5.65a) and (5.65c) using equations of overall equilibrium.

6.9. The hyperbolic paraboloid shell shown in Figure 6–21 is supported on four piers which are at the same elevation as the corners. The center point O is raised up by 20 and the shell has a load of 100, taken as uniform over the horizontal projection.

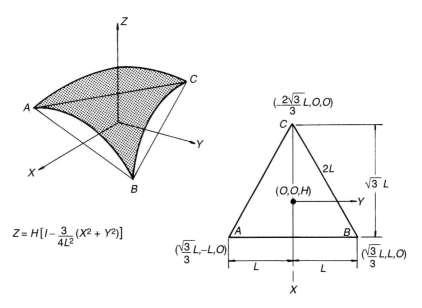

FIGURE 6–20   Exercise 6.7

(a) Compute the membrane theory shear stress in the shell
(b) Compute the forces in the edge beams
(c) Find the forces in the piers

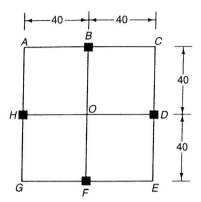

FIGURE 6–21 Plan layout for the hypar shell of Exercise 6.9. (Adapted from Kelkar and Sewell, Ref. 24)

6.10. The hyperbolic paraboloid shown in Figure 2–8(a) may be idealized as shown in Figure 6–22. It consists of four saddle hypars and is subjected to a loading of 50, considered as uniform over the horizontal projection.
(a) Compute the membrane theory shear stress in the shell
(b) Find the forces in the edge beams

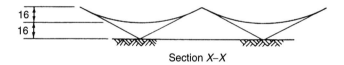

Section X–X

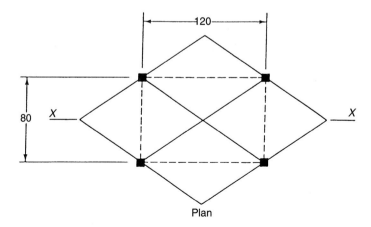

Plan

FIGURE 6–22 Configuration of four adjoining hypar shells to form a single structure, Exercise 6.10. (Adapted from Kelkar and Sewell, Ref. 24.)

7

Deformations

7.1 GENERAL

In the earlier chapters, we developed a geometric description of the middle surface of a shell which proved to be adequate to derive the equations of equilibrium. In turn, a simplified subset of the equilibrium equations formed the basis of the membrane theory of shells, for which many important practical applications were illustrated. Although membrane action in a shell is desirable from the dual standpoints of mathematical simplification and material efficiency, the requisite conditions for corresponding behavior cannot always be simulated in an actual structure. Consequently, to expand our base of understanding of shell behavior, we must develop relationships between the forces and the deformations of the shell. The first step is a description of the displacements, where we follow the vector approach suggested by Novozhilov.[1]

7.2 DISPLACEMENT

7.2.1 Displacement Vector

In Figure 7–1(a), a differential element of the middle surface is shown before and after deformation. The reference point o located on the middle surface moves to the position o', and this displacement is denoted by Δ. Considering a second point $o(\zeta)$, located a distance ζ from

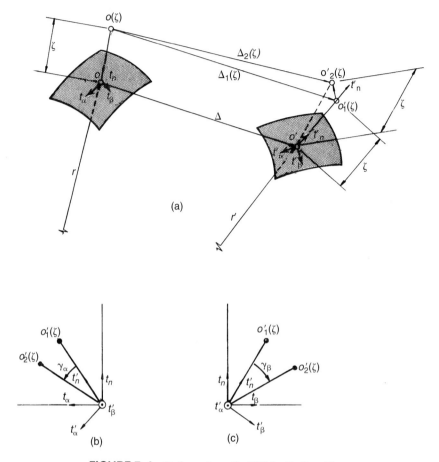

FIGURE 7-1   Deformation of a Middle-Surface Element.

o along the original unit $\mathbf{t_n}$, the corresponding position after deformation is $o_i'(\zeta)$ ($i = 1$ or 2), and the displacement between $o(\zeta)$ and o_i' (ζ) is $\mathbf{\Delta_i}(\zeta)$.

We have shown two possible positions of $o_i'(\zeta)$, $o_1'(\zeta)$ and $o_2'(\zeta)$. Point o_1' (ζ) lies on the unit normal to the deformed middle surface, $\mathbf{t_n'}$, at the same distance ζ from the middle surface. The point $o_1'(\zeta)$ is located on $\mathbf{t_n'}$ because of assumption [3], Table 1–1, and remains ζ from the middle surface because of assumption [4]. On the other hand, point o_2' is also located ζ from the deformed middle surface, but no longer necessarily lies on the normal $\mathbf{t_n'}$; that is, the location of $o_2'(\zeta)$ only requires the enforcement of assumption [4]. Because assumption [3] refers to the suppression of transverse shearing strains, we may surmise that the difference in $o_1'(\zeta)$ and $o_2'(\zeta)$ is the effect of the transverse shearing *strain*. This is clearly shown in Figures 7–1(b) and 7–1(c), where views normal to $\mathbf{t_\beta'}$ and $\mathbf{t_\alpha'}$ are shown with the transverse shearing strains defined as γ_α and γ_β. The positive sense shown is chosen arbitrarily to highlight γ_α and γ_β.

It was stated in the introduction that most classical work in shell and plate theories has been based on the suppression of transverse shearing deformations to achieve mathematical simplification. Currently, however, the analysis of complex plates and shells is frequently carried out by powerful numerical techniques that are not necessarily dependent on this simplification. Also, some authors have expressed an opinion that the inclusion of transverse shearing strains extends the bounds of the theory to include somewhat thicker plates and shells. This is somewhat difficult to quantify, since a true thick-shell theory should account for transverse normal stresses as well, but it appears to be correct in a heuristic sense. In the interest of generality, therefore, we will retain the transverse shearing strains and treat $o'_2(\zeta)$ as the deformed point away from the middle surface, and $\boldsymbol{\Delta}_2(\zeta)$ as the corresponding displacement. Subsequently, we will show that a theory which neglects transverse shearing strains is easily obtained from the more general theory.

From Figure 7–1,

$$\zeta \mathbf{t_n} + \boldsymbol{\Delta}_2(\zeta) = \boldsymbol{\Delta} + \zeta \mathbf{t'_n} + \zeta \gamma_\alpha \mathbf{t'_\alpha} + \zeta \gamma_\beta \mathbf{t'_\beta} \tag{7.1}$$

or

$$\boldsymbol{\Delta}_2(\zeta) = \boldsymbol{\Delta} + \zeta(\mathbf{t'_n} - \mathbf{t_n}) + \zeta(\gamma_\alpha \mathbf{t'_\alpha} + \gamma_\beta \mathbf{t'_\beta}) \tag{7.2}$$

Equation (7.2) expresses the deformation of a point off the middle surface in terms of the deformation of the corresponding middle surface point and the unchanging distance between the middle surface and the point in question. This establishes the pattern for subsequent developments: to relate the behavior of a point *away* from the middle surface to the behavior of the corresponding point *on* the middle surface.

7.2.2 Displacements in Terms of Middle-Surface Parameters

Now we express the displacement vectors in terms of the unit tangent vectors of the undeformed middle surface:

$$\boldsymbol{\Delta} = D_\alpha \mathbf{t_\alpha} + D_\beta \mathbf{t_\beta} + D_n \mathbf{t_n} \tag{7.3a}$$

$$\boldsymbol{\Delta}_2(\zeta) = D_\alpha(\zeta)\mathbf{t_\alpha} + D_\beta(\zeta)\mathbf{t_\beta} + D_n(\zeta)\mathbf{t_n} \tag{7.3b}$$

Here, D_α, D_β, and D_n are the components of the middle-surface displacement in the respective directions, and $D_\alpha(\zeta)$, $D_\beta(\zeta)$, and $D_n(\zeta)$ are the corresponding components away from the middle surface. Also, the position vector to the deformed middle surface is expressed in terms of the displacement vector by

$$\begin{aligned} \mathbf{r'} &= \mathbf{r} + \boldsymbol{\Delta} \\ &= \mathbf{r} + D_\alpha \mathbf{t_\alpha} + D_\beta \mathbf{t_\beta} + D_n \mathbf{t_n} \end{aligned} \tag{7.4}$$

In equation (7.2), the unit vectors of the deformed middle surface, $\mathbf{t'_\alpha}$, $\mathbf{t'_\beta}$, and $\mathbf{t'_n}$, are present. In line with equation (7.4) for the vector $\mathbf{r'}$, we seek to write $\mathbf{t'_\alpha}$, $\mathbf{t'_\beta}$, and $\mathbf{t'_n}$ in terms of the unit vectors of the *undeformed* middle surface. Because

$$\mathbf{t'_n} = \mathbf{t'_\alpha} \times \mathbf{t'_\beta} \tag{7.5}$$

it is sufficient to develop suitable expressions for $\mathbf{t'_\alpha}$ and $\mathbf{t'_\beta}$.

By analogy with the definitions in equations (2.16) and (2.8), we have

$$t'_\alpha = \frac{r'_{,\alpha}}{A'} \tag{7.6a}$$

$$t'_\beta = \frac{r'_{,\beta}}{B'} \tag{7.6b}$$

where

$$A' = \pm(r'_{,\alpha} \cdot r'_{,\alpha})^{1/2} \tag{7.7a}$$

$$B' = \pm(r'_{,\beta} \cdot r'_{,\beta})^{1/2} \tag{7.7b}$$

Thus, we have defined A' and B' as the Lamé parameters for the deformed middle surface. Now, taking $\partial/\partial\alpha$ of equation (7.4)

$$r'_{,\alpha} = r_{,\alpha} + (D_\alpha t_\alpha)_{,\alpha} + (D_\beta t_\beta)_{,\alpha} + (D_n t_n)_{,\alpha} \tag{7.8}$$

The first term in equation (7.8) is replaced using equation (2.16a), and the remaining terms are differentiated by the product rule, with the derivatives of the unit tangent vectors obtained from equation (2.17). After carrying out the indicated operations, we have

$$r'_{,\alpha} = A[(1 + \varepsilon_\alpha)t_\alpha + \varepsilon_{\alpha\beta}t_\beta - \psi_\alpha t_n] \tag{7.9}$$

where the coefficients of the unit vectors are given by

$$\varepsilon_\alpha = \frac{1}{A}D_{\alpha,\alpha} + \frac{1}{AB}A_{,\beta}D_\beta + \frac{1}{R_\alpha}D_n \tag{7.10a}$$

$$\varepsilon_{\alpha\beta} = \frac{1}{A}D_{\beta,\alpha} - \frac{1}{AB}A_{,\beta}D_\alpha \tag{7.10b}$$

$$\psi_\alpha = -\left(\frac{1}{A}D_{\alpha,\alpha} - \frac{1}{R_\alpha}D_\alpha\right) \tag{7.10c}$$

Similarly,

$$r'_{,\beta} = B[\varepsilon_{\beta\alpha}t_\alpha + (1 + \varepsilon_\beta)t_\beta - \psi_\beta t_n] \tag{7.11}$$

in which

$$\varepsilon_\beta = \frac{1}{B}D_{\beta,\beta} + \frac{1}{AB}B_{,\alpha}D_\alpha + \frac{1}{R_\beta}D_n \tag{7.12a}$$

$$\varepsilon_{\beta\alpha} = \frac{1}{B}D_{\alpha,\beta} - \frac{1}{AB}B_{,\alpha}D_\beta \tag{7.12b}$$

$$\psi_\beta = -\left(\frac{1}{B}D_{n,\beta} - \frac{1}{R_\beta}D_\beta\right) \tag{7.12c}$$

The coefficients defined in equations (7.10) and (7.12) are extremely important in what follows and are interpreted from a physical standpoint at a later stage of this chapter.

We are ready to substitute equations (7.9) to (7.12) into equations (7.6a) and (7.6b). As we carry out this operation, recall assumption [2] of Table 1–1, which restricted the development to the domain of comparatively small deformations. This limitation enables products of the coefficients defined in equations (7.10) and (7.12) to be neglected, and ensures that the subsequent relationships will be linear. We wish to emphasize at this point that although assumption [2] greatly simplifies the problem from the mathematical standpoint, the justification is based on verification of the magnitudes of the neglected terms for a particular problem or class of problems. In most practical applications, the linear theory has been verified to be adequate, but, if not, a variety of nonlinear theories are available.[2] In any case, all of the consistent nonlinear theories reduce to the identical linear theory for small strains.

Performing the indicated substitutions, we have from equations (7.7a) and (7.9)

$$
\begin{aligned}
A' &= A[(1 + \varepsilon_\alpha)^2 + \varepsilon_{\alpha\beta}^2 + \psi_\alpha^2]^{1/2} \\
&\simeq A(1 + \varepsilon_\alpha)
\end{aligned}
\tag{7.13}
$$

and from equations (7.7b) and (7.11),

$$
\begin{aligned}
B' &= B[\varepsilon_{\beta\alpha}^2 + (1 + \varepsilon_\beta)^2 + \psi_\beta^2]^{1/2} \\
&\simeq B(1 + \varepsilon_\beta)
\end{aligned}
\tag{7.14}
$$

after the product terms are dropped. Then, from equation (7.6a),

$$
\begin{aligned}
\mathbf{t}_\alpha' &= \frac{A[(1 + \varepsilon_\alpha)\mathbf{t}_\alpha + \varepsilon_{\alpha\beta}\mathbf{t}_\beta - \psi_\alpha\mathbf{t}_n]}{A(1 + \epsilon_\alpha)} \\
&\simeq \mathbf{t}_\alpha + \varepsilon_{\alpha\beta}\mathbf{t}_\beta - \psi_\alpha\mathbf{t}_n
\end{aligned}
\tag{7.15}
$$

and, from equation (7.6b),

$$
\begin{aligned}
\mathbf{t}_\beta' &= \frac{B[\varepsilon_{\beta\alpha}\mathbf{t}_\alpha + (1 + \varepsilon_\beta)\mathbf{t}_\beta - \psi_\beta\mathbf{t}_n]}{B(1 + \varepsilon_\beta)} \\
&\simeq \varepsilon_{\beta\alpha}\mathbf{t}_\alpha + \mathbf{t}_\beta - \psi_\beta\mathbf{t}_n
\end{aligned}
\tag{7.16}
$$

where

$$
\frac{\varepsilon_{ij}}{1 + \varepsilon_i} \simeq \varepsilon_{ij} \quad \text{and} \quad \frac{\psi_i}{1 + \varepsilon_i} \simeq \psi_i \begin{pmatrix} i = \alpha, \beta \\ j = \beta, \alpha \end{pmatrix}
$$

Finally, we compute the vector product $\mathbf{t}_\alpha' \times \mathbf{t}_\beta'$, which, after simplification, is

$$
\mathbf{t}_n' = \mathbf{t}_n + \psi_\alpha\mathbf{t}_\alpha + \psi_\beta\mathbf{t}_\beta
\tag{7.17}
$$

Equation (7.17) is the desired expression of the unit normal to the *deformed* middle surface, in terms of the tangent vectors to the *undeformed* middle surface. Substituting equations (7.15), (7.16), and (7.17) into equation (7.2) gives

$$
\begin{aligned}
\mathbf{\Delta}_2(\zeta) &= \mathbf{\Delta} + \zeta[(\psi_\alpha + \gamma_\alpha + \gamma_\beta\varepsilon_{\beta\alpha})\mathbf{t}_\alpha + (\psi_\beta + \gamma_\beta + \gamma_\alpha\varepsilon_{\alpha\beta})\mathbf{t}_\beta \\
&\quad - (\gamma_\alpha\psi_\alpha + \gamma_\beta\psi_\beta)\mathbf{t}_n]
\end{aligned}
\tag{7.18a}
$$

Again, products of the deformation parameters, such as $\gamma_\beta\varepsilon_{\beta\alpha}$, are neglected, so that equation (7.18a) simplifies to

$$\mathbf{\Delta}_2(\zeta) = \mathbf{\Delta} + \zeta[(\psi_\alpha + \gamma_\alpha)\mathbf{t}_\alpha + (\psi_\beta + \gamma_\beta)\mathbf{t}_\beta] \qquad (7.18b)$$

In view of equation (7.3a), equation (7.18b) may be rewritten as

$$\mathbf{\Delta}_2(\zeta) = [D_\alpha + \zeta(\psi_\alpha + \gamma_\alpha)]\mathbf{t}_\alpha + [D_\beta + \zeta(\psi_\beta + \gamma_\beta)]\mathbf{t}_\beta + D_n\mathbf{t_n} \qquad (7.19)$$

so that the components of $\mathbf{\Delta}_2$ (ζ) are

$$D_\alpha(\zeta) = D_\alpha + \zeta(\psi_\alpha + \gamma_a) \qquad (7.20a)$$
$$D_\beta(\zeta) = D_\beta + \zeta(\psi_\beta + \gamma_\beta) \qquad (7.20b)$$
$$D_n(\zeta) = D_n \qquad (7.20c)$$

Here, we have expressed the components of the displacement of an *arbitrary point* on the shell in terms of the displacement of the *corresponding point* on the *middle surface* and the *unchanging separation* of the points along the *unit normal vector*. Equation (7.20c) is, of course, a direct restatement of assumption [4]. At this point, we observe that the popular shell theories that incorporate assumption [3], as well, may be obtained by simply setting $\gamma_\alpha = \gamma_\beta = 0$ in the previous equations.

7.2.3 Rotations

Recalling that the coefficients D_α and D_β represent the respective displacements of the middle surface along the s_α and s_β coordinate lines, the terms ψ_α and ψ_β may be viewed as rotations of the normal $\mathbf{t}'_n$ about the $\mathbf{t}'_\beta$ and $\mathbf{t}'_\alpha$ axes, respectively. To show this, we construct Figures 7–2(a) and 7–2(b), which are, in turn, identical to the views shown in Figures 7–1(b) and 7–1(c). On Figure 7–2, the projections of the unit vectors to the undeformed surface

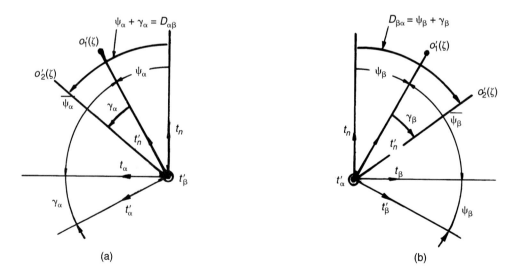

(a)   (b)

FIGURE 7–2 Transverse Shearing Strains.

$\mathbf{t}_\alpha$ and $\mathbf{t}_\beta$ may be considered to remain unit vectors within the scope of this theory; that is, $\mathbf{t}_\alpha \cdot \mathbf{t}_\alpha' \simeq 1$ from equation (7.15). If we scalar multiply equation (7.17) by $\mathbf{t}_\alpha$ and then $\mathbf{t}_\beta$, we get

$$\psi_\alpha = \mathbf{t}_\alpha \cdot \mathbf{t}_n' \tag{7.21a}$$

and

$$\psi_\beta = \mathbf{t}_\beta \cdot \mathbf{t}_n' \tag{7.21b}$$

Equation (7.21a) is interpreted in Figure 7–2(a). Continuing,

$$\mathbf{t}_\alpha \cdot \mathbf{t}_n' = \cos \overline{\psi}_\alpha$$
$$= \sin\left(\frac{\pi}{2} - \overline{\psi}_\alpha\right) \tag{7.21c}$$

For small rotations

$$\sin\left(\frac{\pi}{2} - \overline{\psi}_\alpha\right) \simeq \frac{\pi}{2} - \overline{\psi}_\alpha = \psi_\alpha \tag{7.21d}$$

or, ψ_α is the rotation of the normal to the deformed middle surface about the $\mathbf{t}_\beta'$ axis. Similarly, from Figure 7–2(b), ψ_β is the rotation of the normal about the $\mathbf{t}_\alpha'$ axis. Within the scope of the small deformation theory, ψ_α and ψ_β may be visualized as rotations about $\mathbf{t}_\beta$ and $\mathbf{t}_\alpha$ as well.

We now consider the transverse shearing strains, initially defined in Figures 7–1(b) and 7–1(c), superimposed on the rotations ψ_α and ψ_β in Figures 7–2(a) and 7–2(b). We see that the terms $\psi_\alpha + \gamma_\alpha$ and $\psi_\beta + \gamma_\beta$, which initially appeared in equation (7.19), may be interpreted as rotations. If we consider a line connecting a point away from the *deformed* middle surface, o_2', and the corresponding middle surface point, and a second line along the normal to the *undeformed* middle surface $\mathbf{t}_n$, the angles formed by the intersection of these lines are, respectively,

$$D_{\alpha\beta} = \psi_\alpha + \gamma_\alpha \tag{7.22a}$$

in the α-n plane and

$$D_{\beta\alpha} = \psi_\beta + \gamma_\beta \tag{7.22b}$$

in the β-n plane. We also may observe from Figure 7–2 that if γ_α and $\gamma_\beta = 0$, $D_{\alpha\beta}$ and $D_{\beta\alpha}$ are, in turn, the angles between $\mathbf{t}_\alpha$ and $\mathbf{t}_\alpha'$, and $\mathbf{t}_\beta$ and $\mathbf{t}_\beta'$. $D_{\alpha\beta}$ and $D_{\beta\alpha}$ are termed middle-surface *rotations* and, together with D_α, D_β, and D_n, constitute the set of *generalized displacements* for the shell theory under study.

The complete description of the displacement of any point on or within the shell surface is given in terms of the five generalized displacements. Seemingly, rotation about the normal, in other words, about the axis denoted by $\mathbf{t}_n$ in Figure 5–1, has no utility provided that the structure is constrained against rigid body movement. However, there are occasionally cases where the shell has a slope discontinuity, such as Figure 4–1, or is approximated by an assemblage of flat (facet) surfaces.[3] It may be then convenient to introduce the normal rotation, which is known as the *drilling* degree of freedom.

7.3 STRAIN

7.3.1 Strain on Middle Surface

Referring to Figure 7–3, consider an arc *op* on the s_α coordinate line that has an initial length of

$$ds_\alpha = A\, d\alpha \qquad (7.23)$$

After deformation, the points *o* and *p* move to *o'* and *p'* and the arc length becomes

$$ds'_\alpha = A'\, d\alpha$$
$$= A(1 + \varepsilon_\alpha)\, d\alpha \qquad (7.24)$$

from equation (7.13). The *extensional strain* is defined as the relative increase in length of a curvilinear line element, or the final length minus the initial length, divided by the initial length. So, in the α direction, we have

$$\frac{ds'_\alpha - ds_\alpha}{ds_\alpha} = \frac{A(1 + \varepsilon_\alpha)d\alpha - A\, d\alpha}{A\, d\alpha} \qquad (7.25)$$
$$= \varepsilon_\alpha$$

Thus ε_α, which appeared in the coefficient of $\mathbf{t}_\alpha$ in equation (7.9), and is defined by equation (7.10a), is the strain in the α direction. Correspondingly in the β direction, the strain is given by ε_β, which is defined in equation (7.12a). We have thus found that the extensional strains occur quite naturally in the expressions for the derivatives of the position vector to the deformed middle surface.

The *in-plane shearing strain* is defined as the change in angle between $\mathbf{t}_\alpha$ and $\mathbf{t}_\beta$ during deformation. Referring to Figure 7–4, the shearing strain is

$$\omega = \bar{\varepsilon}_{\alpha\beta} + \bar{\varepsilon}_{\beta\alpha} = \sin\left(\bar{\varepsilon}_{\alpha\beta} + \bar{\varepsilon}_{\beta\alpha}\right)$$
$$= \cos\left(\frac{\pi}{2} - \bar{\varepsilon}_{\alpha\beta} - \bar{\varepsilon}_{\beta\alpha}\right) \qquad (7.26)$$
$$= \mathbf{t}'_\alpha \cdot \mathbf{t}'_\beta$$

We form the indicated scalar product using equations (7.15) and (7.16), which gives

$$\omega = \mathbf{t}'_\alpha \cdot \mathbf{t}'_\beta = \varepsilon_{\beta\alpha} + \varepsilon_{\alpha\beta} + \psi_\alpha\psi_\beta \qquad (7.27)$$

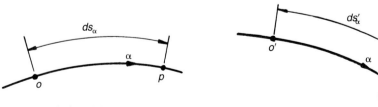

before deformation   after deformation

FIGURE 7–3   Deformation of a Coordinate Line.

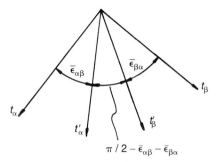

FIGURE 7–4 In-Plane Shearing Strain.

Since the last term is negligible, we have

$$\omega = \varepsilon_{\alpha\beta} + \varepsilon_{\beta\alpha} \tag{7.28}$$

as the shearing strain. Again, $\varepsilon_{\alpha\beta}$ and $\varepsilon_{\beta\alpha}$ are found among the coefficients of the equations defining the derivatives of the position vector of the deformed middle surface as equations (7.10b) and (7.12b). The transverse shearing strains γ_{α} and γ_{β} have previously been defined.

7.3.2   Strain on Parallel Surface

The differential arc length along coordinate line s_{α} of the parallel surface is

$$ds_{\alpha}(\zeta) = A(\zeta)\, d\alpha \tag{7.29}$$

Referring to the normal section, Figure 7–5,

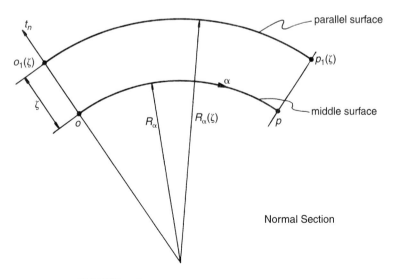

Normal Section

FIGURE 7–5 Displacement of a Parallel Surface.

$$\frac{ds_\alpha(\zeta)}{ds_\alpha} = \frac{R_\alpha(\zeta)}{R_\alpha} \tag{7.30a}$$

Since

$$R_\alpha(\zeta) = R_\alpha + \zeta \tag{7.30b}$$

and

$$ds_\alpha = A \, d\alpha \tag{7.30c}$$

then

$$ds_\alpha(\zeta) = \frac{R_\alpha + \zeta}{R_\alpha} A \, d\alpha$$

$$= A\left(1 + \frac{\zeta}{R_\alpha}\right) d\alpha \tag{7.30d}$$

Comparing equations (7.29) and (7.30d), the Lamé parameter for the parallel surface is

$$A(\zeta) = A\left(1 + \frac{\zeta}{R_\alpha}\right) \tag{7.31}$$

For a parallel surface on a normal section along the s_β coordinate line

$$R_\beta(\zeta) = R_\beta + \zeta \tag{7.32a}$$

$$ds_\beta(\zeta) = B\left(1 + \frac{\zeta}{R_\beta}\right) d\beta \tag{7.32b}$$

and

$$B(\zeta) = B\left(1 + \frac{\zeta}{R_\beta}\right) \tag{7.33}$$

Example 7.3(a)

Returning to the toroidal shell shown in Figure 2–14 and described in Example 2–8, we have from equations (7.30b) and (7.32a)

$$R_1(\zeta) = R_1 + \zeta$$

$$= \frac{r_1}{\cos \theta_2} + r_2 + \zeta \tag{7.34a}$$

$$R_2(\zeta) = R_2 + \zeta$$

$$= r_2 + \zeta \tag{7.34b}$$

Then, from equations (7.31), (7.33), (2.62), and (2.63)

$$A(\zeta) = (r_1 + r_2 \cos \theta_2)\left(1 + \frac{\zeta \cos \theta_2}{r_1 + r_2 \cos \theta_2}\right) \tag{7.35a}$$

$$= r_1 + (r_2 + \zeta) \cos \theta_2$$

$$B(\zeta) = r_2\left(1 + \frac{\zeta}{r_2}\right) \tag{7.35b}$$

We recognize that the parallel surface is described by a system of curvilinear coordinates that is identical to those which describe the middle surface. Therefore, the relationships between the strains and displacements defined for the middle surface in equations (7.10a), (7.10b), (7.12a), and (7.12b) may be written for the parallel surface by replacing R_α, R_β, A, and B by $R_\alpha(\zeta)$, $R_\beta(\zeta)$, $A(\zeta)$, and $B(\zeta)$, respectively; and, by replacing D_α, D_β, and D_n by $D_\alpha(\zeta)$, $D_\beta(\zeta)$, and $D_n(\zeta)$, as defined in equations (7.20).

For the extensional strains, we get

$$\varepsilon_\alpha(\zeta) = \varepsilon_\alpha + \zeta\kappa_\alpha \tag{7.36}$$

$$\varepsilon_\beta(\zeta) = \varepsilon_\beta + \zeta\kappa_\beta \tag{7.37}$$

after dropping terms of order (h/R) compared to 1; the latter order of magnitude comparison is widely used in shell theory and is concisely stated as $O(h/R)$: 1. The coefficients of ζ are

$$\kappa_\alpha = \frac{1}{A} D_{\alpha\beta,\alpha} + \frac{1}{AB} A_{,\beta} D_{\beta\alpha} \tag{7.38}$$

and

$$\kappa_\beta = \frac{1}{B} D_{\beta\alpha,\beta} + \frac{1}{AB} B_{,\alpha} D_{\alpha\beta} \tag{7.39}$$

where $D_{\alpha\beta}$ and $D_{\beta\alpha}$ are defined in equation (7.22). The terms κ_α and κ_β correspond to the familiar curvature terms in linear beam theory. For plates, they are indeed curvatures, but, for shells, they are more properly called *changes in curvature,* because shells by definition are initially curved.

For the in-plane shearing strain $\omega(\zeta)$, we first combine $\varepsilon_{\alpha\beta}$ and $\varepsilon_{\beta\alpha}$ as indicated in equation (7.28). Adding equations (7.10b) and (7.12b), we have

$$\omega = \frac{1}{A} D_{\beta,\alpha} - \frac{1}{AB} A_{,\beta} D_\alpha + \frac{1}{B} D_{\alpha,\beta} - \frac{1}{AB} B_{,\alpha} D_\beta \tag{7.40}$$

Now, referring to the parallel surface, we perform the indicated substitutions for the radii of curvature, Lamé parameters, and displacements, and also drop terms of $O(h/R)$:1. The resulting expression for the in-plane shearing strain on the parallel surface is

$$\omega(\zeta) = \varepsilon_{\alpha\beta} + \zeta\tau_\alpha + \varepsilon_{\beta\alpha} + \zeta\tau_\beta \tag{7.41}$$

where

$$\tau_\alpha = \frac{1}{A} D_{\beta\alpha,\alpha} - \frac{1}{AB} A_{,\beta} D_{\alpha\beta} \tag{7.42}$$

and

$$\tau_\beta = \frac{1}{B} D_{\alpha\beta,\beta} - \frac{1}{AB} B_{,\alpha} D_{\beta\alpha} \tag{7.43}$$

It is convenient to introduce

$$\tau = \frac{1}{2} (\tau_\alpha + \tau_\beta) \tag{7.44}$$

Then, equation (7.41) is written as

$$\omega(\zeta) = \omega + 2\zeta\tau \tag{7.45}$$

where τ is called the twist or the *torsion* of the middle surface.

We have now completely defined each strain on the parallel surface, except for the transverse shearing strain, as a *linear combination of two terms:* (a) the *corresponding strain* on the *middle surface,* which is expressed in terms of the *middle-surface displacements* by one of equations (7.10) or (7.12); and (b) a change in *curvature* or a *twist,* which is, in turn, given in terms of the *middle-surface displacements* by equations (7.38), (7.39), or (7.42)–(7.45), respectively. The transverse shearing strains are also defined in terms of middle-surface displacements, equations (7.22), (7.10c), and (7.12c), but remain *constant* over the shell depth.

Equations (7.36–7.45) suggest that the transverse shearing strains γ_α and γ_β should also vary *linearly* through the thickness. However, Figure 7–6 shows that these strains are *constant* through the thickness. The differences in the variation of the strains through the thickness when transverse shearing strains are included (linear vs. constant) give rise to a curious phenomenon known as *shear locking.* Shear locking is a problem in finite element models when numerical integrations through the thickness are required.[4] If suitable precautions are not taken, the resulting added stiffness can lead to serious errors; this is one reason that solutions incorporating transverse shearing strains have been slow to gain popularity.

These relationships among strains, changes in curvature, and twist on the one hand, and the displacements of the middle surface on the other, are of primary importance in the

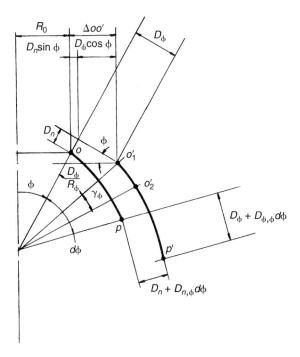

FIGURE 7–6   Displacement of the Meridian of a Shell of Revolution.

theory of shells and are called generalized *strain-displacement* or *kinematic* equations. We have thus established the second basic component of the elasticity problem as outlined in Section 1.1, the conditions of *compatibility* of the strains and displacements, suitably specialized for the shell theory under consideration:

$$\varepsilon_\alpha = \frac{1}{A} D_{\alpha,\alpha} + \frac{1}{AB} A_{,\beta} D_\beta + \frac{1}{R_\alpha} D_n \tag{7.46a}$$

$$\varepsilon_\beta = \frac{1}{B} D_{\beta,\beta} + \frac{1}{AB} B_{,\alpha} D_\alpha + \frac{1}{R_\beta} D_n \tag{7.46b}$$

$$\omega = \frac{1}{A} D_{\beta,\alpha} - \frac{1}{AB} A_{,\beta} D_\alpha + \frac{1}{B} D_{\alpha,\beta} - \frac{1}{AB} B_{,\alpha} D_\beta \tag{7.46c}$$

$$\kappa_\alpha = \frac{1}{A} D_{\alpha\beta,\alpha} + \frac{1}{AB} A_{,\beta} D_{\beta\alpha} \tag{7.46d}$$

$$\kappa_\beta = \frac{1}{B} D_{\beta\alpha,\beta} + \frac{1}{AB} B_{,\alpha} D_{\alpha\beta} \tag{7.46e}$$

$$\tau = \frac{1}{2}\left[\frac{1}{A} D_{\beta\alpha,\alpha} + \frac{1}{B} D_{\alpha\beta,\beta} - \frac{1}{AB}(A_{,\beta} D_{\alpha\beta} + B_{,\alpha} D_{\beta\alpha})\right] \tag{7.46f}$$

$$\gamma_\alpha = \frac{1}{A} D_{n,\alpha} - \frac{1}{R_\alpha} D_\alpha + D_{\alpha\beta} \tag{7.46g}$$

$$\gamma_\beta = \frac{1}{B} D_{n,\beta} - \frac{1}{R_\beta} D_\beta + D_{\beta\alpha} \tag{7.46h}$$

If we choose to neglect transverse shearing strains, we set $\gamma_\alpha = \gamma_\beta = 0$ in equations (7.46g) and (7.46h) respectively, so that $D_{\alpha\beta} = \psi_\alpha$ and $D_{\beta\alpha} = \psi_\beta$. Then, replacing $D_{\alpha\beta}$ and $D_{\beta\alpha}$ in equations (7.46d) to (7.46f) by ψ_α and ψ_β as given by equations (7.10c) and (7.12c), we get the changes in curvatures directly in terms of the middle-surface displacements:

$$\kappa_\alpha = -\frac{1}{A}\left(\frac{1}{A} D_{n,\alpha} - \frac{1}{R_\alpha} D_\alpha\right)_{,\alpha} - \frac{1}{AB} A_{,\beta}\left(\frac{1}{B} D_{n,\beta} - \frac{1}{R_\beta} D_\beta\right) \tag{7.47a}$$

$$\kappa_\beta = -\frac{1}{B}\left(\frac{1}{B} D_{n,\beta} - \frac{1}{R_\beta} D_\beta\right)_{,\beta} - \frac{1}{AB} B_{,\alpha}\left(\frac{1}{A} D_{n,\alpha} - \frac{1}{R_\alpha} D_\alpha\right) \tag{7.47b}$$

$$\tau = \frac{1}{2}\left\{-\frac{1}{A}\left(\frac{1}{B} D_{n,\beta} - \frac{1}{R_\beta} D_\beta\right)_{,\alpha} - \frac{1}{B}\left(\frac{1}{A} D_{n,\alpha} - \frac{1}{R_\alpha} D_\alpha\right)_{,\beta}\right.$$
$$\left. + \frac{1}{AB}\left[A_{,\beta}\left(\frac{1}{A} D_{n,\alpha} - \frac{1}{R_\alpha} D_\alpha\right) + B_{,\alpha}\left(\frac{1}{B} D_{n,\beta} - \frac{1}{R_\beta} D_\beta\right)\right]\right\} \tag{7.47c}$$

Note with respect to equations (7.46) and (7.47) that no force-deformation relationships are involved and, hence, linear material behavior is not a prerequisite for the application of these equations. Also, as previously demonstrated for the equilibrium equations, a considerable simplification of these equations often occurs for specific geometries through the elimination of some of the coupling terms.

7.4 STRAIN-DISPLACEMENT RELATIONS FOR SHELLS OF REVOLUTION

7.4.1 Specialization of Equations

The specialization of the curvilinear coordinate system to the shell of revolution geometry, shown in Figure 2–11, is carried out in Section 3.3.1. The corresponding strain-displacement equations for $\alpha = \phi$ and $\beta = \theta$ are

$$\varepsilon_\phi = \frac{1}{R_\phi}(D_{\phi,\phi} + D_n) \tag{7.48a}$$

$$\varepsilon_\theta = \frac{1}{R_0}(D_{\theta,\theta} + \cos\phi\, D_\phi + \sin\phi\, D_n) \tag{7.48b}$$

$$\omega = \frac{1}{R_\phi}D_{\theta,\phi} + \frac{1}{R_0}(D_{\phi,\theta} - \cos\phi\, D_\theta) \tag{7.48c}$$

$$\kappa_\phi = \frac{1}{R_\phi}D_{\phi\theta,\phi} \tag{7.48d}$$

$$\kappa_\theta = \frac{1}{R_0}D_{\theta\phi,\theta} + \frac{\cos\phi}{R_0}D_{\phi\theta} \tag{7.48e}$$

$$\tau = \frac{1}{2}\left[\frac{1}{R_\phi}D_{\theta\phi,\phi} + \frac{1}{R_0}(D_{\phi\theta,\theta} - \cos\phi\, D_{\theta\phi})\right] \tag{7.48f}$$

$$\gamma_\phi = \frac{1}{R_\phi}(D_{n,\phi} - D_\phi) + D_{\phi\theta} \tag{7.48g}$$

$$\gamma_\theta = \frac{1}{R_0}(D_{n,\theta} - \sin\phi\, D_\theta) + D_{\theta\phi} \tag{7.48h}$$

If the transverse shearing strains are neglected, we have the alternate form

$$\kappa_\phi = -\frac{1}{R_\phi}\left[\frac{1}{R_\phi}(D_{n,\phi} - D_\phi)\right]_{,\phi} \tag{7.49a}$$

$$\kappa_\theta = -\frac{1}{R_0^2}(D_{n,\theta} - \sin\phi D_\theta)_{,\theta} - \frac{\cos\phi}{R_\phi R_0}(D_{n,\phi} - D_\phi) \tag{7.49b}$$

$$\tau = \frac{1}{2}\left\{-\frac{1}{R_\phi}\left[\frac{1}{R_0}(D_{n,\theta} - \sin\phi D_\theta)\right]_{,\phi} - \frac{1}{R_\phi R_0}(D_{n,\phi} - D_\phi)_{,\theta}\right.$$
$$\left. + \frac{\cos\phi}{R_0^2}(D_{n,\theta} - \sin\phi D_\theta)\right\} \tag{7.49c}$$

We may also write the strains and displacements in the Fourier series form that proved expedient in the treatment of nonsymmetric loading on shells of revolution. Following the procedure of Section 5.1.2, we have

$$
\begin{Bmatrix} \varepsilon_\phi \\ \varepsilon_\theta \\ \omega \\ \kappa_\phi \\ \kappa_\theta \\ \tau \\ \gamma_\phi \\ \gamma_\theta \end{Bmatrix} = \sum_{j=0}^{\infty} \begin{Bmatrix} \varepsilon_\phi^j \cos j\theta \\ \varepsilon_\theta^j \cos j\theta \\ \omega^j \sin j\theta \\ \kappa_\phi^j \cos j\theta \\ \kappa_\theta^j \cos j\theta \\ \tau^j \sin j\theta \\ \gamma_\phi^j \cos j\theta \\ \gamma_\theta^j \sin j\theta \end{Bmatrix}
\tag{7.50}
$$

and

$$
\begin{Bmatrix} D_\phi \\ D_\theta \\ D_n \\ D_{\phi\theta} \\ D_{\theta\phi} \end{Bmatrix} = \sum_{j=0}^{\infty} \begin{Bmatrix} D_\phi^j \cos j\theta \\ D_\theta^j \sin j\theta \\ D_n^j \cos j\theta \\ D_{\phi\theta}^j \cos j\theta \\ D_{\theta\phi}^j \sin j\theta \end{Bmatrix}
\tag{7.51}
$$

Equations (7.50) and (7.51) are substituted into equations (7.48) and (7.49) to derive the kinematic equations for general harmonic j:

$$
\varepsilon_\phi^j = \frac{1}{R_\phi} (D_{\phi,\phi}^j + D_n^j)
\tag{7.52a}
$$

$$
\varepsilon_\theta^j = \frac{1}{R_0} (j D_\theta^j + \cos\phi\, D_\phi^j + \sin\phi\, D_n^j)
\tag{7.52b}
$$

$$
\omega^j = \frac{1}{R_\phi} D_{\theta,\phi}^j - \frac{1}{R_0} (j D_\phi^j + \cos\phi\, D_\theta^j)
\tag{7.52c}
$$

$$
\kappa_\phi^j = \frac{1}{R_\phi} D_{\phi\theta,\phi}^j
\tag{7.52d}
$$

$$
\kappa_\theta^j = \frac{1}{R_0} j D_{\theta\phi}^j + \frac{\cos\phi}{R_0} D_{\phi\theta}^j
\tag{7.52e}
$$

$$
\tau^j = \frac{1}{2} \left[\frac{1}{R_\phi} D_{\theta\phi,\phi}^j - \frac{1}{R_0} (j D_{\phi\theta}^j + \cos\phi\, D_{\theta\phi}^j) \right]
\tag{7.52f}
$$

$$
\gamma_\phi^j = \frac{1}{R_\phi} (D_{n,\phi}^j - D_\phi^j) + D_{\phi\theta}^j
\tag{7.52g}
$$

$$
\gamma_\theta^j = -\frac{1}{R_0} (j D_n^j + \sin\phi\, D_\theta^j) + D_{\theta\phi}^j
\tag{7.52h}
$$

or with transverse shearing strains neglected

$$\kappa_\phi^j = -\frac{1}{R_\phi}\left[\frac{1}{R_\phi}(D_{n,\phi}^j - D_\phi^j)\right]_{,\phi} \tag{7.53a}$$

$$\kappa_\theta^j = \frac{j}{R_0^2}(jD_n^j + \sin\phi\, D_\theta^j) - \frac{\cos\phi}{R_\phi R_0}(D_{n,\phi}^j - D_\phi^j) \tag{7.53b}$$

$$\tau^j = \frac{1}{2}\left\{\frac{1}{R_\phi}\left[\frac{1}{R_0}(jD_n^j + \sin\phi\, D_\theta^j)\right]_{,\phi} + \frac{j}{R_\phi R_0}(D_{n,\phi}^j - D_\phi^j)\right.$$

$$\left. - \frac{\cos\phi}{R_0^2}(jD_n^j + \sin\phi\, D_\theta^j)\right\} \tag{7.53c}$$

7.4.2 Physical Interpretation

The first two equations of (7.48a–h) for the extensional strains may be physically interpreted by considering Figure 7–6, which shows a segment of the meridian of a shell of revolution. The first term of equation (7.48a) follows from the basic definition of strain as the change in length of the differential segment op, $D_{\phi,\phi}d\phi$, divided by the initial length $R_\phi d\phi$. The second term of the equation represents the strain from the normal displacement D_n; this is easily visualized as a change in the arc length op caused by a change in R_ϕ, $(R_\phi + D_n)\,d\phi - R_\phi d\phi$, again divided by the initial length $R_\phi d\phi$. In equation (7.48b), the first term is again straightforward. The remaining two terms occur because of the change in the radius of the parallel circle caused by D_ϕ and D_n, which is shown as $\Delta_{oo'}$ on Figure 7–6. The horizontal projection of D_ϕ is $D_\phi \cos\phi$, and that of D_n is $D_n \sin\phi$. Together, we have the change in the length of the circular arc, $(D_\phi \cos\phi + D_n \sin\phi)\,d\theta$, divided by the original length, $R_0 d\theta$.

    Next, we consider equations (7.48d) and (7.48e), the changes of the curvature. These are defined as the change in the surface rotations over the segments $d\phi$ and $d\theta$, respectively, divided by the length of the segments. Equation (7.48d) is the change in the rotation in the ϕ direction, $D_{\phi\theta,\phi}d\phi$, divided by the arc length, $R_\phi d\phi$, or $(1/R_\phi)D_{\phi\theta,\phi}$. We may also interpret this result directly in terms of the displacements by using Figure 7–6. The rotation at the initial point in the segment, point o, as it moves to o_1' is $(1/R_\phi)D_\phi$. From o_1' to o_2', we have the transverse shearing strain contribution of γ_ϕ, where the positive sense was established by Figure 7–2. There is also a rotation at o_1' attributed to D_n. This is illustrated in Figure 7–7, where the same differential segment is shown with an exaggerated D_n displacement. The rotation at o_1' caused by the change in D_n is seen to be $(1/R_\phi)D_{n,\phi}$, in a sense opposite that of the rotation attributed to D_ϕ. Hence the total change in rotation over the segment

$$\left(\frac{1}{R_\phi}D_\phi + \gamma_\phi - \frac{1}{R_\phi}D_{n,\phi}\right)_{,\phi}d\phi$$

is divided by the segment length, $R_\phi d\phi$, to derive the meridional change in curvature. This expression is verified by substituting equation (7.48g) into (7.48d).

    Equation (7.48e) has two components. The first term is the change in the rotation $D_{\theta\phi}, D_{\theta\phi,\theta}d\theta$, divided by the arc length, $R_0 d\theta$, or $(1/R_0)D_{\theta\phi,\theta}$. Again, if a segment of the shell along a parallel circle is examined, this term may be interpreted directly in terms of the displacements. Note that there is a second term present in the expression for κ_θ. This term is

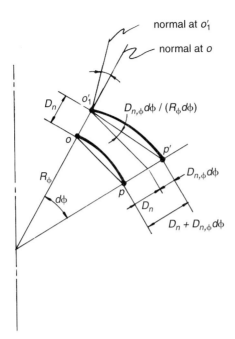

FIGURE 7–7 Normal Displacement of the Meridian.

$(\cos \phi / R_0) D_{\phi\theta}$, which acts about $\mathbf{t}_\phi$. The contribution of rotation $D_{\phi\theta}$ may be followed by referring to Figure 3–6 and replacing the term $N_\theta R_\phi d\phi$ by $D_{\phi\theta}$ on the FORCE element and on view HV1. Then, moving to view MV1, we see the resolution of this vector into the meridional direction and the normal direction. The drilling component, corresponding to rotation about the normal, is not admissible in the present theory. The meridional component, acting in the negative ϕ direction, is the rotation $D_{\phi\theta} \cos \phi$ and represents a change in slope along the parallel circle over the segment. The sign of this component is positive because, as shown in Figure 7–2(b) with $\alpha = \phi$ and $\beta = \theta$, the positive sense of the rotation $D_{\theta\phi}$ and hence κ_θ points in the negative ϕ direction. After division by the arc length, $R_0 d\theta$, we have the second term of equation (7.48e).

The preceding use of an argument from the *equilibrium* equations to explain a *compatibility* relationship is not an isolated coincidence. This is a representative part of a *static-geometric* analogy that exists between the equations of equilibrium and the equations of compatibility. This analogy is sometimes useful in constructing dual solutions. A complete development of the static-geometric analogy may be found in Gol'denviezer.[5]

We now turn our attention to the in-plane shearing strain ω, which is given by equation (7.48c) and is shown on Figure 7–8 as the sum of ω_ϕ and ω_θ. The term ω_ϕ is given by $D_{\phi,\theta} d\theta / (R_0 d\theta) = (1/R_0) D_{\phi,\theta}$. To evaluate ω_θ, note that point p has moved a distance $D_\theta + D_{\theta,\phi} d\phi$ in the θ direction. When the θ displacement of point o, D_θ, is projected onto the corresponding displacement of p to p', it becomes

$$D_\theta \left(\frac{R_0 + R_{0,\phi} d\phi}{R_0} \right) = D_\theta \left(1 + \frac{R_\phi \cos \phi}{R_0} d\phi \right)$$

because of the change in the horizontal radius R_0. The difference,

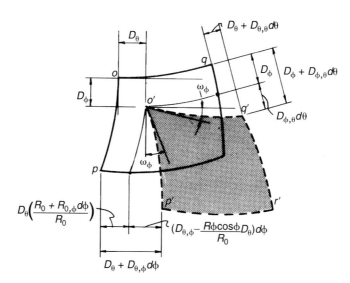

FIGURE 7–8 Shearing Displacements.

$$\frac{[D_{\theta,\phi} - D_\theta(R_\phi/R_0)\cos\phi]\,d\phi}{R_\phi d\phi} = \omega_\theta$$

and the sum of ω_ϕ and ω_θ is ω, as given by equation (7.48c).

Finally, we examine the twist τ, which is given by equations (7.48f). Since τ is obtained from τ_ϕ and τ_θ as defined in equation (7.44), we concentrate on the basic terms τ_θ and τ_θ. Referring to Figures 7–8 and 7–9(a), the twist in the ϕ direction, τ_ϕ, is given by the change in the rotation in the θ direction at o' as ϕ varies $D_{\theta\phi,\phi}d\phi$, divided by the arc length, $R_\phi d\phi$, which is the first term in equation (7.48f). The twist in the θ direction, τ_θ, consists partially of the corresponding change in the rotation in the ϕ direction at o' as θ varies, $D_{\phi\theta,\theta}d\theta$, divided by the arc length, $R_0 d\theta$, as shown in Figure 7–9(b). Another contribution to the twist in the θ direction is attributed to the change in the horizontal radius R_0 as ϕ changes. From equation (7.48h), $D_{\theta\phi}$ is approximately inversely proportional to R_0 so that $D_{\theta\phi}$ is reduced to

$$\frac{R_0}{R_0 + R_{0,\phi}d\phi}D_{\theta\phi}$$

over the segment, or we have a change of

$$-\frac{R_{0,\phi}d\phi}{R_0}D_{\theta\phi} = -\frac{R_\phi}{R_0}\cos\phi D_{\theta\phi}d\phi$$

After division by the arc length over which the change in horizontal radius occurs, $R_\phi d\phi$, we have the third term of equation (7.48f).

We have attempted to provide an explanation of the individual terms of the compatibility equations by using physical arguments for the specialized shell of revolution geometry. In a general shell, the physical illustrations would become untenable; yet, the initial derivation

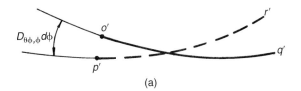

(a)

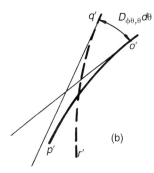

(b)

FIGURE 7–9   Twisting Displacements.

by vector algebra was straightforward. In the past, considerable effort has been expended in deriving compatibility relationships for specialized geometries using strictly physical reasoning.[6] Here, in both the equilibrium and compatibility treatments, we have sought to vary the approach somewhat by initially deemphasizing the physical in favor of the mathematical approach, but then closely supporting the results with specific physical illustrations.

7.5   STRAIN-DISPLACEMENT RELATIONS FOR PLATES

As initially set forth in Section 3.4, the shell equations are transformed into a sufficiently general form for medium-thin plates by letting the radii of curvature approach infinity. From equations (7.46),

$$\varepsilon_\alpha = \frac{1}{A} D_{\alpha,\alpha} + \frac{1}{AB} A_{,\beta} D_\beta \tag{7.54a}$$

$$\varepsilon_\beta = \frac{1}{B} D_{\beta,\beta} + \frac{1}{AB} B_{,\alpha} D_\alpha \tag{7.54b}$$

$$\omega = \frac{1}{A} D_{\beta,\alpha} - \frac{1}{AB} A_{,\beta} D_\alpha + \frac{1}{B} D_{\alpha,\beta} - \frac{1}{AB} B_{,\alpha} D_\beta \tag{7.54c}$$

$$\kappa_\alpha = \frac{1}{A} D_{\alpha\beta,\alpha} + \frac{1}{AB} A_{,\beta} D_{\beta\alpha} \tag{7.54d}$$

$$\kappa_\beta = \frac{1}{B} D_{\beta\alpha,\beta} + \frac{1}{AB} B_{,\alpha} D_{\alpha\beta} \tag{7.54e}$$

$$\tau = \frac{1}{2}\left[\frac{1}{A}D_{\beta\alpha,\alpha} + \frac{1}{B}D_{\alpha\beta,\beta} - \frac{1}{AB}(A_{,\beta}D_{\alpha\beta} + B_{,\alpha}D_{\beta\alpha})\right]$$ (7.54f)

$$\gamma_\alpha = \frac{1}{A}D_{n,\alpha} + D_{\alpha\beta}$$ (7.54g)

$$\gamma_\beta = \frac{1}{B}D_{n,\beta} + D_{\beta\alpha}$$ (7.54h)

When the transverse shearing strains are neglected, we have

$$\kappa_\alpha = -\left[\frac{1}{A}\left(\frac{1}{A}D_{n,\alpha}\right)_{,\alpha} + \frac{A_{,\beta}}{AB^2}D_{n,\beta}\right]$$ (7.55a)

$$\kappa_\beta = -\left[\frac{1}{B}\left(\frac{1}{B}D_{n,\beta}\right)_{,\beta} + \frac{B_{,\alpha}}{A^2B}D_{n,\alpha}\right]$$ (7.55b)

$$\tau = -\frac{1}{2}\left\{\frac{1}{A}\left(\frac{1}{B}D_{n,\beta}\right)_{,\alpha} + \frac{1}{B}\left(\frac{1}{A}D_{n,\alpha}\right)_{,\beta} - \frac{1}{AB}\left[\frac{A_{,\beta}}{A}D_{n,\alpha} + \frac{B_{,\alpha}}{B}D_{n,\beta}\right]\right\}$$ (7.55c)

or

$$\tau = -\frac{1}{AB}\left(D_{n,\alpha\beta} - \frac{1}{A}A_{,\beta}D_{n,\alpha} - \frac{1}{B}B_{,\alpha}D_{n,\beta}\right)$$ (7.55d)

since $A \neq A(\alpha)$ and $B \neq B(\beta)$ for a flat plate.

It has been noted several times that the adoption of assumption [3] and the subsequent suppression of transverse shearing strains generally leads to significant mathematical simplifications. An example of this may be seen for the plate case by comparing equations (7.54d) to (7.54f) with equations (7.55a) to (7.55c). In the latter equations, the curvatures and hence the entire flexural behavior of the shell are dependent on only the normal displacement, D_n, and are essentially uncoupled from the extensional behavior. The form of the curvature-displacement relations is identical to that encountered in elementary beam theory. On the other hand, the retention of transverse shearing strains leads to a considerably more complicated mathematical formulation. This is discussed further in Chapter 10.

7.6 GEOMETRY OF DEFORMED SURFACE

7.6.1 Change in Gaussian Curvature

At the conclusion of Section 2.6, it was remarked that a set of compatibility equations for the deformed middle surface, parallel to equations (2.35), (2.36), and (2.37), may be derived.[7] Of the three relations, the one corresponding to equation (2.37) is notable in the context of characterizing deformed surfaces. The right-hand side of equation (2.37) contains the term $K = 1/R_\alpha R_\beta$, the Gaussian curvature.

We consider the differential of K,

$$d(K) = d\left(\frac{1}{R_\alpha R_\beta}\right) = \frac{1}{R_\alpha}d\left(\frac{1}{R_\beta}\right) + \frac{1}{R_\beta}d\left(\frac{1}{R_\alpha}\right) = -\left(\frac{\kappa_\alpha}{R_\beta} + \frac{\kappa_\beta}{R_\alpha}\right)$$ (7.56)

by the definition of normal curvature κ_i in equation (2.21). Therefore, we define

$$\overline{K} = \frac{\kappa_\alpha}{R_\beta} + \frac{\kappa_\beta}{R_\alpha} \tag{7.57}$$

as the *change* in Gaussian curvature.

The sign is irrelevant here; thus, we observe that just as the Gaussian curvature is an important single parameter for the classification of undeformed surfaces,[8] the change in Gaussian curvature may serve the same role for a deformed surface. It should be reiterated that both terms are scalar point functions so that global generalizations should be guarded; however, locally these expressions may lead to useful insights.

While equation (7.57) may be evaluated from the basic geometric relationships in Chapter 2, it would be computationally expedient to have an expression in terms of deformation variables. Starting with an initially flat surface, Calladine[9] derived the expressions:

$$\overline{K} = -\epsilon_{y,xx} + \omega_{,xy} - \epsilon_{x,yy} \tag{7.58}$$

in Cartesian coordinates. Note that if the surface remains flat, $K = 0$ and equation (7.58) becomes the plane strain compatibility condition familiar from the 2-D theory of elasticity. Calladine[9] further argues that equation (7.58) may be applied to two additional classes of surfaces that are not initially flat:

1. Developable surfaces.
2. Surfaces which may be adequately described by Cartesian coordinates. Developable surfaces qualify since $K = 0$, just as for a totally flat surface. We have identified a class of shells which fits into the second category in Section 2.8.1, shallow shells.

It may be further reasoned from the concept of finite elements, mentioned throughout the book several times, that a surface which is not globally shallow, in the sense of Section 2.8.1, may be closely represented by an assemblage of discrete elements which are individually locally shallow. Given, the local nature of $\overline{K}$, the expression may be computed in local Cartesian coordinates. Many computer codes will provide the necessary information needed to evaluate equation (7.58) at selected points on a shell surface, perhaps in terms of local stresses which may be readily converted to strains.

7.6.2 Inextensional Deformations

In Section 5.2.1, the topics of pure bending deformations and inextensional deformations were introduced. The specific difference between the two states is that in-plane shearing strains, as represented by ω, may still occur with inextensional deformations. In this section, they are regarded as identical: Deformations which neither stretch nor shear the middle surface.[10] Considering equation (7.58), with the middle-surface strains equal to zero, $\overline{K} = 0$. Thus, there is no change in Gaussian curvature when the deformations are inextensional. While equation (7.58) was derived for an initially flat (developable) surface, Calladine[9] has shown that the unchanging Gaussian curvature holds for polygonalised approximations of closed (positive $\overline{K}$) shells and, in the limit, for their doubly curved analogues. As an illustration, consider a circular cylindrical shell, Figure 7–10(a). We evaluate equation (7.57) using the change in curvature expression, equation (7.47a), corresponding to negligible transverse shearing strains. With $\alpha = x$ and $\beta = \theta$, $R_\alpha = \infty$, $R_\beta = a$, $A = 1$, $B = a$; therefore, $A_\theta = 0$ and equation (7.47a) becomes $\kappa_x = -D_{n,xx}$ so that the change in Gaussian curvature is

$$\overline{K} = -\frac{1}{a} D_{n,xx} \tag{7.59}$$

Equating equations (7.59) and (7.58), we have

$$\frac{1}{a} D_{n,xx} = \epsilon_{y,xx} - \omega_{,xy} + \epsilon_{x,yy} \tag{7.60}$$

which may be used to compute displacements for a cylindrical shell when the surface strains are known. This problem is addressed in a somewhat different form in Section 8.3.4.1.

Now we return to the case of inextensional deformation, for which equation (7.60) becomes

$$D_{n,xx} = 0 \tag{7.61}$$

which has a solution

$$D_n(X,\theta) = f_1(\theta) + xf_2(\theta) \tag{7.62}$$

where f_1 and f_2 depend on the boundary conditions in the x-direction.

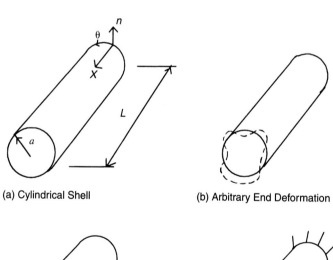

(a) Cylindrical Shell   (b) Arbitrary End Deformation

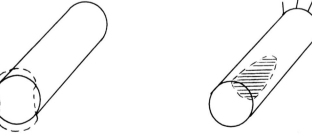

(c) Second Harmonic   (d) Trough Deformation of Fixed Shell

FIGURE 7–10   Inextensional Deformation of a Cylindrical Shell.

Initially, we impose

$$D_n(0,\theta) = D_n(L,\theta) = 0 \tag{7.63}$$

which corresponds to a radial line constraint at each end. This produces $D_n = 0$ throughout. Note that no constraints are placed on the circumferential or longitudinal deformations at either end.

We now evaluate the other components of displacement, D_x and D_θ. From equation (7.46b) with $\varepsilon_\beta = \varepsilon_\theta = 0$ and $D_n = 0$,

$$D_{\theta,\theta} = 0 \tag{7.64}$$

so that

$$D_\theta = f_3(x) \tag{7.65}$$

while from equation (7.46a) with $\varepsilon_\alpha = \varepsilon_x = 0$,

$$D_{x,x} = 0 \tag{7.66}$$

so that

$$D_x = f_4(\theta) \tag{7.67}$$

Then assuming that the in-plane shearing strain ω is also zero, we obtain a connective relationship from equation (7.46c):

$$D_{\theta,x} + \frac{1}{a} D_{x,\theta} = 0 \tag{7.68}$$

or, considering equations (7.65) and (7.67)

$$[f_3(x)]_{,x} + \frac{1}{a}[f_4(\theta)]_{,\theta} = 0 \tag{7.69}$$

which is solved as

$$D_\theta(x) = f_3(x) = C_1 + C_2 x \tag{7.70}$$
$$D_x(\theta) = f_4(\theta) = C_3 - C_2 a\theta \tag{7.71}$$

where C_1, C_2, and C_3 are constants.

Before pursuing inextensional (non-trivial) solutions of equations (7.62), (7.70), and (7.71), we may check some simple cases. If there is a uniform axial displacement $D_x(\theta) = \overline{D}_x$, $C_3 = \overline{D}_x$ and $C_2 = 0$; if there is a uniform circumferential displacement $D_\theta(x) = \overline{D}_\theta$, $C_1 = \overline{D}_\theta$ and $C_2 = 0$. The latter two cases are independent of the boundary condition on D_n.

We now return to the problem at hand and relax the radial constraint at one end of the tube, say at $X = L$. A possible form of a nonzero displacement at $X = L$ is shown in Figure 7–10(b). With $D_n(0,\theta) = 0$ in equation (7.62), $f_1(\theta) = 0$ but $f_2(\theta)$ is undetermined. While $f_2(\theta)$ is somewhat arbitrary because the restrictions on the boundary are too few to completely specify the surface, there are certain restrictions on the function.[9]

1. $f_2(\theta)$ is bounded in magnitude by the small deformation restriction, Assumption [2] Chapter 1, which limits D_n and the applicability of the linear kinematic relationships, equations (7.46).

2. Because D_n, is periodic in θ, $f_2(\theta)$ must be periodic, in other words,

$$f_2 (\theta + 2\pi) = f_2 (\theta) \tag{7.72}$$

3. At any value of x, the hoop strain must be inextensional. Returning to equation (7.46b) for ε_θ but with $D_n \neq 0$,

$$D_n = -D_{\theta,\theta} \tag{7.73}$$

Therefore,

$$\int_0^{2\pi} D_n \, d\theta = D_\theta (0) - D_\theta (2\pi) = 0 \tag{7.74}$$

because D_θ is also periodic. Hence

$$\int_0^{2\pi} f_2 (\theta) \, d\theta = 0 \tag{7.75}$$

Again, following Calladine,[9] we shall seek harmonic solutions of the form $\sin j\theta$ and $\cos j\theta$, familiar in general from Section 5.1 and specifically for cylindrical shells in Section 6.2. Taking a single term of a Fourier series

$$f_2 (\theta) = \frac{D_n^j}{L} \cos j\theta \tag{7.76}$$

where D_n^j is the amplitude of the displacement in harmonic j,

$$D_n = D_n^j \frac{x}{L} \cos j\theta \tag{7.77}$$

from equation (7.62), because $f_1 (\theta) = 0$. Then integrating equations (7.73) and (7.66) gives

$$D_{\theta,\theta} = -D_n^j \frac{x}{L} \cos j\theta \tag{7.78}$$

$$D_\theta = -\frac{1}{j} D_n^j \frac{x}{L} \sin j\theta + f_5 (x) \tag{7.79}$$

and

$$D_x = f_6 (\theta) \tag{7.80}$$

Finally, returning to equation (7.68), we have

$$-\frac{1}{jL} D_n^j \sin j\theta + [f_5 (x)]_{,x} + \frac{1}{a} [f_6 (\theta)]_{,\theta} = 0 \tag{7.81}$$

The middle term $f_{5,x}$ can be at most a constant because there are no other functions of x. However, upon integration with respect to θ, a nonzero constant would contribute to $f_6(\theta)$ as a linear and nonperiodic term, so that f_5 itself is taken as

$$f_5 = 0 \tag{7.82}$$

Continuing the integration of equation (7.81)

$$D_x = \frac{1}{j^2 L} D_n^j \cos j\theta \tag{7.83}$$

in view of equation (7.80).

In Figure 7–10(c), the displacements for $j = 2$ are shown. Along with the obvious D_n displacement, there is a displacement D_x which is proportional to D_n^j circumferentially, but not a variable function of x, and a displacement D_θ which is out of phase circumferentially. It is also of interest to consider the comparative amplitudes of the deformations at $x = L$:

Component	Amplitude	$j = 2$
D_n	D_n^j	D_n^2
D_θ	$1/j\, D_n^j$	$1/2\, D_n^2$
D_x	$a/(j^2 L)D_n^j$	$a/(4L)D_n^2$

While the relationship between D_n and D_θ is determined entirely by the deformation of the inextensional hoop, the amplitude of D_x is also dependent on the a/L (radius/length) ratio of the cylinder.

Obviously, for higher harmonics the longitudinal deformations, D_x, become negligible in comparison to D_n.

The phenomenon described mathematically in the preceding paragraphs may be easily illustrated physically by wrapping a sheet of paper into a cylinder and taping or gluing the seam. To approximate the boundary condition $D_n(0) = 0$, a coin or other thin disk could be used as a mold at one end; alternately, a forefinger may be used as a mandrel, with the tube wrapped just snug below the knuckle. Then, the inextensional mode is easily produced with a slight pinch at the free end. The normal, D_n, and circumferential, D_θ, deformations are easily observed but the longitudinal deformation, D_x, is difficult to see. We may carry the example further by imposing an additional boundary condition $D_x(0,\theta) = 0$. From equation (7.83), $D_n^j = 0$; from equation (7.77), $D_n = 0$; and from equation (7.79), $D_\theta = 0$ as $f_5 = 0$. Seemingly the inextensional deformations are suppressed. But, if the paper cylinder is pushed up the forefinger snug against the top and the cylinder is pinched, the free end seems just as flexible.

Calladine explains this paradox by considering a different type of deformation—a "snapped-through" trough caused by a one-sided disturbance,[11] Figure 7–10(d). He invokes requirements of smoothness and the presence of large deformations that invalidate the small deformation kinematic equations, (7.46) and (7.47), which are the basis of the inextensional deformation theory. He also suggests that perhaps some stretching or shearing could be allowed as a transition between the original (undeformed) and final (inextensional) geometry.

In summary, the free end of a cylindrical shell is inherently flexible in the hoop direction when subjected to normal excitations and tends to oval, regardless of the support condition at the other end. This is also true for other open shells of zero and negative Gaussian curvature. The common preventative measure is the ring stiffener, discussed in a somewhat different context in Section 4.5.1 and shown for some tower-type shells in Figure 2–8(o).

The assumption of inextensional deformations has been explored by several investigators to obtain lower bound solutions for shell instability as briefly mentioned in Section 13.3.7 and illustrated by the curve labeled λ_β in Figure 13–9. Also it is used to analyze a cylindrical tank for the effects of differential settlement in Section 8.3.5.

REFERENCES

1. V. V. Novozhilov, *Thin Shell Theory* [translated from 2nd Russian ed. by P. G. Lowe] (Groningen, The Netherlands: Noordhoff, 1964): 14–27.
2. H. Kraus, *Thin Elastic Shells* (New York: Wiley, 1967), chap. 1.
3. P. Bergan and M. K. Nygard, Nonlinear shell elements with six freedoms per node, *Proc. First World Congress on Computational Mechanics,* University of Texas at Austin, Austin, September 1985.
4. C. K. Choi, Reduced integrated nonconforming plate element, *Journal of Engineering Mechanics,* ASCE 112, No. 4 (April 1986): 370–385.
5. A. L. Gol'denveizer, *Theory of Elastic Thin Shells* [translated from Russian ed.] (New York: Pergamon Press, 1961): 92–96.
6. S. Timoshenko and S. Woinowsky-Krieger, *Theory of Plates and Shells,* 2nd ed. (New York: McGraw-Hill, 1959): 165–173.
7. Novozhilov, *Thin Shell Theory:* 14.
8. C. Calladine, *Theory of Shell Structures* (Cambridge: Cambridge University Press, 1983): 175–177.
9. Calladine, *Theory of Shell Structures:* 146–157.
10. Novozhilor, *Thin Shell Theory:* 151–163.
11. Calladine, *Theory of Shell Structures:* 184.

EXERCISES

7.1. Perform the differentiations indicated in equation (7.8) to verify equation (7.9); also verify equation (7.11).

7.2. Verify the expressions given for κ_α and κ_β in equations (7.38) and (7.39); also verify the expressions for τ_α and τ_β, as given in equations (7.42) and (7.43).

7.3. On an appropriate sketch, show the physical meaning of the first term in equation (7.48e) in terms of the middle-surface displacements.

7.4. Derive the strain-displacement relations for a shell of revolution where the Lamé parameter A is taken to correspond to the axial coordinate Z. Give a physical explanation of each term using appropriate sketches.

7.5. Show that the four terms $\varepsilon_{\alpha\beta}$, $\varepsilon_{\beta\alpha}$, τ_α, and τ_β satisfy the identity

$$\tau_\alpha + \frac{\varepsilon_{\beta\alpha}}{R_\alpha} = \tau_\beta + \frac{\varepsilon_{\alpha\beta}}{R_\beta}$$

$$\text{if } \gamma_\alpha = \gamma_\beta = 0.$$

7.6. Consider the strain-displacement relations for a plate in Cartesian coordinates, equations (7.54 a, b, c) and show that equation (7.58) gives $\overline{K} = 0$.

7.7. For a cylindrical shell with radius a and length L, Figure 7–10 (a), the boundary conditions are

$$At\ x = 0, D_x = D_x^j \cos j\theta; D_n = 0$$

$$At\ x = L,\ Free\ end$$

Show that the components of displacement are

$$D_x = D_x^j \cos j\theta$$

$$D_\theta = D_x^j j\,(x/a)\sin j\theta$$

$$D_n = -D_x^j j^2\,(x/a)\cos j\theta$$

for inextensional deformations.[11]

7.8. For the shell in Exercise 7.7, the boundary condition at $x = 0$ is

$$D_x = 0, D_\theta = D_\theta^j \cos j\theta$$

Show that the components of displacement are

$$D_x = 0$$

$$D_\theta = D_\theta^j \cos j\theta$$

$$D_n = D_n^j \sin j\theta$$

8

Constitutive Laws, Boundary Conditions, and Displacements

8.1 CONSTITUTIVE LAWS

The constitutive law or stress-strain relationship is the third basic component of the elasticity problems. This subject is quite broad, but the detailed examination of various alternatives that are dependent on the characteristics of particular engineering materials is not properly within the scope of this book. Rather, it remains within the purview of the theory of elasticity, as we may accommodate a variety of material laws within our formulation of the shell or plate problem. Initially, we use the basic Hooke's law for isotropic materials, and then we illustrate how some extended material laws can be accommodated.

8.1.1 Isotropic Material Law

8.1.1.1 Strain-Stress Relationship Including Temperature Effects. The strains, as defined in Section 7.3, are related to the stresses, introduced in Section 3.1, by

$$\varepsilon_\alpha(\alpha,\beta,\zeta) = \frac{1}{E}\left(\sigma_{\alpha\alpha} - \mu\sigma_{\beta\beta}\right) + \overline{\alpha}\,T(\alpha,\beta,\zeta) \tag{8.1a}$$

$$\varepsilon_\beta(\alpha,\beta,\zeta) = \frac{1}{E}\left(\sigma_{\beta\beta} - \mu\sigma_{\alpha\alpha}\right) + \overline{\alpha}\,T(\alpha,\beta,\zeta) \tag{8.1b}$$

$$\omega(\alpha,\beta,\zeta) = \frac{1}{G}\sigma_{\alpha\beta} = \frac{1}{G}\sigma_{\beta\alpha} \tag{8.1c}$$

$$\gamma_\alpha(\alpha,\beta) = \frac{1}{G_n}\sigma_{\alpha n} \tag{8.1d}$$

$$\gamma_\beta(\alpha,\beta) = \frac{1}{G_n}\sigma_{\beta n} \tag{8.1e}$$

In equation (8.1), E = Young's modulus, G and G_n = in-plane and normal shearing moduli, and μ = Poisson's ratio. Also, $\bar{\alpha}$ represents the coefficient of thermal expansion, which will be taken as constant, and $T(\alpha,\beta,\zeta)$ = the temperature, measured from a specified reference value. Thermal effects are included because they are very important in some applications.[1]

Matrix 8–1

$$
\begin{Bmatrix} \sigma_\alpha(\alpha,\beta,\zeta) \\ \sigma_{\beta\beta}(\alpha,\beta,\zeta) \\ \sigma_{\alpha\beta}(\alpha,\beta,\zeta) \\ \sigma_{\alpha n}(\alpha,\beta) \\ \sigma_{\beta n}(\alpha,\beta) \end{Bmatrix} = E
\begin{bmatrix}
\frac{1}{1-\mu^2} & \frac{\mu}{1-\mu^2} & 0 & 0 & 0 \\
\frac{\mu}{1-\mu^2} & \frac{1}{1-\mu^2} & 0 & 0 & 0 \\
0 & 0 & \frac{1}{2(1+\mu)} & 0 & 0 \\
0 & 0 & 0 & \frac{1}{2(1+\mu)} & 0 \\
0 & 0 & 0 & 0 & \frac{1}{2(1+\mu)}
\end{bmatrix}
\begin{Bmatrix} \varepsilon_\alpha + \zeta k_\alpha \\ \varepsilon_\beta + \zeta k_\beta \\ \omega + 2\zeta\tau \\ \gamma_\alpha \\ \gamma_\beta \end{Bmatrix} - \varepsilon\bar{\alpha}T(\alpha,\beta,\zeta)
\begin{Bmatrix} \frac{1}{1-\mu} \\ \frac{1}{1-\mu} \\ 0 \\ 0 \\ 0 \end{Bmatrix}
$$

We find it convenient to invert equation (8.1) and substitute the expressions $\varepsilon_\alpha(\alpha,\beta,\zeta)$, $\varepsilon_\beta(\alpha\ \beta\ \zeta)$, and $\omega(\alpha,\beta,\zeta)$ as given by equations (7.36), (7.37), and (7.45), respectively. After performing the indicated substitutions and assuming for the present that $G_n = G = E/[2(1+\mu)]$, we obtain the stress-strain law

$$\{\sigma(\alpha,\beta,\zeta)\} = [C]\{\varepsilon(\alpha,\beta,\zeta)\} - \{\sigma_T(\alpha,\beta,\zeta)\} \tag{8.2}$$

which is given in detail in Matrix 8–1.

We use the matrix representation to illustrate that any other consistent stress-strain relationship may be accommodated by altering the elements of $[C]$ and $\{\sigma_T\}$, which are essentially independent from the reminder of the development.

8.1.1.2 Stress Resultant-Strain and Stress Couple-Curvature Relations

In the surface structure problem, the constitutive equations are required to relate the stress resultants and stress couples, instead of merely stresses, to the corresponding strain and curvatures. These relationships are obtained in a straightforward fashion. We substitute the stress-strain law, Matrix 8–1, into equations (3.6) to (3.9). Consider the expression for N_α in equation (3.6):

Example 8.1(a)

$$N_\alpha = \int_{-h/2}^{h/2} \sigma_{\alpha\alpha} \left(1 + \frac{\zeta}{R_\beta} \right) d\zeta$$

$$= E \int_{-h/2}^{h/2} \left(\frac{\varepsilon_\alpha}{1 - \mu^2} + \frac{\zeta\kappa_\alpha}{1 - \mu^2} + \frac{\mu\varepsilon_\beta}{1 - \mu^2} + \frac{\mu\zeta\kappa_\beta}{1 - \mu^2} - \frac{\overline{\alpha}\,T}{1 - \mu} \right)\left(1 + \frac{\zeta}{R_\beta} \right) d\zeta \qquad (8.3)$$

Because $\zeta/R_\beta \ll 1$, it may be neglected in this integration. We split the integral into

$$N_\alpha = \frac{E}{1 - \mu^2} \int_{-h/2}^{h/2} (\varepsilon_\alpha + \zeta\kappa_\alpha + \mu\varepsilon_\beta + \mu\zeta\kappa_\beta)d\zeta - \frac{E\overline{\alpha}}{1 - \mu} \int_{-h/2}^{h/2} T(\zeta)d\zeta \qquad (8.4)$$

In this development, the cross-section is assumed to be symmetric about the middle plane or surface, so that

$$\int_{-h/2}^{h/2} \zeta d\zeta = 0$$

and the second and fourth terms of the first integral vanish, leaving

$$N_\alpha = \frac{Eh}{1 - \mu^2} (\varepsilon_\alpha + \mu\varepsilon_\beta) - N_{\alpha T} \qquad (8.5)$$

where

$$N_{\alpha T} = \frac{E\overline{\alpha}}{1 - \mu} \int_{-h/2}^{h/2} T(\zeta)d\zeta$$

which may be called the *thermal load,* is left in general form to accommodate the possibility that $T = T(\zeta)$. The remainder of the stress resultant-strain equations are obtained in the same fashion.

Example 8.1(b)

Now, consider the expression for M_α in equation (3.8)

$$M_\alpha = \int_{-h/2}^{h/2} \sigma_{\alpha\alpha}\zeta \left(1 + \frac{\zeta}{R_\beta} \right) d\zeta \qquad (8.6)$$

which, referring to equations (8.3) and (8.4), becomes

$$M_\alpha = \frac{E}{1 - \mu^2} \int_{-h/2}^{h/2} (\zeta\varepsilon_\alpha + \zeta^2\kappa_\alpha + \mu\zeta\varepsilon_\beta + \mu\zeta^2\kappa_\beta)d\zeta - \frac{E\overline{\alpha}}{1 - \mu} \int_{-h/2}^{h/2} T(\zeta)\zeta d\zeta \qquad (8.7)$$

For a symmetric cross-section, the first and third terms of the first integral vanish, leaving

$$M_\alpha = \frac{Eh^3}{12(1 - \mu^2)} (\kappa_\alpha + \mu\kappa_\beta) - M_{\alpha T} \qquad (8.8)$$

where

$$M_{\alpha T} = \frac{E\overline{\alpha}}{1 - \mu} \int_{-h/2}^{h/2} T(\zeta)\zeta d\zeta$$

is called the *thermal moment.* If T is constant through the thickness, there are no thermal moments.

The remainder of the stress couple-curvature equations are obtained in an identical fashion. Collecting the equations, we can develop Matrix 8–2, where

$$N_{\alpha T} = N_{\beta T} = N_T = \frac{E\bar{\alpha}}{1 - \mu} \int_{-h/2}^{h/2} T(\zeta) d\zeta \tag{8.9a}$$

$$M_{\alpha T} = M_{\beta T} = M_T = \frac{E\bar{\alpha}}{1 - \mu} \int_{-h/2}^{h/2} T(\zeta)\zeta \, d\zeta \tag{8.9b}$$

Matrix 8–2

$$
\begin{Bmatrix} N_\alpha \\ N_\beta \\ N_{\alpha\beta} \\ M_\alpha \\ M_\beta \\ M_{\alpha\beta} \\ Q_\alpha \\ Q_\beta \end{Bmatrix} = E
\begin{bmatrix}
\frac{h}{1-\mu^2} & \frac{\mu h}{1-\mu^2} & 0 & 0 & 0 & 0 & 0 & 0 \\
\frac{\mu h}{1-\mu^2} & \frac{h}{1-\mu^2} & 0 & 0 & 0 & 0 & 0 & 0 \\
0 & 0 & \frac{h}{2(1-\mu)} & 0 & 0 & 0 & 0 & 0 \\
0 & 0 & 0 & \frac{h^3}{12(1-\mu^2)} & \frac{\mu h^3}{12(1-\mu^2)} & 0 & 0 & 0 \\
0 & 0 & 0 & \frac{\mu h^3}{12(1-\mu^2)} & \frac{h^3}{12(1-\mu^2)} & 0 & 0 & 0 \\
0 & 0 & 0 & 0 & 0 & \frac{h^3}{12(1+\mu)} & 0 & 0 \\
0 & 0 & 0 & 0 & 0 & 0 & \frac{\lambda h}{2(1+\mu)} & 0 \\
0 & 0 & 0 & 0 & 0 & 0 & 0 & \frac{\lambda h}{2(1+\mu)}
\end{bmatrix}
\begin{Bmatrix} \varepsilon_\alpha \\ \varepsilon_\beta \\ \omega \\ \kappa_\alpha \\ \kappa_\beta \\ \tau \\ \gamma_\alpha \\ \gamma_\beta \end{Bmatrix} -
\begin{Bmatrix} N_{\alpha T} \\ N_{\beta T} \\ 0 \\ M_{\alpha T} \\ M_{\beta T} \\ 0 \\ 0 \\ 0 \end{Bmatrix}
$$

and where the shear moduli are now assumed to be related by $G_n = \lambda G$. λ, sometimes called the *shear correction factor* or the *transverse warping shape factor,* is relatively complex to define explicitly, but is commonly taken as 5/6 for isotropic shells.[2]

Matrix 8–2 in the form

$$\{N\} = [D]\{\varepsilon\} - \{N_T\} \tag{8.10}$$

serves to illustrate again that more complex material properties, expressed through a modification of the stress-strain law [C], can be represented through a corresponding generalization of [D] without affecting the remaining development. Similarly, the $\{N_T\}$ matrix can be suitably altered if $\{\sigma_T\}$ is changed. We consider some of these possibilities in the next section. Furthermore, we must emphasize that two terms of [D], D_{33} and D_{66}, often appear in slightly different forms, modified by a factor of 2. This arises because of the somewhat ambiguous definitions of ω and τ as given in equations (7.28) and (7.44); there is little consensus on whether ω should be defined as in equation (7.28), or as one-half of that quantity. Similarly, the factor of 1/2 introduced into the definition of τ could be omitted. The only requirement is to maintain consistency in the constitutive matrix, and [D], as given in Matrix 8–2, is correct in this regard.

Also, note that for plates with no in-plane loading, the first three rows and columns of Matrix 8–2 may be deleted. If transverse shearing strains are not included in the formulation—either for shells or for plates—the last two rows and columns are not relevant; rather, the transverse shear resultants are obtained from the fourth and fifth equilibrium equations, (3.22a) and (3.22b), once the stress couples are computed.

It is also interesting to note that although the transverse shear resultants are readily computed from the last two equations in Matrix 8–2, the corresponding stresses in Matrix 8–1 violate the stress-free conditions on the faces $\pm h/2$, because γ_α and γ_β are constant through the thickness. This has been a major consideration in the formulation of a consistent theory of plates, as discussed later in Section 11.2.4. However, within the context of our treatment, this violation is only at the point (elasticity) level and may be circumvented by computing the shear stresses directly from the stress resultants, as described in equation (3.10c).

In the preceding equations and throughout this book, the temperature T is taken as a point function of the spatial coordinates which has been evaluated independently of the mechanical characteristics (material, properties, stress, strain, etc.). Hence the thermal terms were attached to the mechanical loading and treated as thermal loads and moments. Although this formulation is adequate for most applications, the temperature effects which produce the loads and moments may be very significant in some shell structures, particularly those susceptible to damage from cracking. A careful analysis of the temperature distributions may be warranted as outlined in Appendix 8A.

8.1.2 Specifically Orthotropic Materials

A generalization of the isotropic condition which is of interest to us is the specifically orthotropic material. The strict definition of this configuration is that elastic properties are specified along two mutually orthogonal *material* axes, whereas the shell still retains the piecewise constant thickness and cross-sectional symmetry presumed so far in this text. Further, it is assumed in our development that the orthogonal material axes

Matrix 8–3

$$[\mathbf{D_{or}}] = \begin{bmatrix} \int_h c_{11}d\zeta & \int_h c_{12}d\zeta & 0 & 0 & 0 & 0 & 0 & 0 \\ \int_h c_{12}d\zeta & \int_h c_{22}d\zeta & 0 & 0 & 0 & 0 & 0 & 0 \\ 0 & 0 & \int_h c_{44}d\zeta & 0 & 0 & 0 & 0 & 0 \\ 0 & 0 & 0 & \int_h c_{11}\zeta^2 d\zeta & \int_h c_{12}\zeta^2 d\zeta & 0 & 0 & 0 \\ 0 & 0 & 0 & \int_h c_{12}\zeta^2 d\zeta & \int_h c_{22}\zeta^2 d\zeta & 0 & 0 & 0 \\ 0 & 0 & 0 & 0 & 0 & \int_h c_{44}\zeta^2 d\zeta & 0 & 0 \\ 0 & 0 & 0 & 0 & 0 & 0 & \lambda_\alpha \int_h c_{55}d\zeta & 0 \\ 0 & 0 & 0 & 0 & 0 & 0 & 0 & \lambda_\beta \int_h c_{66}d\zeta \end{bmatrix}$$

coincide with the coordinate lines of the middle surface of the *shell*. The development of composite matenal technology has made it feasible to produce very efficient shells and plates having such orthotropic properties.

With this as a preface, we introduce a modified matrix $[\mathbf{C}]$ into equation (8.2)

$$[\mathbf{C}_{or}] = \begin{bmatrix} c_{11} & c_{12} & 0 & 0 & 0 \\ c_{12} & c_{22} & 0 & 0 & 0 \\ 0 & 0 & c_{44} & 0 & 0 \\ 0 & 0 & 0 & c_{55} & 0 \\ 0 & 0 & 0 & 0 & c_{66} \end{bmatrix} \tag{8.11}$$

The elements of matrix $[\mathbf{C}_{or}]$ are generally determined experimentally.

$[\mathbf{C}_{or}]$ may be used to derive a corresponding stress resultant-strain matrix $[\mathbf{D}_{or}]$ and a corresponding thermal load vector $\{\mathbf{N}_{Tor}\}$ by following the same procedure described in Section 8.1.1.2. We also redefine the thermal coefficients and transverse warping shape factors to permit different values in each direction. The resulting generalization of equation (8.10) is

$$\{\mathbf{N}_{or}\} = [\mathbf{D}_{or}]\{\boldsymbol{\varepsilon}\} - \{\mathbf{N}_{Tor}\} \tag{8.12}$$

where $[\mathbf{D}_{or}]$ is given in general form as shown in Matrix 8–3, and the elements of $\{\mathbf{N}_{Tor}\}$ are

$$N_{\alpha Tor} = \int_{h} (c_{11}\overline{\alpha}_{\alpha} + c_{12}\overline{\alpha}_{\beta})T(\zeta)d\zeta \tag{8.13a}$$

$$N_{\beta Tor} = \int_{h} (c_{12}\overline{\alpha}_{\alpha} + c_{22}\overline{\alpha}_{\beta})T(\zeta)d\zeta \tag{8.13b}$$

$$M_{\alpha Tor} = \int_{h} (c_{11}\overline{\alpha}_{\alpha} + c_{12}\overline{\alpha}_{\beta})T(\zeta)d\zeta \tag{8.13c}$$

$$M_{\beta Tor} = \int_{h} (c_{12}\overline{\alpha}_{\alpha} + c_{22}\overline{\alpha}_{\beta})T(\zeta)d\zeta \tag{8.13d}$$

For homogeneous shells, the integrals in Matrix 8–3 and equations (8.13a–d) are easily evaluated in explicit form. For multilayered sections, the integrals may be computed by summing over the thickness. Multilayered shells and plates are widely used in applications of high-performance materials. For such layered shells, the shape factors λ_{α} and λ_{β} are determined for specific cases by Dong and Tso,[3] through correlation with selected reference solutions based on the theory of elasticity. A complete bibliography on the calculation of shape factors for layered plates and shells is also given there.

8.1.3   Stiffened Shells and Plates and Reticulated Shells

There are two additional structural systems which are neither shells nor plates by the strict definition, but may be conveniently treated as such. These are (a) shells or plates to which discrete stiffeners, surface undulations, or folds have been added to supplement the strength and/or rigidity of the basic constant thickness section; and, (b) frames composed of intersecting grids of closely spaced members that follow the topography of a curved surface. Such frameworks are termed shell-like structures or *reticulated* shells and combine the favorable structural characteristics of the basic geometric form with the possible use of efficient prefabricated components, requiring little or no falsework for erection.[4] The ancient Pantheon

(Figure 1–3) is a stiffened shell; the contemporary Astrodome in Houston, Texas (Figure 2–8 [m]), is a reticulated shell.

Although it is theoretically possible to model these relatively complex systems as ensembles of the various constituent components, it is often feasible, considerably simpler, and more economical to use surface structure equations. Naturally, the question arises as to the appropriate value for the "thickness" because, by definition, a shell or plate has a piecewise constant or smoothly varying thickness. It appears to be proper to consider this question at the level of the *stress-resultant-strain* relationships, matrix **[D]**, rather than at the *stress-strain* level, matrix **[C]**, since the latter matrix is based on the *stress-at-a-point* focus theory of the theory of elasticity—which obviously has little meaning when we must consider equivalent thicknesses.

A fairly common and quite illustrative example is that of a shell or plate with equi-spaced stiffening ribs running in one direction, as shown in Figure 8–1, or in both directions. A common method of treating this problem is to combine the properties of the basic shell or plate and the stiffener over a repeating interval of the cross-section, such as d_β, and then to compute an equivalent thickness in the direction of the stiffener. This process is often referred to by the colloquial title of *verschmieren,* after the German verb "to smear over." For a shell that behaves essentially in accordance with membrane theory, in other words, resisting transverse loading by in-plane forces, the equivalent thickness is logically computed with respect to the extensional rigidity, whereas for a plate subjected to bending, it would be based on the flexural rigidity.

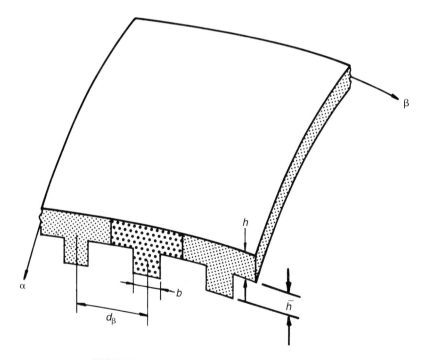

FIGURE 8–1   Shell or Plate with Stiffening Ribs.

Example 8.1(c)

To illustrate such a computation, consider the situation shown in Figure 8–1. For a shell, we compute the area of the repeating cross-section, $hd_\beta + \bar{h}b$, and divide by the spacing, d_β, to get

$$h_\alpha = h + \frac{b\bar{h}}{d_\beta} \tag{8.14}$$

In the other direction, we use $h_\beta = h$. If the stiffened structure is a plate acting primarily in bending, we would compute the moment of inertia of the repeating T section about its centroidal axis, $I_{h\bar{h}}$. Then, the equivalent thickness h_α is found from setting

$$\frac{1}{12} d_\beta h_\alpha^3 = I_{h\bar{h}} \tag{8.15a}$$

or

$$h_\alpha = \left(\frac{12\, I_{h\bar{h}}}{d_\beta} \right)^{1/3} \tag{8.15b}$$

The thickness in the β direction is again taken as $h_\beta = h$.

The equivalent thicknesses are then used to compute the corresponding terms of matrix [**D**]. Referring to the elements of Matrix 8–2, we would use h_α to compute the terms in the first and fourth rows, and h_β for the terms in the second and fifth rows. It has been suggested[5] that the terms corresponding to the shearing rigidity (rows three, seven, and eight) be evaluated by using the basic thickness or, in some cases, be neglected altogether.

This technique may be easily extended to cases in which the stiffeners run in both directions by computing appropriate h_β thicknesses from formulas analogous to equations (8.14) and (8.15). A form of a plate supplemented by beams running in two directions known as a *waffle slab* (Figure 2–8[w]) is quite widely used in reinforced concrete building construction.

This simple illustration of a *verschmieren* procedure raises several questions which we consider at this point. If only one equivalent thickness is defined in each direction, should it be the extensional or the flexural thickness for a shell that may have significant bending? We have opted for the extensional, but will this equivalent thickness adequately represent the flexural characteristics of the stiffened shell? Also, what about the shearing and twisting thicknesses? To include only the basic shell or plate, as mentioned in the preceding paragraph, seems to be a rather crude approximation. Finally, the use of different equivalent thicknesses in the α and β directions for the extensional and bending rigidities would cause matrix [**D**] to be unsymmetric because of the off-diagonal terms in rows one, two, four, and five. This leads to contradictions and computational difficulties that should be avoided.

From the preceding questions, it is obvious that a procedure that embodies more of the characteristics of the actual structure is desirable. This is especially true as the stiffeners become more prominent with respect to the basic cross-section, up to the case of reticulated structures, where they may constitute practically the entire means of resisting the primary forces and moments.

A promising approach to this problem is the *split rigidity* concept introduced by Buchert.[4] The method is a generalization of a development by E. Reissner, who has been one of the most prolific contributors to the theory of shells in the middle- and late-twentieth century.[6] Essentially, different equivalent thicknesses for the extensional, shearing, bending, and twisting terms may be specified. To avoid difficulties with nonsymmetry in [**D**], Poisson's ratio is neglected. This leads to a diagonal matrix $\lceil \mathbf{D_{eq}} \rfloor$, which replaces [**D**] in equation (8.10):

$$\lceil \mathbf{D_{eq}} \rfloor = E \left\lceil h_{\alpha(m)} h_{\beta(m)} \ \frac{h_{\alpha\beta(m)}}{2} \ \frac{h_{\alpha(b)}^3}{12} \ \frac{h_{\beta(b)}^3}{12} \ \frac{h_{\alpha\beta(b)}^3}{12} \ \frac{\lambda h_{\alpha(m)}}{2} \ \frac{\lambda h_{\beta(m)}}{2} \right\rfloor \qquad (8.16)$$

The additional subscripts (m) and (b) indicate membrane or extensional, and bending or flexural, respectively.

To illustrate the split-rigidity concept, we choose a reticulated shell composed of a gridwork of structural members, as shown in Figure 8–2. The members are assumed to have locally constant spacing, d_α and d_β; areas A_α and A_β; moments of inertia I_α and I_β; and St. Venant torsion constants K_α and K_β, in each direction. Any infilled material between the main members is assumed to be either restricted to local loading distribution, and therefore negligible in this analysis, or included in the section properties of the main members.

With these idealizations, we may immediately compute

$$h_{\alpha(m)} = \frac{A_\alpha}{d_\beta} \qquad (8.17)$$

and

$$h_{\beta(m)} = \frac{A_\beta}{d_\alpha} \qquad (8.18)$$

The equivalent in-plane shear thickness is a bit harder to visualize. A first approximation, considering only the primary members of the frame, would be to take the average of the equivalent extensional thicknesses, $1/2[h_{\alpha(m)} + h_{\beta(m)}]$, so that

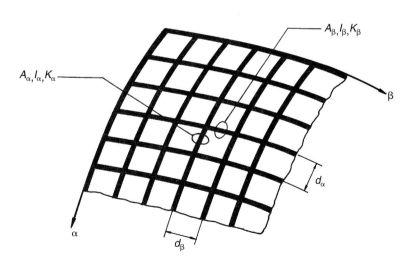

FIGURE 8–2   Reticulated Shell.

$$h_{\alpha\beta(m)} = \frac{1}{2}\left(\frac{A_\alpha}{d_\beta} + \frac{A_\beta}{d_\alpha}\right) \tag{8.19}$$

We note that there is the possibility of increasing the in-plane shear resistance substantially through the integration of the infilling material with the primary frame structure.

Now, we turn to the equivalent flexural thicknesses. Considering the α direction first, we equate the moments of inertia per unit width of the frame member and the equivalent shell

$$\frac{I_\alpha}{d_\beta} = \frac{1}{12} h_{\alpha(b)}^3$$

from which

$$h_{\alpha(b)} = \left(12\,\frac{I_\alpha}{d_\beta}\right)^{1/3} \tag{8.20}$$

Similarly, in the β direction

$$h_{\beta(b)} = \left(12\,\frac{I_\beta}{d_\alpha}\right)^{1/3} \tag{8.21}$$

Last, we turn to the twisting thickness. It is in this mode of resistance that a shell-like structure often differs most from a true shell. Whereas a true shell or plate has relatively great twisting rigidity as compared with the flexural rigidity, the reticulated structure is likely to have comparatively little twisting rigidity, especially if it is comprised of thin-walled members with open cross-section.

In his study of orthotropic plates, Huffington suggested a procedure based on the St. Venant torsional rigidity of the members.[7] This is, by no means, a general solution; however, it serves to illustrate a rational approach applicable to a specific class of problems. It is likely that the twisting rigidity will be of greater significance in plates which behave in a primarily flexural mode, but may have little influence on shell-like forms. Since we are following a general course applicable to shells as well, we take only the average extensional thickness for computing the twisting rigidity. Therefore,

$$h_{\alpha\beta(b)} = \frac{1}{2}\left(\frac{A_\alpha}{d_\beta} + \frac{A_\beta}{d_\alpha}\right) \tag{8.22}$$

Substituting equations (8.17) through (8.22) into equation (8.16), we get

$$\lceil\mathbf{D_{eq}}\rfloor = E\left\lceil\frac{A_\alpha}{d_\beta}\ \frac{A_\beta}{d_\alpha}\ \frac{1}{4}\left(\frac{A_\alpha}{d_\beta} + \frac{A_\beta}{d_\alpha}\right)\ \frac{I_\alpha}{d_\beta}\ \frac{I_\beta}{d_\alpha}\ \frac{1}{96}\left(\frac{A_\alpha}{d_\beta} + \frac{A_\beta}{d_\alpha}\right)^3\ \frac{\lambda A_\alpha}{2d_\beta}\ \frac{\lambda A_\beta}{2d_\alpha}\right\rfloor \tag{8.23}$$

The use of an equivalent stress resultant-strain matrix, such as the one defined here, enables a large variety of curved frames and gridworks to be efficiently analyzed by using solutions based on shell and plate theories, i.e., by using a *continuum* as opposed to a *discrete* approach. We must remember, however, that the stress resultants and couples computed from such solutions are intensities per unit length of middle surface and must subsequently be converted to forces and moments in the individual members. A logical procedure for accomplishing this conversion is to take the average value of the stress resultant or couple over a repeating integral, d_α or d_β, multiplied by the interval.

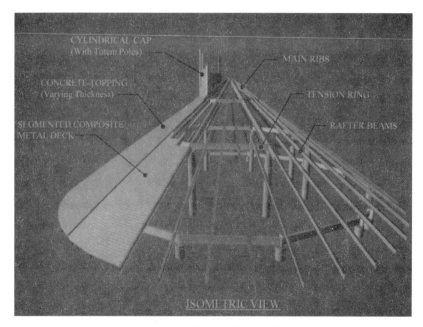

(a) Isometric View of Framing System

(b) Apex Construction

FIGURE 8–3b Roof of Parliament Building, Solomon Islands. (Courtesy Baldridge & Associates Structural Engineering, Inc.)

A modern example of a structure with the form of a continuous shell but which is supported by a discrete structural framework, termed a ribbed dome, is the Parliament Building for the Solomon Islands shown in Figure 2–8(z). The desired spans precluded forming the shell from conventional poured-in-place concrete, and post-tensioning was not feasible because of the locale. As shown in Figure 8–3, the cone shape is approximated by the segmented steel framing. Note that the rafter beams are elevated above the circumferential support beams to match the cone profile. The cone is completed by the variable thickness concrete topping on the composite metal deck to produce a curved surface in the circumferential direction. While the form of the roof was intended to reflect the native culture, the execution demonstrated modern technology adapted to a remote region.[8]

Another modern surface structure supported by a steel framework is the Convention Centre in Hong Kong, Figure 2–8(l), constructed for the historic "handover" to China in 1997. The combination of appealing curved surfaces supported by a framework of discrete members appears to be the contemporary form of signature space structures.

8.2   BOUNDARY CONDITIONS

8.2.1   Role of Boundary Conditions in the Formulation of Shell and Plate Theories

In the preceding section, we have presented the *constitutive* laws that serve to connect the *equilibrium* equations derived in Chapter 3 and the *compatibility* relations developed in Chapter 7. The classical formulation of the solid mechanics problem embodies these three components, which may be collected as a set of *partial differential equations* or as an *energy principle*. We explore both of these formats in depth in the succeeding chapters; however, there is one more ingredient necessary to complete the theory in either case: a suitable set of *boundary conditions*. The possibilities may be classified as follows: (a) *static* (a stress resultant or stress couple is specified); (b) *kinematic* (a displacement or rotation is specified); and (c) *elastic* (a linear combination of a static and a kinematic quantity is specified). This may be represented by a spring-type support.

In the following discussion, we place two restrictions on the boundary conditions to simplify the formulation somewhat: (a) all boundaries coincide with either an s_α or s_β coordinate line and, (b) all kinematic boundary conditions are homogeneous. Exceptions to the first restriction include such problems as skew intersections and cutouts and are best treated using discrete approximation techniques such as the finite element method. Nonhomogeneous boundary conditions are not necessarily more difficult to deal with than the homogeneous variety, but because they occur comparatively infrequently, it is usually expedient to exclude this possibility from the general formulation and to treat such situations that may arise as special cases.

8.2.2   Static and Kinematic Boundary Conditions

We consider two representative boundaries of a surface in Figure 8–4. The boundary coinciding with the s_α coordinate line is located by $\beta = \bar{\beta}$, and the boundary along the s_β coordinate line is designated by $\alpha = \bar{\alpha}$. On the boundary $\alpha = \bar{\alpha}$, there are five static quantities act-

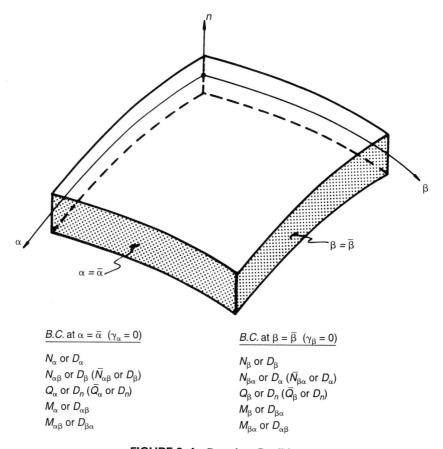

B.C. at $\alpha = \bar{\alpha}$ $(\gamma_\alpha = 0)$

N_α or D_α

$N_{\alpha\beta}$ or D_β $(\bar{N}_{\alpha\beta}$ or $D_\beta)$

Q_α or D_n $(\bar{Q}_\alpha$ or $D_n)$

M_α or $D_{\alpha\beta}$

$M_{\alpha\beta}$ or $D_{\beta\alpha}$

B.C. at $\beta = \bar{\beta}$ $(\gamma_\beta = 0)$

N_β or D_β

$N_{\beta\alpha}$ or D_α $(\bar{N}_{\beta\alpha}$ or $D_\alpha)$

Q_β or D_n $(\bar{Q}_\beta$ or $D_n)$

M_β or $D_{\beta\alpha}$

$M_{\beta\alpha}$ or $D_{\alpha\beta}$

FIGURE 8–4   Boundary Conditions.

ing: N_α, $N_{\alpha\beta}$, Q_α, M_α, and $M_{\alpha\beta}$. Also, five kinematic components are present: D_α, D_β, D_n, $D_{\alpha\beta}$, and $D_{\beta\alpha}$. Note that *each static quantity* performs work through *one and only one kinematic quantity* within the small deformation theory. This restriction is relaxed when instability and finite deformations are considered in Chapters 11 and 13. We group the associated quantities as *static-kinematic correspondents* as listed on Figure 8–3. For the boundary $\beta = \bar{\beta}$, the correspondents are also given on the figure. This pairing is quite important, for at any point on either boundary, only *one of each pair* of static-kinematic correspondents may be specified—except for an elastic boundary represented by a linear combination of the two. In no case, however, may *both* or *neither* of the correspondents be designated on a boundary. Further, when one correspondent is specified, the other must be accommodated by the boundary; for example, if a displacement such as D_α is set equal to 0, the boundary must resist whatever value of N_α is calculated from the subsequent solution; or if the stress resultant $N_\alpha = 0$, the boundary must be free to displace through the computed D_α. In the case of static boundary conditions, nonzero values are often prescribed as a known loading term, for example, in Figure 4–12, $N_z(0) = -N$.

TABLE 8–1   Boundary Conditions at $\alpha = \overline{\alpha}$

Condition Desired	Specific Quantities
Free	$N_\alpha = N_{\alpha\beta} = \underline{Q}_\alpha = M_\alpha = M_{\alpha\beta} = 0$ or
	$N_\alpha = N_{\alpha\beta} = \overline{Q}\alpha = M_\alpha = 0$
Fixed	$D_\alpha = D_\beta = D_n = D_{\alpha\beta} = D_{\beta\alpha} = 0$
Hinged	$M_\alpha = 0; D_\alpha = D_\beta = D_n = D_{\beta\alpha} = 0$
Roller ($\perp$ to middle surface)	$Q_\alpha = M_\alpha = 0; D_\alpha = D_\beta = D_{\beta\alpha} = 0$
Simply supported ($\parallel$ to middle surface)	$N_\alpha = M_\alpha = 0; D_\beta = D_n = D_{\beta\alpha} = 0$
Sliding ($\perp$ to middle surface)	$Q_\alpha = 0; D_\alpha = D_\beta = D_{\alpha\beta} = D_{\beta\alpha} = 0$

Before examining some specialized forms of the boundary conditions, we should mention one further requirement, the elimination of *rigid body* displacements. To accomplish this, it is sufficient that D_α, D_β, D_n, $D_{\alpha\beta}$, and $D_{\beta\alpha}$ each be restrained to vanish at one point on the surface. Seemingly, we would also need to restrict the drilling rotation (about the normal), even though this degree of freedom is not germane to the theory. There are also certain cases when the loading is self-equilibrated, such as those corresponding to the $j > 1$ harmonics in Section 5.3, for which the rigid body displacements do not have to be eliminated explicity.

As specific examples of easily visualized boundary conditions that are frequently encountered, we consider the boundary $\alpha = \overline{\alpha}$ (see Table 8–1).

Recognize that the mechanical terms used to describe the boundaries—e.g., hinged and roller—are essentially two-dimensional in implication and do not fully describe the constraint in the β direction. Consequently, the conditions on D_β, and/or $D_{\beta\alpha}$ may be interchanged with $N_{\alpha\beta}$, and/or $M_{\alpha\beta}$, respectively, if the physical situation seems appropriate. Also, the term *simply supported* is often imprecisely used to also describe the hinged or roller cases. Technically, this condition should be reserved for beam-like situations for which a constraint is not necessary or feasible, such as the barrel shells shown in Figure 2–8(e) which would be restrained longitudinally only by the weak-axis bending of the transverse supporting arch or diaphragm. On the other hand, the lower boundary of the cooling tower shell shown in Figures 2–8(o) and (p) are obviously free between supports, but the condition is more complicated within the support region as discussed in Section 5.3.3. A similar set of conditions can be derived for the boundary $\beta = \overline{\beta}$.

8.2.3   Kirchhoff Boundary Conditions

We have mentioned several times that much of the classical theory of shells and plates is based on the suppression of the transverse shearing strains. From the statement of the admissible boundary conditions presented in the preceding section, we can anticipate that the most general system of governing differential equations is of tenth order in each direction, permitting prescription of five conditions on each of the four boundaries. When transverse shearing strains are not included, the governing equations are reduced to eighth order in each direction, and a contracted set of boundary conditions is required.

The derivation of the static conditions consistent with the suppression of transverse shearing strains is one of the more interesting theoretical problems in the theories of shells and plates. These are commonly known as the *Kirchhoff boundary conditions* after the noted mathematician G. Kirchhoff, who first established the correct relationships for a plate using an energy formulation.[9] Although Kirchhoff's derivation served to clarify the situation somewhat, it was mathematical in basis, and it remained for two distinguished nineteenth-century scientists, Lord Kelvin and P. G. Tait,[10] to provide a physical interpretation of the Kirchhoff conditions.

To illustrate the Kelvin-Tait argument and to provide an extension into the shell formation, refer to the boundary $\alpha = \bar{\alpha}$ as shown on Figure 8–5(a). We identify four adjacent differential half-segments *ab, bo, oc* and *cd,* each of length $\Delta s_\beta/2$, and indicate the resultant forces acting on segment *boc* due to $N_{\alpha\beta}$ and Q_α. These resultants act at point *o*. Also, we show the resultant couples on two adjacent segments, *abo* and *ocd,* due to $M_{\alpha\beta}$. These act at points *b* and *c,* respectively. The directions shown correspond to the sign convention of Figure 3–2. Now, in Figure 8–5(b), we examine a normal section showing the middle surface of the differential segment and replace the moments caused by the stress couples by equivalent forces at *a, o,* and *d*. The replacement of a moment on the boundary by a pair of equal and opposite forces separated by a differential distance is justified by the St. Venant principle,[11] one of the many enduring contributions of this outstanding nineteenth-century scientist. From $M_{\alpha\beta}$, we have the forces $(M_{\alpha\beta}\,\Delta s_\beta)/\Delta s_\beta = M_{\alpha\beta}$ acting at *a* and *o* normal to the chord *ao*. From $M_{\alpha\beta} + M_{\alpha\beta,\beta}\Delta\beta$, we derive a similar pair of forces acting normal to chord *od*. Also shown by dashed vectors are the *effective* in-plane shear $\bar{N}_{\alpha\beta}$ and the *effective* transverse shear $\bar{Q}_\alpha$, which we wish to evaluate at point *o*.

Summing forces in the β direction, we have

$$\bar{N}_{\alpha\beta}\Delta s_\beta = N_{\alpha\beta}\Delta s_\beta + M_{\alpha\beta} \sin\left(\frac{\Delta s_\beta}{2R_\beta}\right) + (M_{\alpha\beta} + M_{\alpha\beta,\beta}\Delta\beta) \sin\left(\frac{\Delta s_\beta}{2R_\beta}\right)$$

Because $M_{\alpha\beta,\beta}\Delta\beta$ is of higher order than the remaining terms and $\sin(\Delta s_\beta/2R_\beta) \simeq \Delta s_\beta/2R_\beta$, the effective in-plane shear is

$$\bar{N}_{\alpha\beta} = N_{\alpha\beta} + \frac{M_{\alpha\beta}}{R_\beta} \tag{8.24}$$

Summing forces in the *n* direction, we have

$$\bar{Q}_\alpha\Delta s_\beta = Q_\alpha\Delta s_\beta - M_{\alpha\beta} \cos\left(\frac{\Delta s_\beta}{2R_\beta}\right) + (M_{\alpha\beta} + M_{\alpha\beta,\beta}\Delta\beta) \cos\left(\frac{\Delta s_\beta}{2R_\beta}\right)$$

Noting that $\cos(\Delta s_\beta/2R_\beta) \simeq 1$, and that $\Delta s_\beta = B\Delta\beta$, the effective transverse shear is

$$\bar{Q}_\alpha = Q_\alpha + \frac{M_{\alpha\beta,\beta}}{B} \tag{8.25}$$

Along the boundary $\beta = \bar{\beta}$, we can develop a similar set of effective shears

$$\bar{N}_{\beta\alpha} = N_{\beta\alpha} + \frac{M_{\beta\alpha}}{R_\alpha} \tag{8.26}$$

and

$$\bar{Q}_\beta = Q_\beta + \frac{M_{\beta\alpha,\alpha}}{A} \tag{8.27}$$

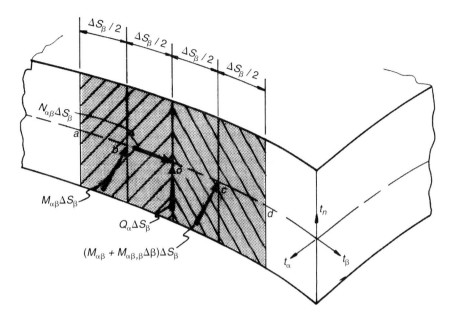

FIGURE 8–5a   Forces and Moments on a Shell Boundary.

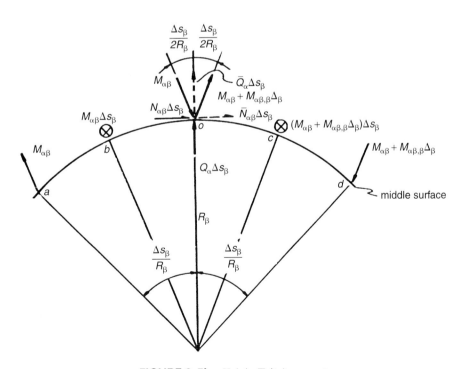

FIGURE 8–5b   Kelvin-Tait Argument.

Now, referring to Figure 8–4, we have the five static boundary conditions contracted to four along each boundary.

For the corresponding kinematic conditions, we first examine equations (7.46g) and (7.46h) with $\gamma_\alpha = \gamma_\beta = 0$. Then $D_{\alpha\beta}$ and $D_{\beta\alpha}$, corresponding to the bending couples, are equal to ψ_α and ψ_β, respectively. For the effective shear stresses, which incorporate the twisting couples, we have D_β and D_n on the $\alpha = \bar{\alpha}$ boundary, and D_α and D_n on the $\beta = \bar{\beta}$ boundary. These are also listed on Figure 8–4.

8.2.4 Boundary Conditions for Plates with No Transverse Shearing Strains

Although the general boundary conditions described in Section 8.2.2 are applicable for plates as well as shells, the Kirchhoff conditions are somewhat simplified for plates. An examination of equations (8.24) and (8.26) reveals R_β and R_α in the denominators of the twisting stress couple terms, $M_{\alpha\beta}$ and $M_{\beta\alpha}$. Because these radii $\to \infty$, the effective in-plane shear is superfluous, and we are left with only the effective transverse shears, as given by equations (8.25) and (8.27).

8.2.5 Boundary Conditions for Shells of Revolution

The shell of revolution geometry was specialized in Section 2.8.2, with the meridians corresponding to the s_α coordinate lines and the parallel circles to the s_β coordinate lines. For this shell form, we only have boundary conditions along the parallel circles. If the meridional coordinate is taken as ϕ, and the circumferential coordinate as θ, a parallel circle boundary corresponds to $\phi = \bar{\phi}$. The Kirchhoff conditions follow from equations (8.24) and (8.25) as

$$\bar{N}_{\phi\theta} = N_{\phi\theta} + \frac{M_{\phi\theta}}{R_\theta} \tag{8.28}$$

and

$$\bar{Q}_\phi = Q_\phi + \frac{M_{\phi\theta,\theta}}{R_0} \tag{8.29}$$

8.2.6 Symmetry

In addition to external boundary conditions such as those we have discussed in the preceding paragraphs, there may be equations of condition that arise because of symmetry properties of both the geometry and the loading. The practical manifestation of the recognition of symmetry conditions is that only a portion of the entire continuum must be analyzed.

A fairly general modern treatment of the topic of symmetry was presented by Glockner,[12] and we focus here on the applications to shells and plates. Of course, we have already used the concept of symmetry extensively in our treatment of shells of revolution as a distinct class of surface structures. Such shells are distinguished *geometrically* by their rotational symmetry. Further, in the analysis by the Fourier series technique in Chapter 5, symmetric and antisymmetric components of the *loading* were identified. Formally, we may state that a loading function $q(x)$ is symmetric about $x = 0$ if $q(-x) = q(x)$, and antisymmetric if $q(-x) = -q(x)$. Furthermore, higher harmonic components, which are self-equilibrated, are said to possess *cyclic* symmetry. This was illustrated by the column-supported shell solution in Section 5.3.3.

Example 8.2(a)

As a first illustration, consider a complete rotational shell, such as a sphere or an ellipsoid, subject to a uniform normal pressure $q_n = q$. Using the notation of Figures 4–2 and 8–4, the following symmetry conditions may be enforced at $\phi = \pi/2$:

$$D_\phi = 0; \qquad Q_\phi = 0$$

These conditions require that only half of the shell be represented in the ensuing analysis. This may not be important in a membrane theory analysis which is statically determinate—for example, equations (4.27) and (4.28)—but may result in considerable savings when an analysis using the general theory is performed.

In some numerical solutions, it also may be necessary to enforce such symmetry conditions to eliminate rigid body displacements.

Example 8.2(b)

Another case of interest is a shell of translation, such as the elliptic paraboloid shown in Figure 6–15, subject to a uniform vertical loading $q_x = q$. We may consider only the quadrant *OEDF* and, using the notation of Figure 6–4, take $\alpha = x$ and $\beta = y$ on Figure 8–4. Then, we have

$$D_x = 0; \qquad N_{xy} = 0; \qquad Q_x = 0; \qquad D_{xy} = 0; \qquad M_{xy} = 0$$

along *OF* and

$$D_y = 0; \qquad N_{yx} = 0; \qquad Q_y = 0; \qquad D_{yx} = 0; \qquad M_{yx} = 0$$

along *OE*. It is possible to affect further savings by defining symmetry along the diagonal *OD*, but because this involves displacements along nonprincipal lines of curvature, it lies outside our present scope. The conditions for the shell of translation are also applicable to uniformly loaded, symmetrically supported plates.

Example 8.2(c)

As a final example, we choose the same elliptic paraboloid illustrated in Figure 6–15 subjected to a hydrostatic loading acting normal to the *X–Y* plane. The distribution in the *Y* direction is given by

$$q_x(X,Y) = q \frac{(b - Y)}{2}$$

and the loading is assumed to be uniform across the *X* direction. This loading is readily decomposed into a constant component ($j = 0$)

$$q_z^0(X,Y) = q\frac{b}{2}$$

and a linear component ($j = 1$)

$$q_z^1(X,Y) = -q\frac{Y}{2}$$

The constant component q_z^0 is symmetrical with respect to both the X and Y axes, and is treated identically to the previous example. The linear component q_z^1 remains symmetric with respect to the Y axis, so the conditions along OF are unaltered. However, this component is antisymmetric with respect to the X axis, since $q_z^1 (X, -Y) = -q_z^1(X, Y)$. The resulting conditions along OE are

$$N_y = 0; \quad D_x = 0; \quad D_n = 0; \quad M_y = 0; \quad D_{xy} = 0$$

which are the correspondents of the conditions on a line of loading symmetry in each case. Again, the antisymmetry conditions are applicable to plates as well.

8.3 MEMBRANE THEORY DISPLACEMENTS

8.3.1 General Aspects

In the preceding sections of this chapter, we have completed the derivation of the individual components required to formulate the general theories of shells and plates. In Chapters 4 to 6, we investigated a very important class of problems, the membrane theory of shells, which requires only the solution of a reduced set of equilibrium equations to determine the in-plane stress resultants. The membrane theory is generally statically determinate as opposed to the general theory, which is statically indeterminate and therefore requires simultaneous consideration of equilibrium, compatibility, and constitutive relations. Nevertheless, we cannot say that the membrane solution is complete until the displacements are determined. As is the usual procedure for statically determinate problems, we may compute these displacements from the equilibrium solution, assuming that the stress field is known.

At the outset, it is instructive to consider the relevance of the membrane theory displacements in shell analysis and design. First, because the design of many shells is based on the membrane theory, the magnitudes of the associated displacements would be useful to verify the applicability of small displacement theory (assumption [2], Table 1–1). Also, knowledge of the expected displacements can be important in providing a supporting system to accommodate the required boundary deformations, as mentioned in the previous section. Further, it is possible to obtain values for the bending stress resultants and couples by substituting the membrane theory displacements into the general strain-displacement and constitutive equations, equations (7.46) or (7.47), and equation (8.10), respectively. At best, the values obtained in this manner may be useful for order-of-magnitude estimation in the regions of the shell dominated by membrane action, but do not account for boundary effects.

Next, even if a solution is obtained using a bending theory, the values of the *in-plane* stress resultants are often not changed significantly from those computed from a membrane theory analysis except, perhaps, in localized regions. Because certain of the stress resultants, such as N_ϕ, in symmetrically and antisymmetrically loading shells of revolution, are dependent on the *overall* equilibrium of the shell, they cannot be altered very much by bending action. This means that the middle-surface displacements, which are primarily functions of the in-plane stress resultants, may be estimated quite closely from the membrane theory solution.

Additionally, we may anticipate that a logical procedure to analyze shells which do not fully meet the requirements of the membrane theory, because of inappropriate boundary conditions, would be to use the flexibility method. An example would be a spherical shell, such as that illustrated in Figure 4–2, with the lower boundary pinned or fixed, as defined in Section 8.2. Briefly, the flexibility method consists of introducing a sufficient number of kinematic releases into the system to make it statically determinate. Then, the conditions of deformation compatibility consistent with the original system constraints are enforced on superposed primary and secondary systems to determine the static correspondents of the kinematic releases, known as the redundants. This procedure is familiar to students of structural mechanics and is described in most standard texts on the analysis of indeterminate structures, so we need not elaborate here. Briefly, we note that the compatibility equations take the form

$$\{\Delta\} = \{\overline{\Delta}\} + [\mathbf{F}]\{\chi\} \tag{8.30}$$

where

   $\{\Delta\}$ = the vector of the displacements corresponding to each of the releases on the original structure. Commonly, $\{\Delta\}$ is null.

   $\{\overline{\Delta}\}$ = the vector of the displacements of each of the releases caused by the applied loading. These displacements are computed on the released (now statically determinate) structure.

   $\{\chi\}$ = the vector of the statical correspondents of the kinematic releases, the redundants, initially unknown. For plates and shells, the elements of $\{\chi\}$ are usually transverse shear forces and bending moments on the boundary.

   $[\mathbf{F}]$ = the matrix of flexibility influence coefficients, each element of which is a displacement of a release resulting from a unit value of one of the elements of $\{\chi\}$. The influence coefficients are computed on the released (statically determinate) structure.

Once $\{\overline{\Delta}\}$ and $[\mathbf{F}]$ are computed and $\{\Delta\}$ is specified, we solve for

$$\{\chi\} = [\mathbf{F}^{-1}][\{\Delta\} - \{\overline{\Delta}\}] \tag{8.31}$$

For shells, $\{\overline{\Delta}\}$ can usually be evaluated with sufficient accuracy by using the membrane theory, as discussed in the previous paragraph. The elements of $[\mathbf{F}]$, being displacements caused by boundary transverse shear forces and bending moments, must still be computed by using the bending theory; however, because no surface loads are involved, only the *homogeneous* bending equations need be considered, which frequently simplifies the solution. A specific illustration of this method is presented in Chapter 12 for the bending analysis of cylindrical shells.

Thus we have suggested a variety of reasons for studying the middle surface displacements of shells in equilibrium under a known internal stress field computed by the membrane theory.

8.3.2 Governing Equations

To connect the middle surface displacements D_α, D_β, and D_n with the stress resultants N_α, N_β, and $N_{\alpha\beta}$, compare equations (7.46a) to (7.46c) with the first three rows of Matrix 8–2. Because Matrix 8–2 gives the stress resultants as functions or strain, we must invert the relations

$$N_\alpha = D_1\varepsilon_\alpha + \mu D_1\varepsilon_\beta - N_{\alpha T} \tag{8.32a}$$

$$N_\beta = \mu D_1\varepsilon_\alpha + D_1\varepsilon_\beta - N_{\beta T} \tag{8.32b}$$

$$S = D_2\omega \tag{8.32c}$$

recalling that $N_{\alpha\beta}$ is replaced by S in the membrane theory. From Matrix 8–2, $D_1 = Eh/(1 - \mu^2)$ and $D_2 = Eh/[2(1 + \mu)]$. Solving for the strains in view of equations (8.1a), (8.1b), and (8.9a), we have

$$\varepsilon_\alpha = \frac{1}{Eh}\left[(N_\alpha - \mu N_\beta) + (1 - \mu)N_T\right] \tag{8.33a}$$

$$\varepsilon_\beta = \frac{1}{Eh}\left[(N_\beta - \mu N_\alpha) + (1 - \mu)N_T\right] \tag{8.33b}$$

$$\omega = \frac{2(1 + \mu)}{Eh}S \tag{8.33c}$$

By equating the rhs of equations (7.46a–c) and equations (8.33a–c), we arrive at the complete set of equations relating the displacements and the stress resultants. With these stress resultants having been determined by using the methods of Chapters 4 to 6, we have three partial differential equations in the three unknown middle surface displacements, D_α, D_β, and D_n.

We may also obtain expressions for the rotations $D_{\alpha\beta}$ and $D_{\beta\alpha}$ in terms of the middle surface displacements from equations (7.46g) and (7.46h), if we neglect the transverse shearing strains. This is consistent with the membrane theory where Q_α and Q_β are assumed to be zero.

Many forms used for shell structures facilitate simplifications in the functional dependence of the geometrical parameters. Therefore, it is expedient to specialize the equations for specific geometrical types.

8.3.3 Shells of Revolution

8.3.3.1 Transformation of Governing Equations.   Since we have already specialized the strain-displacement relations for shells of revolution with $\alpha = \phi$ and $\beta = \theta$ in Section 7.4, we have from equations (7.48a–c) and (8.33a–c)

$$R_\phi\varepsilon_\phi = D_{\phi,\phi} + D_n = \frac{R_\phi}{Eh}\left[(N_\phi - \mu N_\theta) + (1 - \mu)N_T\right] \tag{8.34a}$$

$$R_0\varepsilon_\theta = D_{\theta,\theta} + \cos\phi\, D_\phi + \sin\phi\, D_n = \frac{R_0}{Eh}\left[(N_\theta - \mu N_\phi) + (1 - \mu)N_T\right] \tag{8.34b}$$

$$R_0\omega = \frac{R_0}{R_\phi}D_{\theta,\phi} + D_{\phi,\theta} - \cos\phi\, D_\theta = \frac{2(1 + \mu)R_0}{Eh}S \tag{8.34c}$$

Because the stress resultants and thermal loads are both presumed to be known at this stage, we drop the thermal terms, which can simply be absorbed into N_ϕ and N_θ if present.

Now, we return to the technique of Section 4.3.1, whereby a judiciously chosen set of auxiliary variables was used to reduce the set of equations. Again, we follow Novozhilov[13] by introducing

$$\bar{\psi} = \frac{D_\theta}{R_0} \tag{8.35a}$$

$$\bar{\xi} = \frac{D_\phi}{\sin \phi} \tag{8.35b}$$

Next, we substitute equations (8.35a) and (8.35b) into equations (8.34a–c) and eliminate D_n between the first two equations, giving

$$\bar{\xi},_\phi - \frac{R_\theta}{\sin \phi}\, \bar{\psi},_\theta = \frac{1}{Eh \sin \phi}\left[(R_\phi + \mu R_\theta)N_\phi - (R_\theta + \mu R_\phi)N_\theta\right] \tag{8.36a}$$

$$\frac{R_\theta^2 \sin \phi}{R_\phi}\, \bar{\psi},_\phi + \bar{\xi},_\theta = \frac{2(1 + \mu)}{Eh}\, R_\theta S \tag{8.36b}$$

These equations may be compared with the transformed equations, (4.6b) and (4.6a), respectively. We see that the lhs of both would be identical if we were to interchange ψ and $\bar{\psi}$, and ξ and $\bar{\xi}$; whereas, the rhs, in both cases, consist entirely of known functions. Therefore, whatever solution strategies were deduced for the membrane theory equilibrium equations are applicable for the equations governing the corresponding displacements. Also, we recognize that once $\bar{\xi}$ and $\bar{\psi}$ are determined, D_ϕ and D_θ follow from equation (8.35a), and (8.35b) and D_n may be calculated from equation (8.34a) or (8.34b).

As a practical matter, when we attempt to use the various solutions already derived in Chapters 4 through 6 for determination of the corresponding displacements, we soon realize that the rhs of equations (8.34a–c) are likely to be more complicated algebraically than the corresponding functions in the equilibrium equations. This is because the expressions for the stress resultants are already quite involved for some of the cases that we solved in Chapters 4–6. After being multiplied by the radii of curvature expressions on the rhs, integrals that can only be evaluated numerically are often produced. This slight complication notwithstanding, the similarity of the system of governing equations for the membrane theory stress resultants in terms of the known applied loading, on the one hand, and the in-plane displacements and the computed stress resultants on the other, is quite striking. This serves as another example of the static-geometric analogy mentioned in Section 7.4.2.

Once we have evaluated the in-plane displacements, we may compute the rotations from equations (7.48g) and (7.48h) with $\gamma_\phi = \gamma_\theta = 0$:

$$D_{\phi\theta} = -\frac{1}{R_\phi}\,(D_{n,\phi} - D_\phi) \tag{8.37a}$$

$$D_{\theta\phi} = -\frac{1}{R_0}\,(D_{n,\theta} - \sin \phi\, D_\theta) \tag{8.37b}$$

We now find it convenient to again consider the axisymmetric ($j = 0$) and nonsymmetric ($j > 0$) cases separately.

8.3.3.2 Axisymmetric Displacements.

We drop the θ-dependent terms from equations (8.36) and (8.37) and integrate the first two of the set. Following Section 4.3.2.1 and using equation (8.34b), we find

$$\bar{\xi} = \frac{D_\phi}{\sin \phi} \tag{8.38a}$$

$$= \frac{1}{E} \int \frac{1}{h \sin \phi} \left[(R_\phi + \mu R_\theta)N_\phi - (R_\theta + \mu R_\phi)N_\theta \right] d\phi \tag{8.38a}$$

$$\bar{\psi} = \frac{D_\theta}{R_0} = 2 \frac{(1 + \mu)}{E} \int \frac{R_\phi}{hR_0} S \, d\phi \tag{8.38b}$$

$$D_n = \frac{R_\theta}{Eh} (N_\theta - \mu N_\phi) - \cot \phi \, D_\phi \tag{8.38c}$$

$$D_{\phi\theta} = -\frac{1}{R_\phi} (D_{n,\phi} - D_\phi) \tag{8.38d}$$

$$D_{\theta\phi} = \frac{D_\theta}{R_\theta} \tag{8.38e}$$

The indefinite integrals in equations (8.38a) and (8.38b) may be written in the alternate form

$$\bar{\xi} = \frac{D_\phi}{\sin \phi} = \bar{\xi}(\phi''') + \frac{1}{E} \int_{\phi'''}^{\phi} \frac{1}{h \sin \phi} \left[(R_\phi + \mu R_\theta)N_\phi - (R_\theta + \mu R_\phi)N_\theta \right] d\phi \tag{8.39a}$$

$$\bar{\psi} = \frac{D_\theta}{R_0} = \bar{\psi}(\phi''') + \frac{2(1 + \mu)}{E} \int_{\phi'''}^{\phi} \frac{R_\phi}{hR_0} S \, d\phi \tag{8.39b}$$

where ϕ''' and ϕ'''' are the boundaries at which D_ϕ and D_θ are specified. We have used ϕ''' and ϕ'''' to emphasize that these boundaries are *not* the same boundaries where the corresponding stress resultants are specified, ϕ' and ϕ'', as identified in equations (4.10) and (4.11). Recalling the arguments of Section 8.2.2, we cannot specify N_ϕ and D_ϕ, nor S and D_θ on the same boundary. Moreover, because we have presumably specified the boundary values $N_\phi(\phi')$ and $S(\phi'')$ in the equilibrium solution, D_ϕ and D_θ must be imposed at the other boundary in each case. Thus, we have little latitude in the choice for ϕ''' and ϕ''''.

When R_θ is not finite, as for $\phi = 0$ on a toroidal shell, equation (8.38c) is not applicable, but an alternate equation for D_n can be found from equation (8.34a).

For a dome, we again encounter indeterminate forms for D_ϕ and D_θ at the pole. It is easily shown using L'Hospital's rule that $D_\phi(0) = D_\theta(0) = 0$. Then, equations (8.38a–c) give

$$D_n(0) = \frac{R_\theta(0)}{Eh} [N_\theta(0) - \mu N_\phi(0)] \tag{8.40}$$

Example 8.3(a)

As an example, we investigate a spherical dome under dead load, using the solution for stresses derived in Section 4.3.2.2 and referring to Figure 4–2 with $\phi_t = 0$. We wish to compute the displacement normal to the middle surface, D_n, at the lower boundary $\phi = \phi_\beta$.

Examining equation (8.38c) we observe that since $D_\phi(\phi_b) = 0$ as dictated by the requirement to develop $N_\phi(\phi_\beta)$ fully, we need only the values of N_ϕ and N_θ at $\phi = \phi_b$. From equations (4.20) and (4.21), we have

$$N_\phi(\phi_b) = \frac{-qa}{1 + \cos\phi_b} \tag{8.41}$$

and

$$N_\theta(\phi_b) = qa\left(\frac{1}{1 + \cos\phi_b} - \cos\phi_b\right) \tag{8.42}$$

We substitute these values into equation (8.38c) with $R_\theta = a$, to get

$$D_n(\phi_b) = \frac{qa^2}{Eh}\left(\frac{1 + \mu}{1 + \cos\phi_b} - \cos\phi_b\right) \tag{8.43}$$

At the pole, the normal displacement is found using equations (4.22) and (8.40):

$$D_n(0) = \frac{-qa^2}{2Eh}(1 - \mu) \tag{8.44}$$

Example 8.3(b)

As a further example, we consider the displacements of a hyperboloidal shell under self-weight load, previously analyzed in Section 4.3.2.3. Referring to equations (8.38a) and (8.38b), because $S = 0$, then $\overline{\psi}$ is 0. Next, $\overline{\xi}$ is evaluated from equation (8.39a), with equations (4.40) and (4.41) substituted for N_ϕ and N_θ and ϕ''' taken as ϕ_b. The resulting integral is quite complicated, but it is easily evaluated using a numerical algorithm such as the trapezoidal method. Finally, D_n is found from equation (8.38c).

A study of the displacements for a parametric range of hyperboloidal shell dimensions, similar to that shown on Figure 4–8 for the stress resultants, is given in Figure 8–6. The nondimensionalizing parameters are indicated on the figure. Also a complete tabulation is available in Gould.[14]

8.3.3.3 Nonsymmetric Loading. To complete the analogous treatment of the equations of equilibrium and compatibility, we apply to equations (8.36b) and (8.36a), respectively, the reduction initially used to obtain equation (5.1) from equations (4.6a) and (4.6b) in Section 5.1.1. The result is

$$\frac{1}{R_\phi R_\theta \sin\phi}\left[\frac{R_\theta^2 \sin\phi}{R_\phi}\overline{\psi}_{,\phi}\right]_{,\phi} + \frac{1}{R_\phi \sin^2\phi}\overline{\psi}_{,\theta\theta} = \frac{1}{Eh\,R_\phi R_\theta \sin\phi}$$

$$\times \left\{2(1 + \mu)(R_\theta S)_{,\phi} + \left(\frac{1}{\sin\phi}[(R_\phi + \mu R_\theta)N_\phi - (R_\theta + \mu R_\phi)N_\theta]\right)_{,\theta}\right\} \tag{8.45}$$

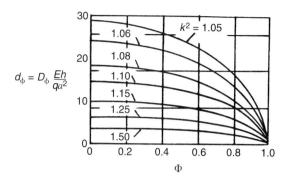

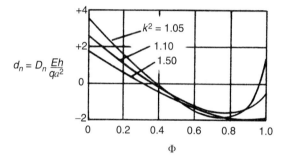

FIGURE 8–6   Nondimensional Self-Load Displacements for a Hyperboloidal Shell with $a/t = 0.90$ and $a/s = 0.55$. (Source: P. L. Gould and S. H. Lee, Hyperbolic cooling towers under seismic design load, *Journal of the Structural Division, ASCE 93,* no. ST3 (June 1967): 95.)

which is in the same form as equation (5.1).

All strains and displacements have previously been expressed in harmonic form by equations (7.50) and (7.51), and the corresponding strain-displacement equations have been written as equations (7.52). With the stress resultants having been expanded in Fourier series in equation (5.2b), equations (8.34a–c) are easily written in separated form. Further, we take the variables $\bar{\psi}$ and $\bar{\xi}$ as

$$\left\{\begin{matrix} \bar{\psi} \\ \bar{\xi} \end{matrix}\right\} = \sum_{j=0}^{\infty} \left\{\begin{matrix} \bar{\psi}^j \cos j\theta \\ \bar{\xi}^j \sin j\theta \end{matrix}\right\} \tag{8.46}$$

and the separated form of equation (8.45), analogous to equation (5.4), is

$$\frac{1}{R_\phi R_\theta \sin \phi} \left[\frac{R_\theta^2 \sin \phi}{R_\phi} \bar{\psi}^j_{,\phi} \right]_{,\phi} - \frac{j^2}{R_\phi \sin^2 \phi} \bar{\psi}^j = \frac{1}{Eh \, R_\phi R_\theta \sin \phi}$$

$$\times \left\{ 2(1 + \mu)(R_\theta S^j)_{,\phi} - j \left(\frac{1}{\sin \phi} [(R_\phi + \mu R_\theta) N^j_\phi - (R_\theta + \mu R_\phi) N^j_\theta] \right) \right\} \tag{8.47}$$

The final major equation for the displacements corresponds to the antisymmetrical case, $j = 1$, earlier discussed in Section 5.2. Following the same steps used in Section 5.2.1, we introduce the variable

$$\overline{Y}(\phi) = \overline{\psi}^1(\phi)R_\theta \sin \phi \tag{8.48}$$

into equation (8.47) with j taken as 1. The equation reduces to

$$\left(\frac{1}{R_\phi \sin \phi} \overline{Y}_{,\phi}\right)_{,\phi} = \frac{1}{Eh\, R_\theta \sin \phi}$$

$$\times \left\{ 2(1 + \mu)(R_\theta S^1)_{,\phi} - \left(\frac{1}{\sin \phi}\left[(R_\phi + \mu R_\theta)N^1_\phi - (R_\theta + \mu R_\phi)N^1_\theta\right]\right)\right\} \tag{8.49}$$

which may be solved as indicated in Section 5.2.

We now turn to the asymmetric case, $j > 1$, which is described by equation (8.47). For spherical and cylindrical shells, it is possible to arrive at solutions for the asymmetric displacements that are analogous to the stress solutions given in Section 5.3.2 and Section 5.3.4, respectively.[15] For more complicated geometries, the transformation applied in Section 5.3.6 may be of some value; however, little work has apparently been done in this area in terms of analytical solutions. Rather, for the most part, numerical techniques are used to evaluate displacements for asymmetrically loaded shells of revolution.

Now we have completed in brief form the displacement analysis for shells of revolution. Once equation (8.47) is solved for $\overline{\psi}^j$ to determine D^j_θ, ξ^j is computed from equation (8.36b) to get D^j_θ and then D^j_n is found from equation (8.34b).

It does not seem particularly useful to evaluate the membrane displacements for specific shell geometries in the same detail that is devoted to stress analysis, because the remaining treatment is self-evident on the basis of the analogy established. However, it is of general interest to summarize some results of various studies that can be consulted by the interested reader.

First, for closed shells as shown in Figure 4–1, certain relationships must exist among the displacements at the pole to avoid singularities in the solution. These have been derived in Brombolich and Gould[16] and are summarized in Table 8–2. The derivation of these relationships is discussed further in Chapter 13.

Next, because hyperboloidal shells under seismic loading were considered in much detail, note that the nondimensionalized displacements for the M distribution are tabulated in Gould and Lee.[17]

TABLE 8–2 Displacement Constraints for Closed Shells

Harmonic Number j	Continuous Dome @ $\phi = 0$ (Figure 4–1[a])	Discontinuous Meridian @ $\phi = \phi_t$ (Figure 4–1[b])
0	$D_\phi = D_\theta = D_{\phi\theta} = D_{\theta\phi} = 0$	$D_\theta = D_{\phi\theta} = D_{\theta\phi} = 0;$ $D_\theta = -\tan \phi_t D_n$
1	$D_n = 0$ $D_\phi + D_\theta = 0; D_{\phi\theta} + D_{\theta\phi} = 0$	$D_{\phi\theta} = D_{\theta\phi} = 0; D_\phi = -\cos \phi_t D_\theta;$ $D_n = -\sin \phi_t D_\theta$
>1	$D_\phi = D_\theta = D_n = D_{\phi\theta} = D_{\theta\phi} = 0$	

Last, we examine the general influences of the various harmonic components of the radial displacements for $j = 0$, $j = 1$, and $j > 1$ on a rotational shell of the tower type. It is convenient to do this with respect to the R–Z Cartesian coordinates system shown on Figure 2–11.

For the $j = 0$ case, the displacement will be principally an elongation or shortening of the shell in the Z (axial) direction, with any R (radial) deformation being a uniform expansion or contraction of the parallel circles, labeled as D^0 on Figure 8–7(a).

For the $j = 1$ case, the displacement will consist of a uniform lateral movement in the R direction, illustrated in Figure 8–7(b), along with flexural-like elongation or shortening

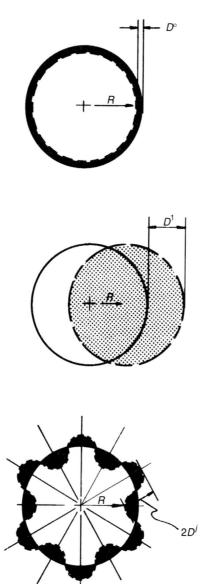

FIGURE 8–7 Harmonic Components of the Displacement of a Shell of Revolution.

across the cross-section in the Z direction. The maximum lateral displacement D^1 must be checked to verify the applicability of the small displacement theory (assumption [2], Table 1–1).

As discussed in Chapters 4 and 5, both the $j = 0$ and $j = 1$ cases may be solved from the elementary theories of axial deformation and beam bending. Consequently, there is no distortion of the cross-section. However, for the $j > 1$ case, the displacements will vary around the circumference in proportion to $\sin j\theta$ and $\cos j\theta$, and the cross-section will distort as shown in Figure 8–7(c). In problems with significant higher harmonic participation, the limits of the small deformation theory are frequently taxed by the circumferential distortions, where the *relative* displacements represented by the difference in the peak amplitudes should be checked. This difference is labeled as $2D^j$ on the figure. Also, significant distortions in the higher harmonics would necessarily be accompanied by circumferential bending, which would weaken the basis of a membrane theory solution.

For an actual loading that may consist of contributions from more than one harmonic, the maximum relative circumferential displacement, after summing all participating harmonics, should be used for the check.

8.3.4 Shells of Translation

8.3.4.1 Cylindrical Shells.   We consider the general strain-displacement relationship, equations (7.46a–c); and the strain-stress resultant relationships, equations (8.33a–c), with the curvilinear coordinates adopted for the cylindrical geometry in Section 6.2. Thus, we take $\alpha = X$, $\beta = \theta, A = 1, \beta = a;$ and $R_\alpha = \infty$, $R_\beta = a$ to get

$$\varepsilon_X = D_{X,X} = \frac{1}{Eh}\left[(N_X - \mu N_\theta) + (1 - \mu)N_T\right] \tag{8.50a}$$

$$a\varepsilon_\theta = D_{\theta,\theta} + D_n = \frac{a}{Eh}\left[(N_\theta - \mu N_X) + (1 - \mu)N_T\right] \tag{8.50b}$$

$$a\omega = aD_{\theta,X} + D_{X,\theta} = \frac{2(1 + \mu)a}{Eh}\,S \tag{8.50c}$$

As explained in Section 8.3.3.1, the thermal terms are conveniently absorbed into the stress resultants and are not carried forth explicitly.

We first integrate equation (8.50a) to get

$$D_X = \frac{1}{Eh}\int(N_X - \mu N_\theta)\,ds + f_3(\theta) \tag{8.51}$$

From equation (8.50c) we write

$$D_\theta = -\frac{1}{a}\int D_{X,\theta}\,dX + \frac{2(1 + \mu)}{Eh}\int S\,dX + f_4(\theta) \tag{8.52}$$

and then from equation (8.50b), we find

$$D_n = \frac{a}{Eh}(N_\theta - \mu N_X) - D_{\theta,\theta} \tag{8.53}$$

The membrane theory rotations are found from equations (7.46g) and (7.46h), with $\gamma_\alpha = \gamma_\beta = 0$:

$$D_{X\theta} = -D_{n,X} \tag{8.54}$$

and

$$D_{\theta X} = -\frac{1}{a}(D_{n,\theta} - D_\theta) \tag{8.55}$$

A ready example is the simply supported cylindrical shell subject to the first harmonic of the Fourier series expansion for the dead load. The stress analysis for this loading was carried out in Section 6.2, and the stress resultants are given by equations (6.10). The boundary conditions must be stated with respect to the displacements. The longitudinal symmetry dictates that $D_X(L/2) = 0$. Because the boundary corresponds to the *simply supported* condition discussed in Section 8.2.2, we also have $D_\theta(0) = 0$. These boundary conditions yield $f_3(\theta) = f_4(\theta) = 0$.

Proceeding, we find by substituting equations (6.10) into equations (8.51) to (8.55) and carrying out the integrations and substitutions,

$$D_X = \frac{4q_d}{\pi Eh}\left(\frac{L}{\pi}\right)\left[\frac{2L^2}{\pi^2 a} - \mu a\right]\cos\frac{\pi X}{L}\cos\theta \tag{8.56a}$$

$$D_\theta = \frac{4q_d}{\pi Eh}\left(\frac{L^2}{\pi^2}\right)\left[\frac{2L^2}{\pi^2 a^2} + 4 + 3\mu\right]\sin\frac{\pi X}{L}\sin\theta \tag{8.56b}$$

$$D_n = -\frac{4q_d}{\pi Eh}\left[(4 + \mu)\frac{L^2}{\pi^2} + \frac{2L^4}{\pi^4 a^2} + a^2\right]\sin\frac{\pi X}{L}\cos\theta \tag{8.56c}$$

$$D_{X\theta} = \frac{4q_d}{\pi Eh}\left(\frac{\pi}{L}\right)\left[(4 + \mu)\frac{L^2}{\pi^2} + \frac{2L^4}{\pi^4 a^2} + a^2\right]\cos\frac{\pi X}{L}\cos\theta \tag{8.56d}$$

$$D_{\theta X} = -\frac{4q_d}{\pi Eh}\left[a - \frac{2\mu L^2}{\pi^2 a}\right]\sin\frac{\pi X}{L}\sin\theta \tag{8.56e}$$

The preceding expressions are useful in the bending analysis of open cylindrical shells, when the membrane theory solution serves as the particular solution. The boundary displacements, as computed from equations (8.56a–c), are corrected by edge forces and moments to satisfy prescribed compatibility conditions on the longitudinal boundaries.[18] The procedure is quite analogous to the classical flexibility method of structural analysis, discussed in Section 8.3.1, and is explored in more detail in Chapter 12.

At this point, we consider an example that represents an exception to the general association of membrane theory analysis with statically determinate systems. Occasionally, boundary conditions are encountered which do not grossly violate the requirements discussed in Section 4.2, and yet do not permit the *a priori* determination of the stress resultants before considering the displacements. We might call this a *statically indeterminate membrane theory* problem, which was mentioned in Section 6.2.

As an illustration, we reconsider the previous example in Section 6.2 with an axial constraint condition

$$D_X(0) = D_X(L) = 0 \tag{8.57a}$$

replacing the condition

$$N_X(0) = N_X(L) = 0 \tag{8.57b}$$

The preceding static condition enabled $f_1(\theta)$ and $f_2(\theta)$ in equations (6.9a) and (6.9b) to be set equal to zero and, subsequently, the explicit expressions for S and N_X to be written as equations (6.10a) and (6.10b). We may retain the other static boundary condition $S(L/2) = 0$, because the symmetry is not altered; thus, $f_1(\theta)$ is still zero. Proceeding, the expressions for the stress resultants in the modified problem are written from equations (6.9) as

$$S = \frac{8q_dL}{\pi^2} \cos \frac{\pi X}{L} \sin \theta \tag{8.58a}$$

$$N_X = -\frac{8q_dL^2}{\pi^3 a} \sin \frac{\pi X}{L} \cos \theta + f_2(\theta) \tag{8.58b}$$

$$N_\theta = -\frac{4q_d a}{\pi} \sin \frac{\pi X}{L} \cos \theta \tag{8.58c}$$

We note from equation (8.58b) that the function $f_2(\theta)$ remains to be determined, obviously from consideration of the displacements. If we substitute equations (8.58a) and (8.58b) into equations (8.51) and (8.52), we have

$$D_X = D_{XS} + \frac{1}{Eh} f_2(\theta)X + f_3(\theta) \tag{8.59a}$$

$$D_\theta = D_{\theta S} - \frac{1}{2aEh} [f_2(\theta)]_{,\theta} X^2 - \frac{1}{a} [f_3(\theta)]_{,\theta} X + f_4(\theta) \tag{8.59b}$$

where D_{XS} and $D_{\theta S}$ represent the corresponding displacements from the simply supported cases, equations (8.56a) and (8.56b), respectively. We now have three functions of integration to be evaluated from the boundary conditions

$$D_X(0) = D_X(L) = 0 \tag{8.60a}$$

together with the condition stated in equation (6.11c)

$$D_\theta(0) = D_\theta(L) = 0 \tag{8.60b}$$

and an additional symmetry condition

$$D_X\left(\frac{L}{2}\right) = 0 \tag{8.60c}$$

These equations form only three independent conditions because of the symmetry of the problem. Because $D_{\theta S}$ and D_{XS} automatically satisfy equations (8.60b) and (8.60c), we find by evaluating $D_X(0)$, $D_\theta(0)$, and $D_X(L/2)$ that

$$D_{XS}(0,\theta) + f_3(\theta) = 0 \tag{8.61a}$$

$$f_4(\theta) = 0 \tag{8.61b}$$

$$\frac{1}{Eh} f_2(\theta) \frac{L}{2} + f_3(\theta) = 0 \tag{8.61c}$$

from which

$$f_2(\theta) = \frac{2Eh}{L} D_{XS}(0,\theta) \tag{8.62a}$$

$$f_3(\theta) = -D_{XS}(0,\theta) \tag{8.62b}$$

$$f_4(\theta) = 0 \tag{8.62c}$$

It is easily verified that the conditions on $D_X(L)$ and $D_\theta(L)$ are similarly satisfied by equations (8.62a–c).

To complete the analysis, we substitute equations (8.62a–c) into equations (8.58) and (8.59) and take D_{XS} and $D_{\theta S}$ from equations (8.56a) and (8.56b). The modified expression for the stress resultants Nx is

$$N_X = -\frac{8q_d}{\pi^2} \left[\frac{L^2}{\pi a} \sin \frac{\pi X}{L} - \left(\frac{2L^2}{\pi^2 a} - \mu a \right) \right] \cos \theta \tag{8.63}$$

with S and N_θ given by equations (8.58a) and (8.58c), respectively. The corresponding displacements are

$$D_X = \frac{4q_d}{\pi Eh} \left(\frac{L}{\pi} \right) \left[\frac{2L^2}{\pi^2 a} - \mu a \right] \left[\cos \frac{\pi X}{L} + \frac{2X}{L} - 1 \right] \cos \theta \tag{8.64a}$$

$$D_\theta = \frac{4q_d}{\pi Eh} \left(\frac{L}{\pi} \right) \left[\frac{L}{\pi} \left(\frac{2L^2}{\pi^2 a^2} + 4 + 3\mu \right) \sin \frac{\pi X}{L} + \left(\frac{2L^2}{\pi^2 a} - \mu a \right) \left(\frac{X^2}{aL} - \frac{X}{a} \right) \right] \sin \theta \tag{8.64b}$$

The remaining displacements D_n, $D_{X\theta}$, and $D_{\theta X}$ are easily evaluated from equations (8.53) to (8.55).

We should remember that the applications of this statically indeterminate membrane analysis are somewhat restricted, because the admissible boundary conditions may not grossly violate the membrane theory requirements as discussed previously in Section 6.2.

8.3.4.2 Shells with Double Curvature.

For doubly curved shells of the form considered in Section 4.4.2, the computation of displacements from the membrane theory stresses is not widely treated in the literature. This is probably because of two main reasons: (a) One of the principal uses for the membrane displacements is to incorporate them into a flexibility-type general solution, as described in the previous section. Although this approach is quite applicable for shells of revolution and for cylindrical shells, it is not particularly suited for doubly curved translational shells because of the lack of homogeneous bending solutions; (b) The integration of the stress resultant-displacement relations generally must be carried out numerically even for the shell of revolution geometry, although the stress resultants may have been evaluated analytically. Rather than deal with partially analytical, partially numerical solutions, it is often expedient to employ a strictly numerical approach. Such a technique for translational shells is described in Hedgren and Billington.[19] Also, the numerical techniques of finite differences[20] and finite elements[21] have been applied to this class of problem.

8.3.5 Base Settlement of Cylindrical Tanks

As mentioned previously, cylindrical shells may be treated either as shells of revolution or as shells of translation. Moreover, because they are so common in application, they are often considered as a distinct class of shells for the purpose of analysis. In fact, we have used both Z and X to describe the axial coordinate of a cylindrical shell depending on whether it is considered a shell of revolution (Chapters 4 and 5) or a shell of translation (Chapter 6). Here, we consider a specialized problem for cylindrical shells using the shell of revolution description, $\alpha = Z$ and $\beta = \theta$.

Cylindrical storage tanks are commonly constructed on shallow foundations and subject to differential settlement under load. The resulting settlements may be measured by established survey procedures and decomposed into uniform translation or settlement; rotation or tilt, and differential settlement. While the first two components may be important in some serviceability considerations, they are essentially rigid body forms with respect to the shell. Differential settlement, however, may produce deformations and stresses in the shell wall.

Large diameter steel tanks for oil or water storage are commonly covered by a floating roof which moves vertically depending on the fluid level. The functioning of such a roof may be affected by distortions in the shell wall. These tanks generally have a thickened section or discrete circumferential stiffening member at the top called the primary wind girder, which may be overstressed by distortions resulting from differential settlement at the base.

The idealized situation is shown in Figure 8–8 after Jonaidi and Ansurian.[22] A comprehensive treatment of the problem, encompassing both membrane and bending theory solutions is given in Kamyab and Palmer[23] and in Palmer.[24] Recently, there have been some simplified approaches to the analysis of a cylindrical tank due to such imposed deformations, based on the number of circumferential waves j in the meridional displacement pattern at the base, $Z = 0$. Considering a typical harmonic in the form of equation 7.51,

$$D_Z(0,\theta) = D_Z^j(0) \cos j\theta \tag{8.65}$$

When the shell moves up or down at the base, lines AA_1 and BB_1 rotate. If the movement is downward as in arc segment AB, points A_1 and B_1 tend to move toward each other. Since the arc segment A_1B_1 is initially circular, this movement can occur by a reduction in the radius of curvature and a corresponding radial displacement.

$$D_n(h,\theta) = D_n^j(h) \cos j\theta \tag{8.66}$$

This radial displacement may proceed without requiring the shell to stretch or shorten, because of the relative flexibility of the arc segment A_1B_1. Rather, the primary effect is circumferential bending at the top, depending on the stiffness of the shell and wind girder. Consequently, the analysis may be approached by the assumption of inextensional deformations, introduced in Section 5.2.1, further developed for cylindrical shells in Section 7.6.2, and applied to this problem by Kamyab and Palmer.[23]

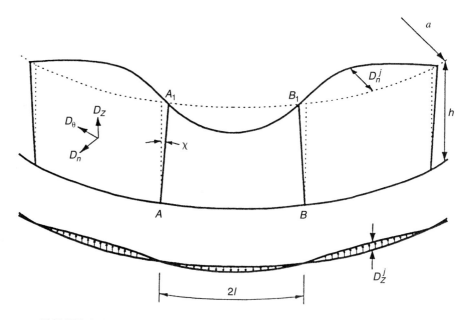

FIGURE 8–8 Shell Deformation Caused by Vertical Base Differential Settlement. (Reprinted with the permission of Elsevier Science, Ltd.)

Considering equation (8.50b)

$$\varepsilon_\theta = \frac{1}{a}(D_{\theta,\theta} + D_n) = 0 \tag{8.67}$$

by the imposition of the condition of inextensional deformations in the circumferential direction. With the circumferential displacement D_θ also in the form of equation (7.51), this leads to

$$D_n^j = -jD_\theta^j \tag{8.68}$$

Now we consider the slope of the shell at the base, which is found as the derivative of D_Z along the coordinate line $s_\theta = a\theta$. Using the notation of Section 2.4,

$$\left.\begin{array}{c} (\ )_{,s_\theta} = (\ )_{,\theta}\theta_{,s_\theta} \\[2mm] s_\theta = a\theta \\[2mm] \theta_{,s_\theta} = \dfrac{1}{a} \end{array}\right\} \tag{8.69a}$$

and

$$\begin{aligned} D_Z(0,\theta)_{,s_\theta} &= \frac{1}{a}D_Z(0,\theta)_{,\theta} \\[2mm] &= -(j/a)\, D_Z^j(0)\sin j\theta \end{aligned} \tag{8.69b}$$

from equation (8.65). The maximum value occurs at $j\theta = \pi/2$ where $D_Z(0,\theta)$, as given by equation (8.66), is equal to zero, which is at point A on Figure 8–8. Therefore, the slope of the straight meridional coordinate line AA_1 is

$$\chi = -(j/a)\, D_Z^j(0) \tag{8.70}$$

At the top point A_1,

$$D_\theta^j(h) = h\chi$$
$$= -(j/a)\, h\, D_Z^j(0) \tag{8.71}$$

and, finally, from equation (8.68)

$$D_n^j(h) = j^2\,(h/a)\, D_Z^j(0) \tag{8.72}$$

This simple relationship between the vertical displacement at the bottom edge and the radial displacement at the top edge of a cylindrical tank is attributed to Kamyab and Palmer.[23] If $D_n^j(h)$ is evaluated analytically, as in equation (8.72), the bending moment in the wind girder may be calculated using the formulas for the bending of circular beams found in Flügge,[25] from which the bending and shear stresses may be evaluated from equations (3.10). In examining equation (8.72), it is obvious that it must be conservative for higher harmonics as $D_n^j(h)$ is unbounded in j, while physically, it must approach zero as j becomes large corresponding to many waves around the circumference.

While several more refined analytical approaches are described in Palmer,[24] Jonaidi and Ansourian,[22] and Hornung and Saal,[26] numerical solutions, namely the finite element method, are the state-of-the-art. Jonaidi and Ansourian[22] have carried out a comparison between results from equations (8.72) and a finite element analysis. For a shell with uniform thickness, the agreement is very good for $j \leq 5$ and $j \geq 10$, while for a graded vertical thickness, the solutions depart for $4 \leq j \leq 12$. For very high values of $j\ (> 20)$, $D_n^j(h)$ does indeed tend to zero and the maximum radial deflection $D_n^j(Z)$ occurs below the top, producing so-called barreling.

When the tank is not anchored, the settlement may produce local uplift if the bottom of the shell does not follow the displacement of the foundation. It was observed by Palmer[24] that local bridging limits the subsequent shell membrane stresses. However, Hornung and Saal[26] found that this was true only for tensile stresses, while the compressive stresses could rise significantly in such cases.

Another cylindrical shell problem in which inextensional deformations may be important is the analysis of unanchored fluid-filled tanks subject to base shaking from an earthquake. This problem was mentioned at the end of Section 5.2.3.

Such tanks have failed due to a meridional buckle at the base which spreads around the circumference as the direction of the shaking changes during the earthquake. This is known as an elephant foot bulge and is attributed to a high concentration of compressive stress, coupled with the effects of internal pressure leading to plastification and the subsequent buckling. This problem is beyond the scope of this book but is relevant in our examination of cylindrical shells which deform inextensionally.

A common approach to the analysis of such tanks is to use a mechanical model in which the shell is considered as rigid but supported by equivalent springs to represent the deformations. This approach is attractive since geometrical and material nonlinearities as well as the elastic flexibility can be introduced through the springs. In a recent study, El-Bkaily and Peek[27] suggest that such models should be used in the post-buckling range only if inextensional deformations are prevented. In the case of the liquid-filled cylindrical tanks, this means that only vessels with roofs can be modeled in this manner since the roof eliminates the inextensional modes.

REFERENCES

1. R. D. Lowry and P. L. Gould, Thermal analysis of orthotropic layered shells of revolution by the finite element method, *Proc. of the IASS Symposium on Shell Structures and Climatic Influences,* University of Calgary, Alberta, Canada, July 1972: 315–325.

2. G. R. Heppler and J. S. Hansen, A mindlin element for thick and deep shells, *Computer Methods in Applied Mechanics and Engineering 54* (1986): 21–47.

3. S. B. Dong and F. K. W. Tso, On a laminated orthotropic shell theory including transverse shear deformations, *Journal of Applied Mechanics, Trans. ASME 39,* series E, no. 4 (December 1972): 1091–1097.

4. K. P. Buchert, *Buckling of Shell and Shell-Like Structures,* (Columbia, Mo.: K. P. Buchert and Associates, 1973): 5–10, 31–34.

5. S. Timoshenko and S. Woinowsky-Krieger, *Theory of Plates and Shells,* 2nd ed., (New York: McGraw-Hill, 1959): 368–369.

6. E. Reissner, Some aspects of the theory of thin elastic shells, *Journal, Boston Society of Engineers,* Boston, Mass., 42, no. 2, (April 1956): 100–133.

7. J. J. Huffington, Jr., Theoretical determination of rigidity properties of orthogonally stiffened plates, *Journal of Applied Mathematics,* ASME, paper no. 55–A–12 (March 1956): 15–20.

8. S. M. Baldridge, Native accents, *Modern Steel Construction,* AISC, Chicago, IL. 34, no. 11 (Nov. 1994): 30–35.

9. G. Kirchhoff, Vorlesungen über mathematische physik, vol 1., *Mechanik,* 1877: 450.

10. Lord Kelvin and P. G. Tait, *Treatise on Natural Philosophy,* vol. 1, pt. 2, 1883: 188.

11. S. Timoshenko and J. N. Goodier, *Theory of Elasticity,* 2nd ed., (New York: McGraw-Hill, 1951): 33.

12. P. G. Glockner, Symmetry in structural mechanics, *Journal of the Structural Division,* ASCE 99, no. ST1 (January 1973): 71–89.

13. V. V. Novozhilov, *Thin Shell Theory* [translated from 2nd Russian ed. by P. G. Lowe] (Groningen, The Netherlands: Noordhof, 1964): 118.

14. P. L. Gould and S. L. Lee, Hyperbolic cooling towers under seismic design load, *Journal of the Structural Division,* ASCE 93, no. ST3, (June 1967): 87–109.

15. W. Flügge, *Stresses in Shells,* 2nd ed., (Berlin: Springer-Verlag, 1973): 85–87, 121–124.

16. L. J. Brombolich and P. L. Gould, Finite element analysis of shells of revolution by minimization of the potential energy functional, *Proc. Conference on Applications of the Finite Element Method in Civil Engineering,* Vanderbilt University, Nashville, Tenn., November 1969: 279–307; L. J. Brombolich and P. L. Gould, A high-precision curved shell finite element, *Synoptic, AIAA Journal* 10, no. 6 (June 1972): 727–728.

17. Gould and Lee, Hyperbolic Cooling Towers under Seismic Design Load.

18. Design of cylindrical concrete roofs, *ASCE Manual of Engineering Practice* 31 (New York: American Society of Civil Engineers, 1952).

19. A. W. Hedgren and D. P. Billington, Numerical analysis of translational shell roofs, *Journal of the Structural Division,* ASCE 92, No. ST1 (February 1966): 223–244.

20. M. Soare, *Application of Finite Difference Equations to Shell Analysis,* (Oxford: Pergamon Press, 1967).

21. R. W. Clough and C. P. Johnson, Finite element analysis of arbitrary thin shells, *Concrete Thin Shells,* ACI publication SP-28 (Detroit: American Concrete Institute, 1971): 333–363.

22. M. Jonaidi and P. Ansourian, Effects of differential settlement on storage tank shells, *Advances in Steel Structures,* Proc. Int. Conf. On Advancement in Steel Structures, Pergamon Press, Dec. 1996: 821–826.

23. H. Kamyab and S. C. Palmer, Analysis of displacement and stresses in oil storage tanks caused by differential settlement, *Proc. Inst. Mech. Engrs,* part c, vol. 203: cl, 1989: 60–70.

24. S. C. Palmer, Stresses in storage tanks caused by differential settlements, *Proc. Inst. Mech. Engrs.,* vol. 208, 1994: 5–16.

25. W. Flügge, *Stresses in Shells,* 2nd ed., Berlin: Springer-Verlag, 1973: 507–510.

26. V. Hornung and H. Saal, Stresses in tank shells due to settlement taking into account local uplift, *Advances in Steel Structures,* vol. II, Proc. Int. Conf. on Advancement in Steel Structures, Pergamon Press, Dec. 1996: 827–832.

27. M. El-Bkaily and R. Peek, Plastic buckling of unanchored roofed tanks under dynamic loads, *Journal of Engineering Mechanics, ASCE,* Vol. 124, No. 6 (June, 1998): 648–657.

28. V. S. Kelkar and R. T. Sewell, *Fundamentals of the Analysis and Design of Shell Structures* (Englewood Cliff, N.J.: Prentice Hall, 1987): 192–193.

29. M. A. Polak. Thermal analysis of reinforced-concrete shells, *Journal of Structural Engineering ASCE,* Vol. 124, No 1 (January, 1998): 105–108.

EXERCISES

8.1. With reference to Section 8.2.2, list the boundary conditions on the boundary $\beta = \bar{\beta}$ for the free, fixed, hinged, roller, and sliding idealizations.

8.2. Generalize the hinged boundary condition on $\alpha = \bar{\alpha}$ to represent elastic displacement and rotation constraints.

8.3. Consider the $\beta = \bar{\beta}$ boundary and derive equations (8.26) and (8.27).

8.4. Consider a shell of revolution geometry and directly derive equations (8.28) and (8.29).

8.5. Determine the normal displacements at points A and B on the toroidal shell shown in Figure 4–9 due to an internal pressure p.

8.6. Determine the decrease in diameter of a complete sphere of radius a and thickness h under an internal suction q.

8.7. For the cylindrical shell under the circumferentially varying loading considered in Section 5.3.4, derive expressions for the displacements for the cases $j = 0$, $j = 1$, and $j > 1$. Sketch the deflections on an elevation and on a cross-section for each harmonic case.

8.8. Write equations (8.34a–c) in harmonic form and consider separately the cases $j = 0$, $j = 1$, and $j > 1$. Verify the pole conditions for D_ϕ, D_θ, and D_n.

8.9. Determine the membrane theory displacements for an open cylindrical shell, as shown in Figure 6–1, subject to a uniform loading q_u, as given by equations (6.12), for the following cases:
(a) Load uniformly distributed in X direction.
(b) Load harmonically distributed in X direction.

8.10. Derive the governing equations for the membrane theory displacements of a hyperbolic paraboloid, as shown in Figure 6–5, for a uniform live load $q_z = p$. If the shell is supported by two vertical arches spanning between the corners of the shell above A and D, and B and C, respectively, state the corresponding boundary conditions.

8.11. A toroidal knuckle transition is introduced at the junction of the cylindrical and conical segments as shown in Figure 8–9. The cylinder and the cone have thicknesses of 0.5 while the toroidal knuckle has a thickness of 1.25. For Poisson's ratio = 0.3, E = 30,000,000 and an internal pressure of 100,
(a) Compute the stresses in the three parts of the shell using membrane theory
(b) Find the differential deflection at the cylindrical-toroidal junction, Point C.

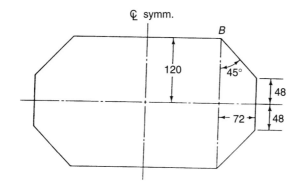

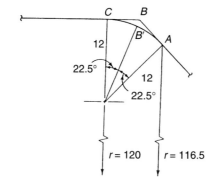

FIGURE 8–9 Toroidal knuckle used as a transition between conical and cylindrical portions of a pressure vessel. (Adapted from Kelkar and Sewell, Ref. 28.)

APPENDIX 8A

Thermal Analysis of Reinforced Concrete Shells

Polak[29] has presented an equation for the temperature T at a point within the shell surface, ζ, as a function of time, t.

$$T = T(\zeta) + T(\zeta,t) \tag{8.73}$$

$$T(\zeta) = T_1^f + (T_2^f - T_1^f)\zeta/h \tag{8.74}$$

$$T(\zeta,t) = \frac{2}{\pi} \sum_{n=1}^{\infty} \left(\left\{ \frac{[T_2^f \cos(n\pi) - T_1^f] - [T_2^i \cos(n\pi) - T_1^i]}{n} \right\} \times \sin \frac{n\pi\zeta}{h} \times \exp \left(\frac{-kn^2\pi^2 t}{h^2} \right) \right) \tag{8.75}$$

where

h = thickness of shell at the point

k = thermal diffusivity of the concrete

ζ = thickness coordinate measured from inside surface of shell, positive toward outside of shell

t = elapsed time

$T_1^i, T_2^i, T_1^f, T_2^f$ = Initial (i) and final (f) temperatures on outside (2) and inside (1) surfaces of shell

A detailed application of this equation is presented in reference 29 where the thermal diffusivity of concrete is taken as $k = 0.774$ mm^2/sec.

9

Energy and Approximate Methods

9.1 GENERAL

In solid mechanics, an energy formulation is often viewed as an alternative to the differential equation statement. A direct connection exists between the energy and the differential equation approaches through the principle of virtual displacements and through various extremum principles. However, we will not pursue this connection, as it is beyond our immediate scope.

Our interest in introducing energy considerations here is twofold. First, energy principles are the basis for many powerful numerical methods for solving shell and plate problems. Second, energy-based solutions are an important resource even for the classical formulations that we stress in this book. We illustrate the latter in some detail in Chapters 11 and 13, and the materials in this chapter may be deferred until the appropriate sections are encountered without loss in continuity.

In this chapter, we highlight some of the more important aspects of energy methods. We primarily stress those items essential to support the ensuing applications. Also, we introduce some approximations that lead to the development of numerically based solution techniques.

9.2   STRAIN ENERGY

The strain energy for an elastic body U_ε may be written in terms of the strain energy density dU_ε as

$$U_\varepsilon = \int_V dU_\varepsilon \tag{9.1}$$

where V represents the volume of the continuum.

In terms of the orthogonal curvilinear coordinates, as defined in Chapter 2, and the differential thickness $d\zeta$, as shown on Figure 3–1 (a), the strain energy density follows from the linear theory of elasticity:

$$\begin{aligned}
dU_\varepsilon = [&\sigma_{\alpha\alpha}(\alpha,\beta,\zeta)\varepsilon_\alpha(\alpha,\beta,\zeta) + \sigma_{\beta\beta}(\alpha,\beta,\zeta)\varepsilon_\beta(\alpha,\beta,\zeta) \\
&+ \sigma_{nn}(\alpha,\beta,\zeta)\varepsilon_n(\alpha,\beta,\zeta) + \sigma_{\alpha\beta}(\alpha,\beta,\zeta)\omega(\alpha,\beta,\zeta) \\
&+ \sigma_{\alpha n}(\alpha,\beta)\gamma_\alpha(\alpha,\beta) + \sigma_{\beta n}(\alpha,\beta)\gamma_\beta(\alpha,\beta)] \, dV
\end{aligned} \tag{9.2}$$

The stress terms are introduced in Section 3.1; the strains are defined in Section 7.3.

For a shell or plate, the differential volume dV may be expressed in terms of the differential area dS and thickness $d\zeta$ as

$$\begin{aligned}
dV &= dS \, d\zeta \\
&= A(\zeta) \, d\alpha \, B(\zeta) \, d\beta \, d\zeta \\
&= A\left(1 + \frac{\zeta}{R_\alpha}\right) d\alpha \, B\left(1 + \frac{\zeta}{R_\beta}\right) d\beta \, d\zeta
\end{aligned} \tag{9.3a}$$

or, after dropping terms of $O(h/R) : 1$,

$$dV = AB \, d\alpha \, d\beta \, d\zeta \tag{9.3b}$$

It is generally preferable to write the strain energy in terms of a single field variable, in other words, stress, strain, or displacement. To retain generality, we choose the strains, which are connected to each of the others by a single set of equations, (8.1) and (7.46), respectively.

We proceed by noting that $\sigma_{nn} = 0$ (assumption [4], Table 1–1), and by rewriting equation (9.1) in the matrix notation introduced in equation (8.2):

$$U_\varepsilon = \frac{1}{2} \int_V \lfloor \boldsymbol{\sigma} \rfloor \{\boldsymbol{\varepsilon}\} \, dV \tag{9.4}$$

Now, substituting equation (8.2) into (9.4), we get

$$U_\varepsilon = \frac{1}{2} \int_V [[\mathbf{C}]\{\boldsymbol{\varepsilon}\} - \{\boldsymbol{\sigma_T}\}]^T \{\boldsymbol{\varepsilon}\} \, dV \tag{9.5}$$

Keeping in mind that the elements of $\{\boldsymbol{\sigma_T}\}$ may be regarded as known, equation (9.5) is an expression of the strain energy in terms of strain-type quantities alone. We may now multiply out the terms in equation (9.5). Because the arithmetic is quite lengthy and straightforward, it is omitted here. After integration through the thickness and some rearrangement, the resulting expression is

$$
U_\varepsilon = \frac{E}{2(1 - \mu^2)} \int_S \left\{ h\left[(\varepsilon_\alpha + \varepsilon_\beta)^2 - 2(1 - \mu)\left(\varepsilon_\alpha \varepsilon_\beta - \frac{\omega^2 + \gamma_\alpha^2 + \gamma_\beta^2}{4} \right) \right] \right.
$$

$$
+ \frac{h^2}{4} \left[(\varepsilon_\alpha + \varepsilon_\beta)(\kappa_\alpha + \kappa_\beta) - (1 - \mu)(\varepsilon_\alpha \kappa_\beta + \varepsilon_\beta \kappa_\alpha - 2\omega\tau) \right]
$$

$$
+ \frac{h^3}{12} \left[(\kappa_\alpha + \kappa_\beta)^2 - 2(1 - \mu)(\kappa_\alpha \kappa_\beta - \tau^2) \right]
$$

$$
- \overline{\alpha}(1 + \mu)(\varepsilon_\alpha + \varepsilon_\beta) \int_{-h/2}^{h/2} T(\zeta)d\zeta
$$

$$
\left. - \overline{\alpha}(1 + \mu)(\kappa_\alpha + \kappa_\beta) \int_{-h/2}^{h/2} T(\zeta)\zeta d\zeta \right\} AB\, d\alpha\, d\beta
$$

(9.6)

where the volume integral has been replaced by a surface integral over S. Modified expressions for U_ε which correspond to other stress-strain laws may be written in the same manner using the appropriate $[\mathbf{C}]$, as discussed in Chapter 8.

If, for the moment, we neglect the temperature-dependent terms, we may group equation (9.6) as

$$
U_\varepsilon = \frac{E}{2(1 - \mu^2)} \int_S \left\{ h[\mathbf{I}] + \frac{h^2}{4}[\mathbf{II}] + \frac{h^3}{12}[\mathbf{III}] \right\} AB\, d\alpha\, d\beta
$$

(9.7)

It is convenient to interpret equation (9.7) with respect to the three component expressions $[\mathbf{I}, \mathbf{II}, \mathbf{III}]$, which are, respectively, linear, quadratic, and cubic in the thickness h. First, the linear component $Eh/[2(1 - \mu^2)][\mathbf{I}]$ represents extensional and shearing energy, which is predominant in shells that behave primarily in accordance with the membrane theory. The quadratic component $Eh^2/[8(1 - \mu^2)][\mathbf{II}]$ is a coupling between extensional and shearing, and twisting and bending terms, and enters into some geometrically nonlinear formulations. Finally, the cubic component $Eh^3/[24(1 - \mu^2)][\mathbf{III}]$ contains both bending and twisting energy which are the primary resistance mechanisms in the flexure of thin plates. In many cases, only one or two of the three components are required.

9.3 POTENTIAL ENERGY OF THE APPLIED LOADS

In equation (3.11e), the surface loading vector, $\mathbf{q}$ is defined in terms of loading intensities per unit area of middle surface $q_\alpha(\alpha,\beta)$, $q_\beta(\alpha,\beta)$ $q_n(\alpha,\beta)$. Here, we may also admit loads which act along a coordinate line. These *line loads* are defined in the same fashion as the surface loading vectors

$$
\hat{\mathbf{q}}(\alpha,\overline{\beta}) = \hat{q}_\alpha(\alpha,\overline{\beta})\mathbf{t}_\alpha + \hat{q}_\beta(\alpha,\overline{\beta})\mathbf{t}_\beta + \hat{q}_n(\alpha,\overline{\beta})\mathbf{t}_n
$$

(9.8a)

for a load acting along the s_α coordinate line corresponding to $\beta = \overline{\beta}$, and

$$\hat{\mathbf{q}}(\alpha,\beta) = \hat{q}_\alpha(\alpha,\beta)\mathbf{t}_\alpha + \hat{q}_\beta(\alpha,\beta)\mathbf{t}_\beta + \hat{q}_n(\alpha,\beta)\mathbf{t_n} \tag{9.8b}$$

for a load acting along the s_β coordinate line corresponding to $\alpha = \overline{\alpha}$. Boundary reactions are frequently represented by such line loads. Furthermore, we may accommodate concentrated loads defined by

$$\overline{\mathbf{q}}(\overline{\alpha},\overline{\beta}) = \overline{q}_\alpha(\overline{\alpha},\overline{\beta})\mathbf{t}_\alpha + \overline{q}_\beta(\overline{\alpha},\overline{\beta})\mathbf{t}_\beta + \overline{q}_n(\overline{\alpha},\overline{\beta})\mathbf{t_n} \tag{9.9}$$

for a load acting at $(\overline{\alpha},\overline{\beta})$. No body forces distributed through the thickness are explicitly included in this development. Rather, for a thin continuum, it is generally sufficient to refer such forces to the middle surface, as we did in Chapter 3 for gravity loading.

Corresponding to each of the loading terms are the components of the middle surface displacement vector $\mathbf{\Delta}(\alpha,\beta)$. These components have been defined in equation (7.3a) as $D_\alpha(\alpha,\beta)$, $D_\beta(\alpha,\beta)$, $D_n(\alpha,\beta)$.

It is convenient to define the change in potential energy of the applied loading U_q as the product of each loading component and the corresponding displacement component. Further, when a positive displacement occurs, U_q decreases so that the product of the correspondents is given a *negative* sign. Thus, we have

$$\begin{aligned}
U_q = -\Bigg\{ &\int_S \mathbf{q}(\alpha,\beta)\cdot\mathbf{\Delta}(\alpha,\beta)\, A(\alpha,\beta)\, B(\alpha,\beta)\, d\alpha\, d\beta \\
&+ \sum \left[\int_{s_\alpha} \hat{\mathbf{q}}(\alpha,\overline{\beta})\cdot\mathbf{\Delta}(\alpha,\overline{\beta})\, A(\alpha,\overline{\beta})\, d\alpha \right. \\
&\left. + \int_{s_\beta} \hat{\mathbf{q}}(\overline{\alpha},\beta)\cdot\mathbf{\Delta}(\overline{\alpha},\beta)\, B(\overline{\alpha},\beta)\, d\beta \right] + \sum \overline{\mathbf{q}}(\overline{\alpha},\overline{\beta})\cdot\mathbf{\Delta}(\overline{\alpha},\overline{\beta}) \Bigg\}
\end{aligned} \tag{9.10a}$$

or, in terms of the components,

$$\begin{aligned}
U_q = -\Bigg\{ &\int_S q_i(\alpha,\beta)D_i(\alpha,\beta)\, A(\alpha,\beta)\, B(\alpha,\beta)\, d\alpha\, d\beta \\
&+ \sum \left[\int_{s_\alpha} \hat{q}_i(\alpha,\overline{\beta})\, D_i(\alpha,\overline{\beta})\, A(\alpha,\overline{\beta})\, d\alpha \right. \\
&\left. + \int_{s_\beta} \hat{q}_i(\overline{\alpha},\beta)\, D_i(\overline{\alpha},\beta)\, B(\overline{\alpha},\beta)\, d\beta \right] \\
&+ \sum \overline{q}_i(\overline{\alpha},\overline{\beta})\, D_i(\overline{\alpha},\overline{\beta}) \Bigg\} \quad (i = \alpha,\beta,n)
\end{aligned} \tag{9.10b}$$

In equations (9.10a) and (9.10b), the second term signifies that the load potentials for each line load acting on the shell surface are summed, and the third term indicates that load potentials for each concentrated load are similarly summed.

Note that equations (9.8) and (9.9) can be generalized to include distributed, line, and concentrated couples corresponding to the rotations $D_{\alpha\beta}$ and $D_{\beta\alpha}$.

9.4 ENERGY PRINCIPLES AND RAYLEIGH-RITZ METHOD

9.4.1 Principles of Virtual Displacements

Consider an elastic body subject to a set of surface tractions composed of the distributed load vectors $\mathbf{q}(\alpha,\beta)$, as defined in equations (3.11e), along with line load vectors $\hat{\mathbf{q}}(\alpha,\overline{\beta})$ and $\hat{\mathbf{q}}(\overline{\alpha},\beta)$ and concentrated load vectors $\overline{\mathbf{q}}(\overline{\alpha},\overline{\beta})$. If we assume that there are no thermal or inertial effects, then the law of conservation of energy requires that the work done by the surface tractions be equal to the strain energy stored in the material. Now a change $\delta\Delta$ is imposed on the displacement field Δ, which moves to a new position $\Delta + \delta\Delta$. $\delta(\ )$ is the variational operator and, for our purposes, $\delta\Delta$ may be regarded as a *linear increment* of the vector Δ. Also, $\delta\Delta$ is assumed to be compatible with the constraints of the system, although this restriction is not necessary but only convenient.[1] The source of $\delta\Delta$ is not specified; in other words, it does not necessarily result from any particular loading system. Hence, $\delta\Delta$ is called a *virtual* displacement.

Next, we write the energy balance corresponding to the virtual displacement. The virtual work performed by the surface tractions is given by

$$
\delta\overline{U}_q = \int_S \mathbf{q}(\alpha,\beta)\cdot\delta\Delta(\alpha,\beta)\, A(\alpha,\beta)\, B(\alpha,\beta)\, d\alpha\, d\beta
$$

$$
+ \sum_S \left[\int_{S_\alpha} \hat{\mathbf{q}}(\alpha,\overline{\beta})\cdot\delta\Delta(\alpha,\overline{\beta})\, A(\alpha,\overline{\beta})\, d\alpha \right. \tag{9.11}
$$

$$
+ \left. \int_{S_\beta} \hat{\mathbf{q}}(\overline{\alpha},\beta)\cdot\delta\Delta(\overline{\alpha},\beta)\, B(\overline{\alpha},\beta)\, d\beta \right] + \sum_S \overline{\mathbf{q}}(\overline{\alpha},\overline{\beta})\cdot\delta\Delta(\overline{\alpha},\overline{\beta})
$$

and the strain energy is changed by dU_ε, which is written from equation (9.1) as

$$
\delta U_\varepsilon = \delta \int_V dU_\varepsilon \tag{9.12}
$$

The rhs of equation (9.12)

$$
\delta \int_V dU_\varepsilon
$$

may be evaluated from equation (9.5); however, because equation (9.5) is not yet written as an explicit function of the displacement Δ, and because variations or increments of Δ are being considered, we choose to leave the rhs of equation (9.12) in the general form.

Then, the energy balance is written by equating equations (9.11) and (9.12):

$$
\delta\overline{U}_q = \delta \int_V dU_\varepsilon \tag{9.13}
$$

which is a statement of the principle of virtual displacements. This principle is a special form of the principle of virtual work, which is regarded by many contemporary mechanicians as the cornerstone of solid mechanics. (For a complete discussion of the interrelation of the various energy principles of solid mechanics, the interested reader is referred to Washizu.[2])

The principle of virtual displacements, as stated in this section, may be taken as an alternate statement of the condition of static equilibrium, as derived in Chapter 3. This interpretation is used freely in the ensuing sections.

9.4.2 Principle of Minimum Total Potential Energy

It is convenient to introduce some further assumptions at this point. First, we specify that the volume V does not change during the virtual displacement and that the surface tractions also do not vary. Next, the variational operator $\delta(\ )$ is restricted to variations or increments in *displacement* only. We now recognize that $\delta \overline{U}_q$, as defined in equation (9.11), is simply $-\delta U_q$, where the load potential U_q was defined in equation (9.10a). Hence, we may rewrite equation (9.13) as

$$\delta U_\tau = \delta U_\varepsilon + \delta U_q = \delta(U_\varepsilon + U_q) = 0 \qquad (9.14)$$

where U_τ is known as the total potential energy. Bear in mind that U_τ itself is an integral function of various algebraic functions of the dependent variables, as demonstrated by equations (9.6) and (9.10). Such a function is called a *functional* in the terminology of the calculus of variations. Equation (9.14) states that *the total potential energy of an elastic system in equilibrium must be stationary with respect to all displacements satisfying the boundary conditions.* Technically, the term stationary corresponds to the first variation of the functional, in this case δU_τ vanishing, and may indicate an absolute or relative maximum or minimum; however, for our applications here, the stationary condition may be regarded as an absolute or relative minimum.

From the standpoint of solution strategy, a dual interpretation of the principle is helpful: We seek a displacement field $\Delta(\alpha, \beta)$ which satisfies the kinematic boundary conditions and makes U_τ a minimum. Then, the stress resultants and stress couples computed from Δ via the strain-displacement and constitutive laws will satisfy the equilibrium considerations.

The classical solution for the problem of finding the minimum of a functional is the fundamental problem of the calculus of variations and is not within our scope here. The interested reader is referred to Forray[3] for an introductory treatment of this subject. For our purposes, we wish to pursue the application of one of the *direct methods* of the calculus of variations for the solution of equation (9.14). Of the several possibilities available, the Rayleigh-Ritz method is perhaps the best known and simplest to understand, and thus we consider this technique first.

However, before proceeding, we might reflect on the first sentence of this chapter, where we noted that an energy formulation may be regarded as an alternative to a differential equation statement. Historically, the latter was called the method of *effective causes,* while the former was known as the method of *final causes.* The renowned mathematician Euler, who was the inventor of the calculus of variations, observed in an argument which concisely blends science and theology: "Since the fabric of the universe is most perfect, and is the work of a most wise Creator, nothing whatsoever takes place in the universe in which some relation of maximum and minimum does not appear. Wherefore there is absolutely no doubt that every effect in the universe can be explained as satisfactorily from final causes, by the aid of the method of maxima and minima, as it can from the effective causes themselves. . . ."[4] The resounding versatility and success of the contemporary energy-based finite element method appears to leave even Euler's lofty claim understated.

9.4.3 Rayleigh-Ritz Method

Consider the displacement vector $\Delta(\alpha,\beta)$ which was defined originally in equation (7.3a) to include D_α, D_β, and D_n. Because we may have additional generalized displacements, such as $D_{\alpha\beta}$ and $D_{\beta\alpha}$ as defined in equations (7.22a) and (7.22b), and because in some cases not all of the displacement components are included, we write Δ in the general form

$$\{\Delta\} = \{\Delta_1\ \Delta_2 \ldots \Delta_k \ldots \Delta_m\} \tag{9.15}$$

where the index m can be set according to the problem at hand.

The next step is to assume that each Δ_k ($k = 1, \ldots, m$) in equation (9.15) is of the form

$$\Delta_k = c_{k1}\phi_{k1} + c_{k2}\phi_{k2} + \cdots c_{kl}\phi_{kl} + \cdots c_{kn}\phi_{kn} \quad (k = 1, m)$$

$$= \sum_{l=1}^{n} c_{kl}\phi_{kl} \tag{9.16}$$

In equation (9.16), each ϕ_k represents a linearly independent coordinate function, and each c_{kl} is an undetermined *constant* coefficient. It is not necessary that each coordinate function satisfies all of the kinematic boundary conditions, but only that the sequence Δ_k meets this requirement. The truncation index for the sequence n may be set at a different value for Δ_k, subject to some guidelines mentioned further on, and it is anticipated that improved results will be obtained by increasing n, which means including more terms in the sequence. Commonly, polynomials or elementary trigonometric functions are selected for the coordinate functions ϕ_{kl}, but a wide variety of possibilities are available according to the problem.

Now, the expressions for Δ_k are substituted into the strain and the change in potential energy terms in equation (9.14). In the strain energy U_ε, as given by equation (9.6), we must first evaluate the corresponding strains by using the appropriate strain-displacement relationships. After the integrations indicated in equation (9.6) and (9.10) are carried out, U_τ becomes an *algebraic* rather than an *integral* function. Therefore, the extremum problem of the calculus of variations is transformed into the maximum-minimum problem of differential calculus. The later problem, in turn, is solved by satisfying

$$\frac{\partial U_\tau}{\partial c_{kl}} = 0 \quad \begin{pmatrix} k = 1,m \\ l = 1,n \end{pmatrix} \tag{9.17}$$

which produces a set of simultaneous algebraic equations for the coefficients c_{kl}. Following the solution of these equations, the displacement functions Δ_k can readily be used to find strains and curvatures, from which stress resultants and couples may be computed.

The convergence of this method in a rigorous mathematical sense is considered to be a difficult theoretical question.[5] In practice, improved results are obtained in two ways: first, by selecting coordinate functions which closely resemble the actual displacement field—the value of physical intuition in this regard is obvious; second, by using an increasing number of terms in each sequence. Of course, this increases the number of simultaneous equations and the computational effort required.

The careful reader may have noted that no specific reference was made to the *a priori* satisfaction of the internal compatibility or (strain-displacement) conditions by the chosen coordinate functions. This illustrates one of the advantages of a displacement formulation. For reasonably chosen coordinate functions, the ability to compute the strains from the strain-displacement conditions in the course of evaluating U_ε represents, in fact, the satisfaction of the compatibility requirements, so that little difficulty is encountered in this regard.

Another point of general interest is to note the limitations of the method. Obviously, the requirement of the displacement function sequences satisfying the boundary conditions can be quite restrictive in the case of irregularly shaped continua. The application of the Rayleigh-Ritz method in a *piecewise* fashion to subdivisions or finite elements of the entire medium, with the undetermined coefficients chosen to enforce continuity requirements along interelement boundaries as well as to satisfy the external boundary conditions, has become one of the most powerful techniques in applied mechanics: the *finite element method.* A wide variety of applications of this technique are presented by Zienkiewicz,[6] who has contributed greatly to its popularity.

Even before the emergence of the finite element method in the form presently familiar to engineers, Timoshenko suggested that the Rayleigh-Ritz technique—originated by Nobel laureate Lord Rayleigh in his epic treatise "Theory of Sound" and elaborated by W. Ritz, who considered applications to the analysis of thin plates—has spurred more research in the strength of materials and the theory of elasticity than any other single mathematical tool.[7] The vast literature accompanying the finite element method resoundingly reinforces this claim.

In regard to the use of the finite element method to model shells, the *explicit* inclusion of rigid body modes has proved to be troublesome due to the curvilinear coordinates. It has been suggested by Heppler and Hansen[8] that the more convenient implicit representation is best implemented with approximations of equal degree for the translational degrees of freedom contained in $\{\Delta\}$, equation (9.15). This implies a single value of n for all such variables.

Also, we should mention that the Rayleigh-Ritz method may be generalized by taking each of the c_{kl} coefficients in equation (9.16) as an unknown *function* of one of the independent variables, rather than as simply a *constant.* In this case, equation (9.16) is generalized to

$$\Delta_k = \phi_{k0}(\alpha,\beta) + \sum_{l=1}^{n} c_{kl}(\alpha \text{ or } \beta)\phi_{kl}(\alpha,\beta) \qquad (9.18)$$

It is usually convenient to choose ϕ_{k0} to satisfy all of the boundary conditions; then, each term of the sequence $c_{kl}\phi_{kl}(l = 1,n)$ must vanish on the boundaries.

This extension of the basic Rayleigh-Ritz technique is known as the *Kantorovich method.* If applied skillfully, it can reduce the numerical work considerably for some problems.[9]

9.5 GALERKIN METHOD

The Galerkin method is quite similar in execution to the Rayleigh-Ritz method in that the solution is postulated to be represented by a sequence of coordinate functions, such as equation (9.16). The most apparent operational difference is that in the Galerkin technique, these coordinate functions are tested in the *governing differential equations* rather than in an energy expression.

As an illustration, assume that the system has been reduced to a single equation of the form

$$\mathbf{B}(\Delta_k) = f_k(\alpha,\beta,n) \tag{9.19}$$

where Δ_k is one element of the general displacement vector $\{\Delta\}$, equation (9.15), and $\mathbf{B}$ is a linear differential operator. For example, the Laplacian operator in Cartesian coordinates, obtained by setting

$$\mathbf{B}(\ ) = (\ )_{,XX} + (\ )_{,YY} \tag{9.20}$$

is frequently encountered in plate problems. The rhs of equation (9.19) represents known quantities, for example, the applied surface loading. When equation (9.16) is substituted into equation (9.19), we have

$$\mathbf{B}\left(\sum_{i=1}^{n} c_{ki}\phi_{ki}\right) - f_k(\alpha,\beta,n) = R_k(\alpha,\beta,n) \tag{9.21}$$

where R_k is a residual error function. Recall that Δ_k has been postulated to satisfy the boundary conditions of the problem. The essence of the Galerkin method is to force Δ_k to also satisfy the governing equation as closely as possible. This means that the residual error R_k should be minimized.

The condition of minimizing R_k is to require that R_k be orthogonal to each ϕ_{kl} over the domain of the problem.[9] That is,

$$\int_V R_k\phi_{kl}dV = 0 \tag{9.22a}$$

or

$$\int_V \left[\mathbf{B}\left(\sum_{i=1}^{n} c_{ki}\phi_{ki}\right) - f_k(\alpha,\beta,n)\right]\phi_{kl}\,dV = 0 \quad (l = 1,n) \tag{9.22b}$$

This produces a system of n linear simultaneous algebraic equations in the n unknown coefficients c_{kl}.

Obviously, this procedure becomes increasingly involved when there are multiple equations and unknowns in the system, instead of just a single equation in Δ_k as we have considered here. But, for a large class of second-order ordinary differential equations, the Galerkin method leads to an identical set of coefficients to those produced by the Rayleigh-Ritz method[10] and may be preferable if it is more convenient to work with the governing equations rather than with the energy functional. Moreover, there are problems for which no satisfactory variational principle has been formulated, but for which a set of governing differential equations is available. This suggests that the Galerkin method may be even broader in applicability than the Rayleigh-Ritz method.

Also, it is interesting to note that recent work by Lim and Liew has applied a seemingly powerful generalization of the Rayleigh-Ritz method, the so-called Ritz pb-2 approach, to plates and shallow shells. Through the use of orthogonal shape functions which are based on complete polynomials and which *a priori* satisfy a variety of kinematic boundary conditions, accurate and efficient solutions have been reported for static loading, free vibration, and elastic buckling.[11]

Applications of the Rayleight-Ritz and Galerkin methods are found in Chapters 10 to 13.

REFERENCES

1. F. L. DiMaggio, Principle of virtual work in structural analysis, *Journal of the Structural Division,* ASCE 86, no. ST11 (November 1960): 65–77.

2. K. Washizu, *Variational Methods in Elasticity and Plasticity,* (New York: Pergamon Press, 1968).

3. M. J. Forray, *Variational Calculus in Science and Engineering,* (New York: McGraw-Hill, 1968), Chaps. 1 and 2.

4. S. Timoshenko, *History of Strength of Materials,* (New York: McGraw-Hill, 1953), p. 31. Copyright, 1953 by McGraw-Hill, Inc. and used with permission of McGraw-Hill Book Co.

5. R. Courant and D. Hilbert, *Methods of Mathematical Physics,* vol. 1, (New York: Interscience Publishers, 1966): 175–176.

6. O. C. Zienkiewicz, *The Finite Element Method in Engineering Science,* (London: McGraw-Hill, 1971).

7. Timoshenko, *History of Strength of Materials:* 338–339, 399–401.

8. G. R. Heppler and J. S. Hansen, A mindlin element for thick and deep shells, *Computer Methods in Applied Mechanics and Engineering* 54 (1986): 21–47.

9. Forray, 199–201.

10. Forray, 189–199.

11. C. W. Lim and K. M. Liew A pb-2 ritz formulation for flexural vibration of shallow cylindrical shells of rectangular planform, *Journal of Sound and Vibration,* 173.3 (1994): 343–375.

EXERCISES

9.1 Generalize equations (9.8) to include the possibility of surface loading consisting of distributed, line, and concentrated couples.

9.2 Consider a simple supported beam of length L, with a uniformly distributed load w.
  (a) Compute the deflected shape $D(x)$, by using a sequence of coordinate functions

$$D(x) = c_1 + c_2 \sin \frac{\pi x}{L} + c_3 \sin \frac{2\pi x}{L} + \cdots$$

and observe the convergence. Use both the Rayleigh-Ritz and Galerkin methods.
  (b) Repeat the procedure, using a sequence of polynomials of the form

$$D(x) = c_1 + c_2 x(L - x) + c_3 x^2(L - x) + c_4 x(L - x)^2 + \cdots$$

For the polynomials, recall that the sequence must satisfy the boundary conditions

9.3 Consider equations (9.10) and generalize the load potential to include distributed, line, and concentrated applied moments.

10

Bending of Plates: Displacement Formulation

10.1 GOVERNING EQUATIONS

10.1.1 General Formulation

The equilibrium equations for initially flat plates are stated as equations (3.25a–e); the strain-displacement relations are given by equations (7.54) or, with transverse shearing strains neglected, as equations (7.55). Taken together with the stress resultant-strain relationships in the form of equation (8.10), the requisite boundary conditions discussed in Section 8.2, and specifically the Kirchhoff conditions equations (8.25) and (8.27), the elements of a quite general plate theory are available and substantiated.

It is convenient to consider the development of the theory of plates in several stages, initially neglecting transverse shearing strains and in-plane stress resultants and also assuming isotropic material properties. Then, the theory will be generalized as necessary.

In the absence of in-plane loading q_α and q_β and when the displacements are small in accordance with assumption [2], Table 1–1, the in-plane stress resultants may also be neglected. Then equations (3.25a) and (3.25b) are removed, leaving

$$[(BQ_\alpha)_{,\alpha} + (AQ_\beta)_{,\beta}] + q_n AB = 0 \tag{10.1a}$$

$$(BM_{\alpha\beta})_{,\alpha} + (AM_\beta)_{,\beta} + B_{,\alpha}M_{\beta\alpha} - A_{,\beta}M_\alpha - Q_\beta AB = 0 \tag{10.1b}$$

$$(BM_\alpha)_{,\alpha} + (AM_{\beta\alpha})_{,\beta} + A_{,\beta}M_{\alpha\beta} - B_{,\alpha}M_\beta - Q_\alpha AB = 0 \tag{10.1c}$$

Because this theory is based on the displacements being small and because the displacements are computed based on this theory, full justification for and the limits of applicability of the linear theory must await the solution of a more general problem admitting finite deformations. This is considered in Sections 11.3.5 and 11.3.6, but meanwhile it is of great practical interest to fully develop the linear theory since the serviceability of many plate structures would be diminished by significant displacements, even though they may possess adequate strength.

Continuing along with the elimination of the in-plane forces, we also disregard D_α, D_β, and the extensional and in-plane shearing strains ε_α, ε_β, and ω. With transverse shearing strains also suppressed, the complete set of kinematic equations is given by equations (7.55). Of course, there remain extensional and in-plane shearing strains away from the middle plane as computed from equations (7.36), (7.37), and (7.45):

$$\varepsilon_\alpha(\zeta) = \zeta \kappa_\alpha \tag{10.2a}$$

$$\varepsilon_\beta(\zeta) = \zeta \kappa_\beta \tag{10.2b}$$

$$\omega(\zeta) = 2\zeta \tau \tag{10.2c}$$

where κ_α and κ_β are the curvatures and τ is the twist of the middle plane.

Recalling equations (7.55a–c), we find

$$\kappa_\alpha = -\left[\frac{1}{A}\left(\frac{1}{A}D_{n,\alpha}\right)_{,\alpha} + \frac{A_{,\beta}}{AB^2}D_{n,\beta}\right] \tag{10.3a}$$

$$\kappa_\beta = -\left[\frac{1}{B}\left(\frac{1}{B}D_{n,\beta}\right)_{,\beta} + \frac{B_{,\alpha}}{A^2B}D_{n,\alpha}\right] \tag{10.3b}$$

$$\tau = -\frac{1}{2}\left\{\frac{1}{A}\left(\frac{1}{B}D_{n,\beta}\right)_{,\alpha} + \frac{1}{B}\left(\frac{1}{A}D_{n,\alpha}\right)_{,\beta} - \frac{1}{AB}\left[\frac{A_{,\beta}}{A}D_{n,\alpha} + \frac{B_{,\alpha}}{B}D_{n,\beta}\right]\right\} \tag{10.3c}$$

We note that these terms are functions of the normal displacement D_n only.

The stress couples are easily expressed in terms of D_n by substituting equations (10.3a–c) into the appropriate constitutive laws. For isotropic materials, we have, from Matrix 8–2 and equations (10.3a–c),

$$M_\alpha = -D\left\{\frac{1}{A}\left(\frac{1}{A}D_{n,\alpha}\right)_{,\alpha} + \frac{A_{,\beta}}{AB^2}D_{n,\beta} + \mu\left[\frac{1}{B}\left(\frac{1}{B}D_{n,\beta}\right)_{,\beta} + \frac{B_{,\alpha}}{A^2B}D_{n,\alpha}\right]\right\} \tag{10.4a}$$

$$M_\beta = -D\left\{\frac{1}{B}\left(\frac{1}{B}D_{n,\beta}\right)_{,\beta} + \frac{B_{,\alpha}}{A^2B}D_{n,\alpha} + \mu\left[\frac{1}{A}\left(\frac{1}{A}D_{n,\alpha}\right)_{,\alpha} + \frac{A_{,\beta}}{AB^2}D_{n,\beta}\right]\right\} \tag{10.4b}$$

$$M_{\alpha\beta} = -D\frac{(1-\mu)}{2}\left\{\frac{1}{A}\left(\frac{1}{B}D_{n,\beta}\right)_{,\alpha} + \frac{1}{B}\left(\frac{1}{A}D_{n,\alpha}\right)_{,\beta} - \frac{1}{AB}\left[\frac{A_{,\beta}}{A}D_{n,\alpha} + \frac{B_{,\alpha}}{B}D_{n,\beta}\right]\right\} \tag{10.4c}$$

in which D is the flexural rigidity of the plate, given by

$$D = \frac{Eh^3}{12(1-\mu^2)} \tag{10.4d}$$

We will assume that $M_{\alpha\beta} = M_{\beta\alpha}$, with the justification following the same argument offered in taking $N_{\alpha\beta} = N_{\beta\alpha}$ in Section 4.1. Note that when the transverse shearing strains are excluded, the transverse shear stress resultants can no longer be calculated from the corresponding constitutive relationship, given by the last two rows of Matrix 8–2. Rather, equations (10.1b) and (10.1c) are solved for

$$Q_\beta = \frac{1}{AB} [(BM_{\alpha\beta})_{,\alpha} + (AM_\beta)_{,\beta} + B_{,\alpha}M_{\beta\alpha} - A_{,\beta}M_\alpha] \tag{10.5a}$$

and

$$Q_\alpha = \frac{1}{AB} [(BM_\alpha)_{,\alpha} + (AM_{\beta\alpha})_{,\beta} + A_{,\beta}M_{\alpha\beta} - B_{,\alpha}M_\beta] \tag{10.5b}$$

so that Q_α and Q_β can be computed once the stress couples have been determined. This is analogous to the elementary theory of beams, in which the shear force is evaluated from equilibrium considerations rather than from a constitutive law. Only transverse shearing *strains* can be suppressed; the corresponding *forces* are required for equilibrium.

The formulation of the plate-bending problem is completed as follows:

1. Introduce equations (10.4) into equations (10.5).
2. Differentiate equations (10.5a) and (10.5b) by β and α, respectively.
3. Substitute into equation (10.1a).

The resulting equation expresses equilibrium in the normal direction in terms of the *single displacement* D_n and constitutes the governing equation of the system. This is a classical displacement formulation. Once D_n is evaluated, the stress couples may be computed by differentiation from equations (10.5a) and (10.5b).

We also have the effective transverse shears, $\overline{Q}_\alpha$ and $\overline{Q}_\beta$,

$$\overline{Q}_\alpha = Q_\alpha + \frac{M_{\alpha\beta,\beta}}{B} \tag{10.6a}$$

$$\overline{Q}_\beta = Q_\beta + \frac{M_{\beta\alpha,\alpha}}{A} \tag{10.6b}$$

which we obtain from equations (8.25) and (8.27). Also, the rotations $D_{\alpha\beta}$ and $D_{\beta\alpha}$ are found from equations (7.54g) and (7.54h) with $\gamma_\alpha = \gamma_\beta = 0$:

$$D_{\alpha\beta} = \psi_\alpha = -\frac{1}{A} D_{n,\alpha} \tag{10.7a}$$

$$D_{\beta\alpha} = \psi_\beta = -\frac{1}{B} D_{n,\beta} \tag{10.7b}$$

It is evident from the complexity of the foregoing expressions that the algebra involved in carrying out the steps outlined in the previous paragraphs becomes quite involved. For our purposes, it is sufficient to specialize the further development for two predominant cases: Cartesian coordinates and polar coordinates.

10.1.2 Cartesian Coordinates

For Cartesian coordinates, shown on Figure 10–1, $\alpha = X$, $\beta = Y$, $n = Z$, and $A = B = 1$. The symbol M_x^+ means $M_x + M_{x,x}\,dx$, etc. Then equations (10.4a–c) become

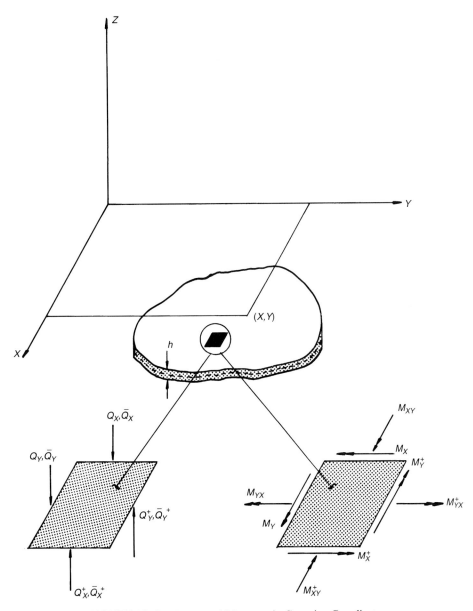

FIGURE 10–1   Forces and Moments in Cartesian Coordinates.

$$M_X = -D(D_{Z,XX} + \mu D_{Z,YY}) \tag{10.8a}$$

$$M_Y = -D(D_{Z,YY} + \mu D_{Z,XX}) \tag{10.8b}$$

$$M_{XY} = M_{YX} = -D(1 - \mu)D_{Z,XY} \tag{10.8c}$$

Likewise, equations (10.5a) and (10.5b) reduce to

$$Q_Y = [M_{XY,X} + M_{Y,Y}]$$
$$= -D[D_{Z,YYY} + \mu D_{Z,XXY} + (1 - \mu)D_{Z,XXY}] \tag{10.9a}$$
$$= -D[D_{Z,YYY} + D_{Z,XXY}]$$

$$Q_X = [M_{X,X} + M_{YX,Y}]$$
$$= -D[D_{Z,XXX} + \mu D_{Z,XYY} + (1 - \mu)D_{Z,XYY}] \tag{10.9b}$$
$$= -D[D_{Z,XXX} + D_{Z,XYY}]$$

Next, we take $\partial/\partial Y$ of equation (10.9a) and $\partial/\partial X$ of equation (10.9b) and substitute into equation (10.1a) to get

$$Q_{X,X} + Q_{Y,Y} + q_Z = 0 \tag{10.10}$$

or

$$-D[D_{Z,XXXX} + 2D_{Z,XXYY} + D_{Z,YYYY}] + q_Z = 0 \tag{10.11}$$

Equation (10.11) may be written concisely as

$$\nabla^2(\nabla^2 D_Z) = \nabla^4 D_Z = \frac{q_Z}{D} \tag{10.12}$$

where

$$\nabla^2(\ ) = (\ )_{,XX} + (\ )_{,YY} \tag{10.13a}$$

and

$$\nabla^4(\ ) = \nabla^2[\nabla^2(\ )] = (\ )_{,XXXX} + 2(\ )_{,XXYY} + (\ )_{,YYYY} \tag{10.13b}$$

$\nabla^2(\ )$ and $\nabla^4(\ )$ are the *Laplacian* and *biharmonic operators*, respectively,[1] and (10.12) is commonly known as *the plate equation*.

The homogeneous part of the governing equation of the engineering theory of medium thin plates is the *biharmonic equation*

$$\nabla^4 D_Z = 0 \tag{10.14}$$

The recognition of this form is important for at least two reasons. First, there is a considerable reservoir of mathematical knowledge dealing with biharmonic equations which can be transferred to our plate theory.[2] Second, because $\nabla^4(\ )$ is invariant, we may obtain the corresponding equation for any other system of orthogonal curvilinear coordinates by simply transforming the biharmonic operator. We will verify this for polar coordinates in the next section and utilize the invariant property of the ∇^2 and ∇^4 operators repeatedly.

Of particular interest with respect to the preceding formulation is the reduction of the system to an equation with a single dependent variable. Stepping back through the deviation, we find that the key step was the expression of the extensional and shearing strains in terms of the normal displacement D_n only, equations (10.2) and (10.3). In order to interpret this from a physical standpoint, we examine a segment of the middle plane along a normal section traced by the X–Z plane as shown in Figure 10–2.

We show the segment deformed in the positive Z direction, where both the displacement D_Z and the derivative $D_{Z,X}$ are positive. Also, we recall the basic assumptions listed in Table 1–1. We focus on an arbitrary point ① which initially lies at a distance ζ from the undeformed middle plane. Because all geometry is referred to the middle plane, we locate the corresponding point on the undeformed middle plane, point ②. During deformation, point ② displaces a distance D_Z parallel to the *original* Z direction (assumption [2]) to point to ③. Because we are neglecting transverse shearing contributions, the deformed position of point ① must lie on the normal to the deformed middle plane, $D_{Z,X}$, erected at point ③ (assumption [3]) Likewise, the thickness remains constant during deformation (assumption [4]), so that the point must remain a distance ζ from the deformed middle plane. The final position is denoted by ④.

Note that point ④ also has a displacement in the X direction denoted by D_X. Because D_X is negative corresponding to the positive sense of $D_{Z,X}$, as shown on Figure 10–2,

$$D_X = -\zeta D_{Z,X} \tag{10.15}$$

Because the Lamé parameters are constant, the strain in the X-direction is found from equation (7.54a), as simply

$$\varepsilon_X(\zeta) = D_{X,X}$$
$$= -\zeta D_{Z,XX} \tag{10.16}$$

which corresponds to equation (10.2a), after κ_α in equation (10.3a) is specialized for Cartesian coordinates.

An identical argument referred to a normal section traced by the YZ plane gives

$$D_Y = -\zeta D_{Z,Y} \tag{10.17}$$

and

$$\varepsilon_Y(\zeta) = -\zeta D_{Z,YY} \tag{10.18}$$

corresponding to equation (10.2b), with equation (10.3b) suitably specialized. Equations (10.15) and (10.17) clearly illustrate the coupling of the *in-plane* displacements to the *normal* displacement, which is the key to the relatively simple form of the plate equation.

The in-plane shearing strain, adapted from equation (7.54c), is

$$\omega = D_{X,Y} + D_{Y,X}$$
$$= -2\zeta D_{Z,XY} \tag{10.19}$$

which matches equation (10.2c) with τ specialized for Cartesian coordinates.

To complete the formulation, we also list the effective transverse shears $\bar{Q}_X$ and $\bar{Q}_Y$. From equations (10.6), (10.8c), and (10.9).

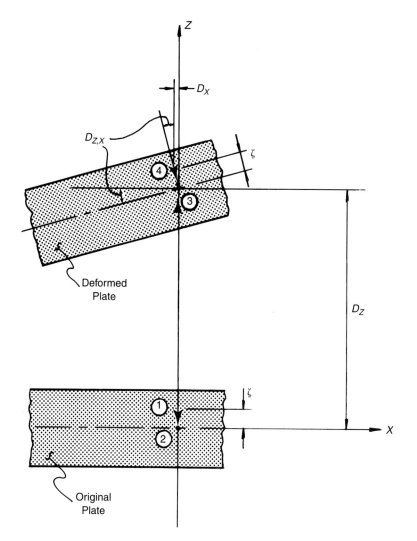

FIGURE 10–2   Displacements for a Plate.

$$\overline{Q}_X = Q_X + M_{XY,Y}$$
$$= -D[D_{Z,XXX} + (2 - \mu)D_{Z,XYY}]$$
(10.20a)

and

$$\overline{Q}_Y = Q_Y + M_{XY,X}$$
$$= -D[D_{Z,YYY} + (2 - \mu)D_{Z,XXY}]$$
(10.20b)

The rotations D_{XY}, and D_{YX} are found from equation (10.7) as

$$D_{XY} = \psi_X = -D_{Z,X} \qquad (10.21a)$$

$$D_{YX} = \psi_Y = -D_{Z,Y} \qquad (10.21b)$$

It is notable that equation (10.11) is the two-dimensional counterpart of the familiar Bernoulli-Euler equation, which is the cornerstone of beam theory. Taking $D_Z = D_Z(X)$ or $D_Z(Y)$ only in equation (10.11) and neglecting Poisson's ratio in the elastic constant D, we have the governing differential equation for a beam of unit width and depth h.

Equation (10.11) also serves to articulate the difference between a plate and a two-dimensional gridwork of beams. Whereas the first and third terms, $D_{Z,XXXX}$ and $D_{Z,YYYY}$ clearly describe *bending* resistance in the respective directions, the second term, $2D_{Z,XXYY}$, represents the contribution of the *twisting* rigidity to the load resistance. This is easily verified by examining equations (10.8) in which the directional derivatives $D_{Z,XX}$ and $D_{Z,YY}$ form the bending stress couples, and the mixed derivative $D_{Z,XY}$ produces the twisting stress couple.

The presence of the twisting couples discerns between two-dimensional flexural behavior and plate behavior. In this regard, it is interesting to note that in 1789, J. Bernoulli proposed approximating a plate with a system of perpendicular intersecting beams, based on an earlier model of a flexible membrane by Euler. S. Germain, aided by Lagrange, incorporated the twisting term to find the homogeneous part of equation (10.11), and the correct strain energy expression was derived by Poisson. A comprehensive plate theory, including loading terms and boundary conditions, was finally advanced by Navier in 1820[3], although the final resolution of the free-edge boundary conditions in terms of effective forces appeared somewhat later, as discussed in Section 8.2.3. We shall thoroughly explore Navier's approach later in this chapter.

10.1.3   Polar Coordinates

For polar coordinates, shown on Figure 10–3, $\alpha = R$, $\beta = \theta$, and $n = Z$. Obviously, $A = 1$, but considering the s_θ coordinate line,

$$ds_\theta = R\, d\theta \qquad (10.22)$$

so that $B = R$. Because $B = B(\alpha)$ in the general formulation, the $B_{,\alpha}$ term is $R_{,R} = 1$ and $A_{,\beta} = 0$. Also shown in Figure 10–3 are the stress resultants and stress couples corresponding to the polar coordinates, as adapted from Figure 3–2.

Equations (10.4a–c) become

$$M_R = -D\left[D_{Z,RR} + \frac{\mu}{R}\left(\frac{1}{R}D_{Z,\theta\theta} + D_{Z,R}\right)\right] \qquad (10.23a)$$

$$M_\theta = -D\left[\frac{1}{R^2}D_{Z,\theta\theta} + \frac{1}{R}D_{Z,R} + \mu D_{Z,RR}\right] \qquad (10.23b)$$

$$M_{R\theta} = M_{\theta R} = -D(1-\mu)\left[\frac{1}{R}\left(D_{Z,R\theta} - \frac{1}{R}D_{Z,\theta}\right)\right] \qquad (10.23c)$$

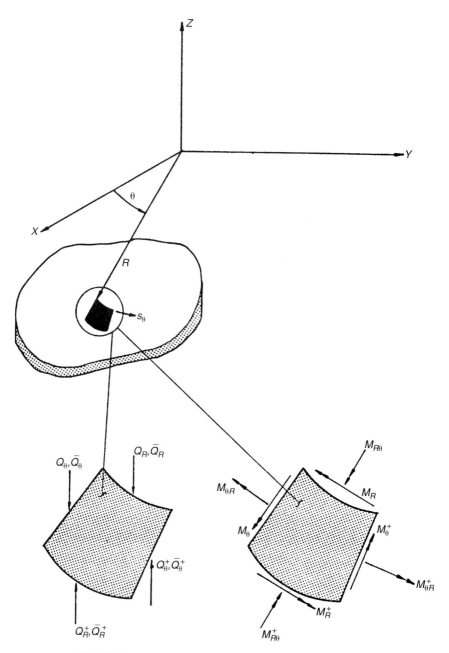

FIGURE 10–3   Forces and Moments in Polar Cooordinate.

Likewise, equations (10.5) specialize to

$$Q_\theta = \frac{1}{R}[(RM_{R\theta}),_R + M_{\theta,\theta} + M_{R\theta}]$$

$$= \frac{1}{R}[2M_{R\theta} + RM_{R\theta,R} + M_{\theta,\theta}] \tag{10.24a}$$

$$= -D\frac{1}{R}\left[\frac{1}{R}D_{Z,R\theta} + D_{Z,RR\theta} + \frac{1}{R^2}D_{Z,\theta\theta\theta}\right]$$

and

$$Q_R = \frac{1}{R}[(RM_R),_R + M_{R\theta,\theta} - M_\theta]$$

$$= \frac{1}{R}[M_R + RM_{R,R} + M_{R\theta,\theta} - M_\theta] \tag{10.24b}$$

$$= -D\left[D_{Z,RRR} + \frac{1}{R}D_{Z,RR} - \frac{1}{R^2}D_{Z,R} + \frac{1}{R^2}D_{Z,R\theta\theta} - \frac{2}{R^3}D_{Z,\theta\theta}\right]$$

To express equation (10.1a) in polar coordinates,

$$(RQ_R),_R + Q_{\theta,\theta} + Rq_Z = 0 \tag{10.25}$$

in terms of D_Z, we take the appropriate derivatives of equations (10.24) to get

$$D_{Z,RRRR} + \frac{2}{R}D_{Z,RRR} - \frac{1}{R^2}(D_{Z,RR} - 2D_{Z,RR\theta\theta})$$

$$+ \frac{1}{R^3}(D_{Z,R} - 2D_{Z,R\theta\theta}) + \frac{1}{R^4}(4D_{Z,\theta\theta} + D_{Z,\theta\theta\theta\theta}) = \frac{q_Z}{D} \tag{10.26}$$

It may be verified that equation (10.26) can be written as the biharmonic equation

$$\nabla^4 D_Z = \nabla^2[\nabla^2 D_Z] = \frac{q_Z}{D} \tag{10.27}$$

where, in polar coordinates,

$$\nabla^2(\ ) = (\ ),_{RR} + \frac{1}{R}(\ ),_R + \frac{1}{R^2}(\ ),_{\theta\theta} \tag{10.28a}$$

or

$$\nabla^2(\ ) = \frac{1}{R}[R(\ ),_R],_R + \frac{1}{R^2}(\ ),_{\theta\theta} \tag{10.28b}$$

Comparing equations (10.12) and (10.27) demonstrates the invariant property of the biharmonic operator. That is, equation (10.12) may be directly transformed into any other coordinate system by transforming $\nabla^4(\ ) = \nabla^2[\nabla^2(\ )]$ accordingly.

It is sometimes convenient to write

$$Q_R = -D[\nabla^2 D_Z]_{,R} \tag{10.29a}$$

and

$$Q_\theta = -D\frac{1}{R}[\nabla^2 D_Z]_{,\theta} \tag{10.29b}$$

which are easily verified.

We also have the effective transverse shears

$$\overline{Q}_R = Q_R + \frac{M_{R\theta,\theta}}{R} \tag{10.30a}$$

$$\overline{Q}_\theta = Q_\theta + M_{R\theta,R} \tag{10.30b}$$

and the rotations

$$D_{R\theta} = \psi_R = -D_{Z,R} \tag{10.30c}$$

$$D_{\theta R} = \psi_\theta = -\frac{1}{R}D_{Z,\theta} \tag{10.30d}$$

A preponderance of circular plate problems encountered in engineering are axisymmetrically loaded, in the same manner as the axisymmetrically loaded shell of revolution treated in Section 4.3.2. For such cases, all functions of θ drop out of the preceding equations, so that

$$M_R = -D\left(D_{Z,RR} + \frac{\mu}{R}D_{Z,R}\right) \tag{10.31a}$$

$$M_\theta = -D\left(\frac{1}{R}D_{Z,R} + \mu D_{Z,RR}\right) \tag{10.31b}$$

$$M_{R\theta} = 0 \tag{10.31c}$$

Also,

$$Q_\theta = 0 \tag{10.32a}$$

$$Q_R = D\left(D_{Z,RRR} + \frac{1}{R}D_{Z,RR} - \frac{1}{R^2}D_{Z,R}\right) \tag{10.32b}$$

and

$$\overline{Q}_R = Q_R \tag{10.32c}$$

Finally, the governing equation simplifies to

$$D_{Z,RRRR} + \frac{2}{R}D_{Z,RRR} - \frac{1}{R^2}D_{Z,RR} + \frac{1}{R^3}D_{Z,R} = \frac{q_Z}{D} \tag{10.33a}$$

or

$$\nabla_a^4 D_Z = \frac{q_Z}{D} \tag{10.33b}$$

where $\nabla_\alpha^4(\ )$ signifies the axisymmetric biharmonic operator.

10.1.4   Force and Moment Transformations

Once we have selected the α and β coordinates, we may evaluate the transverse shear resultants and stress couples from the preceding equations only on planes parallel to the s_α and s_β coordinate lines. Frequently, we are interested in values on other planes, because the maximum transverse shear and/or bending moment may act in directions other than those coinciding with the coordinate lines.

The approach is similar to that used to evaluate principal stresses in linear elasticity. In Figure 10–4(a), a second set of orthogonal axes (ξ, η), which represent arbitrary directions for which we wish to evaluate the forces and moments, are defined. These axes are oriented by the angle γ measured clockwise in the $\alpha - \beta$ plane. On the differential rectangular element in the inset, bound by s_α and s_β coordinate lines, we superimpose the rotated coordinates and consider shaded triangle ①, which is enlarged in Figure 10–4(b). The positive senses of the stress resultants acting on the element are obtained from Figure 3–2.

Summing forces in the Z direction, we have

$$-Q_\alpha \Delta s_\xi \sin \gamma - Q_\beta \Delta s_\xi \cos \gamma + Q_\eta \Delta s_\xi = 0$$

or

$$Q_\eta = Q_\alpha \sin \gamma + Q_\beta \cos \gamma \tag{10.34}$$

From the shaded triangle ②, we may find

$$Q_\xi = Q_\alpha \cos \gamma - Q_\beta \sin \gamma \tag{10.35}$$

Next, we treat the stress couples acting on the triangular element ①, as shown in Figure 10–4(c). Again, the positive senses of the couples are set from Figure 3–2. From moment equilibrium about the ξ axis, we have

$$-M_\eta \Delta s_\xi + M_\alpha \sin^2 \gamma \Delta s_\xi + M_\beta \cos^2 \gamma \Delta s_\xi + M_{\beta\alpha} \cos \gamma \sin \gamma \Delta s_\xi + M_{\alpha\beta} \sin \gamma \cos \gamma \Delta s_\xi = 0$$

With $M_{\beta\alpha} = M_{\alpha\beta}$,

$$M_\eta = M_\alpha \sin^2 \gamma + M_\beta \cos^2 \gamma + 2M_{\alpha\beta} \sin \gamma \cos \gamma \tag{10.36a}$$

This equation may be written in terms of the double angle 2γ as

$$M_\eta = \frac{1}{2}(M_\alpha + M_\beta) - \frac{1}{2}(M_\alpha - M_\beta) \cos 2\gamma + M_{\alpha\beta} \sin 2\gamma \tag{10.36b}$$

Moment equilibrium about the η axis yields

$$M_{\eta\xi} \Delta s_\xi - M_\alpha \sin \gamma \cos \gamma \Delta s_\xi + M_\beta \cos \gamma \sin \gamma \Delta s_\xi + M_{\alpha\beta} \sin^2 \gamma \Delta s_\xi - M_{\beta\alpha} \cos^2 \gamma \Delta s_\xi = 0$$

from which

$$M_{\eta\xi} = (M_\alpha - M_\beta) \sin \gamma \cos \gamma - M_{\alpha\beta}(\sin^2 \gamma - \cos^2 \gamma) \tag{10.37a}$$

or, in terms of the double angle,

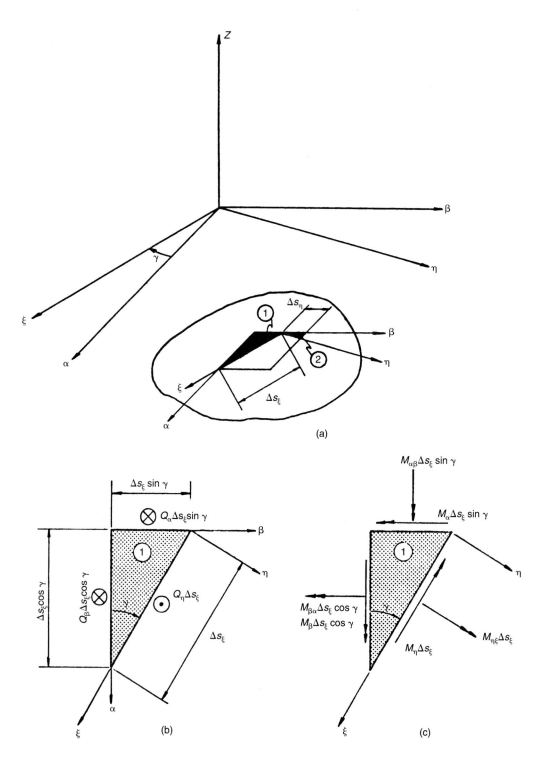

FIGURE 10–4 Force and Moments in Rotated Coordinates.

$$M_{\eta\xi} = \frac{1}{2}(M_\alpha - M_\beta)\sin 2\gamma + M_{\alpha\beta}\cos 2\gamma \tag{10.37b}$$

A similar calculation for triangle ② gives

$$M_\xi = M_\alpha \cos^2\gamma + M_\beta \sin^2\gamma - 2M_{\alpha\beta}\sin\gamma\cos\gamma \tag{10.38a}$$

or

$$M_\xi = \frac{1}{2}(M_\alpha + M_\beta) + \frac{1}{2}(M_\alpha - M_\beta)\cos 2\gamma - M_{\alpha\beta}\sin 2\gamma \tag{10.38b}$$

and

$$M_{\xi\eta} = M_{\eta\xi} \tag{10.39}$$

Equations (10.34) to (10.39) enable the transverse shear resultants and the stress couples to be calculated on any specified section of the plate. They play the same role in the theory of plates as the familiar Mohr's circle transformations occupy in the analysis of stresses at a point in the theory of elasticity.

10.1.5 Alternate Formulation

In the preceding sections, we have almost routinely referred to the biharmonic equation as *the* plate equation and, indeed, this is common terminology. Nevertheless there are alternative formulations, one of which may be anticipated when we reflect on our previously discussed analogies to beam theory. Just as in beam theory, we can separate the fourth-order governing equation, *deflection = f(load)*, into two coupled second-order equations, *deflection = f (bending moment)* and *bending moment = f (load)*. The latter form is directly applicable to statically determinate beams and is the basis for some methods of analyzing statically indeterminate beams as well.

For the plate problem, we add equations (10.36b) and (10.38b) to get

$$M_\xi + M_\eta = M_\alpha + M_\beta \tag{10.40}$$

Equation (10.40) states that the sum of the flexural stress couples acting about two mutually orthogonal axes is *invariant* with respect to coordinate transformation. Thus, we may proceed using the specialized Cartesian coordinate expressions for $M_X + M_Y$ while retaining complete generality. Summing equations (10.8a) and (10.8b),

$$M_X + M_Y = -D(1 + \mu)(D_{Z,XX} + D_{Z,YY})$$
$$= -D(1 + \mu)\nabla^2 D_Z \tag{10.41}$$

Equations (10.40) and (10.41) constitute a formal proof of the *invariant* property of ∇^2.

We now set

$$\tilde{M} = \frac{M_X + M_Y}{1 + \mu} = -D\nabla^2 D_Z \tag{10.42}$$

so that

$$\nabla^2 D_Z = -\frac{1}{D}\tilde{M} \tag{10.43}$$

Then, from equations (10.12) and (10.43),

$$\nabla^4 D_Z = \nabla^2(\nabla^2 D_Z) = -\frac{1}{D}\nabla^2\tilde{M} = \frac{q_Z}{D} \tag{10.44}$$

or

$$\nabla^2\tilde{M} = -q_Z \tag{10.45}$$

Equations (10.43) and (10.45) comprise the alternate formulation and allow the plate problem to be solved in an analogous manner as a uniformly stretched, laterally loaded membrane. The membrane problem is regarded as somewhat simpler to treat numerically.[4]

10.1.6 Boundary Conditions

The boundary conditions corresponding to the theory of plates without transverse shearing strains are easily synthesized from Section 8.2.2 and Figure 8–4. Because no extensional forces are considered, three basic conditions apply: fixed or clamped, simply supported, and free. These are listed for boundaries $\alpha = \bar{\alpha}$ and $\beta = \bar{\beta}$ in Table 10–1. They are stated in homogeneous form but may be generalized, as described at the conclusion of this section.

     With respect to the *fixed* condition, often called a *clamped* boundary, note from equations (10.7) that $D_{\alpha\beta} = 0$ and $D_{\beta\alpha} = 0$ imply $D_{Z,\alpha} = 0$ and $D_{Z,\beta} = 0$, respectively.

     The *simply supported* boundary may be written in an alternate form for Cartesian coordinates when the boundaries are coincident with the coordinate lines. We consider first a boundary defined by $X = \bar{X}$, as shown, for example, in Figure 10–5. The condition of simple support states that $D_Z(\bar{X}, Y) = 0$. Further, it is implied that the boundary is straight, and therefore, $D_{Z,YY}(\bar{X},Y) = 0$. A similar argument with respect to a boundary $Y = \bar{Y}$ leads to $D_{Z,XX}(X,\bar{Y}) = 0$. Equations (10.8a) and (10.8b) for simply supported boundaries become

$$M_X(\bar{X},Y) = -DD_{Z,XX}(\bar{X},Y) = 0 \tag{10.46a}$$

TABLE 10–1   Boundary Conditions for Plates

Condition and Symbol		Boundary	
		$\alpha = \bar{\alpha}$	$\beta = \bar{\beta}$
Fixed or clamped		$D_Z = D_{\alpha\beta} = 0$ or $D_Z = D_{Z,\alpha} = 0$	$D_Z = D_{\beta\alpha} = 0$ or $D_Z = D_{Z,\beta} = 0$
Simply supported		$D_Z = 0;\ M_\alpha = 0$ or $D_Z = D_{Z,\alpha\alpha} = 0$	$D_Z = 0;\ M_\beta = 0$ or $D_Z = D_{Z,\beta\beta} = 0$
Free		$\bar{Q}_\alpha = M_\alpha = 0$	$\bar{Q}_\beta = M_\beta = 0$

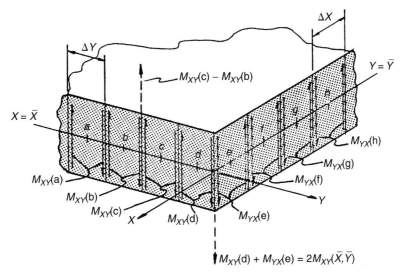

FIGURE 10–5   Forces at Right-angle Corner.

and

$$M_Y(X,\overline{Y}) = -DD_{Z,YY}(X,\overline{Y}) = 0 \qquad (10.46b)$$

which, in view of the preceding discussion, generalize to

$$M_X(\overline{X},Y) = -D\nabla^2 D_Z(\overline{X},Y) = 0 \qquad (10.47a)$$

$$M_Y(X,\overline{Y}) = -D\nabla^2 D_Z(X,\overline{Y}) = 0 \qquad (10.47b)$$

Thus, on each boundary, if

$$\text{If } M_X = 0 \rightarrow D_{Z,XX} = 0 \rightarrow \nabla^2 D_Z = 0$$

and if

$$M_Y = 0 \rightarrow D_{Z,YY} = 0 \rightarrow \nabla^2 D_Z = 0$$

The statement of the simply supported boundary condition as $\nabla^2 D_Z = 0$ is predicated on the edge being straight, as well as undeflecting. In polar coordinates, the second term of equation (10.28a) would not vanish on a simply supported boundary, since it is proportional to the rotation $D_{R\theta}$, equation (10.30c), which is unconstrained. Therefore, $M_R = 0$ does *not* imply $\nabla^2 D_Z = 0$.

Both idealizations of supported boundaries for plates, *fixed* and *simply supported*, may be difficult to achieve in practice. First of all, the provision of a support transverse to the plate denoted by $D_Z = 0$, which is required in both cases, often introduces a constraint in the plane of the plate. In turn, this would lead to in-plane extensional and/or shearing forces, similar to those present in a shell. Further, the fully clamped boundary corresponding to the *fixed* condition makes the release of the in-plane constraint even more difficult. On the other hand, the presence of in-plane constraints and the corresponding forces in transversely loaded plates may impart considerable additional strength into a plate which has been proportioned based on elementary plate theory. Of course, the in-plane forces cannot be mobilized to resist transverse loading until the plate has undergone significant transverse displacement, when the resistance mechanism becomes more like a shell. These deformations may render the plate unserviceable (i.e., an initially flat plate should remain reasonably flat under load) so that the in-plane forces are useful only for reserve capacity.

Now consider the *free-edge* condition. In Section 8.2.3, we presented the contraction of the general boundary conditions into the Kirchhoff conditions as a result of the suppression of transverse shearing strains. For the plate problem, equations (10.6) indicate that the transverse shear and the twisting stress couple are combined into an effective transverse shear force which is prescribed to vanish on a free edge.

In addition to the free-edge situation, the effective transverse shear forces have some interesting implications for edges that are supported. In Figure 10–5, we show two boundaries of a simply supported rectangular plate, where the Kelvin-Tait representation developed on Figures 8–4 and 8–5 is used to replace the twisting stress couples by closely spaced forces. The positive sense of these couples is established from Figure 10–1.

We first study two adjacent differential segments, say b and c, on the $X = \overline{X}$ boundary. At the junction, we have a *net* contribution to the effective shear equal to

$$M_{XY}(c) - M_{XY}(b) \tag{10.48}$$

Again, the positive sense is established from the direction of Q_X in Figure 10–1(a). If M_{XY} is a continuous function with *continuous* derivatives, equation (10.48) becomes

$$M_{XY,Y}\Delta Y \tag{10.49}$$

which combines with $Q_X \Delta Y$ to give the effective shear $\overline{Q}_X$, as written in equation (10.20a) after the ΔY terms have been canceled. On the other hand, consider the case in which M_{XY} may be continuous, but *not* the first derivative in the Y direction. This corresponds to a jump discontinuity in M_{XY} along the boundary which produces a concentrated force at the point of discontinuity, with the magnitude given by equation (10.48). Conversely, if a concentrated force of a given magnitude is *applied* along the boundary, the twisting stress couple must undergo a jump equal to the magnitude of the applied force.

We carry the argument one step further by considering the intersection of the two boundaries $X = \overline{X}$ and $Y = \overline{Y}$ at the corner. Here, there is definitely a jump discontinuity producing a concentrated force of magnitude

$$M_{XY}(d) + M_{YX}(e) \tag{10.50a}$$

Because $M_{YX} = M_{XY}$, we have the *corner* force

$$R_{\text{corner}} = 2M_{XY} \tag{10.50b}$$

The concept of the corner force is not limited to the intersection of two free boundaries, where it obviously must vanish. In general, any right-angle corner where at least one of the intersecting boundaries can develop M_{XY} or M_{YX} will have a corner force. Physically, a plate having twisting stress couples acting as shown in Figure 10–5 would require that the force

$2M_{XY} (\overline{X},\overline{Y})$ be developed at the corner to prevent the corner from uplifting. The case of non-right-angle intersections and the detailed distribution of the effective transverse shear along the entire boundary will be examined for some specific examples in the following sections.

It should also be mentioned that the homogeneous boundary conditions discussed in this section may be generalized to include prescribed edge displacements and forces. For example, within the definition of fixed and hinged conditions, a nonzero value of the transverse displacement D_Z may be accommodated. This may be of interest in support settlement problems. Also, a specified edge moment may be inserted in place of M_α or $M_\beta = 0$ for the hinged and free conditions. Depending on the solution routine, known edge moments and edge forces may be grouped with the applied loading terms, but should properly be regarded as static boundary conditions. Several illustrations of this generalization are found in the ensuing sections.

Furthermore, a linear contribution of a static-kinematic correspondent pair may be encountered in the description of an elastic support. For example, $M_\alpha (\overline{\alpha},\beta) = kD_{\alpha\beta} (\overline{\alpha},\beta)$ would represent a linear rotational spring with spring constant k along the boundary $\alpha = \overline{\alpha}$.

10.2 RECTANGULAR PLATES

10.2.1 Bending Under Edge Loading

10.2.1.1 General Solution.   Consider a rectangular plate subject to the uniformly distributed edge moments $M_X = M_1$, and $M_Y = M_2$, as shown in Figure 10–6. The sign convention for the moments is consistent with Figure 3–2(b). Also, the deflected middle plane is shown on the figure.

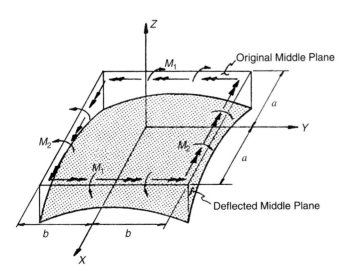

FIGURE 10–6   Rectangular Plate Loaded by Edge Moments.

For this problem, it is expedient to shortcut the formal procedure of solving the governing differential equation, (10.12). Rather, we surmise that the state of bending is uniform throughout the plate. That is,

$$M_X(X,Y) = M_1 \tag{10.51a}$$

$$M_Y(X,Y) = M_2 \tag{10.51b}$$

Then, we may write from equations (10.8a) and (10.8b)

$$M_X = M_1 = -D(D_{Z,XX} + \mu D_{Z,YY}) \tag{10.52a}$$

$$M_Y = M_2 = -D(D_{Z,YY} + \mu D_{Z,XX}) \tag{10.52b}$$

We eliminate $D_{Z,YY}$ to get

$$D_{Z,XX} = -\frac{(M_1 - \mu M_2)}{D(1 - \mu^2)} \tag{10.53}$$

which may be integrated to give

$$D_{Z,X} = -\frac{(M_1 - \mu M_2)}{D(1 - \mu^2)} X + f_1(Y) + C_1 \tag{10.54}$$

and

$$D_Z(X,Y) = -\frac{(M_1 - \mu M_2)}{2D(1 - \mu^2)} X^2 + X f_1(Y) + f_2(Y) + C_1 X + C_2 \tag{10.55}$$

We may proceed further by noting from equations (10.51) to (10.53) that $D_{Z,YY}$ must also be constant. Therefore,

$$D_{Z,YY} = X f_1(Y)_{,YY} + f_2(Y)_{,YY} = \text{constant} \tag{10.56}$$

Because there is no X in the second term of equation (10.56) to cancel the variable X in the first term, $f_1(Y)_{,YY} = 0$ and

$$f_1(Y) = C_3 Y + C_4 \tag{10.57a}$$

$$f_2(Y) = C_5 Y^2 + C_6 Y + C_7 \tag{10.57b}$$

and

$$D_{Z,YY} = 2C_5 \tag{10.57c}$$

We then substitute equation (10.55) into equation (10.52b), taking into account equations (10.53) and (10.57c), to find

$$M_2 = -D\left[2C_5 - \frac{\mu}{D} \frac{(M_1 - \mu M_2)}{(1 - \mu^2)}\right] \tag{10.58a}$$

from which

$$C_5 = -\frac{(M_2 - \mu M_1)}{2D(1 - \mu^2)} \tag{10.58b}$$

Also, from equations (10.8c), (10.54), and (10.57a),

$$M_{XY} = -D(1 - \mu)C_3 \tag{10.59}$$

Upon substituting equation (10.57a) and (10.57b) into equations (10.55), we may combine the term XC_4, which arises from $Xf_1(Y)$, with the term C_1X to produce a term C_8X, and also absorb the constant C_7 into C_2 to make C_9. Then, from equations (10.55) to (10.59), we have the general solution

$$D_Z = \frac{-1}{2D(1 - \mu^2)} [(M_1 - \mu M_2)X^2 + (M_2 - \mu M_1)Y^2] \tag{10.60}$$
$$+ C_3 XY + C_8 X + C_6 Y + C_9$$

At this point, we inquire further into the nature of the constants in equation (10.60). Constant C_3 multiplies the term XY and thus will contribute to M_{XY} via equation (10.8c), as shown in equation (10.59). The remaining constants C_8, C_6, and C_9 are coefficients of linear or constant terms; therefore, they cannot contribute to the stress couples or to the transverse shear stress resultants which are dependent on quadratic and cubic terms respectively, as is evident from equations (10.8) and (10.9). These are *rigid-body* terms which depend only on the location of the origin and the orientation of the coordinate system, but *not* on the applied loading.

With the origin of the coordinate system placed in the center of the plate, as shown in Figure 10–6, double symmetry is created. Therefore, $D_Z(a, Y) = D_Z(-a,Y)$, from which

$$C_3 = C_8 = 0 \tag{10.61a}$$

Also, $D_Z(X,b) = D_Z(X,-b)$ so that

$$C_6 = 0 \tag{10.61b}$$

Because $C_3 = 0$, equation (10.59) gives

$$M_{XY} = 0 \tag{10.62}$$

This leaves only C_9 still to be determined.

Because we have no readily apparent kinematic boundary conditions to constrain the displacement, we may arbitrarily select a point on the deflected plate as the reference for measuring D_Z. It is convenient to choose the origin $X = 0$, $Y = 0$, as a point of *zero* displacement, i.e.,

$$D_Z(0,0) = 0 \tag{10.63}$$

meaning that Z is measured from the middle surface of the *deflected* plate. Once again, we stress that this arbitrary choice does not affect the stress resultants or couples, since C_9 is a rigid-body term. Imposing equation (10.63) on equation (10.60) gives $C_9 = 0$ so that, in view of equations (10.61),

$$D_Z(X,Y) = \frac{-1}{2D(1 - \mu^2)} [(M_1 - \mu M_2)X^2 + (M_2 - \mu M_1)Y^2] \tag{10.64}$$

Finally, because $D_{Z,XX}$ and $D_{Z,YY}$ are constant, $Q_X = Q_Y = 0$ from equation (8.9). Likewise, with $M_{XY} = 0$, $\overline{Q}_X = 0$, and $\overline{Q}_Y = 0$ from equation (10.20), and there are no corner forces.

We have treated the rectangular plate subject to edge moments in detail. It is noteworthy that the computational effort was substantially reduced by starting with a physically plausible state of constant bending in each direction and then proceeding to derive a consistent solution for the problem at hand. Although this approach may appear to be *ad hoc* and is obviously not completely general, it is often productive. It is in the spirit of the semi-inverse method, attributed to St. Venant,[5] that is so widely applied in the theory of elasticity.

It is also important to reflect on the role of the preceding solution in the engineering application of the theory of plates. First, we will use some special cases to demonstrate that certain loading combinations are valuable for experimental verification of the theory. Second, we will find that the preponderance of readily obtainable analytic solutions exist for plates with simply supported boundaries. The well-known flexibility method of structural analysis, outlined in Section 8.3.1, suggests that solutions for plates with fixed boundaries can be obtained by combining the corresponding solution for a simply supported boundary condition with a solution resulting from edge moments of appropriate magnitude, so as to satisfy the zero-rotation condition at the boundaries.

10.2.1.2 Parabolic Bending.   Referring to Figure 10–6, we take $M_1 = M_2 = M$, which represents edge moments of an equal intensity applied uniformly around the boundary of the plate. The solution, equation (10.64), reduces to

$$D_Z(X,Y) = -\frac{M}{2D(1 + \mu)}(X^2 + Y^2)$$ (10.65)

which is the equation of a paraboloid of revolution. We previously encountered a surface *initially* in this form in our study of translational shells, equation (6.66).

10.2.1.3 Cylindrical Bending.   Another interesting case is to determine what combination of edge moments would be required to produce a cylindrical deflected shape. Because a cylindrical surface is only singly curved, we want $D_Z = D_Z(X)$ or $D_Z = D_Z(Y)$. From equation (10.64), it appears that $M_1 = \mu M_2$ or $M_2 = \mu M_1$ would be required to achieve these conditions. We illustrate $D_Z = D_Z(Y)$ on Figure 10–7 and choose $M_2 = M$ and $M_1 = \mu M$. The deflected shape is given by

$$D_Z = -\frac{M}{2D}Y^2$$ (10.66)

It is instructive to interpret this solution from a physical standpoint. At first thought, one might surmise that a plate could be bent into a cylindrical shape by simply applying equal couples to two parallel edges and leaving the other edges unloaded. Closer scrutiny reveals that the Poisson effect under this loading would produce a contraction of the top fibers and an extension of the bottom fibers, with a corresponding curvature as shown in the cross-section on Figure 10–7. Thus, the moment μM is required on the other edges to produce compensating extension in the top and contraction in the bottom so as to restore the rectangular cross-section. Parenthetically, this simple problem motivates a possible experimental means of determining Poisson's ratio, μ. We consider a plate initially deflected by a known couple M_2 acting on the edges parallel to the X-axis, and then loaded incrementally by a cou-

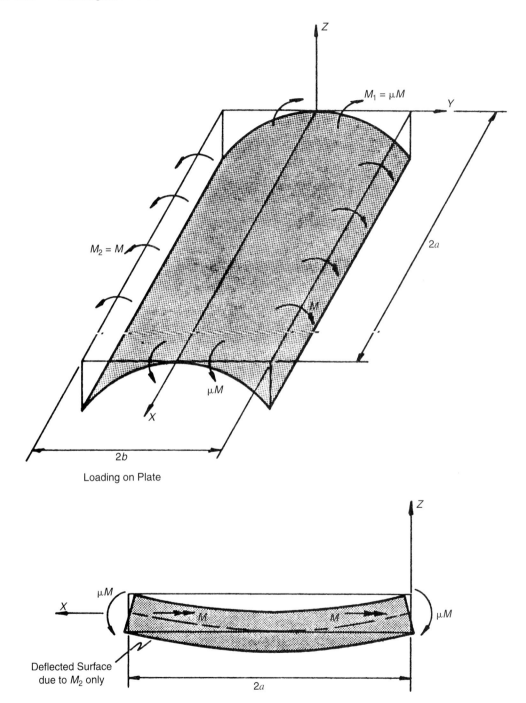

Loading on Plate

Cross-Section at $Y = 0$

FIGURE 10–7   Cylindrical Bending of Rectangular Plate.

ple M_1 along the edges parallel to the Y-axis. Observing the deflection in the X direction, the ratio of M_1/M_2 for which a line on the surface parallel to the X-axis becomes straight would give the effective value of μ. As mentioned in Chapter 8, the determination of the appropriate elastic constants to insert into a solution is often a difficult proposition. The utilization of experiments based on known analytical solutions in which the constants are indirectly determined is often productive in this regard.

10.2.1.4 Anticlastic Bending.

The loading case $M_1 = -M_2 = M$ is shown in Figure 10–8(a). Substituting into equation (10.64), we have

$$D_Z(X,Y) = \frac{-M}{2D(1-\mu)}(X^2 - Y^2) \tag{10.67}$$

which we recognize from equation (6.23) as the equation of the hyperbolic paraboloid, describing an anticlastic surface.

We are intrigued by this result because we know that the two sets of real characteristics should occur as straight lines on the deflected surface. In our study of shells of this shape, we found the families of straight lines quite important in the physical understanding of the load-resistance mechanism.

If we restrict our inquiry to square plates, we find from Section 6.4.1 that the straight lines are oriented at $45°$ to the X-Y coordinates, as shown in Figure 10–8(b). We wish to examine the state of stress along these straight lines, which correspond to a rotation of $\gamma = +45°$ as defined in Figure 10–4. We take ξ and η as $\tilde{X}$ and $\tilde{Y}$, respectively, as shown in Figure 10–8(b), and evaluate $M_{\tilde{X}\tilde{X}}$, $M_{\tilde{X}\tilde{Y}}$, and $M_{\tilde{Y}\tilde{Y}}$ from equations (10.36b), (10.37b), and (10.38b). With $M_\alpha = M_X = M$, $M_\beta = M_Y = -M$, and $M_{\alpha\beta} = M_{XY} = 0$, we find

$$M_{\tilde{X}\tilde{X}} = M_{\tilde{Y}\tilde{Y}} = 0 \tag{10.68}$$

and

$$M_{\tilde{X}\tilde{Y}} = M \tag{10.69}$$

Constructing the Kelvin-Tate argument, Figure 10–8(c), we see that the state of stress on the portion of the plate bound by the $45°$ straight lines which connect the midpoints of each side is equivalent to that obtained by taking a square plate with sides of length $\sqrt{2}a$, loaded only with opposing corner forces of magnitude $2M$. Naturally, the moments within the perimeter of the straight lines on the original plate are identical to those on the corner-loaded square plate.

The bending of a plate into an anticlastic surface by pairs of opposed corner forces was first investigated by Lamb,[6] who is renown for his work in hydrodynamics. It is obvious that the situation depicted in Figure 10–8(c) is relatively simple to simulate physically, and classical experiments based on this configuration were performed by Nádai to verify the flexural theory of plates.[7]

10.2.2 Navier Solution for Simply Supported Rectangular Plates

The governing equation for the bending of a rectangular plate under an arbitrary distributed loading $q_Z(X,Y)$, shown in Figure 10–9, is written from equation (10.12) as

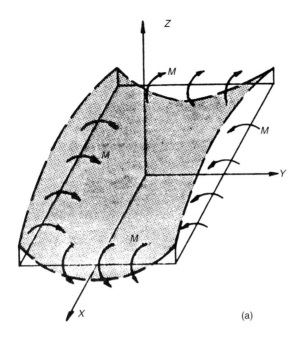

(a)

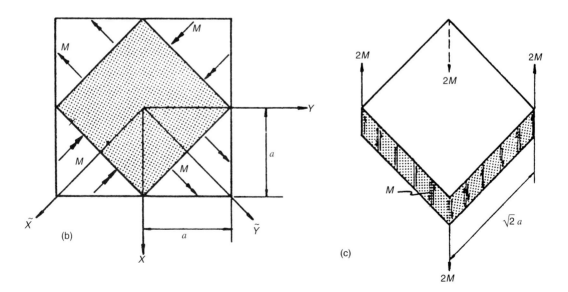

(b)

(c)

FIGURE 10–8   Anticlastic Bending of Rectangular Plate.

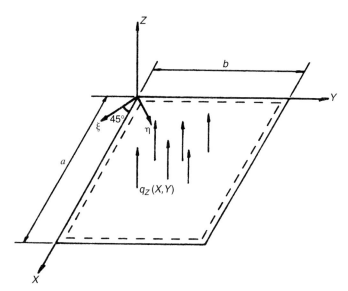

FIGURE 10–9 Arbitrary Tansverse Loading on Rectangular Plate.

$$\nabla^4 D_Z(X,Y) = \frac{q_Z(X,Y)}{D} \tag{10.70}$$

10.2.2.1 General Solution. Among his many accomplishments, Navier is credited with developing the first satisfactory theory of plates, in the form of equation (10.70).[3] A widely used solution technique, which bears his name, is to expand both D_Z and q_Z into double Fourier sine series, to solve the resulting equations for a general harmonic, and to obtain the complete solution by superimposing a sufficient number of harmonic components to attain an acceptable precision. We may recall that Fourier series were introduced for shells of revolution in Section 5.1.2. Here, sine series are chosen because the simply supported boundary conditions are satisfied identically, as we will show later.

Proceeding, we assume

$$D_Z(X,Y) = \sum_{j=1}^{\infty} \sum_{k=1}^{\infty} D_Z^{jk} \sin j\pi \frac{X}{a} \sin k\pi \frac{Y}{b} \tag{10.71}$$

where D_Z^{jk} is the Fourier coefficient for the displacement in the general harmonics j and k. Then,

$$\nabla^4 D_Z = \sum_{j=1}^{\infty} \sum_{k=1}^{\infty} \left[\left(\frac{j\pi}{a} \right)^4 + 2 \left(\frac{j\pi}{a} \right)^2 \left(\frac{k\pi}{b} \right)^2 + \left(\frac{k\pi}{b} \right)^4 \right] D_Z^{jk} \sin j\pi \frac{X}{a} \sin k\pi \frac{Y}{b}$$

$$= \pi^4 \sum_{j=1}^{\infty} \sum_{k=1}^{\infty} \left[\left(\frac{j}{a} \right)^2 + \left(\frac{k}{b} \right)^2 \right]^2 D_Z^{jk} \sin j\pi \frac{X}{a} \sin k\pi \frac{Y}{b} \tag{10.72}$$

The rhs of equation (10.70) is expanded as

$$\frac{q_Z(X,Y)}{D} = \frac{1}{D} \sum_{j=1}^{\infty} \sum_{k=1}^{\infty} q_Z^{jk} \sin j\pi \frac{X}{a} \sin k\pi \frac{Y}{b} \tag{10.73}$$

where q_Z^{jk} is the Fourier coefficient for the loading.

We proceed to evaluate q_Z^{jk} by the standard method. To keep the notation consistent, j and k should be regarded as *specific*, though general, indices in the following operations; l and m are introduced as the corresponding *variable* indices:

1. Multiply both sides by $\sin m\pi \ (Y/b) \ dY$ and integrate from 0 to b.
2. Note that

$$\int_0^b \sin k\pi \frac{Y}{b} \sin m\pi \frac{Y}{b} \, dY = \begin{cases} 0 & (m \neq k) \\ \frac{b}{2} & (m = k) \end{cases} \tag{10.74a}$$

3. Multiply both sides by $\sin l\pi \ (X/a) \ dX$ and integrate from 0 to a.
4. Note that

$$\int_0^a \sin j\pi \frac{X}{a} \sin l\pi \frac{X}{a} \, dX = \begin{cases} 0 & (l \neq j) \\ \frac{a}{2} & (l = j) \end{cases} \tag{10.74b}$$

Steps (2) and (4) demonstrate the *orthogonality* property that greatly facilitates the evaluation of the Fourier coefficients. Therefore,

$$q_Z^{jk} = \frac{4}{ab} \int_0^a \int_0^b q_Z(X,Y) \sin j\pi \frac{X}{a} \sin k\pi \frac{Y}{b} \, dY \, dX \tag{10.75}$$

We now substitute equations (10.72) and (10.73) into equation (10.70) and write the equation for a specific j and k:

$$\pi^4 \left[\left(\frac{j}{a}\right)^2 + \left(\frac{k}{b}\right)^2 \right]^2 D_Z^{jk} = \frac{q_Z^{jk}}{D} \tag{10.76}$$

where q_Z^{jk} is given by equation (10.75). We then solve equation (10.76) for D_Z^{jk} and substitute into equation (10.71) to write the complete solution as

$$D_Z(X,Y) = \frac{1}{\pi^4 D} \sum_{j=1}^{\infty} \sum_{k=1}^{\infty} q_Z^{jk} \frac{1}{\left[\left(\frac{j}{a}\right)^2 + \left(\frac{k}{b}\right)^2 \right]^2} \sin j\pi \frac{X}{a} \sin k\pi \frac{Y}{b} \tag{10.77}$$

It remains to be shown that equation (10.77) satisfies the simply supported boundary conditions. Referring to Section 10.1.6, we require $D_Z(0, Y) = D_Z(a, Y) = D_Z(X,0) = D_Z(X, b) = 0$ which are obviously satisfied by equation (10.77); and $M_X(0, Y) = M_X(a, Y) = M_Y(X, 0) = M_Y(X, b) = 0$. The latter conditions can be written as $\nabla^2 D_Z(0, Y) = \nabla^2 D_Z(a, Y) = \nabla^2 D_Z(X, 0) = \nabla^2 D_Z(X, b) = 0$ from equation (10.47). Obviously, $[( )_{,XX} + ( )_{,YY}]$ operating on equation (10.77) contains the product $\sin j\pi(X/a) \sin k\pi(Y/b)$ which vanishes along the boundaries. Thus, the identical satisfaction of the simply supported boundary conditions by the Navier double sine series solution is verified. Unfortunately, it is not as easy to find appropriate functions to satisfy other boundary conditions identically.

Although we claim to have found the general solution as equation (10.77), the analysis is not complete. We must proceed through the tedious differentiations required to compute the stress couples and transverse shear resultants. However, we can regard the operations as being routine, because the displacement function is obviously continuous and repeatedly differentiable. This ensures that we do not violate the requirements of internal compatibility, which is obviously a positive feature of the *displacement* formulation selected. Also, it is noteworthy that here we were able to derive a solution sufficiently general to apply for all harmonics; whereas, for example, in the analysis of doubly curved shells of revolution, we found it necessary to treat the symmetrical and antisymmetrical harmonics separately from the general case.

Because we have obtained a quite general solution, we will summarize the various rotations, stress resultants, and couples. These are obtained from equations (10.8), (10.9), (10.20), and (10.21), and all have the form

$$F(X,Y) = F(D_Z) = F_1 \sum_{j=1}^{\infty} \sum_{k=1}^{\infty} q_Z^{jk} \frac{F_2}{\left[\left(\frac{j}{a}\right)^2 + \left(\frac{k}{b}\right)^2\right]^2} F_3\left(j\pi \frac{X}{a}\right) F_4\left(k\pi \frac{Y}{b}\right) \quad (10.78)$$

where the various functions are collected in Table 10–2.

10.2.2.2 Uniformly Distributed Loading.   For a uniformly distributed load,

$$q_Z(X,Y) = q_0 \quad (10.79)$$

equation (10.75) is

$$q_Z^{jk} = \frac{4q_0}{ab} \int_0^a \int_0^b \sin j\pi \frac{X}{a} \sin k\pi \frac{Y}{b} \, dY \, dX$$

$$= \frac{16}{\pi^2 jk} q_0 \quad (j \text{ and } k \text{ odd})$$

or

$$= 0 \quad (j \text{ and } k \text{ even}) \quad (10.80)$$

from which we find

$$D_Z(X,Y) = \frac{16q_0}{\pi^6 D} \sum_{j=1,3\ldots}^{\infty} \sum_{k=1,3\ldots}^{\infty} \frac{1}{jk\left[\left(\frac{j}{a}\right)^2 + \left(\frac{k}{b}\right)^2\right]^2} \sin j\pi \frac{X}{a} \sin k\pi \frac{Y}{b} \quad (10.81)$$

from equation (10.77).

From a computational standpoint, it is instructive to examine the convergence of this series. For simplicity, we take a square plate, $a = b$, and consider the deflection in the center, $X = Y = a/2$.

$$D_Z\left(\frac{a}{2},\frac{a}{2}\right) = \frac{16q_0 a^4}{\pi^6 D} \sum_{j=1,3,\ldots}^{\infty} \sum_{k=1,3,\ldots}^{\infty} \left[\frac{\sin j\frac{\pi}{2} \sin k\frac{\pi}{2}}{jk(j^2 + k^2)^2}\right] \quad (10.82)$$

The term in the square brackets takes on the following values:

TABLE 10–2 Functions for Double Sine Series Solution

$$F(D_Z) = F_1 \sum_{j=1}^{\infty} \sum_{k=1}^{\infty} q_Z^{jk} \frac{F_2}{\left[\left(\dfrac{j}{a}\right)^2 + \left(\dfrac{k}{b}\right)^2\right]^2} F_3\left(j\pi\,\frac{X}{a}\right) F_4\left(k\pi\,\frac{Y}{b}\right)$$

$F(D_Z)$	F_1	F_2	F_3	F_4
D_Z	$\dfrac{1}{\pi^4 D}$	1	$\sin$	$\sin$
$D_{Z,X} = -D_{XY}$	$\dfrac{1}{\pi^3 D}$	$\dfrac{j}{a}$	$\cos$	$\sin$
$D_{Z,Y} = -D_{YX}$	$\dfrac{1}{\pi^3 D}$	$\dfrac{k}{b}$	$\sin$	$\cos$
$D_{Z,XX}$	$\dfrac{1}{\pi^2 D}$	$\left(\dfrac{j}{a}\right)^2$	$-\sin$	$\sin$
$D_{Z,XY}$	$\dfrac{1}{\pi^2 D}$	$\left(\dfrac{j}{a}\right)\left(\dfrac{k}{b}\right)$	$\cos$	$\cos$
$D_{Z,YY}$	$\dfrac{1}{\pi^2 D}$	$\left(\dfrac{k}{b}\right)^2$	$\sin$	$-\sin$
$D_{Z,XXX}$	$\dfrac{1}{\pi D}$	$\left(\dfrac{j}{a}\right)^3$	$-\cos$	$\sin$
$D_{Z,XXY}$	$\dfrac{1}{\pi D}$	$\left(\dfrac{j}{a}\right)^2\left(\dfrac{k}{b}\right)$	$-\sin$	$\cos$
$D_{Z,XYY}$	$\dfrac{1}{\pi D}$	$\left(\dfrac{j}{a}\right)\left(\dfrac{k}{b}\right)^2$	$\cos$	$-\sin$
$D_{Z,YYY}$	$\dfrac{1}{\pi D}$	$\left(\dfrac{k}{b}\right)^3$	$\sin$	$-\cos$
M_X	$\dfrac{1}{\pi^2}$	$\left(\dfrac{j}{a}\right)^2 + \mu\left(\dfrac{k}{b}\right)^2$	$\sin$	$\sin$
M_Y	$\dfrac{1}{\pi^2}$	$\mu\left(\dfrac{j}{a}\right)^2 + \left(\dfrac{k}{b}\right)^2$	$\sin$	$\sin$
M_{XY}	$-\dfrac{1}{\pi^2}$	$(1-\mu)\left(\dfrac{j}{a}\right)\left(\dfrac{k}{b}\right)$	$\cos$	$\cos$
Q_X	$\dfrac{1}{\pi}$	$\left(\dfrac{j}{a}\right)\left[\left(\dfrac{j}{a}\right)^2 + \left(\dfrac{k}{b}\right)^2\right]$	$\cos$	$\sin$
Q_Y	$\dfrac{1}{\pi}$	$\left(\dfrac{k}{b}\right)\left[\left(\dfrac{j}{a}\right)^2 + \left(\dfrac{k}{b}\right)^2\right]$	$\sin$	$\cos$
$\overline{Q}_X$	$\dfrac{1}{\pi}$	$\left(\dfrac{j}{a}\right)\left[\left(\dfrac{j}{a}\right)^2 + (2-\mu)\left(\dfrac{k}{b}\right)^2\right]$	$\cos$	$\sin$
$\overline{Q}_Y$	$\dfrac{1}{\pi}$	$\left(\dfrac{k}{b}\right)\left[(2-\mu)\left(\dfrac{j}{a}\right)^2 + \left(\dfrac{k}{b}\right)^2\right]$	$\sin$	$\cos$

j	k	[ ]
1	1	+0.2500
1	3	−0.0033
3	1	−0.0033
1	5	+0.0003

This indicates quite rapid convergence, so that even the first term of the series should be quite close to the actual displacement. However, in this regard, we recall that the bending and twisting stress couples are computed from the second derivatives, and the transverse shear resultants from the third derivatives of D_Z.

From Table 10–2, it is obvious that each differentiation adds a j or k to the numerator, thereby reducing the convergence rate. Therefore, although the displacements may converge quite quickly, many more terms of the series may be required to evaluate the stress couples and transverse shears accurately.

10.2.2.3 Approximate Analysis for Uniformly Distributed Loading.

Because the $j = 1$, $k = 1$ term of the series gives a very close approximation for the displacement caused by the uniform load q_0, we may study the essential characteristics of the transverse shear resultants and the stress couples using a single harmonic approximation of a uniformly distributed loading, as shown in Figure 10–10. From equation (10.80), we compute

$$q_Z^{11} = \frac{16}{\pi^2} q_0 \tag{10.83}$$

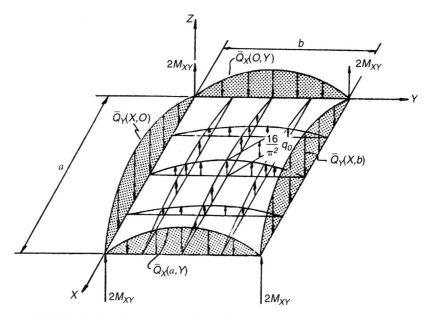

FIGURE 10–10   Single Harmonic Distributed Loading on Rectangular Plate.

Substituting equation (10.83) into equation (10.77), we have

$$D_Z(X,Y) = \left(\frac{16q_0}{\pi^2}\right)\left(\frac{1}{\pi^4 D}\right)\frac{1}{\left(\dfrac{1}{a^2}+\dfrac{1}{b^2}\right)^2}\sin\pi\frac{X}{a}\sin\pi\frac{Y}{b} \qquad (10.84)$$

from which we find the effective transverse shears $\overline{Q}_X$ and $\overline{Q}_Y$ from Table 10–2:

$$\overline{Q}_X(X,Y) = \left(\frac{16q_0}{\pi^3}\right)\frac{\left[\dfrac{1}{a^2}+(2-\mu)\dfrac{1}{b^2}\right]}{a\left(\dfrac{1}{a^2}+\dfrac{1}{b^2}\right)^2}\cos\pi\frac{X}{a}\sin\pi\frac{Y}{b} \qquad (10.85a)$$

and

$$\overline{Q}_Y(X,Y) = \left(\frac{16q_0}{\pi^3}\right)\frac{\left[(2-\mu)\dfrac{1}{a^2}+\dfrac{1}{b^2}\right]}{b\left(\dfrac{1}{a^2}+\dfrac{1}{b^2}\right)^2}\sin\pi\frac{X}{a}\cos\pi\frac{Y}{b} \qquad (10.85b)$$

On Figure 10–10, we show the effective transverse shears on the boundaries, $\overline{Q}_X$ (0, Y), $\overline{Q}_X$ (a, Y), $\overline{Q}_Y$ (X, 0), and $\overline{Q}_Y$ (X, b) The positive sense of these forces is found by referring to Figure 10–1.

We now consider equilibrium in the Z direction and define the total applied load

$$Q_{Z1} = \int_0^a \int_0^b q_Z \, dY \, dX \qquad (10.86a)$$

With the load intensity given by

$$q_Z(X,Y) = (16/\pi^2)q_0 \sin\pi(X/a)\sin\pi(Y/b) \qquad (10.86b)$$

as specified by equations (10.73) and (10.83),

$$Q_{Z1} = \frac{16}{\pi^2}q_0 \int_0^a \int_0^b \sin\pi\frac{X}{a}\sin\pi\frac{Y}{b}\,dY\,dX$$

$$= \frac{64}{\pi^4}q_0\,ab \qquad (10.86c)$$

which acts in the positive Z direction.

Considering now the effective transverse shears on the boundaries, we have the resultant Q_{Z2} acting in the positive Z direction given by

$$Q_{Z2} = -\int_0^b [|\overline{Q}_X(0,Y)| + |\overline{Q}_X(a,Y)|]dY$$

$$-\int_0^a [|\overline{Q}_Y(X,0)| + |\overline{Q}_Y(X,b)|]dX \qquad (10.87)$$

After considerable algebra we find

$$Q_{Z2} = -\frac{64}{\pi^4} q_0 \frac{ab}{(a^2 + b^2)^2} [a^4 + 2(2 - \mu)a^2b^2 + b^4] \tag{10.88}$$

Multiplying the numerator and denominator of equation (10.86c) by $(a^2 + b^2)^2$ and adding it to equation (10.88), we have

$$Q_{Z1} + Q_{Z2} = -128 \frac{(1 - \mu)}{\pi^4} q_0 \frac{a^3b^3}{(a^2 + b^2)^2} \tag{10.89}$$

as an apparently unbalanced force. But, we must also include the corner forces. From equation (10.50b) and Figure 10–5, we see that there are corner forces of magnitude $2M_{XY}$ as shown on Figure 10–10. Referring to Table 10–2 and equation (10.83), we find

$$M_{XY}(X,Y) = -\left(\frac{16}{\pi^2} q_0\right)\left(\frac{1}{\pi^2}\right) \frac{(1 - \mu)}{ab} \frac{1}{\left[\left(\frac{1}{a}\right)^2 + \left(\frac{1}{b}\right)^2\right]^2} \cos \pi \frac{X}{a} \cos \pi \frac{Y}{b} \tag{10.90}$$

In evaluating R_{XY} at each corner, we note from Figure 10–5 that a *positive* M_{XY} produces a corner force in the *negative* Z direction. At each corner, $(X = 0, a; Y = 0, b)$ M_{XY} is negative, so that the corner forces are directed in the positive Z direction. Each force is given by

$$R_{XY} = -2|M_{XY}(0,0)| = -2\left[\frac{16}{\pi^4} q_0 \frac{a^3b^3}{(a^2 + b^2)^2} (1 - \mu)\right] \tag{10.91}$$

producing a total force in the *positive Z* direction of

$$Q_{Z3} = -4R_{XY}$$
$$= 128 \frac{(1 - \mu)}{\pi^4} q_0 \frac{a^3b^3}{(a^2 + b^2)^2} \tag{10.92}$$

Now, adding equation (10.92) to equation (10.89) we get

$$Q_{Z1} + Q_{Z2} + Q_{Z3} = 0 \tag{10.93}$$

Thus, we demonstrate the contribution of the corner forces to the overall equilibrium of a rectangular plate. As an order of magnitude estimate, each corner force will be about 10% of the total load for a square plate.

It may be shown that for non-right-angle corners of simply supported polygonal plates, there is no concentrated corner force, but, as the included angle approaches 90°, the intensities of the effective transverse shears on the boundaries, $\overline{Q}_\xi$ and $\overline{Q}_\eta$, increase rapidly near the corner. This may be demonstrated for a triangular plate which is discussed in Section 10.4.2.

We now consider the stress couples, which are easily computed using Table 10–2 and equation (10.83):

$$M_X(X,Y) = \left(\frac{16}{\pi^2} q_0\right)\left(\frac{1}{\pi^2}\right) \frac{\left(\frac{1}{a}\right)^2 + \mu\left(\frac{1}{b}\right)^2}{\left[\left(\frac{1}{a}\right)^2 + \left(\frac{1}{b}\right)^2\right]^2} \sin \pi \frac{X}{a} \sin \pi \frac{Y}{b} \tag{10.94a}$$

and

$$M_Y(X,Y) = \left(\frac{16}{\pi^2} q_0\right)\left(\frac{1}{\pi^2}\right) \frac{\mu\left(\frac{1}{a}\right)^2 + \left(\frac{1}{b}\right)^2}{\left[\left(\frac{1}{a}\right)^2 + \left(\frac{1}{b}\right)^2\right]^2} \sin \pi \frac{X}{a} \sin \pi \frac{Y}{b} \qquad (10.94b)$$

It is obvious that both functions will follow the load distribution shown in Figure 10–10 with maxima at the center. $X = a/2$, $Y = b/2$.

Example 10.2(a)

In order to assess the relative magnitude of the bending moments at the center of a square plate where $b = a$, we evaluate

$$M_X\left(\frac{a}{2},\frac{a}{2}\right) = M_Y\left(\frac{a}{2},\frac{a}{2}\right) = \frac{4(1 + \mu)}{\pi^4} q_0 a^2 \approx (0.04 \sim 0.05) q_0 a^2 \qquad (10.95)$$

as the maximum moment.

For comparison, consider a beam of unit width carrying the same sinusoidal loading over the span a. With a loading

$$q(X) = \frac{16}{\pi^2} q_0 \sin \frac{X}{a} \qquad (10.96a)$$

the differential equation for a prismatic beam

$$EID_{Z,XXXX} = q \qquad (10.96b)$$

has a solution

$$D_Z(X) = \frac{1}{EI}\left(\frac{16q_0}{\pi^2}\right)\left(\frac{a}{\pi}\right)^4 \sin \pi \frac{X}{a} \qquad (10.96c)$$

from which

$$M(X) = EID_{Z,XX}$$
$$= \frac{16q_0}{\pi^4} a^2 \sin \pi \frac{X}{a} \qquad (10.96d)$$

with a maximum value of

$$M\left(\frac{a}{2}\right) = \frac{16}{\pi^4} q_0 a^2 \approx 0.17 q_0 a^2 \qquad (10.96e)$$

By comparing equations (10.95) and (10.96e), it is apparent that the plate is a far more efficient flexural member than the equivalent beam because of the two-way flexural action and the twisting rigidity. Equation (10.95), which gives the maximum bending moment for a uniformly loaded square plate, also confirms the observations of the seventeenth-century French physicist Mariotte, who surmised that the *total* load Q_0 corresponding to the *maximum* moment, in this case $Q_0 = q_0 a^2$, should remain constant and independent of the size of the plate if the thickness is not changed.[8]

It is interesting to compute the moments in the vicinity of the corners by considering rotated axes ξ, η at $45°$ to the $X - Y$ coordinates, as shown on Figure 10–9. Referring to Figure 10–4(c) and taking $\alpha = X$, $\beta = Y$, and $\gamma = 45°$ in equations (10.36b), (10.38b), and (10.37b), respectively, give

$$M_\eta = M_{XY} = \frac{R}{2}$$

$$M_\xi = -M_{XY} = -\frac{R}{2}$$

$$M_{\eta\xi} = M_{\xi\eta} = 0$$

Thus, in the vicinity of the corner, we find a state of anticlastic bending as described in Section 10.2.1.4. The moments $\pm M_{XY}$ or $\pm R/2$ are the same order magnitude as the moments in the center of the plate, that is, from equations 10.90 and 10.95, $M_{XY} = [(1 - \mu)/(1 + \mu)]M_{max}$ for a square plate.[9] M_η is essentially a clamping moment, similar to that developed at a fixed boundary.

Another manifestation of the corner force phenomenon may be observed by visualizing a rectangular plate under self-weight set on a continuous knife edge. Seemingly, this boundary would correspond to an ideal simple support condition. However, for an applied loading in the downward (negative Z) direction, the preceding solution would indicate corner forces acting downward on the plate. If the plate simply rests on the knife edge with no tiedowns, the corners will tend to rise away from the support. It may also be remarked that the lack of restraint to develop the corner forces will remove the clamping effect of M_η and thereby cause the maximum moments at the center to increase,[9] perhaps by 35% for a square plate.

10.2.2.4 Rectangular Patch Loading.

In Figure 10–11, we show a loading uniformly distributed over a rectangular patch on the plate. The total magnitude of the load is Q_0, so that

$$q_Z = \frac{Q_0}{cd} \left(\begin{array}{l} \xi - \dfrac{c}{2} \le X \le \xi + \dfrac{c}{2} \\[2mm] \eta - \dfrac{d}{2} \le Y \le \eta + \dfrac{d}{2} \end{array} \right) \tag{10.97}$$

$$= 0 \quad \text{otherwise}$$

ξ and η denote the X and Y coordinates, respectively, of the center of the patch.

Substituting equation (10.97) into equation (10.75), we have

$$q_Z^{jk} = \left(\frac{4}{ab}\right)\left(\frac{Q_0}{cd}\right) \int_{\xi-c/2}^{\xi+c/2} \int_{\eta-d/2}^{\eta+d/2} \sin j\pi \frac{X}{a} \sin k\pi \frac{Y}{b} \, dY \, dX$$

$$= \frac{16}{\pi^2 cdjk} Q_0 \sin j\pi \frac{\xi}{a} \sin k\pi \frac{\eta}{b} \sin j\pi \frac{c}{2a} \sin k\pi \frac{d}{2b} \tag{10.98}$$

As a check, we select the uniformly distributed load considered in section 10.2.2.2. For that case, we have $Q_0 = q_0 ab$; $c = a$, $d = b$; and $\xi = a/2$, $\eta = b/2$. Substituting these values into

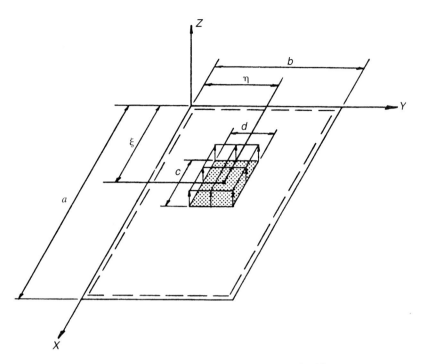

FIGURE 10–11   Patch Loading on Rectangular Plate.

equation (10.98), we get the identical result to that obtained in equations (10.80). The solution for the deflection function and the subsequent stress resultants and couples follow from Table 10Ð2.

10.2.2.5 Concentrated Loading. The solution just obtained for the patch loading may be used to describe a concentrated load at any point on the plate. To illustrate, we consider Q_0 acting at $X = \xi$ and $Y = \eta$, as shown on Figure 10Ð12. We take equation (10.98) with c and d approaching zero and get

$$q_z^{jk} = \lim_{\substack{c \to 0 \\ d \to 0}} \frac{16}{\pi^2 jk} Q_0 \sin j\pi \frac{\xi}{a} \sin k\pi \frac{\eta}{b} \left(\frac{\sin j\pi \dfrac{c}{2a}}{c} \right) \left(\frac{\sin k\pi \dfrac{d}{2b}}{d} \right) \tag{10.99}$$

The limit is evaluated by LÕHospitalÕs rule, which gives, for the last two terms,

$$\lim_{\substack{c \to 0 \\ d \to 0}} \frac{\sin j\pi \dfrac{c}{2a}}{c} \frac{\sin k\pi \dfrac{d}{2b}}{d} = \frac{j\pi}{2a} \frac{k\pi}{2b}$$

reducing equation (10.99) to

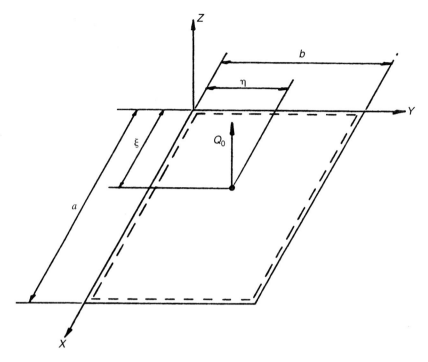

FIGURE 10–12   Concentrated Load on Rectangular Plate.

$$q_Z^{jk} = \frac{4Q_0}{ab} \sin j\pi \frac{\xi}{a} \sin k\pi \frac{\eta}{b} \tag{10.100}$$

From equation (10.77), we write the complete solution as

$$D_Z(X,Y;\xi,\eta) = Q_0 \frac{4}{\pi^4 abD} \sum_{j=1}^{\infty} \sum_{k=1}^{\infty} \frac{\sin j\pi \dfrac{\xi}{a} \sin k\pi \dfrac{\eta}{b}}{\left[\left(\dfrac{j}{a}\right)^2 + \left(\dfrac{k}{b}\right)^2\right]^2} \sin j\pi \frac{X}{a} \sin k\pi \frac{Y}{b} \tag{10.101}$$

$$= Q_0 G(X,Y;\xi,\eta)$$

We introduce the notation $D_Z(X, Y; \xi, \eta)$ to emphasize that D_Z is dependent on *two* sets of coordinates: (a) the *point of observation* of the deflection, (X, Y), and (b) *the point of application* of the applied load Q_0, (ξ, η). The coefficient of Q_0 in equation (10.101) is defined as the Green's function, $G(X, Y; \xi, \eta)$, and represents the displacement at the point of observation (X, Y) due to a *unit* load at the point of application (ξ, η). Graphically, the Green's function for the displacement of the plate may be visualized as an *influence surface* representing the deflected shape of the entire plate caused by a unit load at a particular point.

Once we establish the Green's function for a system, we may readily determine the deflection at any point of observation due to a prescribed surface loading $q_Z = q_Z(\xi, \eta)$. Then, by superposition we have

$$D_Z(X,Y) = \int\int_{\text{Area}} q_Z(\xi,\eta)G(X,Y;\xi,\eta)d\xi\,d\eta \tag{10.102}$$

The Green's function approach is widely used in solid mechanics.[10]

If we consider the solution given by equation (10.101) in more detail, we expect the series for D_Z to converge rapidly because of the j^4 and k^4 terms in the denominator. However, the series expressing the stress couples and transverse shear resultants are found to be divergent right at the point of application of the concentrated load, for example, $X = \xi$, $Y = \eta$. Of course, the series for the load itself will not converge. Thus, this solution, although simple in form, is of limited value for numerical computations and the approach described in the following section is more useful.

Example 10.2(b)

It is informative to compute D_Z for a square plate, $b = a$, with a concentrated load Q_0 at the center. The first four terms of the series for this case give

$$D_Z\!\left(\frac{a}{2},\frac{a}{2}\right) \simeq 0.01\,\frac{Q_0}{D}\,a^2 \tag{10.103}$$

To compare equation (10.103) with a beam solution, we take the entire plate as the beam width a. Then $(EI)_{\text{beam}} = aD$ and

$$D_Z\!\left(\frac{a}{2}\right) = \frac{Q_0 a^2}{48D} \simeq 0.02\,\frac{Q_0}{D}\,a^2 \tag{10.104}$$

again indicating the relative rigidity of the plate as compared to a flexural member of equal span.

10.2.3 Lévy-Nádai Solution for Rectangular Plates

10.2.3.1 General Technique.   Although the double sine series solution developed in the previous section is straightforward, the resulting expressions often require that many terms be included in the summation to attain acceptable precision. Of course, this objection is largely anachronistic in the era of the computer; nevertheless, it is interesting from a fundamental standpoint to explore the possibility of generating more efficient solutions. Also, the Navier solution is well suited for the simply supported boundary only, and the desire to treat other boundary conditions gave rise to the work of Lévy, a pupil of St. Venant, and Estanave.[6] The ensuing single series approach was exploited by Nádai as well and carries the name of the Lévy-Nádai solution.

We introduce this technique by taking the displacement function D_Z as the sum of (a) D_{Z1}, a beam solution which satisfies equilibrium and the boundary conditions in one direction; and, (b) D_{Z2}, a homogeneous solution which can be used to enforce the boundary conditions in the other direction, while not violating the boundary conditions in the first direction.

We refer to Figure 10–13 where the X axis is located along the center line of the plate, for reasons which will become obvious. We select $D_{Z1} = D_{Z1}(X)$ as the beam bending solu-

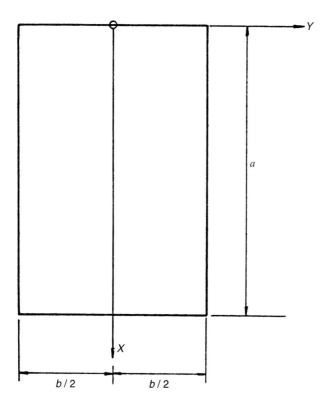

FIGURE 10–13   Coordinates for Lévy-Nádai Solution.

tion for a loading q_Z which satisfies the boundary conditions at $X = 0$ and $X = a$. For example, a simply supported beam under a uniform load $q_Z = q_0$ has a deflection function

$$D_Z(X) = \frac{q_0 X}{24EI} (a^3 - 2aX^2 + X^3) \tag{10.105}$$

A number of similar formulae for various loadings and boundary conditions are collected in the *AISC Manual of Steel Construction.*[11]

If we make the adjustment for the Poisson effect and change the term EI to D, equation (10.105) will satisfy the *plate equation*, equation (10.12), as well as the simply supported boundary conditions at $X = 0$ and $X = a$. However, the boundary conditions at $Y \pm b/2$ are obviously not addressed because $D_{Z1} = D_{Z1}(X)$ only. To correct this, we need D_{Z2}, which satisfies the boundary conditions in *both* directions. Because D_{Z1} already will balance the loading, D_{Z2} should be a homogeneous solution of equation (10.12).

10.2.3.2 Simply Supported Plate Under Uniform Load.   We consider the case of a uniformly loaded, simply supported plate, previously treated in Section 10.2.12. Using equation (10.105) for D_{Z1}, with EI replaced by D, we require the functions to satisfy

$$\nabla^4 D_Z = \nabla^4 D_{Z1} + \nabla^4 D_{Z2} = \frac{q_0}{D} + 0 = \frac{q_0}{D} \tag{10.106}$$

For D_{Z2}, we choose a *single* series solution

$$D_{Z2} = \sum_{j=1}^{\infty} \psi^j(Y) \sin j\pi \frac{X}{a} \tag{10.107}$$

where ψ^j is a function of Y alone. This expression for D_{Z2} automatically satisfies the boundary conditions in the X direction. Substituting equation (10.107) into equation (10.12), we have

$$\nabla^4 D_{Z2} = \sum_{j=1}^{\infty} \left[\left(\frac{j\pi}{a}\right)^4 \psi^j - 2\left(\frac{j\pi}{a}\right)^2 \psi^j_{,YY} + \psi^j_{,YYYY} \right] \sin j\pi \frac{X}{a} = 0 \tag{10.108}$$

Equation (10.108) is satisfied for all j if the terms within $[\ ] = 0$, leading to the linear ordinary differential equation for $\psi^j (Y)$

$$\psi^j_{,YYYY} - 2\left(\frac{j\pi}{a}\right)^2 \psi^j_{,YY} + \left(\frac{j\pi}{a}\right)^4 \psi^j = 0 \tag{10.109}$$

which may be solved by classical means as

$$\psi^j(Y) = C^j_1 \cosh j\pi \frac{Y}{a} + C^j_2 j\pi \frac{Y}{a} \sinh j\pi \frac{Y}{a} + C^j_3 \sinh j\pi \frac{Y}{a}$$
$$+ C^j_4 j\pi \frac{Y}{a} \cosh j\pi \frac{Y}{a} \tag{10.110}$$

The selection of the origin at the middle of the side parallel to the Y-axis now becomes useful. Because the loading is uniform, we recognize that the plate deflection is symmetrical about the X-axis. This being the case, only *even* functions of Y, $[f(-Y) = f(Y)]$, can participate in the solution and the coefficients of the odd terms in equation (10.110), $[f(-Y) = -f(Y)]$, drop out. Therefore, we have C^j_3 and $C^j_4 = 0$, and the complete solution to equation (10.106) reduces to

$$D_Z(X,Y) = D_{Z1} + D_{Z2} \tag{10.111}$$

$$= \frac{q_0 X}{24D} (a^3 - 2aX^2 + X^3) + \sum_{j=1}^{\infty} \left[C^j_1 \cosh j\pi \frac{Y}{a} + C^j_2 j\pi \frac{Y}{a} \sinh j\pi \frac{Y}{a} \right] \sin j\pi \frac{X}{a}$$

We have the integration constants C^j_1 and C^j_2 available to satisfy the simply supported boundary conditions at $Y = \pm b/2$. Because we have already invoked the requirement of symmetry to suppress the odd terms in equation (10.110), the conditions at $Y = +b/2$,

$$D_Z\left(X,\frac{b}{2}\right) = D_{Z,YY}\left(X,\frac{b}{2}\right) = 0 \tag{10.112}$$

are sufficient. It is necessary to expand the expression for D_{Z1} in a Fourier series so that the constants can be chosen in harmonic form. This series is found routinely[9] as

$$D_{Z1} = \frac{4q_0 a^4}{\pi^5 D} \sum_{j=1,3,\text{odd}}^{\infty} \frac{1}{j^5} \sin j\pi \frac{X}{a} \tag{10.113}$$

and, after substitution of equation (10.113), equation (10.111) can be written in the form

$$D_Z(X,Y) = \frac{q_0 a^4}{D} \sum_{j=1,3,\text{odd}}^{\infty} \left[\frac{4}{\pi^5 j^5} + C_5^j \cosh j\pi \frac{Y}{a} + C_6^j j\pi \frac{Y}{a} \sinh j\pi \frac{Y}{a} \right] \sin j\pi \frac{X}{a} \qquad (10.114)$$

where the original constants C_1^j and C_2^j have been modified by the common factor. Now, the constants are evaluated by applying equation (10.112):

$$C_5^j = \frac{-2(\lambda^j \sinh \lambda^j + 2 \cosh \lambda^j)}{\pi^5 j^5 \cosh^2 \lambda^j} \qquad (10.115a)$$

and

$$C_6^j = \frac{2}{\pi^5 j^5 \cosh \lambda^j} \qquad (10.115b)$$

where

$$\lambda^j = \frac{j\pi b}{2a} \qquad (10.115c)$$

The rotations, stress couples, and transverse shear stress resultants can be obtained directly from equations (10.21), (10.8), and (10.9).

In retrospect, it is apparent that the "beam" solution could have been taken in the form of a series initially. This would lead directly to equation (10.114) without encountering equation (10.111).

A detailed study of this solution, including the convergence properties, is contained in Timoshenko and Woinowsky-Krieger.[9] We have already examined this problem in detail in Section 10.2.2.2, and the results found there can be confirmed with the single series solution. Also available are parametric tables for the various functions for uniform and hydrostatic loading.[9] Additionally, single series solutions are developed for some additional cases including the bending moments under a concentrated load,[12] a somewhat difficult computational exercise as noted in Section 10.2.2.5.

Some results of the comprehensive study in Reference 12 are of general interest. The single series Lévy-Nádai solution is usually more rapidly converging than the double series Navier solution. Heuristically, this indicates that starting with the known beam function as a first approximation leads to efficiency in the computational algorithm. Particularly since smaller computers are becoming increasingly popular in engineering, the savings achieved by using a solution based on a well-founded initial trial may be significant.

An interesting observation is that when the aspect ratio $b/a \geq 3$, a plate behaves essentially as a one-way flexural member spanning in the *short* direction, except near the edges. This may be appreciated by noting the denominators of equations (10.115a) and (10.115b), which get large with increasing b/a. With the influence of D_{Z2} diminishing, the deflection function is dominated by the beam solution D_{Z1}.

10.2.3.3 Other Single-Series Solutions.

Another classical application of the Lévy-Nádai approach is shown in Figure 10-14, where a rectangular plate is loaded by constant edge moments. If $M_{Y1}(X) \neq M_{Y2}(X)$, the solution is facilitated by the superposition of a symmetrical and an antisymmetrical case. The details are left as an exercise.

Also, we observe that rectangular plates which have only two opposite sides simply supported may be treated by inserting a beam solution with the appropriate boundary conditions for D_{Z1} in equation (10.111) or the equivalent series in equation (10.114) and proceed-

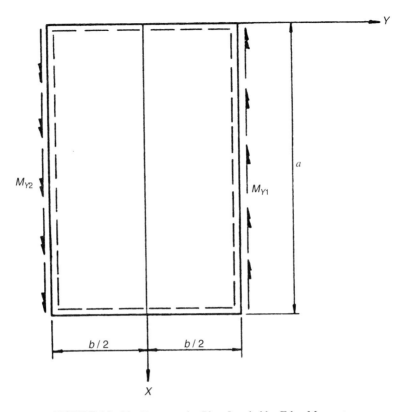

FIGURE 10–14   Rectangular Plate Loaded by Edge Moments.

ing accordingly. This will be adequate for plates which have two sides simply supported, two sides clamped; three sides simply supported, one side free; and two sides simply supported, one side clamped, one side free.

More complicated boundary conditions are treated with Lévy-Nádai type solutions in Timoshenko and Woinowsky-Krieger.[13,14] Also, some additional solution procedures are discussed in Kiattikomol, Keer and Dundurs,[15] and extensive tabulated results are provided by Szilard.[16] These analytical approaches become rather involved, and one must soon turn to numerical techniques. The classical solutions remain valuable, however, as a basis for incisively evaluating the results of numerical solutions through quantitative comparisons and order-of-magnitude bounds.

10.3   CIRCULAR PLATES

10.3.1   Axisymmetric Bending

10.3.1.1   Governing Equation.   Axisymmetric bending of circular plates is described by equations (10.33). Using equation (10.27) and (10.28b) with the θ terms omitted, we may rewrite equation (10.33a) in the form

$$\frac{1}{R}\left[R\left(\frac{1}{R}[R(D_Z),_R],_R\right),_R\right],_R = \frac{q_Z}{D} \tag{10.116}$$

which is quite convenient for solution.

10.3.1.2 Homogeneous Solution. We first consider the homogeneous part of equation (10.116) and integrate consecutively as follows:

$$\frac{1}{R}\left[R\left(\frac{1}{R}[R(D_Z),_R],_R\right),_R\right],_R = 0$$

$$\left[R\left(\frac{1}{R}[R(D_Z),_R],_R\right),_R\right],_R = 0$$

$$\left[R\left(\frac{1}{R}[R(D_Z),_R],_R\right),_R\right] = C_1$$

$$\left(\frac{1}{R}[R(D_Z),_R],_R\right),_R = \frac{C_1}{R}$$

$$\left(\frac{1}{R}[R(D_Z),_R],_R\right) = C_1 \ln R + C_2$$

$$[R(D_Z),_R],_R = C_1 R \ln R + C_2 R$$

$$[R(D_Z),_R] = C_1\left(\frac{R^2}{2}\ln R - \frac{R^2}{4}\right) + C_2\frac{R^2}{2} + C_3$$

$$= C_1 R^2 \ln R + C_2 R^2 + C_3$$

where C_1, and C_2 have been redefined. Continuing

$$(D_Z),_R = C_1 R \ln R + C_2 R + C_3 \frac{1}{R}$$

$$D_Z = C_1\left(\frac{R^2}{2}\ln R - \frac{R^2}{4}\right) + \frac{C_2 R^2}{2} + C_3 \ln R + C_4$$

and, finally,

$$D_Z = D_{Zh} = C_1 R^2 \ln R + C_2 R^2 + C_3 \ln R + C_4 \tag{10.117}$$

where C_1 and C_2 have again been redefined. The terms composing equation (10.117) are known as *biharmonic* functions, because they *individually* and *collectively* satisfy the biharmonic equation in polar coordinates.

It is informative to compute the rotation, stress couples, and transverse shear corresponding to D_{Zh}. These are found from equations (10.30c), (10.31), and (10.32b) and are listed in Table 10–3. We note the following characteristics of the tabulated functions:

1. $R^2 \ln R$ is singular at $R = 0$ and $R = \infty$, and is the only term which contributes to Q_R.
2. R^2 is singular at $R = \infty$; for this function $M_R = M_\theta$, (homogeneous bending).
3. $\ln R$ is singular at $R = 0$.
4. 1 is regular throughout

TABLE 10-3 Solution Components for a Circular Plate

Solution Component	$D_Z(R)$	$D_{R\theta}(R)$ $= -D_{Z,R}$	$M_R(R)/D$ $= -\left[D_{Z,RR} + \dfrac{\mu}{R} D_{Z,R} \right]$	$M_\theta(R)/D$ $= -\left[\dfrac{1}{R} D_{Z,R} + \mu D_{Z,RR} \right]$	$\dfrac{Q_R(R)}{D}$ $= D_{Z,RRR}$ $+ \dfrac{1}{R} D_{Z,RR}$ $- \dfrac{1}{R^2} D_{Z,R}$	$\dfrac{q_Z(R)}{D}$ $= \nabla_a^4 D_Z$
Functions of D_Z						
Homogeneous $D_{Zh}(R)$						
C_1 ×	$R^2 \ln R$	$-R(2\ln R + 1)$	$-[2(1+\mu)\ln R + 3 + \mu]$	$-[2(1+\mu)\ln R + 1 + 3\mu]$	$-4/R$	0
C_2 ×	R^2	$-2R$	$-2(1+\mu)$	$-2(1+\mu)$	0	0
C_3 ×	$\ln R$	$-\dfrac{1}{R}$	$(1-\mu)/R^2$	$-(1-\mu)/R^2$	0	0
C_4 ×	1	0	0	0	0	0
Particular $q_Z(R) = q_0$						
$q_0/64D$ ×	R^4	$-4R^3$	$-4(3+\mu)R^2$	$-4(1+3\mu)R^2$	$-32R$	64

These characteristics are useful for expediting solutions to various problems, as we will see later.

10.3.1.3 Solid Plates.

Consider the case of a constant edge moment, M, applied around the circumference of a simply supported circular plate with radius a, as shown in Figure 10–15.

The boundary conditions are

$$D_Z(a) = 0 \tag{10.118a}$$

$$M_R(a) = M \tag{10.118b}$$

Of course, we may write equation (10.118b) as a function of D_Z, but, as we shall see, this is unnecessary. Two additional boundary conditions are required because we have a fourth-order system. These conditions follow from the requirement that the solution remains finite at $R = 0$. This is a familiar situation, which we first encountered in the treatment of domes in Chapter 4. In our preceding characterization of the various biharmonic terms, we noted the singularity of the $\ln R$ and $R^2 \ln R$ terms at $R = 0$, so that we must set $C_1 = C_3 = 0$. Then, we find, by substituting the remaining terms in equation (10.117) into equations (10.118) with the help of Table 10–3, that

$$C_2 a^2 + C_4 = 0 \tag{10.119a}$$

$$-C_2 2(1 + \mu) = \frac{M}{D} \tag{10.119b}$$

from which

$$C_2 = \frac{-M}{2D(1 + \mu)} \tag{10.120a}$$

$$C_4 = \frac{Ma^2}{2D(1 + \mu)} \tag{10.120b}$$

and

$$D_Z(R) = \frac{M}{2(1 + \mu)D} (a^2 - R^2) \tag{10.121}$$

Also, we find

$$M_R(R) = M_\theta(R) = -2C_2 D(1 + \mu) = M \tag{10.122}$$

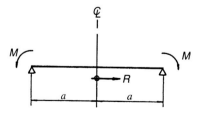

FIGURE 10–15   Edge Moments on Circular Plate.

which indicates that the bending moment is constant throughout the plate in all directions. A state of homogeneous bending was previously encountered for a rectangular plate in Section 10.2.1.2.

Example 10.3(a)

The first logical case of applied surface loading is a uniform load $q_Z = q_0$, as shown on Figure 10–16. It is easily verified by direct substitution into equation (10.116) that the particular solution is

$$D_{Zp}(R) = \frac{q_0}{64D} R^4 \tag{10.123}$$

For convenience, the various functions corresponding to this particular solution are recorded in Table 10–3.

Now, combining equation (10.117) with equation (10.123) after setting $C_1 = C_3 = 0$ as previously discussed, we have

$$D_Z(R) = C_2 R^2 + C_4 + \frac{q_0}{64D} R^4 \tag{10.124}$$

with the boundary conditions corresponding to the simple support at $R = a$

$$D_Z(a) = M_R(a) = 0 \tag{10.125}$$

With the aid of Table 10–3, we write

$$C_2 a^2 + C_4 + \frac{q_0}{64D} a^4 = 0 \tag{10.126a}$$

and

$$-2(1 + \mu)C_2 - \frac{q_0}{64D}[4(3 + \mu)a^2] = 0 \tag{10.126b}$$

from which

$$C_2 = -\frac{q_0}{64D}\left[\frac{2(3 + \mu)}{(1 + \mu)}\right]a^2 \tag{10.127a}$$

$$C_4 = \frac{q_0}{64D}\left[\frac{2(3 + \mu)}{(1 + \mu)} - 1\right]a^4$$

$$= \frac{q_0}{64D}\left[\frac{5 + \mu}{1 + \mu}\right]a^4 \tag{10.127b}$$

and then

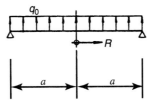

q_0

R

a a

FIGURE 10–16   Uniform Load on Circular Plate.

$$D_Z(R) = \frac{q_0}{64D}\left[\frac{-2(3+\mu)}{(1+\mu)}a^2R^2 + \left[\frac{5+\mu}{1+\mu}\right]a^4 + R^4\right]$$

$$= \frac{q_0}{64D}(a^2 - R^2)\left\{\left[\frac{5+\mu}{1+\mu}\right]a^2 - R^2\right\} \tag{10.128}$$

We may also find the moments from Table 10–3 as

$$M_R(R) = \frac{q_0}{64}\left[\frac{-2(3+\mu)}{(1+\mu)}a^2[-2(1+\mu)] - 4(3+\mu)R^2\right]$$

$$= \frac{q_0}{16}(3+\mu)(a^2 - R^2) \tag{10.129a}$$

$$M_\theta(R) = \frac{q_0}{64}\left[\frac{-2(3+\mu)}{(1+\mu)}a^2[-2(1+\mu)] - 4(1+3\mu)R^2\right]$$

$$= \frac{q_0}{16}[(3+\mu)a^2 - (1+3\mu)R^2] \tag{10.129b}$$

At the boundary, $R = a$,

$$M_R = 0; \quad M_\theta = \frac{q_0}{8}(1-\mu)a^2 \tag{10.130}$$

while at the center, $R = 0$,

$$M_R = M_\theta = \frac{q_0}{16}(3+\mu)a^2 \tag{10.131}$$

so that the maximum moment occurs at the center of the plate. At the pole, $M_R(0) = M_\theta(0)$, which again illustrates the isotrophy condition first encountered in the study of axisymmetrically loaded shells of revolution, Section 4.3.2.1.

Example 10.3(b)

We now investigate a clamped plate under uniform load, as shown in Figure 10–17. There are two evident procedures: (a) We may superimpose the solutions for the cases shown in Figures 10–16 and 10–15 and enforce the compatibility condition $D_{R\theta}(a) = 0$ to compute $M_R(a)$; or, (b) we may take the solution as equation (10.124), with the constants determined from the boundary conditions

$$D_Z(a) = D_{R\theta}(a) = 0 \tag{10.132}$$

We select (a), the superposition approach, and enter Table 10–3 with constants from equations (10.120) and (10.127) to find

$$D_{R\theta}(a) = \frac{q_0}{64D}\left[\frac{-2(3+\mu)}{(1+\mu)}a^2(-2a) - 4a^3\right] - \frac{M}{2(1+\mu)D}(-2a)$$

$$= 0 \tag{10.133}$$

from which

$$M = M_R(a) = -\frac{q_0 a^2}{8} \tag{10.134}$$

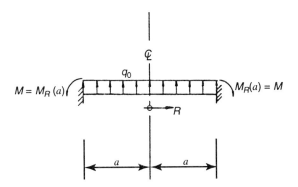

FIGURE 10-17 Clamped Circular Plate.

The deflection function for the clamped plate is found by superimposing equations (10.128) and equation (10.121) with $M = M_R (a)$ as given by equation (10.134).
and may be written in the perfect square form

$$D_Z(R) = \frac{q_0}{64D} (a^2 - R^2)^2 \tag{10.135}$$

We may consisely plot the stress couples for both the simply supported and the clamped boundary conditions on a common graph, Figure 10–18, because the edge moment $M_R (a)$ produces a constant value of M_R and M_θ throughout.

10.3.1.4 Plates with Annular Openings.

Another case of interest is a circular plate with an annular opening, as illustrated in Figure 10–19. We first consider the plate under an exterior edge moment M_1 and an interior edge moment M_2, both uniformly distributed around the circumference. Only the exterior boundary is constrained against displacement in the Z direction. We have the homogeneous solution, equation (10.117), subject to the boundary conditions

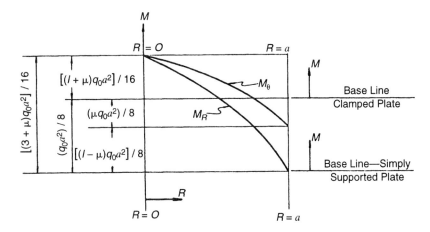

FIGURE 10-18 Moments in Circular Plates.

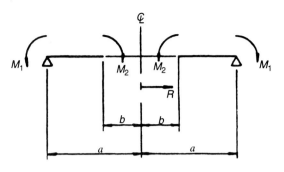

FIGURE 10–19 Circular Plate with Annular Opening.

$$D_Z(a) = 0 \quad (a)$$
$$M_R(a) = M_1 \quad (b)$$
$$Q_R(b) = 0 \quad (c)$$
$$M_R(b) = M_2 \quad (d)$$

(10.136)

A check of Table 10–3 reveals that only the C_1 term contributes to $Q_R(R)$. From equation (10.136c), $Q_R(b) = 0$ so that $C_1 = 0$. The remaining conditions, equations (10.136a), (10.136b), and (10.136d), lead to the complete solution.[17]

$$D_Z(R) = \frac{1}{D(a^2 - b^2)} \left[\frac{a^2 M_1 - b^2 M_2}{2(1 + \mu)} (a^2 - R^2) - \frac{a^2 b^2 (M_1 - M_2)}{(1 - \mu)} \ln \frac{R}{a} \right] \quad (10.137)$$

Example 10.3(c)

A special case of equation (10.137) is $M_2 = 0$ and $M_1 = M$, as shown in Figure 10–20. Equation (10.137) then reduces to

$$D_Z(R) = \frac{Ma^2}{D(a^2 - b^2)} \left[\frac{(a^2 - R^2)}{2(1 + \mu)} - \frac{b^2}{(1 - \mu)} \ln \frac{R}{a} \right] \quad (10.138)$$

We calculate the circumferential bending moment M_θ from Table 10–3, where it is indicated that only the terms R^2 and $\ln R$ contribute. The constants C_2 and C_3 are infered from equations (10.138) as

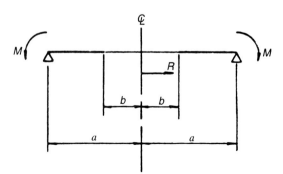

FIGURE 10–20 Annular Plate with Peripheral Edge Moment.

$$C_2 = \frac{Ma^2}{D(a^2 - b^2)}\left[\frac{-1}{2(1 + \mu)}\right] \tag{10.139a}$$

and

$$C_3 = \frac{Ma^2}{D(a^2 - b^2)}\left[\frac{-b^2}{1 - \mu}\right] \tag{10.139b}$$

whereupon

$$
\begin{aligned}
M_\theta(R) &= \frac{Ma^2}{a^2 - b^2}\left\{\frac{-1}{2(1 + \mu)}[-2(1 + \mu)] - \frac{b^2}{(1 - \mu)}\left[\frac{-(1 - \mu)}{R^2}\right]\right\} \\
&= \frac{Ma^2}{a^2 - b^2}\left(1 + \frac{b^2}{R^2}\right)
\end{aligned} \tag{10.139c}
$$

The maximum value of M_θ is at $R = b$,

$$M_\theta(b) = \frac{2Ma^2}{a^2 - b^2} \tag{10.140}$$

Now, if we let the hole shrink, for example, $b \to 0$, then $M_\theta (b) \to 2M$, indicating that there will be a *stress concentration* around the hole approaching twice the value of the moment applied on the outer edge.

The preceding examination of the stress concentration may be generalized. If we consider a plate of arbitrary shape which has a state of homogeneous bending, M, and a circular hole of radius d, as shown in Figure 10–21, the solution in polar coordinates should contain the same biharmonic terms as equation (10.138),

$$D_Z(R) = C_2 R^2 + C_3 \ln R + C_4 \tag{10.141}$$

The influence of the $\ln R$ term on the moments diminishes rapidly away from the hole, whereas the R^2 term produces homogeneous bending and C_4 is a rigid body term which may be ignored here. If we use the boundary conditions

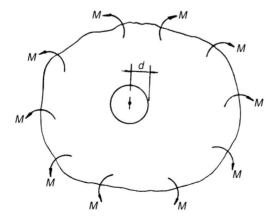

FIGURE 10–21 Arbitrary Plate with Circular Hole.

$$M_R(d) = 0 \tag{10.142a}$$

$$M_R(R \to \infty) = M \tag{10.142b}$$

we find from Table 10–3 that

$$M_R(d) = -2DC_2(1 + \mu) + \frac{DC_3(1 - \mu)}{d^2} = 0 \tag{10.143a}$$

and

$$M_R(\infty) = -2DC_2(1 + \mu) = M \tag{10.143b}$$

from which

$$D_Z(R) = -\frac{M}{D}\left[\frac{1}{2(1 + \mu)}R^2 + \frac{d^2}{(1 - \mu)}\ln R\right] \tag{10.144}$$

and

$$M_\theta(R) = M\left(1 + \frac{d^2}{R^2}\right) \tag{10.145}$$

so that for $R = d$, $M_\theta \to 2M$.

Now consider a simply supported open circular plate with a total load P uniformly distributed around the inner circumference, so that

$$Q_R(b) = -\frac{P}{2\pi b} \tag{10.146}$$

as shown in Figure 10–22. The algebraic sign of Q_R is established in accordance with Figure 10–3. We note from Table 10–3 that only the term $C_1R^2 \ln R$ of the displacement function can contribute to Q_R. Hence,

$$C_1\left(-\frac{4}{b}\right) = -\frac{P}{2\pi bD} \tag{10.147a}$$

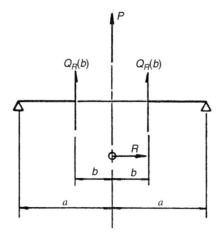

FIGURE 10–22 Uniform Transverse Shear on Annular Plate.

or

$$C_1 = \frac{P}{8\pi D} \tag{10.147b}$$

The remaining constants are determined from the boundary conditions

$$M_R(b) = 0 \tag{10.148a}$$

$$D_Z(a) = 0 \tag{10.148b}$$

$$M_R(a) = 0 \tag{10.148c}$$

and the complete solution finally takes the form[17]

$$D_Z(R) = \frac{P}{8\pi D} R^2 \left(\ln \frac{R}{a} - 1 \right) - \frac{C_5}{4} R^2 - C_6 \ln \frac{R}{a} + C_7 \tag{10.149}$$

where

$$C_5 = \frac{P}{4\pi D} \left[\frac{1 - \mu}{1 + \mu} - \frac{2b^2}{a^2 - b^2} \ln \frac{b}{a} \right] \tag{10.150a}$$

$$C_6 = -\frac{P}{4\pi D} \left[\frac{(1 + \mu)}{(1 - \mu)} \frac{a^2 b^2}{(a^2 - b^2)} \ln \frac{b}{a} \right] \tag{10.150b}$$

$$C_7 = \frac{Pa^2}{8\pi D} \left[1 + \frac{(1 - \mu)}{2(1 + \mu)} - \frac{b^2}{a^2 - b^2} \ln \frac{b}{a} \right] \tag{10.150c}$$

If we now let $b \to 0$, noting that

$$\lim_{b \to 0} \frac{\ln b}{\dfrac{1}{b^2}} = 0 \tag{10.151}$$

by L'Hospital's rule, we find

$$D_Z(R) = \frac{P}{8\pi D} \left[R^2 \ln \frac{R}{a} + \frac{(3 + \mu)}{2(1 + \mu)} (a^2 - R^2) \right] \tag{10.152}$$

which is the solution for a *solid* plate under a *concentrated* load, P, at the center.

It is instructive to review the *order* of the limiting operations which were employed in the preceding paragraphs. In going from equation (10.149) to equation (10.152), the effect of the hole was eliminated because we treated the *general* function $D_Z(R)$ rather than a *specific* value. On the other hand, the limiting operation based on equation (10.140) retained the effect of the hole, because it was performed on a specific value $M_\theta(b)$ rather than the general function, $M_\theta(R)$.

There are a number of additional cases that can be handled with the solutions developed here. For example, a plate with a rigid annular insert or plug is shown in Figure 10–23. If the plug is subjected to an arbitrary axisymmetric loading, $q_Z(R)$, the total load on the plug is

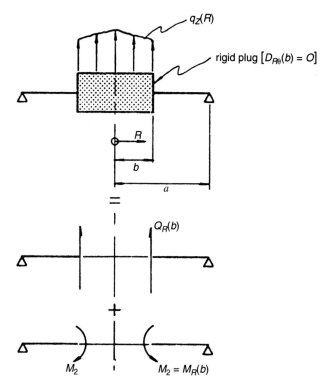

FIGURE 10–23   Annular Plate with Rigid Plug.

$$P = 2\pi \int_0^b q_Z(R)R \, dR \tag{10.153}$$

We may then represent the region of the plate $R \geq b$ by the superposition of (a) a solution corresponding to Figure 10–22 with $Q_R = -P/(2\pi b)$; and (b) a solution based on Figure 10–19 with $M_1 = 0$ and $M_2 = M_R(b)$, as shown in Figure 10–23. Then, M_R (b) is evaluated from the compatibility condition, $D_{R\theta}(b) = 0$. There are numerous variations of the rigid insert problem which are important in machine design. A compilation of these solutions may be found in Timoshenko and Woinowsky-Krieger[17] and in Szilard.[18]

Example 10.3(d)

In Figure 10–24, we show a plate with an annular opening which is axisymmetrically loaded outside the opening. It is convenient to first solve the case of the same loading applied to a solid plate, Figure 10–24(b). In this first solution, we may take $q_Z(R)$ in the region $0 \leq R \leq b$ as any function of R. For convenience, we show $q_Z(R) = q_Z$ (b) for $0 \leq R \leq b$ on Figure 10–24(b). For the solid plate, we evaluate $Q_R(b)$ and $M_R(b)$. For example, with a uniform load $q_Z(R) = q_0$, we would have

$$Q_R = -q_0 \frac{\pi b^2}{2\pi b} = -q_0 \frac{b}{2} \tag{10.154a}$$

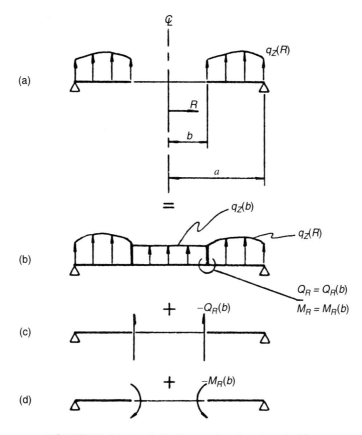

FIGURE 10–24   Radially Varying Load on Annular Plate.

and also calculate $M_R(b)$ from equation (10.129a) as

$$M_R(b) = \left(\frac{q_0}{16}\right)(3 + \mu)(a^2 - b^2) \tag{10.154b}$$

For more complex loadings, the corresponding values of $Q_R(b)$ and $M_R(b)$ would be determined by combining the homogeneous solution, equation (10.117), with an appropriate particular solution. Once $Q_R(b)$ and $M_R(b)$ are evaluated for the solid plate, we correspondingly apply a force and a moment of equal magnitude and opposite sense to the open plate, as shown in Figures 10–24(c) and (d). These latter solutions correspond to the cases shown in Figures 10–22 and 10–19 respectively, and the complete solution for $b \leq R \leq a$ is found by superposition.

It should be noted that the preceding solutions can also represent a clamped plate with the addition of a solution for $M_R(a) = M_1$, as shown in Figure 10–19. The value of M_1 would be determined from the compatibility condition, $D_{R\theta}(a) = 0$.

The loading case shown in Figure 10–22, a uniform transverse shear on an annular plate, can be generalized to a *solid* plate where the transverse shear becomes a ring load around the circle with radius b.[17] If the magnitude of the ring load is given by $Q_R(b) = P$, displacement functions may be derived for the outer ring $r \geq b$ and the inner plate, $r \leq b$.

For the outer ring, Timoshenko and Woinowski-Krieger[17] give

$$D_Z(R) = D_{Z1} = \frac{P}{8\pi D} \left\{ (a^2 - R^2) \left[1 + \frac{1}{2} \left(\frac{1-\mu}{1+\mu} \right) \left(\frac{a^2 - b^2}{a^2} \right) \right] + (b^2 + R^2) \ln \frac{R^2}{a} \right\} \quad (10.155a)$$

For the inner plate, the deflection is

$$D_Z(R) = D_{Z1}(b) + D_{Z2}(R) \quad (10.155b)$$

where

$$D_{Z1}(b) = \frac{P}{8\pi D} \left\{ (a^2 - b^2) \left[1 + \frac{1}{2} \left(\frac{1-\mu}{1+\mu} \right) \left(\frac{a^2 - b^2}{a^2} \right) \right] + 2b^2 \ln \frac{b}{a} \right\} \quad (10.155c)$$

$$D_{Z2}(R) = \frac{P}{8\pi D} \frac{(b^2 - R^2)}{(1 + \mu)} \left[\frac{(1-\mu)(a^2 - b^2)}{2a^2} - (1 + \mu) \ln \frac{b}{a} \right] \quad (10.155d)$$

D_{Z2} represents the deflection due to pure bending of the inner plate. Timoshenko and Woinowski-Kreiger[17] also solve the ring load case for clamped boundaries.

For the outer ring,

$$D_Z(R) = \frac{P}{8\pi D} \left[(a^2 - R^2) \frac{a^2 + b^2}{2a^2} + (b^2 + R^2) \ln \frac{R}{a} \right] \quad (10.155e)$$

and for the inner plate,

$$D_Z(R) = \frac{P}{8\pi D} \left[(b^2 + R^2) \ln \frac{b}{a} + \frac{(a^2 + R^2)(a^2 - b^2)}{2a^2} \right] \quad (10.155f)$$

These solutions may be taken as Green's functions to generate solutions for distributed ring loads on circular plates.

10.3.2 GENERAL BENDING

10.3.2.1 Solution of Governing Equation

We return to the general biharmonic equation, equation (10.26), and seek a set of biharmonic functions which satisfy the θ dependence as well. The solution to the homogeneous equation may be written in the separated form attributed to the noted mechanician Clebsch[19]

$$D_{Zh} = F^0(R) + \sum_{j=1}^{\infty} F^j(R) \cos j\theta \quad (10.156)$$

This approach is identical to that employed for the shell of revolution in Section 5.1.2. From a physical standpoint, this is plausible because the circular plate is the degenerate case of the shell of revolution as $R_\phi \to \infty$.

The function $F^0(R)$ is the solution of the axisymmetric case as given by equation (10.117). For $j \geq 1$, we substitute equation (10.156) into equation (10.26) to get

$$F^j_{,RRRR} + \frac{2}{R} F^j_{,RRR} - \frac{(1 + 2j^2)}{R^2} F^j_{,RR} + \frac{(1 + 2j^2)}{R^3} F^j_{,R}$$
$$- \frac{j^2(4 - j^2)}{R^4} F^j = 0 \qquad (10.157)$$

Just as in the case of shells of revolution, the solution for $j = 1$ is somewhat simpler and is given by

$$F^1(R) = C^1_1 R \ln R + C^1_2 R^3 + C^1_3 R + C^1_4 \frac{1}{R} \qquad (10.158)$$

For $j > 1$,

$$F^j(R) = C^j_1 R^j + C^j_2 R^{-j} + C^j_3 R^{j+2} + C^j_4 R^{-(j-2)} \qquad (10.159)$$

These homogeneous solutions may be combined with appropriate particular solutions to solve specific loading cases, with the integration constants determined from the boundary conditions. In each case, the loading function and hence the particular solution should be described in a Fourier cosine series to conform to D_{Zh}, as given by equation (10.156).

10.3.2.2 Linearly Varying Load.   We investigate a circular plate under a linearly varying load, as shown in Figure 10–25. This represents hydrostatic pressure and may be resolved into symmetric and antisymmetric components, as shown on the figure. We write the symmetric component as

$$q^0_Z = \frac{q_1 + q_2}{2} \qquad (10.160)$$

the antisymmetric component as

$$q^1_Z = \frac{q_1 - q_2}{2} \qquad (10.161)$$

and the displacement function as

$$D_Z = D^0_Z + D^1_Z \qquad (10.162)$$

For the symmetric component, q^0_Z, we have already obtained solutions for simply supported and clamped boundaries in Section 10.3.1.3. To apply the previously obtained solutions to this problem, q^0_Z is inserted for q_0 in equation (10.128) for the simply supported boundary, or in equation (10.135) for the clamped boundary.

For the antisymmetric component, q^1_Z, we write

$$q_Z(R,\theta) = q^1_Z \frac{X}{a} \qquad (10.163)$$

With $X = R \cos \theta$,

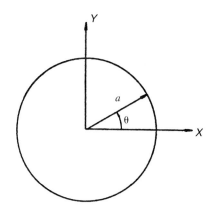

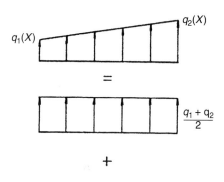

FIGURE 10–25   Hydrostatically Loaded Circular Plate.

$$q_Z(R,\theta) = q_Z^1 \frac{R}{a} \cos\theta \qquad (10.164)$$

Because the governing equation, equation (10.26), requires four derivatives with respect to R, we select a fifth-order function for the particular solution

$$D_{Z_p}^1 = C_5^1 R^5 \cos\theta \qquad (10.165)$$

Substituting equation (10.165) into equation (10.26), we find

$$C_5^1\{120 + 2(60) - [20 + 2(20)] + [5 + 2(5)]$$
$$+ (-4 + 1)\} R \cos\theta = \frac{q_Z^1}{D} \frac{R}{a} \cos\theta \qquad (10.166a)$$

or

$$192 C_5^1 R \cos\theta = \frac{q_Z^1}{D} \frac{R}{a} \cos\theta \qquad (10.166b)$$

from which

$$C_5^1 = \frac{q_Z^1}{192Da} \tag{10.167}$$

and

$$D_{Zp}^1(R,\theta) = \frac{q_Z^1}{192D} \frac{R^5}{a} \cos\theta \tag{10.168}$$

Combining equation (10.168) with the homogeneous solution $D_{Zh}^1 = F^1 \cos\theta$, where F^1 is given by equation (10.158), we have

$$D_Z^1(R,\theta) = \left[C_1^1 R \ln R + C_2^1 R^3 + C_3^1 R + C_4^1 \frac{1}{R} + \frac{q_Z^1}{192D} \frac{R^5}{a} \right] \cos\theta \tag{10.169}$$

So that the solution remains finite at $R = 0$, we take $C_1^1 = C_4^1 = 0$. The remaining constants are evaluated from the appropriate boundary conditions.

For a clamped outer boundary,

$$D_Z^1(a,\theta) = D_{Z,R}^1(a,\theta) = 0 \tag{10.170}$$

which leads to the constants

$$C_2^1 = -\frac{2q_Z^1}{192D} a; \quad C_3^1 = \frac{q_Z^1}{192D} a^3 \tag{10.171}$$

and the deflection function

$$D_Z^1(R,\theta) = \frac{q_Z^1}{192D} \frac{R}{a} (a^2 - R^2)^2 \cos\theta \tag{10.172}$$

For a simply supported outer boundary,

$$D_Z^1(a,\theta) = M_R^1(a,\theta) = 0 \tag{10.173}$$

which gives the constants

$$C_2^1 = -\frac{q_Z^1}{192D} \frac{2(5 + \mu)}{(3 + \mu)} a; \quad C_3^1 = \frac{q_Z^1}{192D} \frac{(7 + \mu)}{(3 + \mu)} a^3 \tag{10.174}$$

and the deflection function

$$D_Z^1(R,\theta) = \frac{q_Z^1}{192(3 + \mu)D} \left[7 + \mu - (3 + \mu)\left(\frac{R}{a}\right)^2 \right] Ra(a^2 - R^2) \cos\theta \tag{10.175}$$

A detailed solution for the antisymmetric loading condition with a simply supported boundary is presented in Timoshenko and Woinowsky-Krieger,[19] where the stress couples, the transverse shear forces, and the locations and magnitudes of the maximum moments are established. Also, the problem of a clamped circular plate under an eccentric concentrated load and some rather extensive generalizations thereof are examined in Timoshenko and Woinowsky-Krieger[19] and in Michell.[20] Among other things, the latter solutions provide an illustration of the Maxwell-Betti reciprocal theorem for a plate.[21]

10.4 PLATES OF OTHER SHAPES

10.4.1 General Approach

There are a number of solutions for plates with other than rectangular or circular planform which can be expressed in Cartesian or polar coordinates. In Cartesian coordinates, solutions for triangular and elliptical plates are found by starting with a deflection function D_Z that is proportional to the equation of the boundary of the plate, which ensures that $D_Z = 0$ on the boundary. The function may be augmented in order to satisfy either a simply supported or a clamped condition. In polar coordinates, a sector shape can readily be treated using the general solution described in Section 10.3.2.1. Some representative illustrations are presented in the following sections. Also, a variety of tabulated results for various shaped plates are contained in Szilard.[16,18]

10.4.2 Triangular Plates

Considering the equilateral triangular plate shown in Figure 10–26, the coordinate system is located with the origin at the centroid of the triangle. With respect to this coordinate system, the equation of each side is shown on the figure and the product of the three terms is,

$$D_{Z1} = -\frac{1}{3}\left[X^3 - 3XY^2 - a(X^2 + Y^2) + \frac{4}{27}a^3\right] \tag{10.176}$$

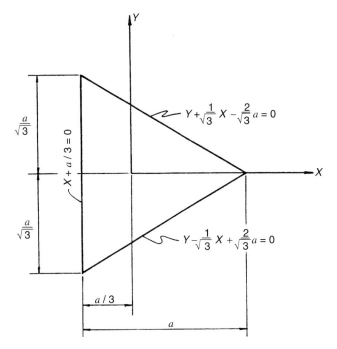

FIGURE 10–26   Triangular Plate.

gives us a start toward a deflection function, because it will satisfy the condition $D_Z = 0$ on all boundaries. The function D_{Z1} must be modified to satisfy the equilibrium condition $\nabla^4 D = 0$ and the remaining boundary conditions.

For an edge moment M uniformly distributed around the boundary, the complete solution is

$$D_Z = -\frac{3M}{4aD} D_{Z1} = \frac{M}{4aD}\left[X^3 - 3XY^2 - a(X^2 + Y^2) + \frac{4}{27}a^3\right] \qquad (10.177)$$

To verify this solution we must show that (a) $\nabla^4 D_Z = 0$ on the entire plate; (b) $D_Z = 0$ on the boundaries; (c) $M_X(-a/3, Y) = M$ (the other boundaries may also be checked but this will not be necessary because of the symmetry).

Proceeding, we consider

1. $\nabla^4 D_Z = 0$ throughout: From equations (10.13) we note that any nonzero term remaining after the application of $\nabla^2 (\ )$ must be at least quadratic in X *and* Y. No such terms are present in equation (10.177).

2. $D_Z = 0$ on boundary: This is obviously satisfied from the derivation of D_{Z1}.

3. $M_X(-a/3, Y) = M$: From equation (10.8a)

$$M_X = -D(D_{Z,XX} + \mu D_{Z,YY})$$

$$= -D\frac{M}{4aD}[6X - 2a + \mu(-6X - 2a)]$$

$$M_X(-a/3,Y) = -\frac{M}{4a}[-2a - 2a + \mu(2a - 2a)] = M$$

and the solution is verified.

For a uniformly distributed load q_0 on a simply supported triangular plate, the deflection function D_{Z1}, is modified to[22]

$$D_Z = \frac{q_0}{64aD}\left[X^3 - 3XY^2 - a(X^2 + Y^2) + \frac{4}{27}a^3\right]\left(\frac{4}{9}a^2 - X^2 - Y^2\right) \qquad (10.178)$$

This solution may be verified in a similar manner to that for equation (10.177). It is instructive to perform a detailed stress analysis on this plate, following the general computational format presented in Section 10.2.2.3. Among other things, this will confirm the absence of concentrated corner forces at a non-right-angle corner. This is left to the exercises.

A solution for a simply supported isosceles right triangular plate is also given by Timoshenko and Woinowsky-Krieger.[22]

10.4.3 Elliptical Plates

A clamped elliptical plate is shown in Figure 10–27 and the equation of the boundary is used to write

$$D_{Z1} = -\frac{X^2}{a^2} - \frac{Y^2}{b^2} + 1 \qquad (10.179)$$

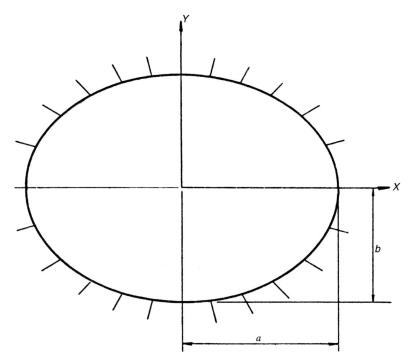

FIGURE 10–27   Elliptical Plate.

For a uniform load q_0 the solution may then be taken as[23]

$$D_{Z2} = C(D_{Z1})^2 \tag{10.180a}$$

where

$$C = \frac{q_0}{8D\left(\dfrac{3}{a^4} + \dfrac{3}{b^4} + \dfrac{2}{a^2b^2}\right)} \tag{10.180b}$$

     A complete stress analysis may be performed in the format of Section 10.2.2.3 and is available in Timoshenko and Woinowsky-Krieger.[23]

10.4.4   Circular Sector Plates

A plate which is a slice of a complete circle may be solved using a procedure similar to that described in Section 10.3.2.1, except that the displacement function for the harmonics $j \geq 1$ is taken as a Fourier sine series. The arithmetic is rather involved, and the interested reader is referred to Szilard.[24]

     The relatively meager array of available solutions for irregularly shaped plates points to the tremendous breakthrough made possible when the finite element technique became available. This method enables almost any planform to be realistically modeled and solved.

REFERENCES

1. R. Szilard, *Theory and Analysis of Plates*, Prentice-Hall, Inc., Englewood Cliffs, NJ, 1974: 36.

2. M. Filonenko-Borodich, *Theory of Elasticity*, [trans. From Russian], (New York: Dover Publications): 260–268.

3. S Timoshenko, *History of the Strength of Materials*, (New York: McGraw-Hill, 1953): 119–122.

4. S Timoshenko and S. Woinowsky-Krieger, *Theory of Plates and Shells*, 2nd ed., (New York: McGraw-Hill, 1959): 92–97.

5. S. Timoshenko and J. N. Goodier, *Theory of Elasticity*, 2nd ed., (New York: McGraw-Hill, 1951): 316–317.

6. S. Timoshenko, *History of the Strength of Materials*: 333–340.

7. A. Nádai, *Die Elastichen Platten*, (Berlin: Springer-Verlag, 1925): 42.

8. S. Timoshenko, *History of the Strength of Materials*: 110–113.

9. S. Timoshenko and W. Woinowsky-Krieger, *Theory of Plates and Shells*: 113–135.

10. R. Courant and D. Hilbert, *Methods of Mathematical Physics*, Vol. I, (New York: Interscience Publishers, 1966): 351–396.

11. *Manual of Steel Construction, Load and Resistance Factor Design*, 1st ed., (New York: American Institute of Steel Construction, 1986): 3-130–3-141.

12. S. Timoshenko and S. Woinowsky-Krieger, *Theory of Plates and Shells*, 2nd ed,: 136–162.

13. Ibid., chap. 6.

14. Ibid., chap. 7.

15. K. Kiattikomol, L. M. Keer, and J. Dundurs, Application of dual series to rectangular plates," *Journal of the Engineering Mechanics Division*, ASCE, vol. 100, no. EM 2, Proc. Paper 10486, April 1974: 433–444.

16. R. Szilard, *Theory and Analysis of Plates*: 650–673.

17. S. Timoshenko and S. Woinowsky-Krieger, *Theory of Plates and Shells*, 2nd ed: 58–65.

18. R. Szilard, *Theory and Analysis of Plates*: 613–649.

19. S. Timoshenko and S. Woinowsky-Krieger, *Theory of Plates and Shells*, 2nd ed: 282–293.

20. J. H. Mitchell, On the flexure of circular plates, *Proc. London Math. Soc.*, 34, 1902: 223–228.

21. N. J. Hoff, *The Analysis of Structures*, (New York: Wiley, 1956): 373–377.

22. S. Timoshenko and S. Woinowsky-Krieger, *Theory of Plates and Shells*, 2nd ed: 313–319.

23. Ibid., 310–313.

24. R. Szilard, *Theory and Analysis of Plates*, 126–129.

EXERCISES

The numerical problems are given without specific units to allow English, metric, or **SI** unit dimensions to be selected.

10.1. Using an appropriate free-body diagram and the definition

$$\varepsilon_Y(\xi) = D_{Y,Y}$$

verify that for a plate described in Cartesian coordinates,

$$D_Y = -\xi D_{Z,Y}$$

and

$$\varepsilon_Y(\xi) = -\xi D_{Z,YY}$$

10.2. A rectangular plate rests on an elastic foundation with modulus K. The units of K are force/surface area/distance in the transverse direction. Derive the governing equations for the bending of this plate.

10.3. Using an appropriate free-body diagram and the Kelvin-Tait argument, verify the effective shear expression in polar coordinates as given by equations (10.30).

10.4. Consider triangle ② in Figure 10–4 and derive equations (10.38) and (10.39).

10.5. The simply supported plate shown in Figure 10–28 is half-submerged in a fluid of unit weight γ.

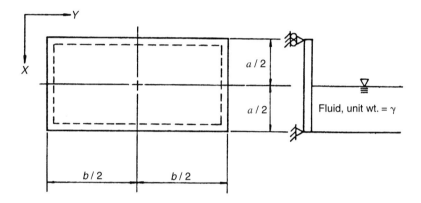

FIGURE 10–28   Exercise 10.5

(a) Obtain a solution for the displacement function using both the Navier and Lévy-Nádai methods.

(b) Using one of the preceding solutions:
  (1) Determine the midspan deflections.
  (2) Determine the maximum value of M_X and the location on the plate.
  (3) Plot the variations of (a) M_X and M_Y along the center lines of the plate and (b) $\overline{Q}_X$ and $\overline{Q}_Y$ along each boundary.
  (4) Compute the corner forces.

10.6. Consider the plate shown in Figure 10–9. The loading $q_Z(X, Y) = q_0$ constant. All edges are simply supported, except the edge along the Y-axis at $X = 0$, which is clamped. Using the Lévy-Nádai approach, derive the deflection function and compare the maximum moments in the X and Y directions to those found when all sides are simply supported. To facilitate the numerical work, a square plate $a = b$ may be used as the basis of comparison.

10.7. Consider the axisymmetrically loaded, simply supported circular plate as shown in Figure 10–29.
  (a) Obtain the solutions for the displacement, the radial moment, and the circumferential moment.
  (b) Evaluate the maximum radial and circumferential moments.
  (c) Verify the solution for a concentrated load at the center of the plate of magnitude $P_0 = q_0 \pi b^2$, given by equation (10.152). Hint: Consider equations 10.155 (a)–(d).

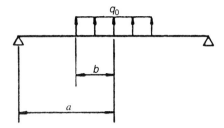

FIGURE 10–29   Exercise 10.7

10.8. Consider the clamped circular plate shown in Figure 10–30 and re-solve Exercise 10.7. Hint: Consider: equations 10.155 (e) and (f).

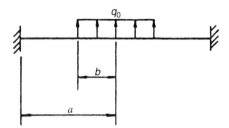

FIGURE 10–30   Exercise 10.8

10.9. Consider the circular plate as shown in Figure 10–31. The plate is subjected to (a) a uniformly distributed load q_0; and (b) a hydrostatic loading with maximum intensity q_1, and is supported on a knife-edge circular support which has a radius b.

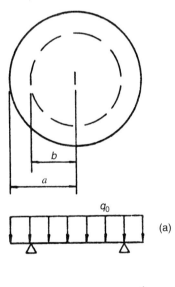

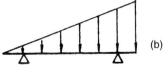

FIGURE 10–31   Exercise 10.9

Derive the general solution for the deflection function for this plate. The solution should consist of a diagram of the superposition representation (if required) plus a careful statement of the appropriate conditions required to evaluate the integration constants. It is suggested that each loading case be treated separately.

10.10. Consider the circular plate with a rigid insert as shown in Figure 10–32. The plate is subjected to a uniformly distributed load of intensity q_0 on the insert and q_1 on the annulus.

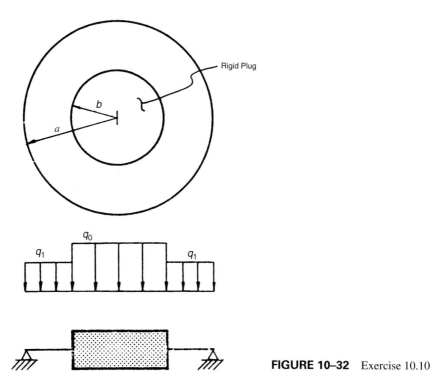

FIGURE 10–32   Exercise 10.10

(a) Derive the general solution for the deflection function for this plate. The solution should consist of a diagram of the superposition representation (if required) plus a careful statement of the appropriate conditions required to evaluate the integration constants.

(b) Compute the expressions for the radial and circumferential moments and determine the maximum values and locations.

10.11. Consider the triangular plate shown in Figure 10–26 subject to a uniformly distributed load q_0.

(a) Verify the solution as given by equation (10.178).

(b) Plot the moments and the transverse shear on a typical edge.

(c) Verify that the total edge shear balances the applied loading and hence, that no corner forces are required or present.

10.12. In Example 10.2(a), the maximum bending moment in a square simply supported plate under a uniform load was shown to be proportional to the total load on the plate, $q_0 a^2$.

(a) Investigate the variation of the maximum normal displacement as the radius a increases while $q_0 a^2$ remains constant.

(b) Investigate this proportionality for a case where two opposite sides are fixed, whereas the other opposite sides are simply supported.

10.13. For a square plate, verify the value of the moment in the vicinity of the corner as discussed in Section 10.2.2.3. Use $\mu = 0.3$ to compute the maximum bending moment in the center of the plate and M_η at the corner.

10.14. Consider the circular plate shown in Figure 10–17, with the addition of a point support at $R = 0$. Using the method of consistent deformations, obtain the deflection function.

11

Bending and Stability of Plates

11.1 ENERGY METHOD SOLUTIONS

11.1.1 Strain Energy for Plates in Flexure

Refer to the general expression for U_ε as given by equation (9.6). We are presently considering no extensional strains, and, in accordance with our previous development, we choose to neglect transverse shearing strains. Therefore, we have only the component **III** in equation (9.7) remaining along with the bending thermal term.

$$
U_\varepsilon = \frac{D}{2} \int_S \left\{ [(\kappa_\alpha + \kappa_\beta)^2 - 2(1 - \mu)(\kappa_\alpha \kappa_\beta - \tau^2)] \right.
$$
$$
\left. - \bar{\alpha}(1 + \mu)(\kappa_\alpha + \kappa_\beta) \int_{-h/2}^{h/2} T(\zeta)\zeta \, d\zeta \right\} AB \, d\alpha \, d\beta
\tag{11.1}
$$

Because we are interested in a displacement formulation, we may substitute the strain-displacement relationships, as given by equations (7.55), into equation (11.1). This will be specialized for the two coordinate systems that we have considered—Cartesian and polar.

For Cartesian coordinates, we take $\alpha = X$, $\beta = Y$, $n = Z$, and $A = B = 1$ in equation (7.55) to get

$$\kappa_X = -D_{Z,XX} \tag{11.2a}$$

$$\kappa_Y = -D_{Z,YY} \tag{11.2b}$$

$$\tau = -D_{Z,XY} \tag{11.2c}$$

Then, substituting equations (11.2a–c) into equation (11.1), we have

$$U_\varepsilon = \frac{D}{2} \int_S \left\{ (D_{Z,XX} + D_{Z,YY})^2 - 2(1 - \mu)[D_{Z,XX}D_{Z,YY} - (D_{Z,XY})^2] \right.$$
$$\left. + \overline{\alpha}(1 + \mu)(D_{Z,XX} + D_{Z,YY})\int_{-h/2}^{h/2} T(\zeta)\zeta \, d\zeta \right\} dY \, dX \tag{11.3}$$

We may rewrite equation (11.3) as

$$U_\varepsilon = \frac{D}{2} \int_S \left\{ (\boldsymbol{\nabla}^2 D_Z)^2 - 2(1 - \mu)[D_{Z,XX}D_{Z,YY} - (D_{Z,XY})^2] \right.$$
$$\left. + \overline{\alpha}(1 + \mu)\boldsymbol{\nabla}^2 D_Z \int_{-h/2}^{h/2} T(\zeta)\zeta \, d\zeta \right\} dY \, dX \tag{11.4a}$$

or

$$U_\varepsilon = U_{\varepsilon 1} + U_{\varepsilon 2} + U_\tau \tag{11.4b}$$

Because the strain energy density dU_ε, which is the integrand of equations (11.4), is obviously independent of the choice of coordinate axes and because $\boldsymbol{\nabla}^2 D_Z$ has been shown to be invariant, the term

$$D_{Z,XX}D_{Z,YY} - (D_{Z,XY})^2 \tag{11.5}$$

also must be invariant. This observation can be useful for transforming U_ε to other coordinate systems.

For polar coordinates, we may repeat the preceding calculations with $\alpha = R$, $\beta = \theta$, $n = Z$, $A = 1$, and $B = R$. However, for variety, we will use the invariant property of dU_ε as discussed in the preceding paragraph because $\boldsymbol{\nabla}^2 D_Z$ is already available in polar coordinates and is given by equation (10.28a) or (10.28b). The thermal term will be omitted because it is proportional to $\boldsymbol{\nabla}^2 D_Z$. Therefore, we need only transform the term stated in equation (11.5).

Referring to Figure 10–3, we have

$$R^2 = X^2 + Y^2 \tag{11.6a}$$

$$R_{,X} = \frac{X}{R} = \cos\theta \tag{11.6b}$$

$$R_{,Y} = \frac{Y}{R} = \sin\theta \tag{11.6c}$$

Also, taking $\partial/\partial X$ of both sides of $X = R\cos\theta$, we find

$$1 = \cos\theta\, R_{,X} - (R\sin\theta)\theta_{,X}$$

from which

$$\theta_{,X} = \frac{\cos\theta\, R_{,X} - 1}{R\sin\theta} = -\frac{\sin\theta}{R} \tag{11.7}$$

using equation (11.6b). Similarly,

$$\theta_{,Y} = \frac{\cos\theta}{R} \tag{11.8}$$

Equations (11.6a–c), (11.7), and (11.8) constitute the basic relations required to transform equation (11.5) into polar coordinates.

We now evaluate the required derivatives using the chain rule:

$$D_{Z,X} = D_{Z,R}R_{,X} + D_{Z,\theta}\theta_{,X} = D_{Z,R}\cos\theta - D_{Z,\theta}\frac{\sin\theta}{R}$$

$$D_{Z,XX} = \cos\theta(D_{Z,RR}R_{,X} + D_{Z,R\theta}\theta_{,X}) + D_{Z,R}(-\sin\theta)(\theta_{,X})$$

$$-\frac{\sin\theta}{R}(D_{Z,\theta R}R_{,X} + D_{Z,\theta\theta}\theta_{,X})$$

$$-D_{Z,\theta}\left[\frac{\cos\theta(\theta_{,X})}{R} - \frac{\sin\theta(R_{,X})}{R^2}\right] \tag{11.9}$$

$$= \cos^2\theta\, D_{Z,RR} - \frac{2\sin\theta\cos\theta}{R}D_{Z,R\theta} + \frac{\sin^2\theta}{R^2}D_{Z,\theta\theta}$$

$$+\frac{\sin^2\theta}{R}D_{Z,R} + \frac{2\sin\theta\cos\theta}{R^2}D_{Z,\theta}$$

Similarly, we compute

$$D_{Z,YY} = \sin^2\theta\, D_{Z,RR} + 2\frac{\sin\theta\cos\theta}{R}D_{Z,R\theta} + \frac{\cos^2\theta}{R^2}D_{Z,\theta\theta}$$

$$+\frac{\cos^2\theta}{R}D_{Z,R} - \frac{2\sin\theta\cos\theta}{R^2}D_{Z,\theta} \tag{11.10}$$

The sum of equations (11.9) and (11.10) checks with equation (10.28a) for $\nabla^2 D_Z$. Finally, we evaluate

$$D_{Z,XY} = \cos\theta(D_{Z,RR}R_{,Y} + D_{Z,R\theta}\theta_{,Y}) - \sin\theta(\theta_{,Y})D_{Z,R}$$

$$- \frac{\sin\theta}{R}(D_{Z,\theta R}R_{,Y} + D_{Z,\theta\theta}\theta_{,Y})$$

$$- \left[\frac{\cos\theta(\theta_{,Y})}{R} - \frac{\sin\theta(R_{,Y})}{R^2}\right]D_{Z,\theta} \qquad (11.11)$$

$$= \sin\theta\cos\theta\, D_{Z,RR} + \frac{(\cos^2\theta - \sin^2\theta)}{R}D_{Z,R\theta}$$

$$- \frac{\sin\theta\cos\theta}{R^2}D_{Z,\theta\theta} - \frac{\sin\theta\cos\theta}{R}D_{Z,R}$$

$$- \frac{(\cos^2\theta - \sin^2\theta)}{R^2}D_{Z,\theta}$$

We may then substitute equations (11.9) to (11.11) into equation (11.4) to get the strain energy in polar coordinates. This is rather involved in the general case, but we can write the axisymmetric expression fairly concisely. We first evaluate equation (11.5), dropping the θ-dependent term. This reduces to $D_{Z,XX}D_{Z,YY} - (D_{Z,XY})^2 = (1/R)D_{Z,R}D_{Z,RR}$, which, along with the first two terms in equations (10.28a), gives

$$U_\varepsilon = \frac{D}{2}\int_S\left[\left(D_{Z,RR} + \frac{1}{R}D_{Z,R}\right)^2 - \frac{2(1-\mu)}{R}D_{Z,R}D_{Z,RR}\right]R\,dR\,d\theta \qquad (11.12)$$

for equation (11.4) with the thermal term omitted. For the common case of a solid circular plate with radius a, equation (11.12) becomes

$$U_\varepsilon = \pi D\int_0^a\left[\left(D_{Z,RR} + \frac{1}{R}D_{Z,R}\right)^2 - \frac{2(1-\mu)}{R}D_{Z,R}D_{Z,RR}\right]R\,dR \qquad (11.13)$$

11.1.2   Simply Supported Plate Under Concentrated Load

We reconsider the problem illustrated in Figure 10–12 and attempt to confirm the solution using the virtual work approach derived in Section 9.4.1. Our first step is to take the solution in the form of a Navier-type double series, as given by equation (10.71), and to evaluate U_ε. With

$$D_Z = \sum_{j=1}^\infty\sum_{k=1}^\infty D_Z^{jk}\sin j\pi\frac{X}{a}\sin k\pi\frac{Y}{b} \qquad (11.14)$$

we have

$$D_{Z,XX} = -\sum_{j=1}^\infty\sum_{k=1}^\infty D_Z^{jk}\left(\frac{j\pi}{a}\right)^2\sin j\pi\frac{X}{a}\sin k\pi\frac{Y}{b} \qquad (11.15a)$$

$$D_{Z,YY} = -\sum_{j=1}^\infty\sum_{k=1}^\infty D_Z^{jk}\left(\frac{k\pi}{b}\right)^2\sin j\pi\frac{X}{a}\sin k\pi\frac{Y}{b} \qquad (11.15b)$$

$$D_{Z,XY} = -\sum_{j=1}^\infty\sum_{k=1}^\infty D_Z^{jk}\left(\frac{j\pi}{a}\right)\left(\frac{k\pi}{b}\right)\cos j\pi\frac{X}{a}\cos k\pi\frac{Y}{b} \qquad (11.15c)$$

When we substitute equations (11.15a–c) into equations (11.4), terms of the form $\sum\limits_{j=1}^{\infty}\sum\limits_{k=1}^{\infty}$ are generated. However, because of the orthogonality relationships, only those terms corresponding to the same j and the same k will remain after integration, and the double summations over j and k reduce to single summations. Therefore, omitting the terms that subsequently drop out, we have

$$U_{\varepsilon 1} = \frac{D}{2}\int_0^a\int_0^b\left\{\sum_{j=1}^{\infty}\sum_{k=1}^{\infty}(D_Z^{jk})^2\left[\left(\frac{j\pi}{a}\right)^2+\left(\frac{k\pi}{b}\right)^2\right]^2\sin^2 j\pi\frac{X}{a}\sin^2 k\pi\frac{Y}{b}\right\}dY\,dX \qquad (11.16)$$

Noting that, $\int_0^b\sin^2(k\pi Y/b)\,dY = b/2$ and $\int_0^a\sin^2(j\pi X/a)\,dX = a/2,$ equation (11.16) reduces to

$$U_{\varepsilon 1} = \frac{\pi^4 abD}{8}\sum_{j=1}^{\infty}\sum_{k=1}^{\infty}(D_Z^{jk})^2\left[\left(\frac{j}{a}\right)^2+\left(\frac{k}{b}\right)^2\right]^2 \qquad (11.17)$$

We now consider

$$U_{\varepsilon 2} = D(1-\mu)\int_0^a\int_0^b\sum_{j=1}^{\infty}\sum_{k=1}^{\infty}(D_Z^{jk})^2\left[\left(\frac{j\pi}{a}\right)^2\left(\frac{k\pi}{b}\right)^2\sin^2 j\pi\frac{X}{a}\sin^2 k\pi\frac{Y}{b}\right.$$
$$\left.-\left(\frac{j\pi}{a}\right)^2\left(\frac{k\pi}{b}\right)^2\cos^2 j\pi\frac{X}{a}\cos^2 k\pi\frac{Y}{b}\right]dY\,dX \qquad (11.18)$$

Because

$$\int_0^b\sin^2 k\pi\frac{Y}{b}\,dY = \int_0^b\cos^2 k\pi\frac{Y}{b}\,dY = \frac{b}{2}$$

and

$$\int_0^a\sin^2 j\pi\frac{X}{a}\,dX = \int_0^a\cos^2 j\pi\frac{X}{a}\,dX = \frac{a}{2}$$

the entire expression for $U_{\varepsilon 2}$ vanishes and $U_\varepsilon = U_{\varepsilon 1}$, with no thermal effects.

We are now prepared to impart a virtual displacement to the system. Recalling the derivation in Section 9.4.1, the virtual displacement is required to conform to the constraints of the system, which in this case means the *boundary* conditions. If the virtual displacement is twice differentiable, the strain displacement conditions, equations (7.55a–d), will automatically satisfy compatibility. Therefore, it is logical to take the virtual displacement in the form of equation (11.14).

We proceed for a single general term of the series and choose for the virtual displacement

$$\delta D_Z = \delta D_Z^{jk}\sin j\pi\frac{X}{a}\sin k\pi\frac{Y}{b} \qquad (11.19)$$

where δD_Z^{jk} represents the amplitude of the virtual displacement. Also, note that the *source* of the virtual displacement need not be specified.

Referring to Figure 10–12, the virtual work done by the external load Q_0 acting at $X = \xi$ and $Y = \eta$ is

$$\delta \overline{U}_q = Q_0 \delta D_Z(\xi, \eta) = Q_0 \delta D_Z^{jk} \sin j\pi \frac{\xi}{a} \sin k\pi \frac{\eta}{b} \tag{11.20}$$

The corresponding change in strain energy is

$$\delta U_\varepsilon = \frac{\partial U_\varepsilon}{\partial D_Z} \delta D_Z \tag{11.21}$$

For U_ε we take $U_{\varepsilon 1}^{jk}$, the single general term of the series given by equation (11.17), and evaluate

$$\frac{\partial U_\varepsilon}{\partial U_Z} = \frac{\partial U_{\varepsilon 1}^{jk}}{\partial D_Z^{jk}} = \frac{\pi^2 Dab}{8} \left[\left(\frac{j}{a}\right)^2 + \left(\frac{k}{b}\right)^2 \right]^2 (2 D_Z^{jk}) \tag{11.22}$$

Substituting equation (11.22) into equation (11.21),

$$\delta U_\varepsilon = \frac{\pi^4 abD}{4} \left[\left(\frac{j}{a}\right)^2 + \left(\frac{k}{b}\right)^2 \right]^2 D_Z^{jk} \, \delta D_Z^{jk} \tag{11.23}$$

Now, equating equations (11.20) and (11.23) and cancelling the terms δD_Z^{jk}, we find

$$D_Z^{jk} = \frac{4 Q_0}{\pi^4 abD} \frac{\sin j\pi \dfrac{\xi}{a} \sin k\pi \dfrac{\eta}{b}}{\left[\left(\dfrac{j}{a}\right)^2 + \left(\dfrac{k}{b}\right)^2 \right]^2} \tag{11.24}$$

and the total deflection is given by

$$D_Z = \sum_{j=1}^{\infty} \sum_{k=1}^{\infty} D_Z^{jk} \sin j\pi \frac{X}{a} \sin k\pi \frac{Y}{b} \tag{11.25}$$

which checks with equation (10.101).

This example serves to illustrate that the principle of virtual work can serve as an alternate statement of equilibrium.

11.1.3 Clamped Plate Under Uniformly Distributed Loading

11.1.3.1 Approximate Solutions. The solution of a fully clamped plate cannot be readily accomplished with the Navier-type double series solution, and the solution of the Lévy-Nádai procedure is fairly involved.[1] This seems to be an ideal case for the application of the Rayleigh-Ritz or Galerkin method. Because the first steps in either method are common, we will proceed with a unified solution as far as possible. We consider the problem shown in Figure 11–1, with a uniformly distributed q_0 acting in the positive Z direction. Referring to Section 9.4.3, we have here only one generalized displacement D_Z so that in equations (9.15) and (9.16), $m = 1$ and $\Delta_1 = D_Z$. We select a single coordinate function

$$\phi_Z = \phi_1 = (X^2 - a^2)^2 (Y^2 - b^2)^2 \tag{11.26}$$

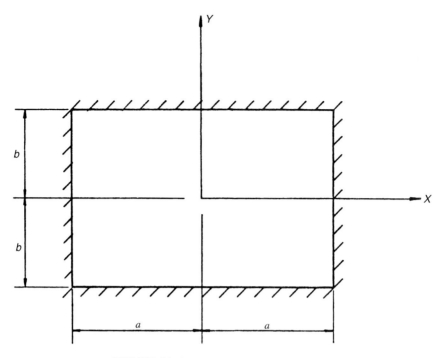

FIGURE 11–1   Clamped Rectangular Plate.

and the one-term approximation for D_Z is

$$D_Z(X,Y) = c(X^2 - a^2)^2(Y^2 - b^2)^2 \qquad (11.27)$$

It is obvious that the coordinate function ϕ_1 satisfies the stated boundary conditions: $D_Z(\pm a, Y) = D_{Z,X}(\pm a, Y) = D_Z(X, \pm b) = D_{Z,Y}(X, \pm b) = 0$. However, it is sometimes difficult to select appropriate coordinate functions when more complicated boundary conditions are encountered. The coordinate function selected here is obviously twice differentiable, so that the compatibility relationships will also be satisfied.

For both the Rayleigh-Ritz and the Galerkin solutions, the Laplacian, $\nabla^2 D_Z$, is required. We compute

$$D_{Z,XX} = 4c(3X^2 - a^2)(Y^2 - b^2)^2 \qquad (11.28a)$$

$$D_{Z,YY} = 4c(X^2 - a^2)^2(3Y^2 - b^2) \qquad (11.28b)$$

$$D_{Z,XY} = 16cXY(X^2 - a^2)(Y^2 - b^2) \qquad (11.28c)$$

and write

$$\begin{aligned} \nabla^2 D_Z &= D_{Z,XX} + D_{Z,YY} \\ &= 4c[(3X^2 - a^2)(Y^2 - b^2)^2 + (X^2 - a^2)^2(3Y^2 - b^2)] \end{aligned} \qquad (11.29)$$

11.1.3.2 Rayleigh-Ritz Solution. We proceed first with the Rayleigh-Ritz solution. Regarding the second term in equations (11.4) as given by equation (11.5), it has been found to be negligible for plates in which the planform is polygonal and the edges remain straight[2] and, in fact, was shown to vanish entirely for the problem considered in the previous section. It may be omitted here as well and

$$U_\varepsilon = \frac{D}{2} \int_{-a}^{a} \int_{-b}^{b} (\nabla^2 D_Z)^2 \, dY \, dX \tag{11.30}$$

where $\nabla^2 D_Z$ is given by equation (11.29).

The potential energy of the applied loads U_q, as defined in equations (9.10), is given by

$$U_q = -\int_{-a}^{a} \int_{-b}^{b} q_0 D_Z \, dY \, dX \tag{11.31}$$

where D_Z is given by equation (11.27).

We then write the total potential energy U_τ as

$$U_\tau = \int_{-a}^{a} \int_{-b}^{b} \left[\frac{D}{2} (\nabla^2 D_Z)^2 - q_0 D_Z \right] dY \, dX \tag{11.32}$$

Now invoking equation (9.17), we set

$$U_{\tau,c} = 0 \tag{11.33}$$

We may simplify the arithmetic by commuting the operations in equation (11.33) and equation (11.32):

$$U_{\tau,c} = \frac{d}{dc} \int_{-a}^{a} \int_{-b}^{b} \left[\frac{D}{2} (\nabla^2 D_Z)^2 - q_0 D_Z \right] dY \, dX$$
$$= \int_{-a}^{a} \int_{-b}^{b} \frac{d}{dc} \left[\frac{D}{2} (\nabla^2 D_Z)^2 - q_0 D_Z \right] dY \, dX = 0 \tag{11.34}$$

Evaluating d/dc of the integrand using equations (11.29) and (11.27),

$$\frac{d}{dc} \left[\quad\quad\quad\quad \right] = Dc\{4[(3X^2 - a^2)(Y^2 - b^2)^2 + (X^2 - a^2)^2(3Y^2 - b^2)]\}^2$$
$$- q_0(X^2 - a^2)^2(Y^2 - b^2)^2 \tag{11.35}$$

Then, substituting this expression into equation (11.34) and solving for c, we have

$$c = \frac{q_0}{D} \frac{\displaystyle\int_{-a}^{a} \int_{-b}^{b} (X^2 - a^2)^2(Y^2 - b^2)^2 \, dY \, dX}{\displaystyle\int_{-a}^{a} \int_{-b}^{b} \{4[(3X^2 - a^2)(Y^2 - b^2)^2 + (X^2 - a^2)^2(3Y^2 - b^2)]\}^2 \, dY \, dX} \tag{11.36}$$

$$= \frac{q_0}{D} \left(\frac{I}{II} \right)$$

The integrations are somewhat tedious and are not given here in detail, but the result is[2]

$$c = \frac{q_0}{Db^4} \left\{ \frac{0.285}{5.20\left[1 + \left(\dfrac{a}{b}\right)^4\right] + 2.97\left(\dfrac{a}{b}\right)^2} \right\}$$ (11.37)

Then $D_Z(X, Y)$ is found by substituting c into equation (11.27).

In comparison to more accurate solutions, the results for the central deflection are found to vary from an error of about 5% for a square plate, $a/b = 1$, to over 30% for $a/b = 0$, representing a long, narrow plate.[2] With such discrepancies on the displacements, the stress couples can be expected to be further in error and intolerable, because these quantities are obtained from twice differentiating the deflection function. In general, when appraising numerical solutions, one should compare the most sensitive meaningful quantities, which are generally those computed by differentiation.

Increased accuracy may be obtained for this problem by adding coordinate functions. A three-term approximation is

$$D_Z(X,Y) = (c_1 + c_2 X^2 + c_3 Y^2)(X^2 - a^2)^2(Y^2 - b^2)^2$$ (11.38)

Only even terms are included because of the double symmetry of the problem.

11.1.3.3 Galerkin Solution.

The one-term Galerkin solution follows from equations (9.19) to (9.22). We first compute the residual error term R_k, defined by equation (9.21), by substituting equation (11.26) into equation (10.12). Continuing from equations (11.28a–c), we find

$$D_{Z,XXXX} = 24c(Y^2 - b^2)^2$$ (11.39a)

$$D_{Z,YYYY} = 24c(X^2 - a^2)^2$$ (11.39b)

$$D_{Z,XXYY} = 16c(3X^2 - a^2)(3Y^2 - b^2)$$ (11.39c)

and

$$R_k = \nabla^4 D_Z - \frac{q_0}{D}$$

$$= 8c[3(Y^2 - b^2)^2 + 4(3X^2 - a^2)(3Y^2 - b^2) + 3(X^2 - a^2)] - \frac{q_0}{D}$$ (11.40)

Then, the orthogonality condition, equation (9.22a), produces

$$\int_{-a}^{a} \int_{-b}^{b} \left\{ 8c[3(Y^2 - b^2)^2 + 4(3X^2 - a^2)(3Y^2 - b^2) + 3(X^2 - a^2)] - \frac{q_0}{D} \right\}$$
$$\cdot \left\{ (X^2 - a^2)^2(Y^2 - b^2)^2 \right\} dY\, dX = 0$$ (11.41)

from which

$$c = \frac{q_0}{D} \frac{I}{III}$$ (11.42)

where I is identical to the numerator of equation (11.36) and

$$III = \int_{-a}^{a} \int_{-b}^{b} 8[3(Y^2 - b^2) + 4(3X^2 - a^2)(3Y^2 - b^2) \\ + 3(X^2 - a^2)](X^2 - a^2)^2(Y^2 - b^2)^2 \, dY \, dX \tag{11.43}$$

In order to show that the one-term Rayleigh-Ritz and Galerkin solutions are identical, we must prove II = III, where II is defined in equation (11.36). We rewrite both II and III in terms of the coordinate function ϕ_Z. In view of equations (11.28a) and (11.28b),

$$II = \int_{-a}^{a} \int_{-b}^{b} (\phi_{Z,XX} + \phi_{Z,YY})^2 \, dY \, dX \\ = \int_{-a}^{a} \int_{-b}^{b} [(\phi_{Z,XX})^2 + 2\phi_{Z,XX}\phi_{Z,YY} + (\phi_{Z,YY})^2] \, dX \, dY \tag{11.44}$$

Also, by comparison with equations (11.39),

$$III = \int_{-a}^{a} \int_{-b}^{b} (\phi_{Z,XXXX} + 2\phi_{Z,XXYY} + \phi_{Z,YYYY})\phi_Z \, dY \, dX \tag{11.45}$$

We now integrate equation (11.44) by parts. For the first term, we have

$$\int_{-a}^{a} \int_{-b}^{b} \phi_{Z,XX}\phi_{Z,XX} \, dY \, dX = \int_{-b}^{b} \phi_{Z,XX}\phi_{Z,X} \Big|_{-a}^{a} \, dY - \int_{-a}^{a} \int_{-b}^{b} \phi_{Z,X}\phi_{Z,XXX} \, dY \, dX \tag{11.46}$$

Integrating the second term by parts again, we have

$$-\int_{-a}^{a} \int_{-b}^{b} \phi_{Z,X}\phi_{Z,XXX} \, dY \, dX = \int_{-b}^{b} \phi_{Z,XXX}\phi_Z \Big|_{-a}^{a} \, dY + \int_{-a}^{a} \int_{-b}^{b} \phi_Z\phi_{Z,XXXX} \, dY \, dX \tag{11.47}$$

If the boundary terms at $X = \pm a$ drop out, we have reduced the first term of equation (11.44) to that of (11.45). Considering the physical boundary conditions discussed in Section 10.1.6, we see that

$$\phi_{Z,XX}\phi_{Z,X} \Big|_{-a}^{a} = 0$$

implies that either the moment or rotation is zero on the boundary. Similarly, for

$$\phi_{Z,XXX}\phi_Z \Big|_{-a}^{a} = 0$$

either the transverse shear or the deflection must vanish. For the clamped plate, the rotation and deflection are zero on $X = \pm a$ so that the boundary terms drop out and the first terms of II and III are identical. Similar integrations by parts reduce the second and third terms of II to the corresponding terms of III, so that the one-term Rayleigh-Ritz and Galerkin solutions are indeed identical.

11.2 EXTENSIONS OF THE ELEMENTARY THEORY

11.2.1 Variable Flexural Rigidity

Consider a simple extension of plate theory where the flexural rigidity D, as defined by equation (10.4d), is generalized to $D(\alpha,\beta)$. Then, the plate equation may be reformulated by taking $D = D(\alpha,\beta)$ in equations (10.4a–c) and following the subsequent steps outlined in Section 10.1.1. To illustrate for Cartesian coordinates, equations (10.8a–c) are written as

$$M_X = -D(X,Y)(D_{Z,XX} + \mu D_{Z,YY}) \tag{11.48a}$$

$$M_Y = -D(X,Y)(D_{Z,YY} + \mu D_{Z,XX}) \tag{11.48b}$$

$$M_{XY} = M_{YX} = -D(X,Y)(1 - \mu)D_{Z,XY} \tag{11.48c}$$

and equations (10.9a) and (10.9b) generalize to

$$Q_Y = -[D(X,Y)(D_{Z,YY} + \mu D_{Z,XX})]_{,Y} + (1 - \mu)[D(X,Y)D_{Z,XY}]_{,X} \tag{11.49a}$$

$$Q_X = -[D(X,Y)(D_{Z,XX} + \mu D_{Z,YY})]_{,X} + (1 - \mu)[D(X,Y)D_{Z,XY}]_{,Y} \tag{11.49b}$$

whereupon the equilibrium equation (10.11) becomes

$$[D(X,Y)(D_{Z,XX} + \mu D_{Z,YY})]_{,XX} + 2(1 - \mu)[D(X,Y)D_{Z,XY}]_{,XY}$$
$$+ [D(X,Y)(D_{Z,YY} + \mu D_{Z,XX})]_{,YY} = q_Z \tag{11.50}$$

Equation (11.50) is used to analyze rectangular plates of variable thickness using the Lévy approach in Timoshenko and Woinowsky-Krieger[3] and in Conway.[4]

Axisymmetrical circular plates with an axisymmetric thickness variation are important in machine part design. The appropriate governing equation is derived in an analogous manner as for the rectangular plate,[5,6] and possible solution schemes and examples are provided in those references.

In practical applications, solutions for plates of variable thickness are often constructed using finite difference or finite element procedures.

11.2.2 Specifically Orthotropic Plates

Another direct extension of the theory of plates is the accommodation of specifically orthotropic properties, as defined in Section 8.1.2. In this case, the material properties are given by $[C_{or}]$, equation (8.11), and specifically by the fourth, fifth, and sixth rows and columns of $[D_{or}]$, Matrix 8–3. In Cartesian coordinates, the generalized moment-curvature expressions are found from equation (8.10) by replacing rows and columns 4 through 6 in Matrix 8–2 with the corresponding elements of $[D_{or}]$; by taking X, Y, Z for α, β, and n; and then by using equations (10.3) for the curvatures κ_X, κ_Y, and τ:

$$M_X = -(D_{or44}D_{Z,XX} + D_{or45}D_{Z,YY}) \tag{11.51a}$$

$$M_Y = -(D_{or55}D_{Z,YY} + D_{or54}D_{Z,XX}) \tag{11.51b}$$

$$M_{XY} = M_{YX} = -D_{or66}D_{Z,XY} \tag{11.51c}$$

The elements of $[\mathbf{D}_{or}]$, D_{orij}, are presumed to be known, as discussed in Section 8.1.2 and $D_{or45} = D_{or54}$.

If we take each D_{orij} as constant, we may write the equilibrium equation by first substituting equations (11.51) into equations (10.9a) and (10.9b) to get

$$Q_Y = -(D_{or55}D_{Z,YYY} + D_{or45}D_{Z,XXY} + D_{or66}D_{Z,XXY}) \tag{11.52a}$$

$$Q_X = -(D_{or44}D_{Z,XXX} + D_{or45}D_{Z,XYY} + D_{or66}D_{Z,XYY}) \tag{11.52b}$$

and then introducing equations (11.52a) and (11.52b) into equation (10.10), which becomes

$$-[D_{or44}D_{Z,XXXX} + 2(D_{or45} + D_{or66})D_{Z,XXYY} + D_{or55}D_{Z,YYYY}] + q_Z = 0 \tag{11.53}$$

To verify the consistency of this derivation, we may check the isotropic case. From Matrix 8–2, $D_{or44} = D_{or55} = D$, $D_{or45} = \mu D$, and $D_{or66} = D(1 - \mu)$. Therefore, $D_{or45} + D_{or66} = D$ and equation (11.53) reduces to equation (10.11).

Equation (11.53) is of interest in the study of stiffened plates. In this application, the term D_{or45} may be neglected, as discussed in Section 8.1.3, and the terms D_{or44}, D_{or55}, and D_{or66} are taken as the corresponding elements of $\lceil \mathbf{D}_{eq} \rfloor$, equation (8.23). Other methods of finding the material constants for specific configurations are discussed in Timoshenko and Woinowsky-Krieger.[7]

For a rectangular plate simply supported on all sides, equation (11.53) may be solved using the Navier approach following Section 10.2.2. Starting with equation (10.71) for D_Z, the lhs of equation (11.53) becomes

$$\sum_{j=1}^{\infty} \sum_{k=1}^{\infty} \left[D_{or44} \left(\frac{j\pi}{a}\right)^4 + 2(D_{or45} + D_{or66}) \left(\frac{j\pi}{a}\right)^2 \left(\frac{k\pi}{b}\right)^2 \right.$$
$$\left. + D_{or55} \left(\frac{k\pi}{b}\right)^4 \right] D_Z^{jk} \sin j\pi \frac{X}{a} \sin k\pi \frac{Y}{b} \tag{11.54}$$

With the rhs expanded in a Fourier series as given in equation (10.73) and q_Z^{jk} evaluated from the specified load distribution $q_Z(X,Y)$ by equation (10.75), we get

$$D_Z^{jk} = \frac{q^{jk}}{\pi^4} \left[\frac{1}{D_{or44}\left(\dfrac{j}{a}\right)^4 + 2(D_{or45} + D_{or66})\left(\dfrac{j}{a}\right)^2\left(\dfrac{k}{b}\right)^2 + D_{or55}\left(\dfrac{k}{b}\right)^4} \right] \tag{11.55}$$

whereupon D_Z is found by entering equation (11.55) into equation (10.71):

$$D_Z(X,Y) = \sum_{j=1}^{\infty} \sum_{k=1}^{\infty} D_Z^{jk} \sin j\pi \frac{X}{a} \sin k\pi \frac{Y}{b} \tag{11.56}$$

Thus, available solutions for simply supported isotropic plates are easily extended to include orthotropic plates, once the material properties are defined.

11.2.3 Multilayered Plates

Another form of anisotropy which is of considerable practical importance is the multilayered plate, which may be composed of two or more bonded layers of isotropic or anisotropic materials. The simplest method of dealing with plates composed of isotropic layers is to use a modified form of the basic plate equation, equation (10.12):

$$\mathbf{\nabla}^4 D_Z = \frac{q_Z}{D_\tau} \qquad (11.57)$$

where D_τ is a transformed flexural rigidity which is computed from the basic properties, E and μ, of the individual layers.[8,9]

One type of layered plate is called a *sandwich plate* and is composed of at least three plies. The outer layers, or skin, are usually relatively thin, but of high strength and resist the flexural and twisting moments by developing couples of opposing in-plane forces; the inner core transmits the shear stresses between the outer layers. This behavior is similar to that of an H-shaped beam, where the two outer layers would represent the flanges and the inner layer, the web. The analysis of this type of plate may be based on a large deflection theory developed by E. Reissner.[10]

11.2.4 Inclusion of Transverse Shearing Deformations

Preceding the development of the Kirchhoff boundary conditions, Section 8.2.3, we saw that the order of governing equations derived by *neglecting* transverse shearing deformations permits only two of the three obvious force conditions to be enforced at a free edge. Also, the presence of concentrated corner forces in rectangular plates is attributable to the suppression of the transverse shearing deformations. For thin plates, these shortcomings appear largely academic and the elementary plate theory is adequate; however, as a plate becomes relatively thicker, transverse shearing effects may be more important. Moreover, eliminating the transverse shearing strain permits the in-plane displacements to be written directly in terms of the normal displacement [see equations (10.15) and (10.17)] and leads to much simplified governing equations.

Although the contradictions incorporated in elementary plate theory have been evident since the time of Kirchhoff, a satisfactory alternative which includes transverse shearing deformations appeared relatively recently (1944) and is attributed to E. Reissner,[11,12] with some significant embellishments by A. Green.[13] More recently, refined plate theories have been classified into *first-* and *higher-order* shearing deformation theories,[14] which carry the respective names of Hencky-Mindlin and Kromm-Reddy. The distinction is drawn because the first-order theory (commonly known as *Mindlin plate theory*) does not satisfy the shear stress-free conditions on the surfaces $\pm h/2$. This is easily seen by referring to the last two equations of Matrix 8–1, where the shear stresses are proportional to the shearing strains which are *constant* through the thickness and, hence, do not necessarily vanish at the surfaces. Ordinarily the lateral (top and bottom) surfaces of a plate, Figure 10–1, would be free of shearing stresses $\sigma_{n\alpha}$ and $\sigma_{n\beta}$. However, symmetry of the stress tensor, $\sigma_{ij} = \sigma_{ji}$, would produce non-zero values based on $\sigma_{\alpha n}$ and $\sigma_{\beta n}$ being constant through the thickness. The higher-order theories attempt to correct this shortcoming by including thickness-dependent factors, at the expense of adding unknowns into the equations. As discussed in Section 8.1.1.2, the deficiency is minor in the context of elementary plate and shell theory. It has also been observed that the effect of shearing deformations is more pronounced in orthotropic than in isotropic plates.[14] The interested reader is referred to Timoshenko and Woinowsky-Krieger[15] and to Reddy[14] for some examples of solutions including transverse shearing deformations. Although the inclusion of transverse shearing deformations complicates the problem considerably for a differential equation formulation, the incorporation of these effects is relatively easy in an energy-based approach. If we consider equations (9.6) and (9.7), we see that the linear component $Eh/[2(1 - \mu^2)]\,[\mathbf{I}]$ contains the transverse shearing strains. Thus, we may add

$$\frac{Eh}{4(1 + \mu)} \int_S (\gamma_\alpha^2 + \gamma_\beta^2)\, AB\, d\alpha\, d\beta \tag{11.58}$$

to equation (11.1) to get the expanded version of U_ε:

$$U_\varepsilon = \int_S \left\{ \frac{D}{2}\left[(\kappa_\alpha + \kappa_\beta)^2 - 2(1 - \mu)(\kappa_\alpha \kappa_\beta - \tau^2) \right] + \frac{Eh}{4(1 + \mu)} \right.$$
$$\left. \cdot (\gamma_\alpha^2 + \gamma_\beta^2) - \bar{\alpha}(1 + \mu)(\kappa_\alpha + \kappa_\beta) \int_{-h/2}^{h/2} T(\zeta)\zeta\, d\zeta \right\} AB\, d\alpha\, d\beta \tag{11.59}$$

When we proceed with the displacement formulation, we substitute the general strain-displacement relationship-equations (7.54a–h) into equation (11.59). For Cartesian coordinates, we have

$$\kappa_X = D_{XY,X} \tag{11.60a}$$

$$\kappa_Y = D_{YX,Y} \tag{11.60b}$$

$$\tau = \frac{1}{2}(D_{YX,X} + D_{XY,Y}) \tag{11.60c}$$

$$\gamma_X = D_{Z,X} + D_{XY} \tag{11.60d}$$

$$\gamma_Y = D_{Z,Y} + D_{YX} \tag{11.60e}$$

and equation (11.59) becomes

$$U_\varepsilon = \int_S \left\{ \frac{D}{2}\left([D_{XY,X} + D_{YX,Y}]^2 \right. \right.$$
$$\left. - 2(1 - \mu)\left[D_{XY,X}D_{YX,Y} - \frac{1}{4}(D_{YX,X} + D_{XY,Y})^2 \right] \right)$$
$$+ \frac{Eh}{4(1 + \mu)}\left([D_{Z,X} + D_{XY}]^2 + [D_{Z,Y} + D_{YX}]^2 \right)$$
$$\left. - \bar{\alpha}(1 + \mu)(D_{XY,X} + D_{YX,Y}) \int_{-h/2}^{h/2} T(\zeta)\zeta\, d\zeta \right\} dX\, dY \tag{11.61}$$

A similar specialization is easily accomplished for polar coordinates. This is left as an exercise.

In anticipation of the use of equation (11.61) in a Rayleigh-Ritz type of solution, it is instructive to compare this generalized form with the version where transverse shearing strains are neglected, equation (11.3). Referring to the procedure described in Section 9.4.3, we see that the index m in equations (9.15) and (9.16) will be equal to 3 instead of 1, corresponding to the generalized displacements

$$\{\mathbf{\Delta}\} = \{D_Z D_{XY} D_{YX}\} \tag{11.62}$$

This means that there will be three times as many coordinate functions and considerably more numerical calculations, but this is not foreboding in the computer age.

There is another, more subtle, difference between the energy expressions which can be important from a computational standpoint. In equation (11.61), the generalized displacements are present only up to the *first* derivative, whereas in the earlier form, equation (11.3), there are *second* derivatives of the dependent variables. The order of the *highest* derivative appearing in the strain energy functional dictates the minimum continuity required at junctions of finite elements.[16] The lower-order continuity necessary for a functional based on equation (11.61) rather than equation (11.3) somewhat compensates for the increased number of generalized displacements and makes the formulation including transverse shearing deformations attractive. The relative ease and efficiency of including these deformations in an energy-based formulation for shells of revolution are demonstrated in Brombolich and Gould.[17] The necessity for employing precautions to avoid overstiffening by "shear locking" was noted in Section 7.3.2.

11.2.5 Folded Plates

Plates are basically shallow flexural members and are somewhat inefficient in flexural action. An appealing procedure to increase the flexural rigidity of a given plate is to introduce undulations or folds in one, or possibly more than one, direction. This increases the section modulus in these directions markedly. Of course, because there is no longer a flat surface, this procedure is impractical for many situations.

A widely used application of this concept is the folded plate, which is used for roofs over large, column-free areas and for many other structures in which a globally curved geometry is best achieved by an assemblage of facets. One such structure, partially designed by the author, is shown in Figure 2–8(x).

A typical configuration for a folded plate roof is shown in Figure 11–2. Basically, this structure is a relatively wide beam with a saw-tooth cross-section of depth H spanning the distance L from support to support. The idealized beam behavior is violated by distortions of the cross-section, which make elementary beam theory, alone, inadequate.

The load-carrying mechanism for folded plates may be conveniently visualized in two parts. The surface loading is resolved into *in-plane* and *transverse* components. Then, the transverse surface loads are resisted by one-way plate bending over a span D. The reactions produced by the transverse loads on the plates are applied at the ridge lines, which act as supports for the plates. Near the ends, some of the loading is directly transferred to the end sup-

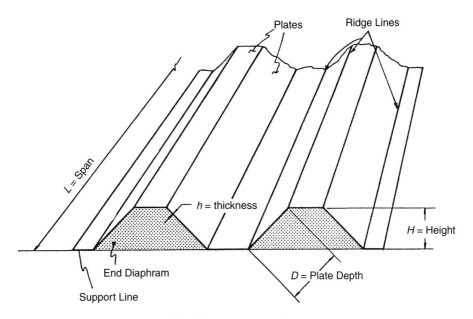

FIGURE 11–2   Folded Plates.

ports, but because $H < L$, the plates are basically one-way and most of the load is carried to the ridge lines. This is called *plate* action.

The ridge lines are subjected to the plate reactions, which are resolved in oblique coordinates to act in the planes of the intersecting plates at each joint. Along with the in-plane component of the surface load on each plate, these forces are resisted by the flexural action of the plate acting as a beam of width h, depth D, and span L. This is termed *diaphragm* action.

The end blocks are generally solid infills or stiff frames, rigid in the vertical plane but flexible in the longitudinal direction, resting on a supporting structure, such as a wall or a line of columns. Away from the ends, the ridge lines deflect in accordance with the diaphragm action of the plates acting as beams with span L, width h, and depth D. Thus, an interaction occurs between the *plate* and the *diaphragm* behavior along the ridge lines, because the supports for the plate action are not unyielding but elastic. This conceptual model is the basis for many of the folded plate theories used in engineering design. Because these design methods are generally developed in terms of planar structural analysis theory rather than in terms of the theory of plates, they are not treated here. The interested reader is referred to Yitzhaki[18] and to Simpson.[19]

The governing equations for folded plates may be written in the context of the theory of plates by combining the equations describing the in-plane forces and corresponding displacements with the basic plate equation, equation (10.12). To do this, we take local Cartesian coordinates such that X and Y are in the plane of each plate and write equations (3.25a) and (3.25b) as

$$N_{X,X} + N_{XY,Y} + q_X = 0 \tag{11.63a}$$

$$N_{XY,X} + N_{Y,Y} + q_Y = 0 \tag{11.63b}$$

Next, we replace the stress resultants by the strains using Matrix 8–2, without thermal effects.

$$\frac{Eh}{1 - \mu^2}\left[(\varepsilon_X + \mu\varepsilon_Y)_{,X} + \frac{1 - \mu}{2}\,\omega_{,Y}\right] + q_X = 0$$

$$\frac{Eh}{1 - \mu^2}\left[\frac{1 - \mu}{2}\,\omega_{,X} + (\varepsilon_Y + \mu\varepsilon_X)_{,Y}\right] + q_Y = 0 \tag{11.64}$$

Finally, we express the strains in terms of the displacements using equations (7.54a–c)

$$\frac{Eh}{1 - \mu^2}\left[(D_{X,X} + \mu D_{Y,Y})_{,X} + \frac{1 - \mu}{2}(D_{Y,X} + D_{X,Y})_{,Y}\right] + q_X = 0 \tag{11.65a}$$

$$\frac{Eh}{1 - \mu^2}\left[\frac{1 - \mu}{2}(D_{Y,X} + D_{X,Y})_{,X} + (D_{Y,Y} + \mu D_{X,X})_{,Y}\right] + q_Y = 0 \tag{11.65b}$$

Equations (11.65a) and (11.65b) together with equation (10.12),

$$\nabla^4 D_Z = \frac{q_Z}{D}$$

form the governing equations for the so-called exact theory of folded plates.[20] Solutions using this theory are presented by several authors,[21–24] and a critical evaluation of various solution procedures is available.[24]

11.2.6 Radial Displacement of a Solid Circular Plate

Often cylindrical tanks and pressure vessels are capped by a circular plate, rather than a spherical or ellipsoidal shell as shown on Figure 4–16. In establishing the compatibility condition at the junction of the cap and the shell, it is usually sufficient to account for only the rotational flexibilities of the shell and the plate since the radial (in-plane) stiffness of the plate is generally much greater than the corresponding normal stiffness of the shell, which would correspond to D_n in Figure 7–7. Thus, the displacement boundary condition for the shell is taken as $D_n = 0$. This is discussed further in Sections 12.3.2 and 13.1.2.4.

  If, however, it is desired to account for the radial flexibility of the plate as well, a solution for the radial displacement in a solid circular plate may be derived from the solution of a thick-walled cylinder, which is based on the Theory of Elasticity. Considering the geometry of the plate from Figure 10–15 with thickness $= h$, radial coordinate $= R$ and outer radius $= a$, and considering a radial line load $Q(a)$ applied axisymmetrically around the circumference, the displacement may be found as[25]

$$D_R(R) = \frac{(1 - \mu)}{Eh}\,Q(a)R \tag{11.66}$$

This solution may be evaluated at $R = a$ and used with the solutions in Section 12.3.2 to formulate a displacement compatibility condition at the junction of the cap and the shell, as discussed in Section 13.1.2.4.

11.3 INSTABILITY AND FINITE DEFORMATION

11.3.1 Modification of Equilibrium Equations

We now direct our interest to the analysis of plates that are subjected to forces acting in the middle plane, along with the transverse loading. If these middle-plane forces are compressive in at least one direction, plate instability is a possibility. Further, if there are no transverse forces but only in-plane compressive forces acting, we have the two-dimensional analogue of the classical Euler column buckling problem. In order to show this, we must relax our basic assumption [2], which enabled the equilibrium equations to be formulated with respect to the *undeformed* middle surface, and now include the effect of the in-plane forces on the equilibrium in the normal direction.

Refer to Figure 3–2(a) and consider a section of the middle surface along an s_α coordinate line with length ds_α. We show two such sections on Figure 11–3, both before and after deformation. Figure 11–3(a) is for N_α, and Figure 11–3(b) pertains to $N_{\beta\alpha}$. The surface rotation $D_{\alpha\beta}$, is denoted in accordance with Figure 7–2 and may be expressed in terms of $D_{n,\alpha}$ by equation (10.7a) if transverse shearing strains are neglected. Note that Figure 11–3 corresponds to the positive sense of $D_{\alpha\beta}$ but to the negative sense of $D_{n,\alpha}$, because the $(+)\mathbf{t}_\alpha$ direction is opposite to that shown on Figure 7–2(a).

We now reexamine the stress vectors $\mathbf{F}_\alpha$ and $\mathbf{F}_\beta$, as defined in equations (3.11a) and (3.11b). First, from Figure 11–3(a), we see that the stress resultant N_α has a normal component $-N_\alpha \sin D_{\alpha\beta} \simeq -N_\alpha D_{\alpha\beta}$ which adds a force $-N_\alpha D_{\alpha\beta}\mathbf{t}_n ds_\beta$ to $\mathbf{F}_\alpha$ in equation (3.11a). The negative sign is in accordance with the sign convention defined on Figure 3–2(a). Next, from Figure 11–3(b), we have the stress resultant $N_{\beta\alpha}$ with a normal component $-N_{\beta\alpha} D_{\alpha\beta}$, which contributes a force $-N_{\beta\alpha} D_{\alpha\beta}\mathbf{t}_n ds_\alpha$ to $\mathbf{F}_\beta$ in equation (3.11b). Similar sections along an s_β coordinate line show the forces $-N_{\alpha\beta} D_{\beta\alpha}\mathbf{t}_n ds_\beta$ adding to $\mathbf{F}_\alpha$ in equation (3.11a) and $-N_\beta D_{\beta\alpha}\mathbf{t}_n ds_\alpha$ going to $\mathbf{F}_\beta$ in equation (3.11b). Because these additional normal forces associate only with Q_α and Q_β, respectively, as coefficients of $\mathbf{t}_n$ in the two equations, they are easily traced through to equations (3.25a–c), the scalar force equilibrium equations for a plate. The in-plane equations are unaffected, and the generalization of equation (3.25c) is

$$[B(Q_\alpha - N_\alpha D_{\alpha\beta} - N_{\alpha\beta} D_{\beta\alpha})]_{,\alpha} + [A(Q_\beta - N_{\beta\alpha} D_{\alpha\beta} - N_\beta D_{\beta\alpha})]_{,\beta} + q_n AB = 0 \qquad (11.67)$$

It is convenient first to expand equation (11.67) and then eliminate certain terms by introducing the in-plane equilibrium equations, (3.25a) and (3.25b). We will proceed for the two coordinate systems which are of interest in the theory of plates, both Cartesian and polar.

For Cartesian coordinates, $\alpha = X$, $\beta = Y$, $n = Z$, and $A = B = 1$. Then, equations (3.25a) and (3.25b) become

$$N_{X,X} + N_{YX,Y} + q_X = 0 \qquad (11.68a)$$

$$N_{XY,X} + N_{Y,Y} + q_Y = 0 \qquad (11.68b)$$

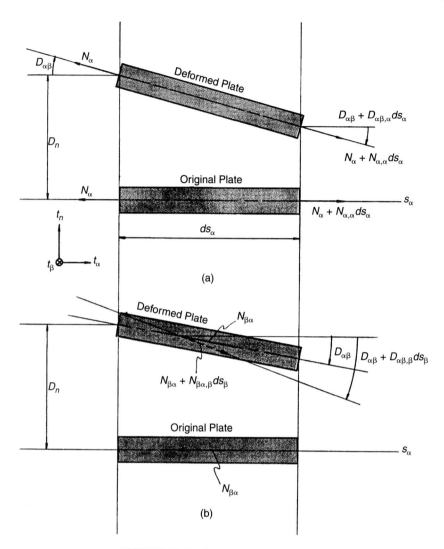

FIGURE 11–3   In-Plane Forces in a Plate.

and equation (11.67) is written as

$$Q_{X,X} + Q_{Y,Y} - D_{XY}(N_{X,X} + N_{YX,Y}) - D_{YX}(N_{XY,X} + N_{Y,Y})$$
$$- N_X D_{XY,X} - N_Y D_{YX,Y} - N_{YX} D_{XY,Y} - N_{XY} D_{YX,X} + q_Z = 0 \qquad (11.69)$$

Substituting equations (11.68a) and (11.68b) into equation (11.69) and setting $N_{XY} = N_{YX}$, we have

$$Q_{X,X} + Q_{Y,Y} - N_X D_{XY,X} - N_Y D_{YX,Y} - N_{XY}(D_{YX,X} + D_{XY,Y})$$
$$+ D_{XY} q_X + D_{YX} q_Y + q_Z = 0 \qquad (11.70)$$

A more familiar form of equation (11.70), applicable for transverse loading only, is found by letting $q_X = q_Y = 0$ and also by suppressing transverse shearing strains, allowing

$$D_{XY} = -D_{Z,X}; \qquad D_{YX} = -D_{Z,Y} \tag{11.71}$$

to be introduced from equations (10.7). The, substituting equations (10.10) to (10.12) into equation (11.70), we have

$$D\nabla^4 D_Z - N_X D_{Z,XX} - N_Y D_{Z,YY} - 2N_{XY} D_{Z,XY} = q_Z \tag{11.72}$$

which becomes the governing equation for the deflection of a plate in the presence of lateral forces.

In polar coordinates, $\alpha = R$, $\beta = \theta$, and $n = Z$ along with $A = 1$ and $B = R$. Equations (3.25a) and (3.25b) become

$$(RN_R)_{,R} + N_{\theta R,\theta} - N_\theta + q_R R = 0 \tag{11.73a}$$

$$(RN_{R\theta})_{,R} + N_{\theta,\theta} + N_{\theta R} + q_\theta R = 0 \tag{11.73b}$$

and equation (11.67) expands to

$$Q_R + RQ_{R,R} + Q_{\theta,\theta} - D_{R\theta}(N_R + RN_{R,R} + N_{\theta R,\theta})$$
$$- D_{\theta R}(N_{R\theta} + RN_{R\theta,R} + N_{\theta,\theta}) - N_R RD_{R\theta,R} \tag{11.74}$$
$$- N_\theta D_{\theta R,\theta} - N_{R\theta} RD_{\theta R,R} - N_{\theta R} D_{R\theta,\theta} + q_Z R = 0$$

Substituting equations (11.73a) and (11.73b) into equation (11.74) and taking $N_{R\theta} = N_{\theta R}$, we have

$$Q_R + RQ_{R,R} + Q_{\theta,\theta} - N_R RD_{R\theta,R} - N_\theta(D_{R\theta} + D_{\theta R,\theta})$$
$$+ N_{R\theta}(D_{\theta R} - RD_{\theta R,R} - D_{R\theta,\theta}) + q_R D_{R\theta} R + q_\theta D_{\theta R} R + q_Z R = 0 \tag{11.75}$$

Now, we let $q_R = q_\theta = 0$ and suppress the transverse shearing strains. From equations (10.30c) and (10.30d), this gives $D_{R\theta} = -D_{Z,R}$ and $D_{\theta R} = -1/R\, D_{Z,\theta}$. Introducing equations (10.24) to (10.27) into equation (11.75), we get

$$D\nabla^4 D_Z - N_R D_{Z,RR} - \frac{N_\theta}{R}\left(D_{Z,R} + \frac{1}{R} D_{Z,\theta\theta}\right) + \frac{2N_{R\theta}}{R}\left(\frac{1}{R} D_{Z,\theta} - D_{Z,R\theta}\right) = q_Z \tag{11.76}$$

for plates with transverse loading only.

For the axisymmetric case, equation (11.76) reduces to

$$D\nabla_a^4 D_Z - N_R D_{Z,RR} - \frac{N_\theta}{R} D_{Z,R} = q_Z \tag{11.77}$$

where $\nabla_a^4(\ ) =$ the axisymmetric biharmonic operator defined in equations (10.33b) or, more conveniently, in equation (10.116). Thus, we establish equations (11.72), (11.76), and (11.77) as the expanded forms of equations (10.12), (10.27), and (10.33b), respectively, when basic assumption [2] is relaxed.

An example using a modified form of equation (11.77) is solved by Timoshenko and Woinowsky-Krieger.[26] Some classical problems in Cartesian coordinates are treated in the subsequent sections. Also, more complex cases including shear loading, orthotropic plates, sandwich plates, and clamped boundaries are investigated by Brush and Almroth.[27]

11.3.2 Modification of Strain Energy

The expression for strain energy, as given by equations (9.6) and (9.7), must be supplemented when coupling of the normal displacements and the in-plane strains is included, because nonlinear strain terms are required. We refer to the basic description of deformation as discussed in Sections 7.2 and 7.3. However, we will not attempt to develop a complete nonlinear theory, but only to retain those higher-order terms containing D_n. This corresponds to a modified finite deformation theory, in which the in-plane displacements remain *small* but the rotations are regarded as *moderate*.

First consider equations (7.13) and (7.14) for A' and B'. Noting from equations (7.10c) and (7.12c), respectively, that ψ_α and ψ_β contain D_n terms, we have

$$A' = A\left[(1 + \varepsilon_\alpha)^2 + \frac{1}{A^2}(D_{n,\alpha})^2\right]^{1/2} \tag{11.78}$$

and

$$B' = B\left[(1 + \varepsilon_\beta)^2 + \frac{1}{B^2}(D_{n,\beta})^2\right]^{1/2} \tag{11.79}$$

Using the binomial theorem, we may simplify these expressions to

$$A' = A\left[1 + \varepsilon_\alpha + \frac{1}{2A^2}(D_{n,\alpha})^2\right] \tag{11.80}$$

and

$$B' = B\left[1 + \varepsilon_\beta + \frac{1}{2B^2}(D_{n,\beta})^2\right] \tag{11.81}$$

We now return to the definition of the middle surface strains in Section 7.3.1. We denote the modified strains as $\bar{\varepsilon}_\alpha$, $\bar{\varepsilon}_\beta$, and $\bar{\omega}$, respectively, and retain the basic definitions of these components of strain. Then, the strain ε_α defined in equation (7.25) becomes

$$\bar{\varepsilon}_\alpha = \varepsilon_\alpha + \frac{1}{2}\left(\frac{D_{n,\alpha}}{A}\right)^2 \tag{11.82}$$

in view of equation (11.80). Considering, equation (11.81), ε_β generalizes to

$$\bar{\varepsilon}_\beta = \varepsilon_\beta + \frac{1}{2}\left(\frac{D_{n,\beta}}{B}\right)^2 \tag{11.83}$$

For the shearing strain, we note the product $\psi_\alpha\psi_\beta$ in equation (7.27), which previously had been neglected. Referring to equations (7.10c) and (7.12c) and retaining only the D_n terms,

$$\psi_\alpha\psi_\beta \simeq \left(\frac{D_{n,\alpha}}{A}\right)\left(\frac{D_{n,\beta}}{B}\right)$$

Therefore, from equation (7.27), we find

$$\overline{\omega} = \omega + \left(\frac{D_{n,\alpha}}{A}\right)\left(\frac{D_{n,\beta}}{B}\right) \tag{11.84}$$

In order to obtain a modified expression for the strain energy, we consider the basic expression, equation (9.4). Assume that the strain energy caused by bending is not changed by the axial forces and remains the cubic component of equation (9.7), in other words $(D/2)$ [**III**]. Also assume that the in-plane stress resultants are caused entirely by applied edge loading in the plane of the plate, in which case they are unchanged during bending. This implies that the external and internal work done by these constant stress resultants acting through the corresponding external and internal *in-plane* displacements will cancel in the energy balance when a virtual transverse displacement is introduced. This may be formally substantiated by rather involved arguments,[26] which are not repeated here. Therefore, the *additional* strain energy $U_{\overline{\varepsilon}}$, is entirely caused by the straining of the middle surface as a result of the bending. With these assumptions, we can write

$$U_{\overline{\varepsilon}} = \int_S [N_\alpha(\overline{\varepsilon}_\alpha - \varepsilon_\alpha) + N_\beta(\overline{\varepsilon}_\beta - \varepsilon_\beta) + N_{\alpha\beta}(\overline{\omega} - \omega)] AB\, d\alpha\, d\beta \tag{11.85}$$

Note that there is no 1/2-coefficient in equation(11.85), because the in-plane stress resultants are already acting when the additional middle-surface strains occur. Substituting equations (11.82) to (11.84) into equation (11.85) gives

$$U_{\overline{\varepsilon}} = \frac{1}{2}\int_S \left[N_\alpha\left(\frac{D_{n,\alpha}}{A}\right)^2 + N_\beta\left(\frac{D_{n,\beta}}{B}\right)^2 + 2N_{\alpha\beta}\left(\frac{D_{n,\alpha}}{A}\right)\left(\frac{D_{n,\beta}}{B}\right)\right] AB\, d\alpha\, d\beta \tag{11.86}$$

and the total strain energy for plate bending in the presence of constant in-plane forces is

$$U_\varepsilon + U_{\overline{\varepsilon}} = U_{\varepsilon+\overline{\varepsilon}} = \frac{D}{2}[\mathbf{III}] + U_{\overline{\varepsilon}} \tag{11.87}$$

where **III** is defined in equations (9.6) and (9.7).

For Cartesian coordinates, equation (11.86) becomes

$$U_{\overline{\varepsilon}} = \frac{1}{2}\int_S [N_X(D_{Z,X})^2 + N_Y(D_{Z,Y})^2 + 2N_{XY}D_{Z,X}D_{Z,Y}]dX\, dY \tag{11.88}$$

and U_ε is given by equation (11.4); for polar coordinates,

$$U_{\overline{\varepsilon}} = \frac{1}{2}\int_S \left[N_R(D_{Z,R})^2 + N_\theta\left(\frac{D_{Z,\theta}}{R}\right)^2 + 2N_{R\theta}D_{Z,R}\left(\frac{D_{Z,\theta}}{R}\right) R\, dR\, d\theta\right] \tag{11.89}$$

For axisymmetric loading on a solid plate of radius a, equation (11.89) simplifies to

$$U_{\overline{\varepsilon}} = \pi\int_0^a N_R(D_{Z,R})^2 R\, dR \tag{11.90}$$

and U_ε is given by equation (11.13).

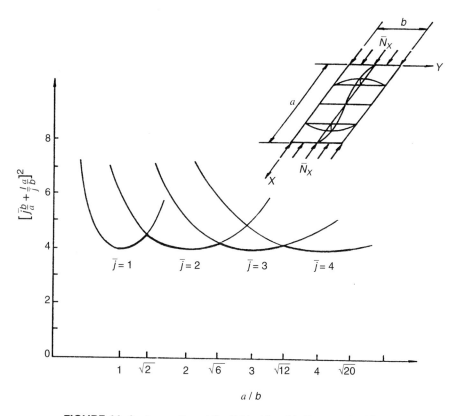

FIGURE 11–4 Lower Bound for Critical Load in Rectangular Plate.

11.3.3 Simply Supported Rectangular Plate Under Transverse and Unidirectional In-Plane Loadings

Consider the rectangular plate shown in Figure 10–9 with the addition of constant in-plane loads $\bar{N}_X(0,Y)$ and $\bar{N}_X(a,Y)$ along the two boundaries parallel to the Y-axis, as shown on the inset in Figure 11–4.

We choose the differential equation approach from Section 11.3.1, whereby equation (11.72) becomes

$$D\boldsymbol{\nabla}^4 D_Z - \bar{N}_X D_{Z,XX} - q_Z = 0 \tag{11.91}$$

Following the procedure used in Section 10.2.2.1 with D_Z given by equation (10.71) and q_Z by equation (10.73), we arrive at the equation for specific harmonics j and k

$$\left\{ \pi^4 D \left[\left(\frac{j}{a}\right)^2 + \left(\frac{k}{b}\right)^2 \right] + \bar{N}_X \pi^2 \left(\frac{j}{a}\right)^2 \right\} D_Z^{jk} = q_Z^{jk} \tag{11.92}$$

from which the Fourier coefficient is

$$D_Z^{jk} = \frac{q_Z^{jk}}{\pi^4 D} \left\{ \frac{1}{\left[\left(\frac{j}{a}\right)^2 + \left(\frac{k}{b}\right)^2\right]^2 + \frac{\overline{N}_X}{\pi^2 D}\left(\frac{j}{a}\right)^2} \right\} \tag{11.93}$$

As before, the Fourier coefficient for the applied surface loading, q_Z^{jk}, is evaluated from equation (10.75). The total plate deflection is then computed from equation (10.71).

If the in-plane load $\overline{N}_X$ is tensile, D_Z^{jk} and hence the deflection D_Z is reduced from the case with no in-plane forces; however, if $\overline{N}_X$ is compressive, the deflection is increased. In fact, there are *critical* values of $\overline{N}_X$, one for each of the specific harmonics $j = \bar{j}$ and $k = \bar{k}$, for which the denominator of { } in equation (11.93) vanishes and D_Z^{jk} becomes infinite:

$$\overline{N}_{Xcr} = -\pi^2 D \left(\frac{a}{\bar{j}}\right)^2 \left[\left(\frac{\bar{j}}{a}\right)^2 + \left(\frac{\bar{k}}{b}\right)^2\right]^2 \tag{11.94}$$

The $\overline{N}_{Xcr}$ are the elastic buckling loads for a uniaxially compressed, simple supported rectangular plate and are unaffected by the transverse load q_Z.

It is evident from equation (11.94) that $\overline{N}_{Xcr}$ is quite similar in form to the Euler buckling formula for columns. It includes the term π^2, the term $D = Eh^3/[12(1 - \mu^2)]$, which is equivalent to the *EI* term in the column formula, and the length terms a and b. However, the length terms are not only in the denominator; rather, the dimension in the direction of loading, a, appears also in the numerator. Also note that the $\bar{j}$ term, corresponding to the harmonic numbers of the buckling modes in the direction of loading, occurs in both the numerator and denominator. The value of $\bar{j}$ which yields the *lowest* critical load is not obvious from equation (11.94). It is quite clear, however, that $\bar{k} = 1$ will give the lowest value of $\overline{N}_{Xcr}$, so that we may write

$$\overline{N}_{Xcr1} = -\pi^2 D \left(\frac{a}{\bar{j}}\right)^2 \left[\left(\frac{\bar{j}}{a}\right)^2 + \left(\frac{1}{b}\right)^2\right]^2 \tag{11.95}$$

which may be conveniently rearranged as

$$\overline{N}_{Xcr1} = \frac{-\pi^2 D}{b^2} \left[\bar{j}\frac{b}{a} + \frac{1}{\bar{j}}\frac{a}{b}\right]^2 \tag{11.96}$$

where a/b is the aspect ratio.

Now, the problem is to determine which value of $\bar{j}$ will produce the lowest N_{cr} for a given aspect ratio. This may be done by plotting the term $[\bar{j} b/a + 1/\bar{j}a/b]^2$ against the aspect ratio a/b for various integer values of $\bar{j}$, as shown in Figure 11–4. It may be observed from this figure that the minimum value

$$\left[\bar{j}\frac{b}{a} + \frac{1}{\bar{j}}\frac{a}{b}\right]^2 \simeq 4 \tag{11.97}$$

is a close lower bound for most cases, so that

$$\overline{N}_{Xcr1} \simeq \frac{-4\pi^2 D}{b^2} \tag{11.98}$$

Of course, the number of waves in the buckled shape will still depend on a/b. The deflected shape for $\bar{j} = 2$, corresponding to $\sqrt{2} \leq a/b < \sqrt{6}$, is shown on the inset.

In contrast to column buckling, note that the lowest critical load is practically independent of the *length* of the member in the direction of the loading. Rather, only the length of the loaded edge, which is a quantity not present in the column problem, is significant.

It should be noted that the single series Lévy-Nádai approach, introduced in Section 10.2.3, is effective for plate instability investigations as well. The latter is the logical alternative to the Navier solution when boundary conditions other than all sides simply supported are encountered. The single-series solution is applied in Brush and Almroth,[27] and an application is suggested in the exercises.

11.3.4 Simply Supported Rectangular Plate under Transverse and Bidirectional In-Plane Loadings

Now consider the rectangular plate with a concentrated load $Q_0 (\xi, \eta)$, as shown in Figure 10–12, with the addition of constant in-plane loadings $\bar{N}_Y(X, 0)$ and $\bar{N}_Y(X,b)$ along the boundaries parallel to the X axis, as well as the $\bar{N}_X$ loads introduced in the preceding section. $\bar{N}_{XY}$ is taken as zero.

We use the virtual work method developed in Section 11.1.2 with D_Z given by equation (11.14) and U_ε by equation (11.17). The first step is to evaluate $U_{\bar{\varepsilon}}$, as given by equation (11.16). Following the procedure detailed in Section 11.3.2, the additional strain energy caused by the in-plane forces is, from equation (11.98),

$$U_{\bar{\varepsilon}} = \frac{1}{2} \int_0^a \int_0^b \left\{ \sum_{j=1}^\infty \sum_{k=1}^\infty (D_Z^{jk})^2 \left[\bar{N}_X \left(\frac{j\pi}{a} \right)^2 \cos^2 j\pi \frac{X}{a} \sin^2 k\pi \frac{Y}{b} \right. \right.$$

$$\left. \left. + \bar{N}_Y \left(\frac{k\pi}{b} \right)^2 \sin^2 j\pi \frac{X}{a} \cos^2 k\pi \frac{Y}{b} \right] dY \, dX \right. \tag{11.99}$$

which integrates to

$$U_{\bar{\varepsilon}} = \frac{\pi^2 ab}{8} \sum_{j=1}^\infty \sum_{k=1}^\infty (D_Z^{jk})^2 \left[\bar{N}_X \left(\frac{j}{a} \right)^2 + \bar{N}_Y \left(\frac{k}{b} \right)^2 \right] \tag{11.100}$$

Adding equation (11.100) to equation (11.17) gives the total strain energy $U_{\varepsilon+\bar{\varepsilon}}$.

Next, we take for $U_{\varepsilon+\bar{\varepsilon}}$ the single general term of the series, $U_{\varepsilon+\bar{\varepsilon}}^{jk}$, and find $\delta U_{\varepsilon+\bar{\varepsilon}}$. From equation (11.100), we have

$$\delta U_{\bar{\varepsilon}} = \frac{\pi^2 ab}{4} \left[\bar{N}_X \left(\frac{j}{a} \right)^2 + \bar{N}_Y \left(\frac{k}{b} \right)^2 \right] D_Z^{jk} \, \delta D_Z^{jk} \tag{11.101}$$

and adding equation (11.101) to equation (11.23), we get

$$\delta U_{\varepsilon+\bar{\varepsilon}} = \frac{\pi^4 abD}{4} \left\{ \left[\left(\frac{j}{a} \right)^2 + \left(\frac{k}{b} \right)^2 \right]^2 + \frac{1}{\pi^2 D} \left[\bar{N}_X \left(\frac{j}{a} \right)^2 + \bar{N}_Y \left(\frac{k}{b} \right)^2 \right] \right\} D_Z^{jk} \, \delta D_Z^{jk} \tag{11.102}$$

The principle of virtual work is applied by equating equations (11.20) and (11.102), from which we find

$$D_Z^{jk} = \frac{4Q_0}{\pi^4 abD} \left\{ \frac{\sin j\pi \frac{\xi}{a} \sin k\pi \frac{\eta}{b}}{\left[\left(\frac{j}{a}\right)^2 + \left(\frac{k}{b}\right)^2 \right]^2 + \frac{1}{\pi^2 D} \left[\overline{N}_X \left(\frac{j}{a}\right)^2 + \overline{N}_Y \left(\frac{k}{b}\right)^2 \right]} \right\} \tag{11.103}$$

Finally, the total deflection may be evaluated by equation (11.25).

We observe that if $\overline{N}_X$ and $\overline{N}_Y$ are both tensile, the deflection is reduced from the case where no in-plane forces are acting. If $\overline{N}_X$ and/or $\overline{N}_Y$ are compressive, we may have instability indicated by the denominator $\to 0$, whereupon

$$\overline{N}_X \left(\frac{j}{a}\right)^2 + \overline{N}_Y \left(\frac{k}{b}\right)^2 = -\pi^2 D \left[\left(\frac{j}{a}\right)^2 + \left(\frac{k}{b}\right)^2 \right]^2 \tag{11.104}$$

Equation (11.104) can lead to several classes of buckling problems, such as (a) $\overline{N}_X$ and $\overline{N}_Y$ are proportional, in other words, $\overline{N}_X = f\overline{N}_{cr}$ and $\overline{N}_Y = g\overline{N}_{cr}$, where f and g are specified constants which permit $\overline{N}_{cr}$ to be evaluated; and, (b) $\overline{N}_X$ (or $\overline{N}_Y$) is a fixed value. Correspondingly, $\overline{N}_{Ycr}$ (or $\overline{N}_{Xcr}$) can be computed. As an example, when $\overline{N}_Y = 0$ we have an identical situation to that discussed in the previous section and covered by equation (11.94). Note also that buckling may occur even if one of the in-plane forces is tensile, although it retards the instability.

Example 11.3(a)

As a simple example of buckling under bidirectional in-plane loading, consider a square plate with $a = b$ and $\overline{N}_X = \overline{N}_Y$. Equation (11.104) reduces to

$$\overline{N}_{cr} = -\frac{\pi^2 D}{b^2} (\bar{j}^2 + \bar{k}^2) \tag{11.105}$$

Obviously the lowest critical load is found for $\bar{j} = \bar{k} = 1$ as

$$\overline{N}_{cr} = -\frac{2\pi^2 D}{b^2} \tag{11.106}$$

This indicates a single wave in each direction.

Example 11.3(b)

Another simple illustration for the square plate is provided by taking $\overline{N}_Y = g\overline{N}_X$. Then, we have

$$\overline{N}_{Xcr} = -\frac{\pi^2 D}{b^2} \frac{(\bar{j}^2 + \bar{k}^2)^2}{(\bar{j}^2 + g\bar{k}^2)} \tag{11.107}$$

$\overline{N}_{Xcr1}$ will again correspond to $\bar{j} = \bar{k} = 1$, so that

$$\overline{N}_{Xcr1} = \frac{-4}{(1 + g)} \frac{\pi^2 D}{b^2} \tag{11.108}$$

The effect of unidirectional tension on the retardation of buckling is demonstrated by considering a negative value of g in equation (11.107). In many practical cases of unidirectional compression, the boundaries in the other direction, parallel to the direction of loading, are restrained in the middle plane. As the plate then deflects, these boundaries develop tensile forces in the middle plane which oppose the onset of instability.

11.3.5 Finite Deformation of Plates

If we consider the modified expressions for the middle-surface strains, equations (11.82) to (11.84), and restrict ourselves for the moment to Cartesian coordinates, we have

$$\bar{\varepsilon}_X = D_{X,X} + \frac{1}{2}(D_{Z,X})^2 \qquad \text{(a)}$$

$$\bar{\varepsilon}_Y = D_{Y,Y} + \frac{1}{2}(D_{Z,Y})^2 \qquad \text{(b)} \qquad (11.109)$$

$$\bar{\omega} = D_{Y,X} + D_{X,Y} + D_{Z,X}D_{Z,Y} \qquad \text{(c)}$$

in view of equations (7.54a–c). Equations (11.109) may be combined into a single compatibility equation in terms of D_Z and the strains.

We form certain second partials, assuming sufficient continuity so that the order of differentiation can be altered:

$$\bar{\varepsilon}_{X,YY} = D_{X,XYY} + (D_{Z,XY})^2 + D_{Z,X}D_{Z,XYY} \qquad \text{(a)}$$

$$\bar{\varepsilon}_{Y,XX} = D_{Y,XXY} + (D_{Z,XY})^2 + D_{Z,Y}D_{Z,XXY} \qquad \text{(b)} \qquad (11.110)$$

$$\bar{\omega}_{,XY} = D_{Y,XXY} + D_{X,XYY} + (D_{Z,XY})^2 + D_{Z,XX}D_{Z,YY} \qquad \text{(c)}$$
$$\qquad + D_{Z,X}D_{Z,XYY} + D_{Z,Y}D_{Z,XXY}$$

Adding equations (11.110a) and (11.110b) and subtracting (11.110c), we obtain

$$\bar{\varepsilon}_{X,YY} + \bar{\varepsilon}_{Y,XX} - \bar{\omega}_{,XY} = (D_{Z,XY})^2 - D_{Z,XX}D_{Z,YY} \qquad (11.111)$$

We next eliminate the strains in favor of the stress resultants. From Matrix 8–2 dropping the thermal terms,

$$N_X = \frac{Eh}{1-\mu^2}(\varepsilon_X + \mu\varepsilon_Y) \qquad \text{(a)}$$

$$N_Y = \frac{Eh}{1-\mu^2}(\varepsilon_Y + \mu\varepsilon_X) \qquad \text{(b)} \qquad (11.112)$$

$$N_{XY} = \frac{Eh}{2(1+\mu)}\omega \qquad \text{(c)}$$

which may be inverted to

$$\varepsilon_X = \frac{1}{Eh}(N_X - \mu N_Y) \qquad \text{(a)}$$

$$\varepsilon_Y = \frac{1}{Eh}(N_Y - \mu N_X) \qquad \text{(b)} \qquad (11.113)$$

$$\omega = \frac{2(1+\mu)}{Eh}N_{XY} \qquad \text{(c)}$$

Now, replacing ε_X, ε_Y, and ω by $\bar{\varepsilon}_X$, $\bar{\varepsilon}_Y$, and $\bar{\omega}$ in equation (11.113a–c) and introducing this equation into equation (11.111), we have

$$\frac{1}{Eh}[N_{X,YY} + N_{Y,XX} - \mu(N_{X,XX} + N_{Y,YY}) - 2(1+\mu)N_{XY,XY}]$$

$$= (D_{Z,XY})^2 - D_{Z,XX}D_{Z,YY} \qquad (11.114)$$

Equation (11.114) and equation (11.72) constitute the compatibility and equilibrium equations, respectively.

Four unknowns still remain in the two equations, so that further refinement is necessary.

If a stress function $\mathscr{F}$ is defined such that

$$N_X = \mathscr{F}_{,YY} \qquad (11.115a)$$

$$N_Y = \mathscr{F}_{,XX} \qquad (11.115b)$$

$$N_{XY} = -\mathscr{F}_{,XY} \qquad (11.115c)$$

and equations (11.115a–c) are introduced into equations (11.114) and (11.72), we obtain

$$\nabla^4\mathscr{F}(X,Y) = Eh[(D_{Z,XY})^2 - D_{Z,XX}D_{Z,YY}] \qquad (11.116)$$

and

$$\nabla^4 D_Z(X,Y) = \frac{1}{D}[q_Z + \mathscr{F}_{,YY}D_{Z,XX} + \mathscr{F}_{,XX}D_{Z,YY} - 2\mathscr{F}_{,XY}D_{Z,XY}] \qquad (11.117)$$

which are known as von Kármán equations for the large deflection of plates, after the famous contemporary mechanician.

The von Kármán equations are coupled and nonlinear. The nonlinearity arises from the relaxation of assumption [2], and the enforcement of this assumption immediately reduces equation (11.117) to the equation of the linear theory, equation (10.12). Also, the equations are written in invariant form and thus may be readily transformed to other coordinate systems. Y. C. Fung[28] observed that the rhs of equation (11.116) is related to the Gaussian curvature of the *deformed* surface. From equation (2.42), we confirm that $(D_{Z,XY})^2 - D_{Z,XX}D_{Z,YY} = \delta$, the discriminant of the *deformed* surface, so that if the plate is bent into a developable surface (zero Gaussian curvature) such as a cylinder, the rhs of equation (11.116) vanishes.

We may investigate this in more detail by referring to the cylindrically bent plate shown in Figure 10–7, which was treated in Section 10.2.1.3. Because of the single curvature, $D_{Z,XX} = D_{Z,YX} = 0$ and the rhs of equation (11.117) reduce to

$$D_{Z,YYYY} = \frac{1}{D}(q_Z + N_Y D_{Z,YY}) \tag{11.118}$$

$D_{Z,YY}$ is left in general form rather than being evaluated from equation (10.66), because the presence of N_Y will modify D_Z somewhat. Equation (11.118) is a fourth-order ordinary linear differential equation with a constant coefficient, which is readily solved by classical methods. The homogeneous solution is written as

$$D_{Zh} = C_1 + C_2 Y + C_3 \sinh \lambda Y + C_4 \cosh \lambda Y \tag{11.119a}$$

where

$$\lambda = \left(\frac{N_Y}{D}\right)^{1/2} \tag{11.119b}$$

with the particular solution depending on the form of q_Z.

For example, if $q_Z = q_0 = $ constant,

$$D_{Zp} = -\frac{q_0}{2N_Y} Y^2 \tag{11.120}$$

With N_Y initially specified, this problem is analogous to an axially loaded beam and may be investigated further for various boundary conditions along the Y-axis. However, when the boundaries parallel to the X-axis are restrained in the middle plane, N_Y is initially unknown. Referring to the cylindrical cross-section in Figure 10–7, this force may be determined from the compatibility condition whereby the *extension* of the plate in the Y-direction produced by N_Y must be equal to the *difference* between the *final* arc length, s_Y, and the *initial* length in the Y-direction, taken as $2b$.

To compute the extension of the plate in the Y-direction, we neglect the contribution of D_Z to $\bar{\varepsilon}_Y$ in equation (11.109b) so that $\bar{\varepsilon}_Y \simeq \varepsilon_Y$. We then substitute $\varepsilon_Y = D_{Y,Y}$ into equation (11.113b) to find

$$D_{Y,Y} = \frac{1}{Eh}(N_Y - \mu N_X) \tag{11.121}$$

Further, we assume that $\varepsilon_X = 0$ in equation (11.113a) and use the resulting $N_X = \mu N_Y$ in equation (11.121) to get

$$D_{Y,Y} = \frac{N_Y(1 - \mu^2)}{Eh} \tag{11.122}$$

from which

$$D_Y = \frac{N_Y(1 - \mu^2)}{Eh} Y \tag{11.123}$$

because $D_Y(0) = 0$.

To express D_Y as the difference between the final arc length and the initial length in the Y direction, refer to equation (2.50) with s_Z replaced by s_Y, dZ by dY, and dR_0 by dD_Z. Because the deformation is symmetric, we need consider only half the plate $(0 < Y < b)$. Then, from equation (2.50),

$$s_Y = \int_0^b [1 + (D_{Z,Y})^2]^{1/2}\, dY \tag{11.124}$$

We use the binomial expansion[27] to simplify equation (11.124) into

$$s_Y \simeq b + \frac{1}{2} \int_0^b (D_{Z,Y})^2\, dY \tag{11.125}$$

so that

$$s_Y - b = \frac{1}{2} \int_0^b (D_{Z,Y})^2\, dY \tag{11.126}$$

Equation (11.123), evaluated at $Y = b$, and equation (11.126) give

$$D_Y(b) = N_Y \frac{(1 - \mu^2)b}{Eh} = \frac{1}{2} \int_0^b (D_{Z,Y})^2\, dY \tag{11.127}$$

Equation (11.127) along with equations (11.119a) and (11.119b) and a particular solution such as equation (11.120) constitute the solution to the problem of cylindrical bending when the in-plane displacements of the middle surface are restrained. An obvious iterative solution algorithm is to begin with an assumed value of N_Y; compute the integration constants in equation (11.119a) using the appropriate boundary conditions on D_Z; complete the expression for D_Z; and finally, check the assumed N_Y using equation (11.127). Solutions of the cylindrical bending problem are available in Timoshenko and Woinowsky-Krieger.[29]

For equations (11.116) and (11.117), a limited number of analytical solutions using the Navier approach are available,[30] but the calculations are quite involved so that numerical solutions are attractive.

An important item which might be investigated with solutions based on the finite deflection theory of plates is the *limit* of the small deflection theory. Specifically, we are interested in two points: (a) how large must the normal displacement D_Z be in order to obtain stress couples significantly different from those calculated using elementary plate theory? and, (b) how significant are the in-plane stress resultants accompanying the finite transverse displacement? These questions are difficult to answer in general but may be studied for circular plates, as we will show in the next section.

11.3.6 Finite Deformation of Axisymmetrically Loaded Circular Plates

Consider an axisymmetrically loaded circular plate. Instead of the general expression equation (10.1), it is convenient to start with the equilibrium equations specialized for axisymmetric loading. Equation (11.73a) gives

$$N_R - N_\theta + R N_{R,R} + q_R R = 0 \tag{11.128}$$

Another relationship is found from generalizing equation (10.32b), which followed from equation (3.25c) through equations (10.1) to (10.5). The first term in the axisymmetric form of equation (11.67) is $[R(Q_R + N_R D_{Z,R})]_{,R}$ when $\alpha = R$, $B = R$, and transverse shearing strains are suppressed. This leads to a modified transverse shear $Q_R + N_R D_{Z,R}$, so that equation (10.32b) becomes

$$Q_R + N_R D_{Z,R} = D\left(D_{Z,RRR} + \frac{1}{R}D_{Z,RR} - \frac{1}{R^2}D_{Z,R}\right) \tag{11.129}$$

From Section 10.3.1.4, we recall that for an axisymmetric loading $q_R(R)$, $Q_R(R)$ can be directly expressed in terms of q_R, so that Q_R may be regarded as a known quantity in what follows.

Now, we use the strain-displacement relations for polar coordinates, which are obtained from equations (11.82) and (11.83) together with (7.54a) and (7.54b), to write

$$\bar{\varepsilon}_R = D_{R,R} + \frac{1}{2}(D_{Z,R})^2 \tag{11.130}$$

and

$$\bar{\varepsilon}_\theta = \frac{D_R}{R} \tag{11.131}$$

Substituting $\bar{\varepsilon}_R$ and $\bar{\varepsilon}_\theta$ in the stress-strain equation (8.10), we find

$$N_R = \frac{Eh}{1-\mu^2}\left[D_{R,R} + \frac{1}{2}(D_{Z,R})^2 + \mu\frac{D_R}{R}\right] \tag{11.132}$$

and

$$N_\theta = \frac{Eh}{1-\mu^2}\left[\frac{D_R}{R} + \mu D_{R,R} + \frac{\mu}{2}(D_{Z,R})^2\right] \tag{11.133}$$

Finally, we substitute equations (11.132) and (11.133) into equations (11.128) and (11.129) to get

$$\begin{aligned}D_{R,RR} = &-\frac{1}{R}D_{R,R} + \frac{1}{R^2}D_R - \frac{(1-\mu)}{2R}(D_{Z,R})^2 \\ &- D_{Z,R}D_{Z,RR} + \frac{(1-\mu^2)}{Eh}q_R\end{aligned} \tag{11.134}$$

and

$$\begin{aligned}D_{Z,RRR} = &-\frac{1}{R}D_{Z,RR} + \frac{1}{R^2}D_{Z,R} \\ &+ \frac{12}{h^2}D_{Z,R}\left[D_{R,R} + \frac{1}{2}(D_{Z,R})^2 + \mu\frac{D_R}{R}\right] + \frac{Q_R}{D}\end{aligned} \tag{11.135}$$

which are again coupled and nonlinear.

These equations are studied for several cases of applied loading and for a variety of boundary conditions in Timoshenko and Woinowsky-Krieger.[31] Of particular interest for our purposes is the case of a simply supported solid plate with edge moment M as shown in Figure 10–15. For this case, $Q_R = q_Z = 0$. Timoshenko and Woinowsky-Krieger present a numerical solution for a plate of thickness h, radius $a \simeq 23h$, and $M = M_R(a) = 2.93 \times 10^{-3}(D/h)$. With respect to the comparison with elementary plate theory as discussed in the previous section, there are two items of major interest: (a) the magnitude of the displacements and moments calculated from the nonlinear theory, as compared to those computed from the elementary theory; and, (b) the significance of the in-plane stress resultants N_R and N_θ, which are not computable from the elementary theory.

Considering first (a) the displacements, the maximum transverse deflection $D_Z \simeq 0.55h$. From equation (10.121), $D_Z(0) = 0.62h$ for $\mu = 0.25$, so that the refined theory gives about a 10% smaller deflection. The elementary theory has stress couples $M_R = M_\theta = M = $ constant throughout the plate, whereas the refined theory yields values for M_R and M_θ at the center which are about 12% less. Thus, the elementary theory appears to be conservative and fairly accurate for deflections up to about one-half the plate thickness.

To assess the significance of the in-plane stress resultants (b), we must refer to actual stresses, rather than the stress resultants. From elementary strength of materials and the definitions in Chapter 3,

$$\sigma_{ii}\left(\pm \frac{h}{2}\right) = \frac{N_i}{h} \pm \frac{6M_i}{h^2} \tag{11.136}$$

The stress resultants N_R and N_θ are both tensile and relatively small in the central portion of the plate. In the outer portion, $N_R \to 0$ and N_θ/h becomes compressive and reaches a magnitude of almost 20% of $(6M/h^2)$, where M is the constant moment of the elementary theory. From this, we conclude that with deflections approaching the half-thickness of the plate, the seemingly simple case of uniform moment may be susceptible to circumferential instability because of the relatively high compressive values of N_θ.

This study has provided us with an opportunity to assess the validity and limitations of one of the important assumptions in the elementary theory of plates, assumption [2].

REFERENCES

1. S. Timoshenko and S. Woinowsky-Krieger, *Theory of Plates and Shells,* 2nd Ed, (New York: McGraw-Hill, 1959): 197–205.

2. M. J. Forray, *Variational Calculus in Science and Engineering,* (New York: McGraw-Hill, 1968): 163–164.

3. S. Timoshenko and S. Woinowsky-Krieger, 2nd Ed. *Theory of Plates and Shells:* 173–179.

4. H. D. Conway, "A Lévy-type solution for a rectangular plate of variable thickness, *Journal of Applied Mechanics,* ASME, Vol. 2J, June 1958: 297–308.

5. S. Timoshenko and S. Woinowsky-Krieger. 2nd Ed. *Theory of Plates and Shells:* 298–308.

6. R. Szilard, *Theory and Analysis of Plates,* (Englewood Cliffs, New Jersey: Prentice-Hall, Inc., 1974): 130–136.

7. S. Timoshenko and S. Woinowsky-Krieger. 2nd Ed. *Theory of Plates and Shells:* 366–369.

8. Ibid.: 390–394.

9. K. S. Pister and S. B. Dong, Elastic bending of layered plates, *Journal of the Engineering Mechanics Division,* ASCE, Vol. 84, No. 10, October 1959: 1–10.

10. E. Reissner, Finite deflections of sandwich plates, *Proc. Of the First U.S. National Congress of Applied Mechanics,* ASME, New York, 1952.

11. E. Reissner, Effect of Shear Deformation on Bending of Elastic Plates, *Journal of Applied Mechanics,* ASME, vol. 12, no. 55, 1947: A69–A77.

12. E. Reissner, On Bending of Elastic Plates, *Quarterly of Applied Mechanics,* vol. 5, 1947: 55–68.

13. A. E. Green, On Reissner's theory of bending of elastic plates, *Quarterly of Applied Mathematics,* vol. 7, 1949: 223–228.

14. J. N. Reddy, *Energy and Variational Methods in Applied Mechanics,* (New York: Wiley, 1984): 354–388.

15. S. Timoshenko and S. Woinowsky-Krieger, *Theory of Plates and Shells:* 165–173.

16. C. S. Desai and J. F. Abel, *Introduction to the Finite Element Method,* (New York: Van Nostrand-Reinhold, 1972): 80–82; 178–181.

17. L. J. Brombolich and P. L. Gould, A high-precision curved shell finite element, *AIAA Journal,* vol. 10, no. 6, June 1972: 727–728.

18. D. Yitzhaki, *Prismatic and Cylindrical Shell Roofs,* (Haifa, Israel: Haifa Science Publishing, 1958).

19. H. Simpson, Design of folded plate roofs, *Journal of the Structural Division,* ASCE, vol. 84, no. 1, January 1958, pp. 1–21.

20. D. P. Billington, *Thin Shell Concrete Structures,* (New York: McGraw-Hill, 1965), chaps. 8 and 9.

21. M. Pultar, *Foundation of Folded Plate Theories, IASS International Colloquium on the Progress of Shell Structures in the Last 10 Years and Its Future Development,* Session VI, Madrid, Spain, September–October 1969: 1–16.

22. V. A. Pulmano and S. L. Lee, Prismatic shells with intermediate Columns. *Journal of the Structural Division,* ASCE, vol. 91, no. ST6, December 1965: 215–237.

23. M. Pultar, D. B. Billington, and J. Riera, Folded plates continuous over flexible supports, *Journal of the Structural Division, ASCE,* vol. 93, no. ST5, October 1967: 253–277.

24. Phase I Report of the Task Committee on Folded Plate Construction, *Journal of the Structural Division, ASCE,* vol. 89, no. ST6, December 1963: 365–406.

25. A. P. Boresi, O. M. Sidebottom, F. B. Seely, and J. O. Smith. *Advanced Mechanics of Materials,* 3rd Ed. (New York, N.Y.: John Wiley and Sons, 1978): 480.

26. S. Timoshenko and S. Woinowsky-Krieger, *Theory of Plates and Shells.* 2nd ed. pp. 380–387.

27. D. O. Brush and B. O. Almroth, *Buckling of Bars, Plates and Shells,* (New York: McGraw-Hill, 1975): 94–112.

28. Y. C. Fung, *Foundations of Solid Mechanics,* (Englewood Cliffs, NJ: Prentice-Hall, 1965): 463–470.

29. S. Timoshenko and S. Woinowsky-Krieger, *Theory of Plates and Shells,* 2nd ed. chap. 1.

30. Ibid.: 421–428.

31. Ibid.: 396–415.

EXERCISES

The numerical problems are given without specific units to allow English, metric, or SI unit dimensions to be selected.

11.1. Re-solve the case of the clamped plate, shown in Figure 11–1 with a uniformly distributed loading q_0, using the virtual work method.

11.2. Consider the clamped rectangular plate subject to a hydrostatic loading as shown in Figure 11–5. Obtain a one-term and a two-term numerical solution for D_Z using the Rayleigh-Ritz method.

11.3. Consider a simply supported rectangular plate under a uniform tension T and compression C, loaded uniformly in one quadrant by a force q, as shown in Figure 11–6.
  (a) Using the Green's function approach, derive the expression for the deflected surface.
  (b) Determine the critical value of C assuming that $T = C/2$.
  (c) Determine the lowest critical value of C and the corresponding buckling mode as a function of the aspect ratio.

11.4. Consider the plate shown in Figure 11–6 subject to the loading q over two quadrants, $0 \le X \le a/2, 0 \le Y \le b$, and re-solve Exercise 11.3 using the virtual work method with $T = C/4$.

11.5. Show that the buckling load for a rectangular plate with dimensions as shown on Figure 11–4 cannot be less than $N_{Xcr} = \pi^2 D/a^2$ if the plate is simply supported on the loaded edges $(0, Y)$ and (a, Y) and the boundary conditions on the other edges are unspecified.

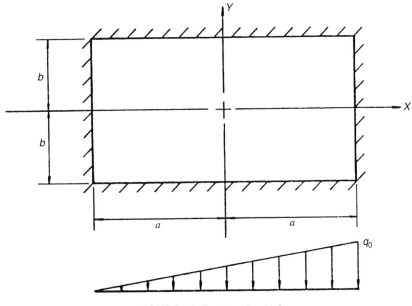

FIGURE 11–5   Exercise 11.2

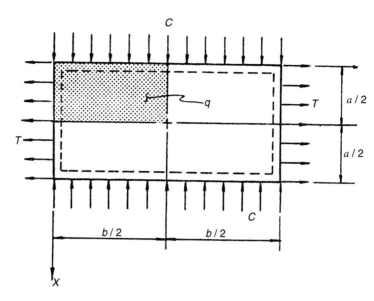

FIGURE 11-6   Exercise 11.4

11.6. Consider the rectangular plate as shown on Figure 11–4.

    (a) Re-solve the all-sides simply supported case treated in Section 11.3.3 using a Lévy-Nádai solution in the form of equations (10.106) and (10.107).

    (b) Determine the critical load for the case when the plate is clamped along the unloaded edges $(X, 0)$ and (X, b).

11.7. Consider the circular plate shown in Figure 10–29 with an added uniform radial compression force N_R. Determine the critical value of N_R.

12

Shell Bending: Applications to Circular Cylindrical Shells

12.1 GENERAL

In earlier chapters, we derived the equilibrium, strain-displacement, and constitutive equations and stated the required boundary conditions for the bending theory of shells, referred to a system of orthogonal curvilinear coordinates. Also, we developed strain energy and potential energy expressions that can be incorporated into an energy formulation of the shell theory. In this chapter, these equations are specialized for various classes of shells, as we have done for the membrane theory equations in Chapters 4 to 6.

Before proceeding, again note that many shells may achieve equilibrium through membrane action alone, provided the requisite conditions are closely approached by the actual shell. For such shells, bending is a secondary phenomenon often confined to narrow regions near boundaries, geometric discontinuities, and concentrated loads. On the other hand, there are shells for which the membrane theory idealization is grossly violated by the physical situation. The bending behavior may alter the stress pattern from that computed by the membrane theory in two ways: (a) significant transverse shearing forces and bending and twisting moments can develop; and, (b) the pattern of the in-plane stress resultants may be altered markedly by the bending deformations. Although a shell may seriously violate the membrane theory requirements, the possibility still remains of resisting *transverse loading* primarily with *in-plane forces,* which is the basic initial attraction of this structural form. It is this latter possibility, whereby the transverse loading may be resisted by a *combination* of in-plane forces and transverse shearing forces, which distinguishes bending of shells from the elementary behavior of plates.

In the study of shell bending, cylindrical and conical shells are frequently considered apart from rotational and translational shells, although they technically may fall into one or both of these classes. The bending behavior of shells with zero Gaussian curvature is quite distinct however, because in the direction where the radius of curvature is infinite, these shells cannot develop any membrane forces to resist the transverse loading. As we have seen in Chapters 4 to 6, the membrane theory solutions for such shells generally show the entire transverse loading being sustained in the curved direction, a one-way resistance pattern. When the membrane boundary conditions in the curved direction are violated and rendered incapable of developing the required reactions, drastic alteration of the stress pattern is often the result because the zero curvature direction is of no help except in a bending mode.

Another important reason for studying cylindrical shells, especially as a distinct class, is that fairly simple analytical solutions may be found, whereas for more complex geometries, such solutions are relatively scarce. Moreover, many of the techniques for solving the governing equations for other geometries involve various approximations and simplifications that cast the equations into a form similar to that of the cylindrical shell. Thus, the cylindrical shell solutions assume added importance from the standpoint of generality.

Finally, we note that in practice, cylindrical shells are probably the most frequently encountered form and certainly merit careful attention.

12.2 SPECIALIZATION OF EQUATIONS

In the discussion of the membrane theory of circular cylindrical shells (Section 4.4.2), we chose $\alpha = Z$, $\beta = \theta$, $A = 1$, and $B = a$, as shown in Figure 4–12(a). For this geometry, $R_\alpha = \infty$ and $R_\beta = a$. Later, in Section 6.2, the axial coordinate was taken as X in deference to its widespread usage in current literature. We should feel comfortable with either choice. For vertical vessels in which the axis of rotation coincides with the global Z axis, $\alpha = Z$ is the logical choice; whereas, for horizontal shells, $\alpha = X$ seems equally suitable. We use $\alpha = X$ in this chapter.

The force equilibrium equations, (3.17a–c), and moment equilibrium equations, (3.22a) and (3.22b), become

$$aN_{X,X} + N_{\theta X,\theta} + q_X a = 0 \tag{12.1a}$$

$$aN_{X\theta,X} + N_{\theta,\theta} + Q_\theta + q_\theta a = 0 \tag{12.1b}$$

$$aQ_{X,X} + Q_{\theta,\theta} - N_\theta + q_n a = 0 \tag{12.1c}$$

$$-aM_{X\theta,X} - M_{\theta,\theta} + Q_\theta a = 0 \tag{12.1d}$$

$$aM_{X,X} + M_{\theta X,\theta} - Q_X a = 0 \tag{12.1e}$$

The strain-displacement relations, equations (7.46a–h), are

$$\varepsilon_X = D_{X,X} \tag{12.2a}$$

$$\varepsilon_\theta = \frac{1}{a}(D_{\theta,\theta} + D_n) \tag{12.2b}$$

$$\omega = D_{\theta,X} + \frac{1}{a}D_{X,\theta} \tag{12.2c}$$

$$\kappa_X = D_{X\theta,X} \tag{12.3d}$$

$$\kappa_\theta = \frac{1}{a}D_{\theta X,\theta} \tag{12.2e}$$

$$\tau = \frac{1}{2}\left[D_{\theta X,X} + \frac{1}{a}D_{X\theta,\theta}\right] \tag{12.2f}$$

$$\gamma_X = D_{n,X} + D_{X\theta} \tag{12.2g}$$

$$\gamma_\theta = \frac{1}{a}(D_{n,\theta} - D_\theta) + D_{\theta X} \tag{12.2h}$$

If transverse shearing strains are neglected, equations (7.47a–c) replace equations (12.2d–h):

$$\kappa_X = -D_{n,XX} \tag{12.3a}$$

$$\kappa_\theta = -\frac{1}{a^2}(D_{n,\theta\theta} - D_{\theta,\theta}) \tag{12.3b}$$

$$\tau = \frac{1}{a}\left(\frac{1}{2}D_{\theta,X} - D_{n,X\theta}\right) \tag{12.3c}$$

There are a number of variations on the strain-displacement relations given by equations (12.1) to (12.3), either with transverse shearing strains included or neglected. These theories are associated with such famous names as Donnell, Mushtari, Love, Timoshenko, Reissner, Naghdi, Sanders, and Flügge, several of whom have been cited earlier in this text and/or will be in Section 12.4.1. The principal differences between the theories are in the expressions for the twisting curvature and in the inclusion of some terms of $O(h/a:1)$ as the shell thickness increases relative to the radius.[1] In relation to the scope of this chapter, the differences in the theories are largely irrelevant, especially if the transverse shearing strains are included as in equations (12.2f–h). However, for the instability and finite deformation problems introduced in Section 13.3, the differences may become significant.

Also, we have the effective shear forces evaluated from equations (8.24) to (8.27):

$$\overline{N}_{X\theta} = N_{X\theta} + \frac{M_{X\theta}}{a} \tag{12.3d}$$

$$\overline{Q}_X = Q_X + \frac{M_{X\theta,\theta}}{a} \tag{12.3e}$$

$$\overline{N}_{\theta X} = N_{\theta X} \tag{12.3f}$$

$$\overline{Q}_\theta = Q_\theta + M_{\theta X,X} \tag{12.3g}$$

The constitutive relations are given by Matrix 8–2, with α and β taken as X and θ, respectively. These are restated here in the matrix form of equation (8.10)

$$\{N\}_{X\theta} = [D]\{\varepsilon\}_{X\theta} - \{N_T\}_{X\theta} \qquad (12.4)$$

with the subscripts serving to remind us of our choice of coordinates.

It is easily verified that the membrane theory equations are recoverable from the preceding expressions. With the bending terms neglected and $N_{\theta X} = N_{X\theta} = S$, equations (12.1a–c) reduce to equations (6.1), and equations (12.2a–c) and the corresponding parts of equation (12.4) are identical to equations (8.50).

12.3 AXISYMMETRICAL LOADING

12.3.1 Displacement Formulation and Solution

Axisymmetrically loaded circular cylindrical shells are employed as pressure vessels and tanks in many industrial applications. Some examples are shown in Figures 2–8(r) and (u). For axisymmetrical loading, the terms q_θ, $N_{\theta X}$, $N_{X\theta}$, $M_{\theta X}$, $M_{X\theta}$, Q_θ, ω, κ_θ, τ, γ_θ, D_θ, $D_{\theta X}$, and all terms differentiated with respect to θ drop out. This leaves the three equilibrium equations

$$N_{X,X} + q_X = 0 \qquad (12.5a)$$

$$Q_{X,X} - \frac{N_\theta}{a} + q_n = 0 \qquad (12.5b)$$

$$M_{X,X} - Q_X = 0 \qquad (12.5c)$$

and the strain-displacement relationships

$$\varepsilon_X = D_{X,X} \qquad (12.6a)$$

$$\varepsilon_\theta = \frac{D_n}{a} \qquad (12.6b)$$

$$\kappa_X = D_{X\theta,X} \qquad (12.6c)$$

$$\gamma_X = D_{n,X} + D_{X\theta} \qquad (12.6d)$$

or, in the absence of transverse shearing strains,

$$\kappa_X = -D_{n,XX} \qquad (12.7)$$

We also have the pertinent constitutive relationships from equation (12.4). These are written explicitly from Matrix 8–2 as

$$N_X = \frac{Eh}{1 - \mu^2}(\varepsilon_X + \mu\varepsilon_\theta) - N_{XT} \qquad (12.8a)$$

$$N_\theta = \frac{Eh}{1 - \mu^2}(\mu\varepsilon_X + \varepsilon_\theta) - N_{\theta T} \qquad (12.8b)$$

$$M_X = D\kappa_X - M_{XT} \tag{12.8c}$$

$$M_\theta = \mu D\kappa_X - M_{\theta T} \tag{12.8d}$$

$$Q_X = \frac{\lambda Eh}{2(1 + \mu)}\,\gamma_X \tag{12.8e}$$

where D has been defined explicitly in equation (10.4d). When transverse shearing strains are suppressed, equation (12.8e) is of no use so that Q_X must be evaluated from equation (12.5c).

We now express the equilibrium equations in terms of the displacements, in the fashion of a classical displacement formulation.

First, we substitute equations (12.6a–d) into equations (12.8a–e) to get the stress resultants and couples in terms of the displacements. Also, in view of equations (8.9), we set $N_{XT} = N_{\theta T} = N_T$ and $M_{XT} = M_{\theta T} = M_T$. Then, we have

$$N_X = \frac{Eh}{1 - \mu^2}\left(D_{X,X} + \mu\,\frac{D_n}{a}\right) - N_T \tag{12.9a}$$

$$N_\theta = \frac{Eh}{1 - \mu^2}\left(\mu D_{X,X} + \frac{D_n}{a}\right) - N_T \tag{12.9b}$$

$$M_X = DD_{X\theta,X} - M_T \tag{12.9c}$$

$$M_\theta = \mu DD_{X\theta,X} - M_T \tag{12.9d}$$

$$Q_X = \frac{\lambda Eh}{2(1 + \mu)}\,(D_{n,X} + D_{X\theta}) \tag{12.9e}$$

In the absence of transverse shearing strains, we use equation (12.7) for κ_X and find

$$M_X = -DD_{n,XX} - M_T \tag{12.10a}$$

$$M_\theta = -\mu DD_{n,XX} - M_T \tag{12.10b}$$

and, from equations (12.5c) and (12.10a),

$$Q_X = -DD_{n,XXX} - M_{T,X} \tag{12.10c}$$

Also, in equation (12.3e), with $M_{X\theta,\theta} = 0$, $\overline{Q}_X = Q_X$.

Now, substituting equations (12.9a–e) into (12.5a–c), we get

$$\frac{Eh}{1 - \mu^2}\left(D_{X,X} + \frac{\mu}{a}D_n\right)_{,X} = -q_X + N_{T,X} \tag{12.11a}$$

$$\frac{Eh}{1 + \mu}\left[\frac{\lambda}{2}(D_{n,X} + D_{X\theta})_{,X} - \frac{1}{a(1 - \mu)}\left(\mu D_{X,X} + \frac{D_n}{a}\right)\right] = -q_n + \frac{N_T}{a} \tag{12.11b}$$

$$DD_{X\theta,XX} - \frac{\lambda Eh}{2(1 + \mu)}\,(D_{n,X} + D_{X\theta}) = M_{T,X} \tag{12.11c}$$

The loading and thermal terms are presumed as known and transposed to the rhs. The resulting set of three equations in the three unknowns D_X, D_n, and $D_{X\theta}$ constitutes the displacement formulation. We will not reduce these equations further, but concentrate on the theory in which transverse shearing strains are neglected.

With $\gamma_X = 0$, $D_{X\theta}$ is expressed in terms of D_n by equation (12.6d). We need retain only the first of equations (12.11), along with a new equation found by substituting equations (12.9b) and (12.10c) into equation (12.5b):

$$\frac{Eh}{1 - \mu^2}\left(D_{X,X} + \frac{\mu}{a}D_n\right)_{,X} = -q_X + N_{T,X} \tag{12.12a}$$

$$-DD_{n,XXXX} - \frac{Eh}{a(1 - \mu^2)}\left(\mu D_{X,X} + \frac{D_n}{a}\right) = -q_n - \frac{N_T}{a} + M_{T,XX} \tag{12.12b}$$

This set may be reduced to a single equation. We first integrate both sides of equation (12.12a), which gives

$$\frac{Eh}{1 - \mu^2}\left(D_{X,X} + \frac{\mu}{a}D_n\right) = -\int q_X\, dX + N_T + C \tag{12.13a}$$

where C is an integration constant.

Comparing equation (12.13a) with equation (12.9a), we find

$$N_X = -\int q_X\, dX + C \tag{12.13b}$$

We may rewrite the integral in equation (12.13b) in the alternate form introduced in Chapter 4 by choosing $C = N_X(0)$. Then, we have

$$N_X(X) = -\int_0^X q_X\, dX + N_X(0) \tag{12.13c}$$

Next, we solve equation (12.13a) for

$$D_{X,X} = \frac{1 - \mu^2}{Eh}\left[-\int_0^X q_X\, dX + N_T + N_X(0)\right] - \frac{\mu}{a}D_n \tag{12.13d}$$

which we substitute into equation (12.12b) to obtain

$$-DD_{n,XXXX} - \frac{Eh}{a(1 - \mu^2)}\left\{\mu\left(\frac{1 - \mu^2}{Eh}\left[-\int_0^X q_X\, dX + N_T + N_X(0)\right]\right.\right.$$
$$\left.\left. - \frac{\mu}{a}D_n\right) + \frac{D_n}{a}\right\} = -q_n - \frac{N_T}{a} + M_{T,XX} \tag{12.14}$$

which consolidates to

$$DD_{n,XXXX} + \frac{Eh}{a^2}D_n = q_n + \left(\frac{1 - \mu}{a}\right)N_T - M_{T,XX} + \frac{\mu}{a}\left[\int_0^X q_X\, dX - N_X(0)\right] \tag{12.15}$$

Equation (12.15) is often written in the form

$$D_{n,XXXX} + 4k^4 D_n = \frac{1}{D}\left(q_n + \frac{(1-\mu)}{a}N_T - M_{T,XX} + \frac{\mu}{a}\left[\int_0^X q_X \, dX - N_X(0)\right]\right) \qquad (12.16a)$$

where

$$k^4 = \frac{3(1-\mu^2)}{a^2 h^2} = \frac{Eh}{4Da^2} \qquad (12.16b)$$

Equation (12.16a) may be recognized as the governing equation for a well-documented problem in mechanics, the bending of a prismatic beam on an elastic foundation. Just as an elastic foundation takes a portion of any transverse loading applied to a beam, the shell parameter Eh/a^2, which appears explicitly in equation (12.15), supplements the flexural rigidity D of the cylindrical shell in resisting the transverse loading q_n. Once the governing equation is solved for D_n, the extensional stress resultants are computed from equations (12.9a) and (12.9b), and the stress couples and transverse shear resultant from equations (12.10a–c).

It is instructive to consider a somewhat specialized case, in which the thermal terms are dropped and the axial load $q_X = 0$. Then, from equation (12.13c), N_X = constant. If one boundary—for example, $X = 0$—is unconstrained against axial deformation, $N_X(0) = 0$ and, therefore, $N_X = 0$ throughout the shell. With these simplifications, equation (12.9a) gives

$$D_{X,X} = -\mu \frac{D_n}{a} \qquad (12.17)$$

and equation (12.9b) becomes

$$N_\theta = \frac{Eh}{a} D_n \qquad (12.18)$$

which is a remarkably simple *algebraic* relationship between the major stress resultant and the primary displacement, in view of the complexity of the system with which we began.

The displacement formulation of the axisymmetrically loaded cylindrical shell may be generalized in a straightforward matter to accommodate shells in which the thickness varies as $h = h(X)$.

Continuing with equation (12.16a), the homogeneous solution is written from the auxiliary equation

$$m^4 + 4k^4 = 0 \qquad (12.19a)$$

with roots

$$m = \pm k(1+i), \pm k(1-i) \qquad (12.19b)$$

as[2]

$$D_{nh} = e^{-kX}(C_1 \cos kX + C_2 \sin kX) + e^{kX}(C_3 \cos kX + C_4 \sin kX) \qquad (12.20)$$

where $C_1 - C_4$ are integration constants.

The particular solution is, of course, dependent on the loading, and the constants $C_1 - C_4$ are found from applying the appropriate boundary conditions, as we see in several examples later.

A comprehensive example of the stress analysis of an axisymmetrically loaded cylindrical shell is presented in Example 13.1(a), Section 13.1.2.4.

12.3.2 Semi-infinite Cylindrical Shells

The complete solution of equation (12.16a) involves four integration constants, as indicated in equation (12.20). This constitutes a two-point boundary value problem. Within the class of axisymmetrically loaded cylindrical shells, however, there are many cases for which the forces at one boundary do not materially affect those at the other boundary. Such shells are termed *semi-infinite.*

We may demonstrate this behavior by referring to equation (12.20). If X is measured from one boundary, the factor e^{kX} will grow very large as X increases unless $C_3 = C_4 = 0$; whereas, the factor e^{-kX} will cause the other terms to attenuate. Thus, the solution simplifies to

$$D_{nh} = e^{-kX}(C_1 \cos kX + C_2 \sin kX) \tag{12.21}$$

with C_1 and C_2 evaluated from the boundary conditions at $X = 0$.

The semi-infinite approach simplifies the ensuing arithmetic considerably. The range of shell parameters for which this assumption is valid may be determined by considering a shell with an arbitrary edge loading and observing the behavior of the solution as the distance from the loaded boundary increases. As shown in Figure 12–1, a transverse shear force Q_0 and bending moment M_0 are applied uniformly around the circumference at $X = 0$. The positive signs are chosen in accordance with Figure 3–2. From equations (12.10a) and (12.10c), we have

$$M_0 = M_X(0) = -DD_{n,XX}(0) \tag{12.22a}$$

$$Q_0 = Q_X(0) = -DD_{n,XXX}(0) \tag{12.22b}$$

from which

$$C_1 = -\frac{1}{2k^3D}(Q_0 + kM_0) \tag{12.23a}$$

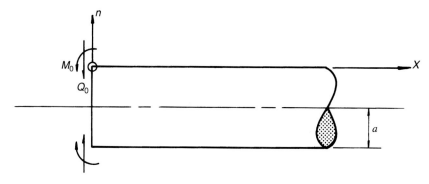

FIGURE 12–1   Semi-infinite Edge-Loaded Cylindrical Shell.

and

$$C_2 = \frac{M_0}{2k^2D} \tag{12.23b}$$

Before we carry the solution further, it is convenient to define the functions

$$F_1(kX) = e^{-kX}(\cos kX + \sin kX) \tag{12.24a}$$

$$F_2(kX) = e^{-kX}(\cos kX - \sin kX) \tag{12.24b}$$

$$F_3(kX) = e^{-kX}\cos kX = \frac{1}{2}(F_1 + F_2) \tag{12.24c}$$

$$F_4(kX) = e^{-kX}\sin kX = \frac{1}{2}(F_1 - F_2) \tag{12.24d}$$

Now, we may express D_n and the various derivatives required to obtain other components of the solution by Matrix 12–1.

Because $F_1(0) = F_2(0) = 1$ and decrease as X increases and because F_3 and F_4 are linear combinations of F_1 and F_2, the behavior of $F_1(X)$ and $F_2(X)$ indicates the propagation of the effects of the boundary loads. These functions are plotted on Figure 12–2 and are tabulated in Timoshenko and Woinowsky-Krieger and Tsui.[2] It is apparent from the figure that the edge loads will produce insignificant effects at about $kX = 3$.

In the literature, it is common to find $L \geq \pi/k$ designated as the length for which a given shell is considered to be a long shell and, therefore, treatable by the semi-infinite approach. L is called the *half-wave length of bending* and may be written as

$$L = \frac{\pi}{k} = \frac{\pi}{\sqrt[4]{3(1 - \mu^2)}}\sqrt{ah} \tag{12.25}$$

using equation (12.16b).

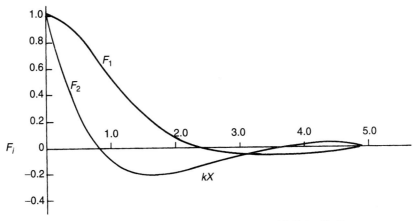

FIGURE 12–2   Solution Functions for Semi-infinite Shells.

Matrix 12–1

$$
\begin{Bmatrix} D_n \\ \\ D_{n,X} \\ \\ D_{n,XX} \\ \\ D_{n,XXX} \end{Bmatrix} = \frac{1}{D} \begin{bmatrix} 0 & -\dfrac{M_0}{2k^2} & -\dfrac{Q_0}{2k^3} & 0 \\ \dfrac{Q_0}{2k^2} & 0 & \dfrac{M_0}{k} & 0 \\ -M_0 & 0 & 0 & -\dfrac{Q_0}{k} \\ 0 & -Q_0 & 0 & 2kM_0 \end{bmatrix} \begin{Bmatrix} F_1(kX) \\ \\ F_2(kX) \\ \\ F_3(kX) \\ \\ F_4(kX) \end{Bmatrix} \quad \begin{matrix} (a) \\ \\ (b) \\ \\ (c) \\ \\ (d) \end{matrix}
$$

Beyond this particular case, L is often used as a measure of the penetration of the effect of various singularities, such as concentrated loads, holes, and discontinuities, into the interior of a shell. Also, Calladine has established a relationship between the extensional and flexural resisting modes of shells based on the change in Gaussian curvature during distortion. In his scheme, the characteristic $\sqrt{ah}$ term plays a central role.[3]

Note that the edge effects discussed here, Q_0 and M_0, are *self-equilibrating* with respect to the overall equilibrium of the shell. The corresponding dissipation of these edge disturbances may be regarded as a demonstration of St. Venant's principle, which we have mentioned in connection with the derivation of the Kirchhoff boundary conditions in Section 8.2.3. Recall that there may be another type of edge effect which is *not* self-equilibrated and readily penetrates to the opposite boundary. An example is the axial line load $\overline{N}$ shown in Figure 4–12(b) and discussed in Section 4.4.2. This is clearly shown by Equation (12.13c) with $N_X(0) = \overline{N}$. We should be careful to reserve the semi-infinite simplification for cases with self-equilibrating edge loads, such as Q_0 and M_0.

Finally, with respect to the problem illustrated in Figure 12–1, we may obtain explicit expressions for the stress resultants from equations (12.10a–c), (12.18), and Matrix 12–1:

$$
M_X(X) = -DD_{n,XX} = M_0 F_1(kX) + \frac{Q_0}{k} F_4(kX) \tag{12.26a}
$$

$$
M_\theta(X) = \mu M_X \tag{12.26b}
$$

$$
Q_X(X) = -DD_{n,XXX} = Q_0 F_2(kX) - 2kM_0 F_4(kX) \tag{12.26c}
$$

$$
N_\theta(X) = \frac{Eh}{a} D_n(X) = \frac{-Eh}{2ak^2 D}\left[M_0 F_2(kX) + \frac{Q_0}{k} F_3(kX) \right] \tag{12.26d}
$$

This solution is utilized frequently in the ensuing examples.

A closely related case is for the cylinder to have an applied end moment of M_0 and a normal displacement $D_n(0) = 0$. This solution is useful for the case where the end of a cylindrical shell is capped by a circular plate. The radial rigidity of the plate would generally be far greater than that of the shell so that the $D_n = 0$ boundary condition would be a reasonable approximation while M_0 would be determined from the rotational compatibility condition (see Section 11.2.6 and Exercise 13.13).

12.3.3 Circumferential Line Loading

Consider a shell subjected to a circumferential line load P (force/length), as shown in Figure 12–3(a), and assume that the shell extends a distance of at least π/k in each direction from the point of application of P. A free-body diagram of the narrow ring under the load, Figure 12–3(b), reveals that the problem reduces to the case of an edge-loaded semi-infinite shell, similar to that treated in the previous section, with boundary conditions

$$Q_0 = Q_X(0) = -\frac{P}{2} \tag{12.27a}$$

$$D_{X\theta}(0) = -D_{n,X}(0) = 0 \tag{12.27b}$$

From equations (12.27a) and (12.27b) and Matrix 12–1(b), we have

$$M_0 = -\frac{Q_0}{2k} = \frac{P}{4k} \tag{12.27c}$$

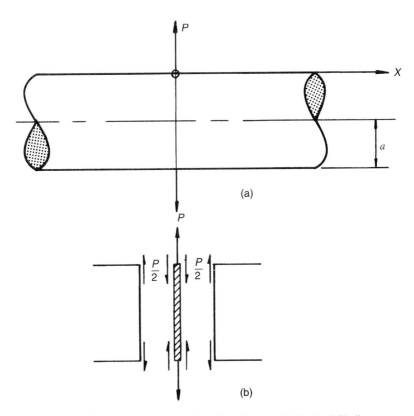

(a)

(b)

FIGURE 12–3   Symmetric Line Loading on a Cylindrical Shell.

The displacement function and stress resultants are written by substituting equations (12.27a) and (12.27c) into Matrix 12–1 and equations (12.26a–d), and then simplifying the ensuing expressions using equations (12.24c) and (12.24d):

$$D_n(X) = \frac{-P}{4k^3D}\left[\frac{1}{2}F_2(kX) - F_3(kX)\right] = \frac{P}{8k^3D}F_1(kX) \tag{12.28a}$$

$$M_X(X) = \frac{P}{2k}\left[\frac{1}{2}F_1(kX) - F_4(kX)\right] = \frac{P}{4k}F_2(kX) \tag{12.28b}$$

$$M_\theta(X) = \mu M_X(X) \tag{12.28c}$$

$$Q_X(X) = -\frac{P}{2}[F_2(kX) + F_4(kX)] = -\frac{P}{2}F_3(kX) \tag{12.28d}$$

$$N_\theta(X) = \frac{Eh}{a}D_n(X) \tag{12.28e}$$

Under the load at $X = 0$, we find the maximum values

$$D_n(0) = \frac{P}{8k^3D} \tag{12.29a}$$

$$M_X(0) = \frac{P}{4k} \tag{12.29b}$$

$$Q_X(0) = -\frac{P}{2} \tag{12.29c}$$

$$N_\theta(0) = \frac{PEh}{8ak^3D} = \frac{Pak}{2} \tag{12.29d}$$

These effects die out with increasing X, as previously indicated.

This solution is of interest in the case of a cylindrical shell with a circular ring stiffener, which we consider later, and also in generating solutions for loading distributed in the X direction by using the Green's function approach.

12.3.4   Axially Distributed Loading

We now consider a radial load uniformly distributed around the circumference but arbitrarily distributed along the X-axis, as shown in Figure 12–4. To treat this problem from a general standpoint, we choose the Green's function technique. In this approach, the response of the shell to a single concentrated load acting at a *point of application, C*, which is arbitrarily located within the loaded region, is studied first. The influence of this load is evaluated at a specified *point of observation*. Then, the effect of the entire distributed load at the point of observation is computed by integrating over the loaded region. In this general example, we must treat two separate cases: (a) point of observation *outside* the loaded region of the shell, such as point A in Figure 12–4; and, (b) point of observation *within* the loaded region of the shell, such as point B in Figure 12–4.

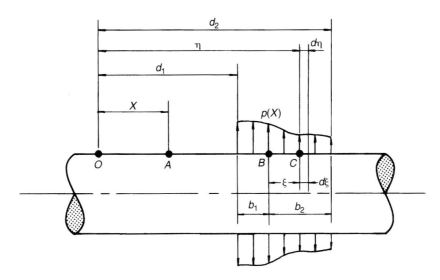

FIGURE 12–4   Distributed Symmetrical Loading on a Cylindrical Shell.

We first select the point of observation as A, outside the loaded region at a distance X from a convenient origin O. We then define the line load produced by $p(\eta)$ acting over the differential length $d\eta$ at the point of application C as $p(\eta)d\eta$. The resulting displacement at A is given by equation (12.28a), with the argument of $F_1(kX)$ taken as the distance between A and C, $\eta - X$:

$$D_n(A,C) = D_n(X,\eta) = \frac{p(\eta)d\eta}{8k^3D} F_1(k[\eta - X]) \tag{12.30}$$

Then, the displacement caused by the entire distributed loading is found as

$$D_n(A) = D_n(X) = \int_{\eta - d_1}^{\eta - d_2} D_n(X,\eta) \tag{12.31}$$

which is easily evaluated once $p(X)$ [or $p(\eta)$] is specified.

Second, locate the point of observation at B, within the loaded region. Point B is a distance $X = d_1 + b_1$, from the origin O, and is a distance ξ from the point of application C. The displacement at B due to the line load $p(\xi)d\xi$ is

$$D_n(B,C) = D_n(X,\xi) = \frac{p(\xi)d\xi}{8k^3D} F_1(k\xi) \tag{12.32}$$

and the displacement of B resulting from the entire load is found to be

$$D_n(B) = D_n(X) = \int_{\xi=0}^{\xi=b_1} D_n(X,\xi) + \int_{\xi=0}^{\xi=b_2} D_n(X,\xi) \tag{12.33}$$

where $b_2 = d_2 - d_1 - b_1$. For both integrals in the preceding equation, the coordinate ξ is positive as measured from point B.

For a point of observation at the edge of the loaded region, A and B are coincident and equations (12.31) and (12.33) should be identical. With $b_1 = 0$, $b_2 = d_2 - d_1$ in equation (12.33), and the identity is confirmed.

The stress resultants and couples for specific loadings $p(X)$ may be evaluated by similar integrations or by subsequent differentiations of the computed $D_n(X)$, as indicated in equations (12.26).

12.3.5 Built-In Shell under Internal Pressure and Temperature Gradient

Next, we examine the cylindrical shell shown in Figure 12–5(a), which is subject to a uniform internal pressure, p, and a linear temperature gradient with T_1 on the outside and T_2 on the inside. The assumption of a linear thermal gradient is generally sufficient for the steady-state condition of thermal conduction but a more elaborate treatment is given in Appendix 8A. Both the pressure and the thermal gradient are taken as constant along the length of the shell and the ends are fixed against translation and rotation. We again assume that L is sufficiently large so that semi-infinite analysis is valid; therefore, only one boundary, say $X = 0$, is considered. A similar problem is considered by Kraus using the more general short-shell solution,[4] but for only a uniform temperature change through the thickness.

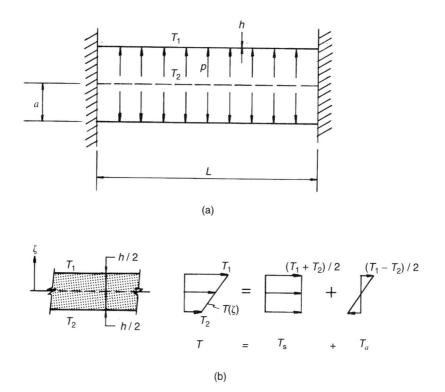

(a)

(b)

FIGURE 12–5   Built-in Cylindrical Shell under Pressure and Thermal Loadings.

Proceeding, we have the homogeneous solution from equation (12.21)

$$D_{nh} = e^{-kX}(C_1 \cos kX + C_2 \sin kX) = C_1 F_3(kX) + C_2 F_4(kX) \qquad (12.34)$$

We now examine equation (12.16a) with respect to the construction of a particular solution and consider each term on the rhs individually.

First, for the pressure term $q_n = p$, we select

$$D_{np1} = \frac{p}{4k^4 D} = \frac{pa^2}{Eh} \qquad (12.35)$$

To find the contribution of the temperature gradient, we first evaluate N_T and M_T from equations (8.9a) and (8.9b). It is helpful to resolve the linear gradient into symmetric and antisymmetric components, as shown in Figure 12–5(b):

$$T(\zeta) = T_s + \frac{\zeta}{h/2} T_a \qquad (12.36a)$$

where

$$T_s = \frac{T_1 + T_2}{2} \qquad (12.36b)$$

$$T_a = \frac{T_1 - T_2}{2} \qquad (12.36c)$$

and ζ = an auxiliary normal coordinate measured from the middle surface. Then, from equation (8.9a) and (8.9b),

$$N_T = \frac{E\bar{\alpha}}{1 - \mu} \int_{-h/2}^{h/2} T(\zeta)d\zeta = \frac{E\bar{\alpha}h}{1 - \mu} T_s \qquad (12.37a)$$

$$M_T = \frac{E\bar{\alpha}}{1 - \mu} \int_{-h/2}^{h/2} T(\zeta)\zeta d\zeta = \frac{E\bar{\alpha}h^2}{6(1 - \mu)} T_a \qquad (12.37b)$$

Now, in equation (12.16a), we have

$$4k^4 D_{np2} = \left[\frac{1 - \mu}{aD}\right]\left[\frac{E\bar{\alpha}h}{1 - \mu} T_s\right] \qquad (12.38)$$

from which

$$D_{np2} = \bar{\alpha}aT_s \qquad (12.39)$$

Because $M_{T,XX} = q_X = 0$ in this case, no particular solutions are required for these terms.

Finally, we have the axial stress resultant term $N_X(0)$, which gives

$$4k^4 D_{np3} = -\frac{\mu}{aD} N_X(0) \qquad (12.40a)$$

or

$$D_{np3} = -\frac{\mu}{4k^4 aD} N_X(0) = -\frac{\mu a}{Eh} N_X(0) \qquad (12.40b)$$

We will evaluate $N_X(0)$ in terms of the temperature gradient later.

Collecting all of the particular solution contributions, we have

$$D_{np} = D_{np1} + D_{np2} + D_{np3} = \frac{pa^2}{Eh} + \bar{\alpha}aT_s - \frac{\mu a}{Eh} N_X(0) \tag{12.41}$$

We may make an interesting observation concerning the particular solution by referring to our treatment of membrane theory displacements for cylindrical shells in Section 8.3.4.1. Specifically, we consider the relevant part of equation (8.53) for the membrane theory normal displacement.

$$D_n = \frac{a}{Eh} (N_\theta - \mu N_X) \tag{12.42a}$$

Now, we substitute the *membrane theory* stress resultant N_θ, as given by equation (4.62d),

$$N_\theta = pa \tag{12.42b}$$

along with

$$N_X = 0 \tag{12.42c}$$

into equation (12.42a), which gives

$$D_n = D_{nm1} = \frac{pa^2}{Eh} \tag{12.42d}$$

which is the same as equation (12.35). For the thermal term, referring to equations (8.9a) and (8.9b), we introduce

$$N_X = N_\theta = N_T = \frac{E\bar{\alpha}h}{1 - \mu} T_s \tag{12.43a}$$

into equation (12.42a). Then,

$$D_n = D_{nm2} = \frac{a}{Eh} \frac{E\bar{\alpha}h}{1 - \mu} T_s(1 - \mu) = \bar{\alpha}aT_s \tag{12.43b}$$

which is identical to equation (12.39). Similarly, with $N_X = N_X(0)$ in equation (12.42a),

$$D_n = D_{nm3} = -\frac{\mu a}{Eh} N_X(0) \tag{12.43c}$$

which matches equation (12.40b). We have thus quantified the earlier assertion that the membrane theory solution frequently serves as a particular solution to the bending theory equations.

We now proceed with the general solution, which is the sum of equations (12.34) and (12.41), by enforcing the boundary conditions

$$D_n(0) = 0 \tag{12.44a}$$

and

$$D_{X\theta}(0) = -D_{n,X}(0) = 0 \tag{12.44b}$$

The first condition gives

$$C_1 + \frac{pa^2}{Eh} + \bar{\alpha}aT_s - \frac{\mu a}{Eh} N_X(0) = 0 \tag{12.45a}$$

while the second condition, after differentiating equation (12.34), yields

$$k(-C_1 + C_2) = 0 \tag{12.45b}$$

whereupon

$$C_1 = -\left[\frac{pa^2}{Eh} + \bar{\alpha}aT_s - \frac{\mu a}{Eh} N_X(0)\right] \tag{12.46a}$$

and

$$C_2 = C_1 \tag{12.46b}$$

To evaluate the term $N_X(0)$ explicitly, we integrate the lhs of equation (12.13d) giving

$$D_X = \int_0^X D_{X,X} dX + D_X(0) \tag{12.47a}$$

with

$$D_{X,X} = \frac{1 - \mu^2}{Eh} [N_T + N_X(0)] \tag{12.47b}$$

when $q_X = 0$. Since the end of the shell is restrained at $X = 0$,

$$D_X(0) = D_{X,X}(0) = 0 \tag{12.47c}$$

and

$$N_X(0) = -N_T = -\frac{E\bar{\alpha}h}{1 - \mu} T_s \tag{12.47d}$$

Then, substituting equation (12.47d) into equations (12.46a) and (12.46b),

$$C_1 = -\left[\frac{pa^2}{Eh} + \left(\frac{1}{1 - \mu}\right)\bar{\alpha}aT_s\right] \tag{12.48a}$$

and

$$C_2 = C_1 \tag{12.48b}$$

Correspondingly, equation (12.41) can be consolidated into

$$D_{np} = \frac{pa^2}{Eh} + \left(\frac{1}{1 - \mu}\right)\bar{\alpha}aT_s \tag{12.49}$$

and together with equation (12.34),

$$D_{nh} = C_1 F_3(kX) + C_2 F_4(kX) \tag{12.50}$$

constitutes the general solution. We also need $D_{X,X}$ which is found from equation (12.13d), in view of equation (12.47d), as

$$D_{X,X} = -\frac{\mu}{a} D_n(X) \tag{12.51}$$

Explicit expressions for the stress resultants and couples may be written by substituting the preceding solution into equations (12.9) and (12.10). This is routine and is omitted for brevity. However, we may surmise from our previous studies of the semi-infinite cylindrical shell that the homogeneous part of the solution D_{nh} will be most influential near the ends and will diminish as X increases, whereas the particular part D_{np}, which is also the membrane theory solution, will predominate away from the ends. Also, note from equations (12.10) that there will be bending in the shell—even after the effects of D_{nh} diminish—because of the constant thermal moment M_T, given by equation (12.37b).

12.3.6 Short Cylindrical Shells

When the distance between boundary points, L, is such that $L < \pi/k$, the semi-infinite assumption is no longer valid and the general solution, (12.20), should be used. We consider the extension of the edge-loaded cylinder problem, shown in Figure 12–1, to the case where line moments and transverse shear forces are applied to both ends (Figure 12–6).

We first rewrite equation (12.20) as

$$D_{nh} = C_1 F_5(kX) + C_2 F_6(kX) + C_3 F_7(kX) + C_4 F_8(kX) \tag{12.52a}$$

where

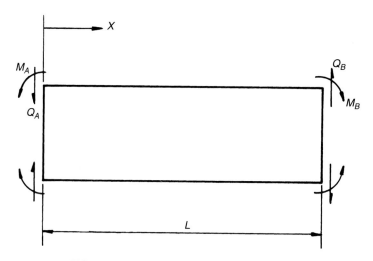

FIGURE 12–6 Edge-Loaded Cylindrical Shell.

$$\left.\begin{array}{l} F_5(kX) = e^{-kX} \cos kX \\ F_6(kX) = e^{-kX} \sin kX \\ F_7(kX) = e^{kX} \cos kX \\ F_8(kX) = e^{kX} \sin kX \end{array}\right\} \tag{12.52b}$$

The boundary conditions for the loading in Figure 12–6 are

$$\left.\begin{array}{l} M_X(0) = -DD_{n,XX}(0) = M_A \\ Q_X(0) = -DD_{n,XXX}(0) = Q_A \\ M_X(L) = -DD_{n,XX}(kL) = M_B \\ Q_X(L) = -DD_{n,XXX}(kL) = Q_B \end{array}\right\} \tag{12.53}$$

We now turn to the tedious calculation of the derivatives of D_{nh}. We first evaluate

$$\begin{aligned} D_{nh,X} = & -C_1 k[F_5(kX) + F_6(kX)] + C_2 k[F_5(kX) - F_6(kX)] \\ & + C_3 k[F_7(kX) - F_8(kX)] + C_4 k[F_7(kX) + F_8(kX)] \end{aligned} \tag{12.54}$$

and then continue with the differentiation in operator notation, for example,

$$\mathbf{D}^n = d^n(\ )/dX^n$$
$$F_5(kX) = F_5$$
$$\mathbf{D}^1 F_5 = -kF_5 - kF_6$$
$$\mathbf{D}^2 F_5 = \mathbf{D}^1(\mathbf{D}^1 F_5) = -k\mathbf{D}^1 F_5 - k\mathbf{D}^1 F_6$$
$$F_6(kX) = F_6$$
$$\mathbf{D}^1 F_6 = -kF_6 + kF_5$$
$$\mathbf{D}^2 F_6 = \mathbf{D}^1(\mathbf{D}^1 F_6) = -k\mathbf{D}^1 F_6 + k\mathbf{D}^1 F_5$$
$$\mathbf{D}^3 F_6 = \mathbf{D}^1(\mathbf{D}^2 F_6) = -k\mathbf{D}^2 F_6 + k\mathbf{D}^2 F_5$$
$$F_7(kX) = F_7$$
$$\mathbf{D}^1 F_7 = kF_7 - kF_8$$
$$\mathbf{D}^2 F_7 = \mathbf{D}^1(\mathbf{D}^1 F_7) = k\mathbf{D}^1 F_7 - k\mathbf{D}^1 F_8$$
$$\mathbf{D}^3 F_7 = \mathbf{D}^1(\mathbf{D}^2 F_7) = k\mathbf{D}^2 F_7 - k\mathbf{D}^2 F_8$$
$$F_8(kX) = F_8$$
$$\mathbf{D}^1 F_8 = kF_8 + kF_7$$
$$\mathbf{D}^2 F_8 = \mathbf{D}^1(\mathbf{D}^1 F_8) = k\mathbf{D}^1 F_8 + k\mathbf{D}^1 F_7$$
$$\mathbf{D}^3 F_8 = \mathbf{D}^1(\mathbf{D}^2 F_8) = k\mathbf{D}^2 F_8 + k\mathbf{D}^2 F_7$$

By repeated substitution, the various derivatives can be written in terms of $F_5 - F_8$:

$$\mathbf{D}^1 F_5 = -k(F_5 + F_6)$$
$$\mathbf{D}^2 F_5 = 2k^2 F_6$$
$$\mathbf{D}^3 F_5 = 2k^3(F_5 - F_6)$$

$$\mathbf{D}^1 F_6 = k(F_5 - F_6)$$
$$\mathbf{D}^2 F_6 = -2k^2 F_5$$
$$\mathbf{D}^3 F_6 = 2k^3(F_5 + F_6)$$
$$\mathbf{D}^1 F_7 = k(F_7 - F_8)$$
$$\mathbf{D}^2 F_7 = -2k^2 F_8$$
$$\mathbf{D}^3 F_7 = -2k^3(F_7 + F_8)$$
$$\mathbf{D}^1 F_8 = k(F_7 + F_8)$$
$$\mathbf{D}^2 F_8 = 2k^2 F_7$$
$$\mathbf{D}^2 F_8 = 2k^3(F_7 - F_8)$$

Matrix 12–2

$$
\begin{bmatrix}
0 & -1 & 0 & 1 \\
k & k & -k & k \\
F_6(kL) & -F_5(kL) & -F_8(kL) & F_7(kL) \\
k[F_5(kL) & k[F_5(kL) & -k[F_7(kL) & k[F_7(kL) \\
-F_6(kL)] & +F_6(kL)] & +F_8(kL)] & -F_8(kL)]
\end{bmatrix}
\begin{Bmatrix}
C_1 \\ C_2 \\ C_3 \\ C_4
\end{Bmatrix}
= \frac{-1}{2k^2 D}
\begin{Bmatrix}
M_A \\ Q_A \\ M_B \\ Q_B
\end{Bmatrix}
$$

Now, we insert the appropriate derivatives into equation (12.53), noting that $F_5(0) = F_7(0) = 1$ and $F_6(0) = F_8(0) = 0$, to get Matrix 12–2, or

$$[\mathbf{F}]\{\mathbf{C}\} = -\frac{1}{2k^2 D}\{\mathbf{B}\} \qquad (12.55a)$$

from which

$$\{\mathbf{C}\} = -\frac{1}{2k^2 D}[\mathbf{F}^{-1}]\{\mathbf{B}\} \qquad (12.55b)$$

Although $\{\mathbf{C}\}$ may be written explicitly in algebraic form, the resulting expressions are cumbersome; also, in the computer era, it is routine to evaluate $\{\mathbf{C}\}$ by specifying numerical values for the shell parameters and performing the inversion and multiplication. We shall regard the problem as essentially solved at this point and consider an application.

12.3.7 Ring-Stiffened Cylindrical Shells

Ring-stiffened cylindrical tubes are commonly used for pressure vessels, submersible vehicles, and rockets. An example is shown in Figure 2–8(r), and an idealization is depicted in Figure 12–7. The rings are often relatively close together, so that the semi-infinite simplification may *not* be valid.

As the cylinder deforms in the radial direction, the ring will retard the expansion or contraction. The resulting contact force between the cylinder and the stiffener is shown as P in Figure 12–7. This problem is somewhat similar in concept to that shown in Figure 4–4, where the ring beam of a dome is analyzed. In that case, the strain incompatibility at the in-

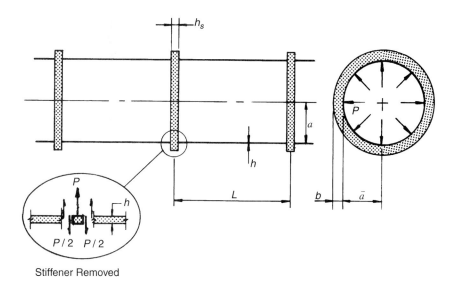

FIGURE 12–7   Ring-Stiffened Cylindrical Shell.

terface of the shell and the beam could not be resolved because of the limitations of the membrane theory. Here, however, we are able to enforce the deformation compatibility between the ring stiffener and the shell.

We assume that the stiffener thickness $h_s \ll L$ so that the reaction can be considered to act on the shell as a radial line load. We take the inside radius of the stiffener as $\bar{a} = a + h/2$, where a is the radius at the shell middle surface and h is the shell thickness. Also, b is the width of the stiffener. In some shells, the stiffeners may be located on the *inside* of the shell, requiring a slight modification of the solution.

The radial deformation of the ring stiffener caused by the line load P (force/length of circumference) is easily computed by considering the stiffener to be a short cylindrical shell with radius = $\bar{a} + b/2$, thickness = b, and length = h_s. Then we apply equation (8.53) which, for the loading and geometry under consideration, reduces to

$$D_n = D_s = \frac{[\bar{a} + (b/2)]}{Eb} N_\theta \tag{12.56a}$$

Assuming the line load to be uniformly distributed over the length h_s, and using equation (4.62c),

$$N_\theta = \frac{P}{h_s}\left(\bar{a} + \frac{b}{2}\right) \tag{12.56b}$$

so that the stiffener deformation in terms of the unknown contact force P is

$$D_s = \frac{P[\bar{a} + (b/2)]^2}{Ebh_s} \tag{12.56c}$$

Now, we examine the forces acting on the shell. We assume the loading is a uniform internal pressure, p, so that the particular solution is given by equation (12.35):

$$D_{np} = D_{np1} = \frac{pa^2}{Eh} \tag{12.57}$$

Also, we have the force P applied to the shell by the ring stiffener. Assuming P is directed outward on the ring, as depicted on the figure, an equal and opposite force reacts on the shell, as shown in the inset. This is the same situation depicted in Figure 12–3, but with the sense reversed. Thus, we may find end conditions from equations (12.27), referring to Figure 12–6 for the correct algebraic signs for $Q_X(0) = Q_A$ and $Q_X(L) = Q_B$:

$$Q_X(0) = \frac{P}{2} \tag{12.58a}$$

$$D_{X\theta}(0) = -D_{n,X}(0) = 0 \tag{12.58b}$$

and

$$Q_X(L) = -\frac{P}{2} \tag{12.58c}$$

$$D_{X\theta}(L) = -D_{n,X}(L) = 0 \tag{12.58d}$$

Substituting equations (12.58a) and (12.58c) into the second and fourth rows of Matrix 12–2, and equations (12.58b) and (12.58d) into equation (12.54) yields four equations in the four unknown integration constants $C_1 - C_4$ and the contact force P. We obtain an additional relationship by equating the radial displacements of the shell and the stiffener:

$$D_n(0) = D_{nh}(0) + D_{np}(0) = D_s \tag{12.59}$$

where D_{nh} is given by equation (12.52a), D_{np} by equation (12.57), and D_s by equation (12.56c).

As mentioned previously, the short-shell solution is best carried out by inserting numerical values pertaining to the case at hand. We may, however, easily proceed for the case where $L > \pi/k$ and the semi-infinite solution is valid. We need only the end conditions given by equations (12.58a) and (12.58b), and may immediately write

$$D_{nh}(0) = -\frac{P}{8k^3D} \tag{12.60}$$

from equation (12.29a) with the appropriate sign change. Correspondingly, equation (12.59) becomes

$$-\frac{P}{8k^3D} + \frac{pa^2}{Eh} = \frac{P[\bar{a} + (b/2)]^2}{Ebh_s} \tag{12.61}$$

from which P is found. Then, the remaining stress resultants and couples are routinely calculated.

A slight variation in this problem is noted in Timoshenko and Woinowsky-Krieger.[2] If the shell is closed at both ends, an axial stress resultant

$$N_X = \frac{p\pi a^2}{2\pi a} = \frac{pa}{2}$$

is possible. This contributes to the particular solution given by equation (12.40b). Thus,

$$D_{np3} = -\frac{\mu a}{Eh} N_X(0) = -p\left(\frac{\mu a^2}{2Eh}\right) \tag{12.62}$$

so that in equation (12.57) and the ensuing equations,

$$D_{np} = \frac{pa^2}{Eh}\left(1 - \frac{\mu}{2}\right) \tag{12.63}$$

For loading cases which are nonaxisymmetrical, circumferential shear forces $N_{X\theta}$ are developed between the shell and the stiffeners. An additional compatibility relationship, equating the circumferential displacements of the shell and the stiffener, is needed to evaluate these contact forces, which may be eccentric to the centroid of the stiffener and also produce circumferential bending moments M_θ. In reinforced concrete barrel shells, discussed in Sections 12.4.1 and 12.4.3, the stiffeners are generally integral with the shell and can be treated using an "effective width" concept.[5]

12.4 GENERAL LOADING

12.4.1 Displacement Formulation and Solution

The analysis of cylindrical shells for nonaxisymmetrical surface loading commands a prominent place in the classical literature, and a comprehensive presentation is beyond our objectives in this text. We should mention, however, that the formulation and application of this theory have attracted the contributions of some of the most prominent twentieth-century mechanicians and engineers to the extensive literature on the subject, including L. H. Donnell, J. Kempner, N. J. Hoff and A. Parme in the United States; H. Schorer of Switzerland; A. Aas Jakobsen in Norway, H. Reissner, W. Flügge, U. Finsterwalder, F. Dishinger, and W. Zerna in Germany; J. E. Gibson and R. S. Jenkins in Great Britain; E. Torroja in Spain; and V. V. Novozhilov, A. I. Lur'e, and V. Z. Vlasov in Russia. Comprehensive historical reviews of these works are found in Flügge[6] from the Western standpoint, and in Novozhilov[7] from the Soviet standpoint. Apparently, the Russian work in this field did not have much impact in the West until the publication of the English language translation of Vlasov's monograph.[8] This formulation is applicable to open, as well as to closed, cylindrical shells. As such, one of the first prominent applications was to the design of so-called barrel shell roofs, Figure 2–8(e). Billington[9] provides an interesting account of the development of this form for shell roofs in Germany by Finsterwalder and his associates and the subsequent technology transfer to the United States by A. Tedesko.

We begin the displacement formulation, with transverse shearing strains neglected, by eliminating the transverse shear stress resultants from equations (12.1b) and (12.1c), using equations (12.1d) and (12.1e). We also take $N_{\theta X} = N_{X\theta}$ and $M_{\theta X} = M_{X\theta}$, leaving the three equilibrium equations

$$aN_{X,X} + N_{X\theta,\theta} + q_X a = 0 \tag{12.64a}$$

$$aN_{X\theta,X} + N_{\theta,\theta} + M_{X\theta,X} + \frac{1}{a}M_{\theta,\theta} + q_\theta a = 0 \tag{12.64b}$$

$$aM_{X,XX} + \frac{1}{a}M_{\theta,\theta\theta} + 2M_{X\theta,X\theta} - N_\theta + q_n a = 0 \tag{12.64c}$$

Next, we substitute equations (12.2a–c) and (12.3a–c) into (12.4) to get the stress resultant-displacement equations

$$N_X = \frac{Eh}{1 - \mu^2}\left[D_{X,X} + \frac{\mu}{a}(D_{\theta,\theta} + D_n)\right] - N_{XT} \tag{12.65a}$$

$$N_\theta = \frac{Eh}{1 - \mu^2}\left[\mu D_{X,X} + \frac{1}{a}(D_{\theta,\theta} + D_n)\right] - N_{\theta T} \tag{12.65b}$$

$$N_{X\theta} = \frac{Eh}{2(1 - \mu)}\left[D_{\theta,X} + \frac{1}{a}D_{X,\theta}\right] \tag{12.65c}$$

$$M_X = -D\left[D_{n,XX} + \frac{\mu}{a^2}(D_{n,\theta\theta} - D_{\theta,\theta})\right] - M_{XT} \tag{12.65d}$$

$$M_\theta = -D\left[\mu D_{n,XX} + \frac{1}{a^2}(D_{n,\theta\theta} - D_{\theta,\theta})\right] - M_{\theta T} \tag{12.65e}$$

$$M_{X\theta} = D\left(\frac{1 - \mu}{a}\right)\left[\frac{1}{2}D_{\theta,X} - D_{n,X\theta}\right] \tag{12.65f}$$

Then, these expressions for the resultants and couples are inserted into equations (12.64a–c) to obtain equilibrium equations in terms of the displacements. Following equations (8.9), we set $N_{XT} = N_{\theta T} = N_T$ and $M_{XT} = M_{\theta T} = M_T$.

$$D_{X,XX} + \left(\frac{1 - \mu}{2a^2}\right)D_{X,\theta\theta} + \left(\frac{1 + \mu}{2a}\right)D_{\theta,X\theta} + \left(\frac{\mu}{a}\right)D_{n,X}$$

$$= \left(\frac{1 - \mu^2}{Eh}\right)(-q_X + N_{T,X}) \tag{12.66a}$$

$$\left(\frac{1 - \mu}{2}\right)D_{\theta,XX} + \left(\frac{1}{a^2}\right)D_{\theta,\theta\theta} + \left(\frac{1 + \mu}{2a}\right)D_{X,X\theta} + \left(\frac{1}{a^2}\right)D_{n,\theta}$$

$$+ \frac{h^2}{12a^2}\left[\left(\frac{1 - \mu}{2}\right)D_{\theta,XX} + \left(\frac{1}{a^2}\right)D_{\theta,\theta\theta} - D_{n,XX\theta} - \left(\frac{1}{a^2}\right)D_{n,\theta\theta\theta}\right] \tag{12.66b}$$

$$= \frac{1 - \mu^2}{Eh}\left(-q_\theta + \frac{N_{T,\theta}}{a} + \frac{M_{T,\theta}}{a^2}\right)$$

$$\frac{h^2}{12}\left[D_{n,XXXX} + \left(\frac{2}{a^2}\right)D_{n,XX\theta\theta} + \left(\frac{1}{a^4}\right)D_{n,\theta\theta\theta\theta}\right]$$

$$-\frac{h^2}{12a^2}\left[\frac{1}{a}D_{\theta,XX\theta} + \left(\frac{1}{a^2}\right)D_{\theta,\theta\theta\theta}\right] + \frac{1}{a^2}\left(\mu a D_{X,X} + D_{\theta,\theta} + D_n\right) \quad (12.66c)$$

$$= \frac{1-\mu^2}{Eh}\left(q_n + \frac{N_T}{a} - M_{T,XX} - \frac{M_{T,\theta\theta}}{a^2}\right)$$

We may simplify equations (12.66b) and (12.66c) by examining the terms multiplied by h^2/a^2. We will drop all such terms as being of $O(h^2/a^2)$:1. At this point, this step involves some presumption. Whereas the first two terms of the $(h^2/12a^2)$ [ ] expression in equation (12.66b) have corresponding terms in the same equation that are of $O(1)$, the remaining terms and all terms contained in $(h^2/12a^2)$ [ ] in equations (12.66c) involve derivatives not found elsewhere in the equation. Nevertheless, this assumption, if not rigorously justified, has been found to give good results for a wide class of problems and greatly simplifies the remaining derivation and solution. A discussion of the possible errors introduced by this step is contained in Kraus.[10]

Insofar as equation (12.66b) is concerned, we may regard the suppression of the $O(h^2/a^2)$ terms as being equivalent to neglecting the influence of the stress couples on the in-plane equilibrium equation. Also, ignoring the $O(h^2/a^2)$ terms in equation (12.66c) is equivalent to replacing the displacement-curvature relationships, equations (12.3), by the corresponding equations for a plate. The latter equations are given by equations (7.55), with

$$\kappa_X = -D_{n,XX} \quad (12.67a)$$

$$\kappa_\theta = -\frac{1}{a^2}D_{n,\theta\theta} \quad (12.67b)$$

$$\tau = -\frac{1}{a}D_{n,X\theta} \quad (12.67c)$$

These interpretations give a reasonable physical basis to the elimination of the $O(h^2/12a^2)$ terms.

We now collect the simplified equations as

$$D_{X,XX} + \left(\frac{1-\mu}{2a^2}\right)D_{X,\theta\theta} + \left(\frac{1+\mu}{2a}\right)D_{\theta,X\theta} + \left(\frac{\mu}{a}\right)D_{n,X} = p_X \quad (12.68a)$$

$$\left(\frac{1-\mu}{2}\right)D_{\theta,XX} + \left(\frac{1}{a^2}\right)D_{\theta,\theta\theta} + \left(\frac{1+\mu}{2a}\right)D_{X,X\theta} + \left(\frac{1}{a^2}\right)D_{n,\theta} = p_\theta \quad (12.68b)$$

$$\frac{h^2}{12}\nabla^4 D_n + \frac{1}{a^2}\left(\mu a D_{X,X} + D_{\theta,\theta} + D_n\right) = p_n \quad (12.68c)$$

where

$$p_X = \frac{1 - \mu^2}{Eh}(-q_X + N_{T,X}) \tag{12.69a}$$

$$p_\theta = \frac{1 - \mu^2}{Eh}\left(-q_\theta + \frac{N_{T,\theta}}{a} + \frac{M_{T,\theta}}{a^2}\right) \tag{12.69b}$$

$$p_n = \frac{1 - \mu^2}{Eh}\left[q_n + \frac{N_T}{a} - \nabla^2(M_T)\right] \tag{12.69c}$$

and

$$\nabla^2(\ ) = (\ )_{,XX} + \frac{1}{a^2}(\ )_{,\theta\theta} \tag{12.69d}$$

is the harmonic operator in the $X - \theta$ cylindrical coordinate system. Kraus[11] has derived a similar set of equations, applicable to noncircular cylindrical shells as well, by neglecting the stress couples in equation (12.64b) and using equations (12.67) for the changes in curvatures at the outset. The only perceptible difference appears to be that the thermal moment gradient term in p_θ, $M_{T,\theta}/a^2$, is not present in his equations.

It is easily verified that for axisymmetric loading, equations (12.68a) and (12.68c) are identical to equations (12.12a) and (12.12b), which reduce ultimately to equations (12.16a) and (12.16b).

The formulation continues by forming a judicious set of mixed partial derivatives as suggested by Kraus:[11] (a) $\partial^2/\partial X^2$ (equation [12.68a]); (b) $\partial^2/\partial\theta^2$ (equation [12.68a]); (c) $\partial^2/\partial X\partial\theta$ (equation [12.68b]); and, (d) solution of (a) and (b) for the mixed partials of D_θ:

$$D_{\theta,XXX\theta} = \frac{2a}{1 + \mu}\left[p_{X,XX} - D_{X,XXXX} - \left(\frac{1 - \mu}{2a^2}\right)D_{X,XX\theta\theta} - \left(\frac{\mu}{a}\right)D_{n,XXX}\right] \tag{12.70a}$$

$$D_{\theta,X\theta\theta\theta} = \frac{2a}{1 + \mu}\left[p_{X,\theta\theta} - D_{X,XX\theta\theta} - \left(\frac{1 - \mu}{2a^2}\right)D_{X,\theta\theta\theta\theta} - \left(\frac{\mu}{a}\right)D_{n,X\theta\theta}\right] \tag{12.70b}$$

and

$$\left(\frac{1 - \mu}{2}\right)D_{\theta,XXX\theta} + \left(\frac{1}{a^2}\right)D_{\theta,X\theta\theta\theta} + \left(\frac{1 + \mu}{2a}\right)D_{X,XX\theta\theta} + \left(\frac{1}{a^2}\right)D_{n,X\theta\theta} - p_{\theta,X\theta} = 0 \tag{12.70c}$$

We now eliminate the first two terms in equation (12.70c) by using equations (12.70a) and (12.70b), and consolidate the remaining terms over the common denominator $a(1 + \mu)$ to get

$$\nabla^4 D_X = -\left(\frac{\mu}{a}\right)D_{n,XXX} + \left(\frac{1}{a^3}\right)D_{n,X\theta\theta} + p_{X,XX} + \frac{2}{a^2(1 - \mu)}p_{X,\theta\theta} - \frac{(1 + \mu)}{a(1 - \mu)}p_{\theta,X\theta} \tag{12.71}$$

Equation (12.71) becomes the governing equation for D_X, once D_n has been determined.

We may derive a similar equation for D_θ by interchanging equations (12.68a) and (12.68b) in the aforementioned mixed partial operations (a), (b), and (c) and solving (a) and (b) for the mixed partials of D_X. This produces

$$D_{X,XXX\theta} = \frac{2a}{1 + \mu} \left[p_{\theta,XX} - \left(\frac{1 - \mu}{2}\right) D_{\theta,XXXX} - \left(\frac{1}{a^2}\right) D_{\theta,XX\theta\theta} - \left(\frac{1}{a^2}\right) D_{n,XX\theta} \right] \tag{12.72a}$$

$$D_{X,X\theta\theta\theta} = \frac{2a}{1 + \mu} \left[p_{\theta,\theta\theta} - \left(\frac{1 - \mu}{2}\right) D_{\theta,XX\theta\theta} - \left(\frac{1}{a^2}\right) D_{\theta,\theta\theta\theta\theta} - \left(\frac{1}{a^2}\right) D_{n,\theta\theta\theta} \right] \tag{12.72b}$$

and

$$D_{X,XXX\theta} + \left(\frac{1 - \mu}{2a^2}\right) D_{X,X\theta\theta\theta} + \left(\frac{1 + \mu}{2a}\right) D_{\theta,XX\theta\theta} + \left(\frac{\mu}{a}\right) D_{n,XX\theta} - p_{X,X\theta} = 0 \tag{12.72c}$$

Eliminating the mixed partials of D_X from equation (12.72c) and clearing, we have

$$\nabla^4 D_\theta = -\left(\frac{2 + \mu}{a^2}\right) D_{n,XX\theta} - \left(\frac{1}{a^4}\right) D_{n,\theta\theta\theta} - \frac{(1 + \mu)}{a(1 - \mu)} p_{X,X\theta}$$

$$+ \left(\frac{2}{1 - \mu}\right) p_{\theta,XX} + \left(\frac{1}{a^2}\right) p_{\theta,\theta\theta} \tag{12.73}$$

which becomes the governing equation for D_θ, once D_n has been determined.

Finally, we may isolate the normal displacement D_n. The procedure is: (a) $\partial/\partial X$ (equation [12.71]); (b) $\partial/\partial\theta$ (equation [12.73]); and, (c) ∇^4 (equation [12.68c]). These operations give

$$(\nabla^4 D_X)_{,X} = \nabla^4(D_{X,X}) = -\left(\frac{\mu}{a}\right) D_{n,XXXX} + \left(\frac{1}{a^3}\right) D_{n,XX\theta\theta} + p_{X,XXX}$$

$$+ \frac{2}{a^2(1 - \mu)} p_{X,X\theta\theta} - \frac{(1 + \mu)}{a(1 - \mu)} p_{\theta,XX\theta} \tag{12.74a}$$

$$(\nabla^4 D_\theta)_{,\theta} = \nabla^4(D_{\theta,\theta})$$

$$= -\left(\frac{2 + \mu}{a^2}\right) D_{n,XX\theta\theta} - \left(\frac{1}{a^4}\right) D_{n,\theta\theta\theta\theta}$$

$$- \frac{(1 + \mu)}{a(1 - \mu)} p_{X,X\theta\theta} + \left(\frac{2}{1 - \mu}\right) p_{\theta,XX\theta} + \left(\frac{1}{a^2}\right) p_{\theta,\theta\theta\theta} \tag{12.74b}$$

and

$$\frac{h^2}{12} \nabla^8 D_n + \frac{1}{a^2} \left[\mu a \nabla^4(D_{X,X}) + \nabla^4(D_{\theta,\theta}) + \nabla^4(D_n) \right] = \nabla^4 p_n \tag{12.74c}$$

After substituting equations (12.74a) and (12.74b) into equations (12.74c), we have the desired equation for D_n:

$$\frac{h^2}{12} \nabla^8 D_n + \left(\frac{1 - \mu^2}{a^2}\right) D_{n,XXXX}$$

$$= \nabla^4 p_n - \frac{1}{a} \left[\mu p_{X,XXX} - \frac{1}{a^2} p_{X,X\theta\theta} + \frac{(2 + \mu)}{a} p_{\theta,XX\theta} + \left(\frac{1}{a^3}\right) p_{\theta,\theta\theta\theta} \right] \tag{12.75}$$

Equations (12.71), (12.73), and (12.75) are among the most famous equations in the theory of thin shells and are commonly referred to as, alternately, *Donnell's equations, Jenkins's equations,* or *Vlasov's equations.* The eighth-order system is uncoupled for the displacements, so that once D_n is determined from equation (12.75), D_X and D_θ may be found by solving equations (12.71) and (12.73), respectively. These equations are also used, with slight elaboration, to study the dynamic response of circular cylindrical shells.[12]

The appropriate boundary conditions for the foregoing equations are established from Figure 8–4, with $\alpha = X$ and $\beta = \theta$. The relationships in which there are no transverse shearing strains included are applicable here. Note that the equations derived pertain both to open and to closed circular cylindrical shells. For closed shells, the boundary conditions in the θ direction are replaced by continuity conditions of the form $f(0) = f(2\pi)$, where $f(\theta)$ is a function which is continuous at $\theta = 0$.

Before considering some specific applications, it is instructive to make some general comments on the analytical solution of the governing equations for this theory, equations (12.71), (12.73), and (12.75), or, alternately, the coupled equations (12.68a–c). Complete solutions are fairly complicated algebraically, but are extensively documented in specialized works, such as Flügge.[13]

Briefly, the solutions of the homogeneous equations for edge loadings at $X = \bar{X} = $ constant are of interest for both open and closed shells and are obtained by taking the Fourier expansions

$$\begin{Bmatrix} D_{Xh} \\ D_{\theta h} \\ D_{nh} \end{Bmatrix} = \sum_{j=0}^{\infty} \begin{Bmatrix} D_{Xh}^j (X) \cos j\theta \\ D_{\theta h}^j (X) \sin j\theta \\ D_{nh}^j (X) \cos j\theta \end{Bmatrix} \tag{12.76}$$

Note that for closed shells, $D_{ih}(0) = D_{ih}(2\pi)(i = X,\theta,n)$, which serve as the continuity or periodicity conditions. Then, for a general harmonic j, the substitution of equations (12.76) into the governing equations gives an eighth-order algebraic system. The solution ultimately takes the general form

$$\begin{aligned}
R^j(X,\theta) = C_G \Big\{ & C_1^j[e^{-\theta_1 x}(f_1^j \cos g_3 x - f_2^j \sin g_3 x)] \\
+ & C_2^j[e^{-\theta_1 x}(f_2^j \cos g_3 x + f_1^j \sin g_3 x)] \\
+ & C_3^j[e^{-\theta_2 x}(f_3^j \cos g_4 x - f_4^j \sin g_4 x)] \\
+ & C_4^j[e^{-\theta_2 x}(f_4^j \cos g_4 x + f_3^j \sin g_4 x)] \\
+ & C_5^j[e^{\theta_1 x}(f_1^j \cos g_3 x + f_2^j \sin g_3 x)] \\
+ & C_6^j[e^{\theta_1 x}(-f_2^j \cos g_3 x + f_1^j \sin g_3 x)] \\
+ & C_7^j[e^{\theta_2 x}(f_3^j \cos g_4 x + f_4^j \sin g_4 x)] \\
+ & C_8^j[e^{\theta_2 x}(-f_4^j \cos g_4 x + f_3^j \sin g_3 x)] \Big\} \begin{pmatrix} \cos j\theta \\ \text{or} \\ \sin j\theta \end{pmatrix}
\end{aligned} \tag{12.77}$$

in which $R^j(X,\theta) = $ a typical displacement, stress resultant, or stress couple for harmonic j; $x = X/a$; $g_1 - g_4$ are quantities dependent on j, the dimensions of the shell, and the material properties; $f_1^j - f_4^j$ are quantities dependent on j, the dimensions of shell, and the material properties for each individual R^j; C_G is a constant dependent on the shell geometry and material properties; and $C_1^j - C_8^j$ are constants of integration.

In the general case, all eight constants are needed, and hence eight boundary conditions in the X direction are required, whereas for a semi-infinite shell, those terms associated with $C_6^j - C_8^j$ are suppressed, and therefore only four boundary conditions are needed.

For an edge loading at $X = \overline{X}$, such as we studied for axisymmetric loading, the integration constants are determined by first substituting the respective displacements, $R^j(X,\theta)$, into the stress resultant-displacement relations, equations (12.65); and then setting those expressions corresponding to the specified edge loadings equal to their boundary values, $N_X(\overline{X})$, $\overline{N}_{X\theta}(\overline{X})$, $\overline{Q}_X(\overline{X})$, and/or $\overline{M}_X(\overline{X})$. Additional equations may be obtained by the specification of kinematic boundary conditions $D_X(\overline{X})$, $D_\theta(\overline{X})$, $D_n(\overline{X})$, and/or $D_{\theta X}(\overline{X})$. For a semi-infinite shell, $\overline{X}$ corresponds to $X = 0$, whereas for a complete shell, $\overline{X}$ stands for $X = L$ as well.

Next, we consider the homogeneous solutions for a shell with an edge loading applied at a boundary $\theta = \overline{\theta} = $ constant. Obviously, this case is pertinent only for open cylindrical shells, such as those discussed in Section 6.2. Here, series solutions of the form

$$\begin{Bmatrix} D_{Xh} \\ D_{\theta h} \\ D_{nh} \end{Bmatrix} = \sum_{k=0}^{\infty} \begin{Bmatrix} D_{Xh}^k(\theta) \cos k\pi \dfrac{X}{L} \\ D_{\theta h}^k(\theta) \sin k\pi \dfrac{X}{L} \\ D_{nh}^k(\theta) \cos k\pi \dfrac{X}{L} \end{Bmatrix} \tag{12.78}$$

are widely used. An eighth-order algebraic system must be evaluated for each harmonic component k from which the solution may be expressed in the same form as equation (12.77), with j replaced by k, and the coordinates X and θ interchanged:

$$\begin{aligned}
R^k = C_G \Big\{ & C_1^k[e^{-g_1\theta}(f_1^k \cos g_3\theta - f_2^k \sin g_3\theta) \\
& + C_2^k[e^{-g_1\theta}(f_2^k \cos g_3\theta + f_1^k \sin g_3\theta)] \\
& + C_3^k[e^{-g_2\theta}(f_3^k \cos g_4\theta - f_4^k \sin g_4\theta) \\
& + C_4^k[e^{-g_2\theta}(f_4^k \cos g_4\theta + f_3^k \sin g_4\theta)] \\
& + C_5^k[e^{g_1\theta}(f_1^k \cos g_3\theta + f_2^k \sin g_3\theta)] \\
& + C_6^k[e^{g_1\theta}(-f_2^k \cos g_3\theta + f_1^k \sin g_3\theta)] \\
& + C_7^k[e^{g_2\theta}(f_3^k \cos g_4\theta + f_4^k \sin g_4\theta)] \\
& + C_8^k[e^{g_2\theta}(-f_4^k \cos g_4\theta + f_3^k \sin g_3\theta)] \Big\}
\end{aligned} \begin{pmatrix} \cos k\pi \dfrac{X}{L} \\ \text{or} \\ \sin k\pi \dfrac{X}{L} \end{pmatrix} \tag{12.79}$$

Again, the effects at opposite boundaries may be uncoupled by using the semi-infinite approach. Because the coordinate lines normal to the loaded boundaries are curved in this case, the effects of the boundary forces may be expected to attenuate more rapidly than those for the previously studied case, in which in-plane forces $\overline{N}_X(\theta,\overline{X})$ applied at $X = \overline{X}$ can propagate along straight lines. After the integration constants are obtained from a specified combination of static and kinematic edge conditions, the homogeneous open-shell solution is completely described. The possible static edge conditions involve

$$N_\theta(\bar\theta), \bar N_{\theta X}(\bar\theta), \bar Q_\theta(\bar\theta), \quad \text{and/or} \quad \bar M_\theta(\bar\theta)$$

whereas the kinematic conditions refer to

$$D_\theta(\bar\theta), D_X(\bar\theta), \quad D_n(\bar\theta), \quad \text{and/or} \quad D_{\theta X}(\bar\theta)$$

Such solutions have been used extensively in the analysis of open cylindrical shell roofs, and extensive tabulations are found in *Design of Cylindrical Concrete Roofs*.[14]

We now consider particular solutions for various applied loadings and thermal effects. A general approach is to take

$$\begin{Bmatrix} D_{Xp} \\ D_{\theta p} \\ D_{np} \end{Bmatrix} = \sum_{j=0}^\infty \sum_{k=0}^\infty \begin{Bmatrix} D_{Xp}^{jk}(\theta) \cos k\pi \dfrac{X}{L} \cos j\theta \\ D_{\theta p}^{jk}(\theta) \sin k\pi \dfrac{X}{L} \sin j\theta \\ D_{np}^{jk}(\theta) \cos k\pi \dfrac{X}{L} \cos j\theta \end{Bmatrix} \tag{12.80}$$

Such solutions may be combined with the edge-load solutions generated from equations (12.76) and (12.78) to solve a wide range of cylindrical shell problems.

Recall from our earlier calculation of the membrane theory displacements for cylindrical shells (Section 8.3.4.1) that the solution for the example considered was precisely in the format of equations (12.80), with $j = k = 1$. This strongly suggests that the membrane theory solution will serve as a particular solution for the bending theory equations in many instances.

Finally, in our general discussion of solution procedures, we outline the technique of Vlasov,[15] which is very attractive for certain problems. Starting with the three coupled equations, (12.68a–c), and using a nondimensional axial coordinate X/L, he introduces a stress function consisting of a homogeneous part and three particular solutions

$$\Phi = \Phi_h + \Phi_{pX} + \Phi_{p\theta} + \Phi_{pn} \tag{12.81}$$

from which the three displacements D_X, D_θ, and D_n, as well as the stress resultants and couples, can be found by differentiation. The particular solutions correspond to the individual surface loading components p_X, p_θ, and p_n, respectively, and the resulting equation is in the same form as equation (12.75) with Φ replacing D_n. He specializes the equation for a normal loading $p_n(X,\theta)$ and further stipulates that the shell is simply supported on all four sides. Referring to Section 8.2.2, this corresponds to boundaries such that where $X = \bar X =$ constant, $N_X = M_X = D_\theta = D_n = 0$; and where $\theta = \bar\theta =$ constant, $N_\theta = M_\theta = D_X = D_n = 0$. Next, he uses the Navier approach, which is familiar from plate solutions, to write

$$\Phi = \sum_{j=1}^\infty \sum_{k=1}^\infty \Phi^{jk} \sin j\pi \frac{\theta}{\bar\theta} \sin k\pi \frac{X}{\bar X} \tag{12.82}$$

This stress function automatically satisfies the simply supported boundary conditions. Φ^{jk} is then determined so as to satisfy the governing equation for each harmonic jk.

The solution by Vlasov's technique is fairly simple to implement, although we do not pursue the details here. An interesting and useful result is the solution for a cylindrical shell under a concentrated load, which isolates the Green's function for such structures.[16]

12.4.2 Column-Supported Cylindrical Shells

We now consider a cylindrical tank resting on a concentric ring of equispaced columns, as shown in Figure 12–8. The shell carries an axisymmetric loading $q(X)$, and the transition between the shell and the columns is assumed to be facilitated by a ring beam. To avoid ambiguity in the subsequent development, it is convenient to set the origin for X at the top of the ring beam and to presume that the ring beam depth $B \ll L$. Thus, the tops of the supporting columns are also located at $X = 0$.

This situation is similar to that shown in Figure 5–11, except that, of course, here the shell is cylindrical. Whereas we were able to treat the spherical geometry by using the membrane theory, a membrane description is inadequate for the column-supported cylindrical shell. This is illustrated in Section 4.4.2 and by Figure 4–12(b), which shows that a force $\overline{N}$ applied to one end simply propagates along the straight meridian in the absence of bending.

Now, we turn to the problem at hand. The representation of the discrete supports is identical to that shown in Figure 5–11, with $\phi_b = \pi/2$ and $R_0(\phi_b) = a$. Here, we have X in place of ϕ for the meridional coordinate. Case (5–11 [b]) of the superposition is described by

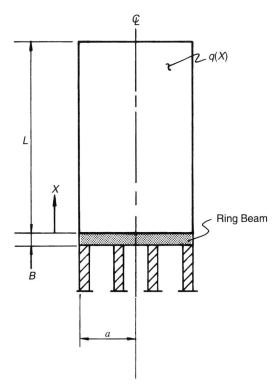

FIGURE 12–8 Column-Supported Cylindrical Shell.

a particular solution, which can be taken as the membrane theory solution, and a homogeneous solution, which satisfies the boundary conditions. The in-plane stress resultants for the particular solution are calculated from equations (4.58) and (4.62a), with the appropriate coordinate modifications:

$$N_{X(b)p}(X) = \frac{Q(X)}{2\pi a} \tag{12.83a}$$

$$N_{\theta(b)p}(X) = q_n(X)a \tag{12.83b}$$

The homogenious solution for case (5–11[b]) is obtained from the treatment of axisymmetrically loaded semi-infinite cylindrical shells in Section 12.3.2 as equation (12.21).

If the tank is supported by a circumferentially rigid ring beam, the appropriate boundary conditions would be

$$D_n(0) = 0 \tag{12.84}$$

and either

$$D_{X\theta}(0) = -D_{n,X}(0) = 0 \tag{12.85a}$$

or

$$M_X(0) = M_0 = 0 \tag{12.85b}$$

depending on whether the ring beam is assumed to prevent or to permit rotations about the circumference. The condition for an actual shell would probably fall somewhere in between the two idealized situations. We will proceed with the former for purposes of illustration.

To find C_1 and C_2 in equation (12.21), the expressions for D_n and $D_{X\theta}$ evaluated at $X = 0$ are substituted into equations (12.84) and (12.85a). In turn, $D_n(0)$ and $D_{X\theta}(0)$ are composed of (a) the particular solution contributions obtained by substituting equations (12.83a) and (12.83b), evaluated at $X = 0$, into equations (8.53) and (8.54); and, (b) the homogeneous solution contributions, found from equation (12.21). Alternately, if boundary condition (12.85b) is selected, $M_X(0)$ comes from equations (12.22a), whereupon C_1 and C_2 may be found.

To complete case (5–11[b]) of the superposition, $M_{X(b)}(X)$, $M_{\theta(b)}(X)$, and $Q_{X(b)}(X)$ are computed from equations (12.26a–c) and $N_{\theta X(b)}(X)$ is the sum of equations (12.83b) and (12.26d). $N_{X(b)}(X)$ remains as the membrane theory value $N_{X(b)p}$, as given in equation (12.83a).

Case (5–11[c]) of the superposition is a homogenous loading condition that is solved following equation (12.76), with the stress resultants, couples, and displacements having the form of equations (12.77). The boundary conditions at $X = 0$ are given by equations (12.84) and (12.85a) or (12.85b), plus two additional equations. First, the ring beam may be regarded as circumferentially rigid, so that

$$D_\theta(0) = 0 \tag{12.86}$$

The final boundary condition at the base is found from the representation of the meridional stress resultants $N_X(0)$ by (a) the negative of the continuous boundary reaction; and, (b) the intensity of the column reactions, as shown on Figure 5–11(c). This condition has been expressed earlier by equation (5.45), which is rewritten for the case at hand as

$$N_{X(c)}(0,\theta) = [-N_{X(b)}(0) + \delta R_{cl}] \tag{12.87}$$

$$\text{where } \delta = 1 \text{ if } m \left(\frac{2\pi}{n_c}\right) - \beta \le \theta \le m \left(\frac{2\pi}{n_c}\right) + \beta$$

$$(m = -n_c/2,\dots,-1,0,1,\dots,n_c/2)$$

$$\delta = 0 \text{ for all other } \theta$$

$$n_c = \text{the total number of columns}$$

$$R_{cl} = \frac{\pi N_{X(b)}(0)}{\beta n_c}$$

As before, equation (12.87) is expanded in a Fourier series in θ

$$N_{X(c)}(0,\theta) = \sum_{j=1}^{\infty} N_{X(c)}^j(0) \cos j\theta \tag{12.88}$$

which gives the Fourier coefficients [see equation (5.50)]

$$N_{X(c)}^j(0) = \frac{2}{j\beta} N_{X(b)}(0) \sin j\beta \quad (j = in_c) \tag{12.89a}$$

and

$$N_{X(c)}^j(0) = 0 \quad (j \ne in_c) \tag{12.89b}$$

Thus, we have expressed the fourth boundary condition at $X = 0$ [equation (12.87)] in the harmonic form

$$N_{X(c)}(0,\theta) = \sum_{j=n_c,2n,\dots}^{\infty} N_{X(c)}^k(0) \cos j\theta \tag{12.90}$$

We presume that the semi-infinite assumption is adequate for our purposes in stress analysis, so that we need not be concerned with the boundary conditions at $X = L$. However, from a practical standpoint, note that open cylindrical tanks are generally stiffened at the top to prevent ovaling caused by the propagation of the concentrated column reactions $N_{X(c)}(0,\theta)$. See Section 7.6.2.

For some specific tank structures, the idealized boundary conditions stated in equations (12.84) and in (12.85a) or (12.85b) might warrant refinement. One possibility is to incorporate the ring stiffener model introduced in Section 12.3.7 to generalize equation (12.84). A second possibility concerns the common use of such a tank to contain a liquid or solid. In such vessels, a circular plate or perhaps a shallow cone or other rotational shell bottom would be attached near the ring beam level. Depending on the connection detail, additional radial and, perhaps, rotational restraints would be introduced. The possibilities are many, depending on the actual case in question, and the realistic incorporation of additional constraining elements into the mathematical model may alter the computed stress pattern significantly.[17] The bottom plate or shell, itself, would behave essentially as an axisymmetric member supported by the ring beam. Such structures have been discussed extensively in the preceding chapters.

If the weight of a portion of the contents of the tank is carried directly to the ring beam by a bottom plate or shell, instead of being transferred through the shell wall, then the resultant force $Q(X)$ in equation (12.83a) would *not* include this portion of the loading for the calculation of $N_{X(b)}(X)$ for $X > 0$; however, for case [5–11(c)] of the superposition, where $N_{X(b)}(0)$ is encountered, the *entire* load from the tank must be included in $Q(0)$ for the calculation of $N_{X(b)}(0)$ and the column reaction R_{cl}. For example, if a circular plate bottom were provided at $X = 0$ in the tank shown in Figure 12–8, all of the weight of the internal contents would be carried directly to the ring beam if no wall friction was assumed. This would leave only the dead weight of the tank and roof in $Q(X)$, $X > 0$, but the weight of the contents would be added for $Q(0)$ in case 5–11(c).

With the boundary conditions for the idealized problem established as equations (12.84), (12.85a) or (12.85b), (12.86), and (12.90), we set the specific expressions for $R^j = D_n^j$, $D_{X\theta}^j$ or M_X^j, D_θ^j, and N_X^j [which are in the form of equation (12.77)] equal to the boundary values at $X = 0$, which are 0, 0, 0, and $N_{X(c)}^j(0)$, respectively. Then the integration constants for harmonic j, $C_1^j - C_4^j$, are found from the simultaneous solution of the four equations. The stress resultants, couples, and displacements for harmonic j are then calculated from the corresponding specific form of equation (12.77).

Convergence is established by the progressive diminution of the results from consecutive participating harmonics ($j = n_c, 2n_c, \ldots$). For the usual cases encountered, four or five harmonics should be adequate. The complete results are obtained by summing the stress resultants and couples and displacements obtained from the specific forms of equation (12.77) over the participating harmonics at various circumferential locations θ, and then adding these values to the case [5–11(b)] results.

Example 12.4(a)

Flügge[18] has considered a case with $a/h = 150$, $n_c = 8$, and $\beta = 6°$. This gives an amplification of $360°/(8 \times 2\beta) = 360/96 = 3.75$ for N_X at the base; that is, $R_{cl} = 3.75 N_{Xb}(0)$. The effect of this amplification and the corresponding change in N_θ dies out at about $X = a$ in a similar fashion as Figure 5–12.

In Gould,[19] the column-supported shell model is generalized to represent a *longitudinally distributed,* as opposed to a *line,* attachment between the column and the shell wall. This study is particularly applicable to elevated storage tanks, such as that shown in Figure 2–8(u).

12.4.3 Multiple Barrel Shells

A multiple barrel shell, composed of adjacent segments of circular cylindrical shells, is depicted in Figure 12–9(a), with the positive sense of the coordinates taken from Figure 6–1. The membrane theory analysis of a single shell of this form was conducted in Sections 6.2 and 8.3.4.1.

First, consider the boundary conditions in the X direction. The great majority of shell roofs of this configuration have been designed assuming simply supported boundaries:

$$N_X = M_X = D_\theta = D_n = 0 \quad \text{at } X = 0 \quad \text{and} \quad X = L \tag{12.91}$$

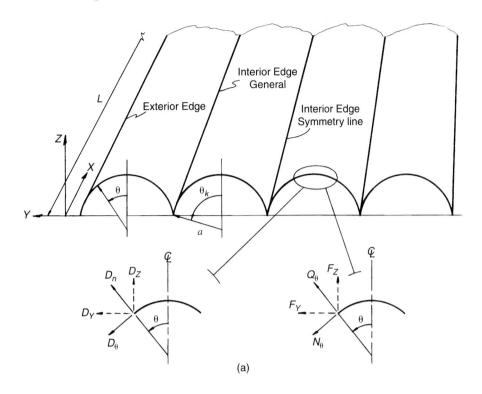

(a)

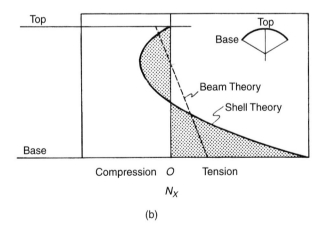

(b)

FIGURE 12–9   Open Cylindrical Shell.

If the support is a wall or frame in the Y–Z plane, it must be sufficiently stiff to prevent displacements in the Y or Z directions; yet, it must be sufficiently flexible to permit displacements along the X-axis and rotations about the Y-axis. Obviously, these requirements are difficult to satisfy exactly in a physical structure, but they are probably the best representation that can be incorporated into an elementary analytical solution.

Now, consider the boundary conditions in the θ direction. Referring to Figure 12–9(a), three situations are identified: exterior, interior-general, and interior-symmetry line. In some cases, the resistance is augmented by stiffening beams spanning the length L, as shown in Figure 2–8(e). Here, we will not consider this situation but only the basic shell; however, the influence of the beams could be incorporated. In a more general form, the exterior edge may be regarded as free so that

$$N_\theta = \overline{N}_{\theta X} = M_\theta = \overline{Q}_\theta = 0 \quad \text{at } \theta = \theta_k \tag{12.92}$$

where $\overline{N}_{\theta X}$ and $\overline{Q}_\theta$ are the effective stress resultants defined in Section 8.2.3. The interior edges are somewhat more complicated to characterize, because they are continuous with the adjacent shell. Clearly, the displacements D_X, D_θ, D_n, and the rotation $D_{\theta X}$ must be compatible between the intersecting shells. Strict enforcement of this condition could result in four simultaneous equations per interior valley line, which greatly complicates the calculations. A simplification, widely used in design, is to treat every interior edge as if it were located on a line of symmetry where the lateral displacement, D_Y, and the rotation about the longitudinal axis, $D_{\theta X}$, vanish. For the coordinate system defined on Figure 12–9(a), these conditions become

$$D_Y(-\theta_k) = 0; \quad D_{\theta X}(-\theta_k) = 0 \tag{12.93a}$$

for the interior edge of the outside barrel shell, and

$$D_Y(\pm\theta_k) = 0; \quad D_{\theta X}(\pm\theta_k) = 0 \tag{12.93b}$$

for all of the interior barrels, where from Figure 12–9(a),

$$D_Y(\theta) = D_\theta \cos\theta + D_n \sin\theta \quad (-\theta_k \le \theta \le \theta_k) \tag{12.93c}$$

If $\gamma_\theta = 0$, the rotation $D_{\theta X}$ is given in terms of D_n and D_θ from equations (12.2h):

$$D_{\theta X} = -\frac{1}{a}(D_{n,\theta} - D_\theta) \tag{12.93d}$$

which is identical to equation (8.55) obtained for the membrane theory rotation.

In addition, we may obtain two static conditions from the symmetry assumption. First, the transverse shear stress resultants $\overline{N}_{\theta X}$ must be equal and opposite on any two coincident edges. When such edges lie along a symmetry line,

$$\overline{N}_{\theta X}(-\theta_k) = 0 \begin{pmatrix} \text{exterior} \\ \text{barrel} \end{pmatrix}; \quad \overline{N}_{\theta X}(\pm\theta_k) = 0 \begin{pmatrix} \text{interior} \\ \text{barrels} \end{pmatrix} \tag{12.94a}$$

is the only possibility. Similarly, any resultant vertical force $\mathbf{F}_Z = F_Z t_Z$ on one edge must have an equal and opposite counterpart on the other edge. Thus, for edges along symmetry lines,

$$F_Z(-\theta_k) = 0 \begin{pmatrix} \text{exterior} \\ \text{barrel} \end{pmatrix}; \quad F_Z(\pm\theta_k) = 0 \begin{pmatrix} \text{interior} \\ \text{barrels} \end{pmatrix} \tag{12.94b}$$

where, from Figure 12–9(a),

$$F_Z = Q_\theta \cos\theta - N_\theta \sin\theta \quad (-\theta_k \le \theta \le \theta_k) \tag{12.94c}$$

Equations (12.91) to (12.94) prescribe the boundary conditions for a simplified bending theory analysis of cylindrical shells, in which only two cases are considered—an exterior and a typical interior barrel. Further, the two cases are uncoupled, thus expeding the calculations.

We first compare the expressions for the membrane theory stress resultants (derived in Section 6.2) and the corresponding displacements (found in Section 8.3.4.1) against these boundary conditions. In particular, consider the case of a uniform dead load q_d (force/area) represented by one term of a Fourier series, as given by equation (6.6b):

$$q_d(X) = \frac{4}{\pi} q_d \sin\pi \frac{X}{L} \tag{12.95}$$

For clarity, all quantities computed from the membrane theory equations will be designated by the subscript m. We check the expressions for the stress resultants, equations (6.10a–c), and find that the static boundary conditions in the X direction, $N_X = M_X = 0$ at $X = 0$ and L, are satisfied. An examination of the membrane theory displacements for this case, given by equations (8.56a–c), reveals that the kinematic boundary conditions in the X direction, $D_\theta = D_n = 0$ at $X = 0$ and L, are similarly satisfied. Thus, the only possible violation of the membrane theory conditions may occur on the $\theta = \pm\theta_k$ boundaries.

Observe from Figure 6–3 that the idealized membrane boundary must develop $N_{\theta m}$ and S_m at $\theta = \pm\theta_k$. This is in clear conflict with the exterior edge condition, equation (12.92), where N_θ and $\overline{N}_{\theta X}$, which corresponds to S_m, are required to vanish. An interior edge can develop only the Y component of $N_{\theta m}$, $F_{Ym} = N_{\theta m}\cos\theta$, as shown on Figure 12–9(a), leaving the Z component, $F_{Zm} = -N_{\theta m}\sin\theta$, unbalanced. With respect to the $N_{\theta X}$ or S_m force, however, a careful examination of Figure 6–3 reveals that the sense of S_m along two adjacent boundaries as computed from the membrane theory solution would be in the same direction, either in the $(+)$ or $(-)$ X direction. On the other hand, two adjacent shells would be expected to develop *equal* and *opposite* shear resultants along their common boundary, because no resultant force along the X axis can exist. Moreover, if the interior valley is a symmetry line as assumed in developing equations (12.94a), then $\overline{N}_{\theta X}$ must be equal to 0. It is apparent, then, that the longitudinal boundary conditions implied for the development of the state of stress predicted by the membrane theory are *grossly* violated.

The basic procedure to rectify these violations consists of starting with the computed membrane theory stress resultants, which are in equilibrium with the applied loading, and the corresponding displacements. Then, corrective edge loads are applied to satisfy the boundary conditions at $\theta = \pm\theta_k$. Mathematically, this is equivalent to adopting the membrane theory solution as the particular solution to the governing equations. This has been shown quantitatively to be a sufficiently close approximation,[20] which is hardly surprising in view of our previous discussions.

With the membrane theory solution taken as the particular solution, we need consider only the homogeneous portion of Donnell's equation for the corrective edge loadings.

Specifically, refer to the solutions given by equations (12.78), which were proposed for loading along edges θ = constant. Notice that these expressions are in the same Fourier series form as our membrane theory solution. This is, of course, no coincidence, but the very reason for taking the membrane theory solutions as Fourier series in the first place. As we stated earlier, the displacements, stress resultants, and couples corresponding to equations (12.78) take the general form of equation (12.79). Now, we further assume that the semi-infinite simplification is applicable, so that the effects at opposite boundaries $\theta = \pm\theta_k$ are *uncoupled*. This leaves only the first half of equation (12.79), with undetermined constants $C_1^k - C_4^k$.

We first consider the exterior edge. The violation of the boundary conditions stated in equation (12.92) are precisely equal to the membrane theory stress resultants evaluated at $\theta = \theta_k$. From equations (6.10a and c),

$$S_m(X,\theta_k) = \frac{8q_dL}{\pi^2} \cos \pi \frac{X}{L} \sin \theta_k \qquad (12.96)$$

and

$$N_{\theta m}(X,\theta_k) = -\frac{4q_da}{\pi} \sin \pi \frac{X}{L} \cos \theta_k \qquad (12.97)$$

Next, from the semi-infinite form of equations (12.79), with $k = 1$, we select expressions for $\overline{N}_{\theta X}^1$, N_θ^1, M_θ^1, and $\overline{Q}_\theta^1$ in terms of the four integration constants $C_1^1 - C_4^1$. Then, we enforce the free edge boundary conditions, equation (12.92), by setting

$$\overline{N}_{\theta X}^1 \left(\frac{X}{L},\theta_k\right) + S_m\left(\frac{X}{L},\theta_k\right) = 0 \qquad (12.98a)$$

$$N_\theta^1 \left(\frac{X}{L},\theta_k\right) + N_{\theta m}\left(\frac{X}{L},\theta_k\right) = 0 \qquad (12.98b)$$

$$M_\theta^1 \left(\frac{X}{L},\theta_k\right) = 0 \qquad (12.98c)$$

$$\overline{Q}_\theta^1 \left(\frac{X}{L},\theta_k\right) = 0 \qquad (12.98d)$$

after which $C_1^1 - C_4^1$ are determined. Note that the functions of (X/L), $\sin \pi(X/L)$ or $\cos \pi (X/L)$, are the same for both the membrane theory and bending theory solutions, so that once the free edge condition is enforced at *one* point in the interval $0 \le X \le L$, it is satisfied at *every* point. This is another advantage of the Fourier series representation.

Once the constants of integration are determined for the exterior edge solution, the stress resultants and couples from the bending theory solution can be evaluated at a sufficient number of coordinates [(X/L), θ] to obtain the stress distribution. Of course, the membrane theory stress resultants are added to $\overline{N}_{\theta X}^1$, N_θ^1, and N_X^1 to get the complete expressions for these functions. Finally, the total displacements may be computed as the sum of the membrane theory values, equations (8.56a–c), and those found from the bending solutions with the constants inserted. This solution is valid in the exterior portion of the shell ($\theta_k \ge \theta \ge 0$).

We have not provided sufficient detail here with respect to the bending solutions to perform calculations on actual shells, but detailed procedures and examples are available (see *Design of Cylindrical Concrete Roofs*[21] and *Billington*[22]). It is often convenient to construct the bending theory corrections to the membrane theory solutions as linear combinations of *unit* load solutions of the type

$$\overline{N}_{\theta X}^1 = 1; N_\theta^1 = M_\theta^1 = \overline{Q}_\theta^1 = 0 \quad \text{at} \left(\frac{X}{L}, \theta_k\right) \tag{12.99a}$$

and

$$N_\theta^1 = 1; \overline{N}_{\theta X} = M_\theta^1 = \overline{Q}_\theta^1 = 0 \quad \text{at} \left(\frac{X}{L}, \theta_k\right) \tag{12.99b}$$

Such solutions are tabulated for a wide range of shell parameters in *Design of Cylindrical Concrete Roofs*.[23]

We now take up the idealized interior edge, referring to the boundary conditions given by equations (12.93) and (12.94). We use an analogous procedure, superimposing the membrane theory solution with bending solutions for edge loadings. The constants of integration are evaluated from the conditions

$$D_Y^1 \left(\frac{X}{L}, \theta_k\right) + D_{Ym} \left(\frac{X}{L}, \theta_k\right) = 0 \tag{12.100a}$$

$$D_{\theta X}^1 \left(\frac{X}{L}, \theta_k\right) + D_{\theta Xm} \left(\frac{X}{L}, \theta_k\right) = 0 \tag{12.100b}$$

$$\overline{N}_{\theta X}^1 \left(\frac{X}{L}, \theta_k\right) + S_m \left(\frac{X}{L}, \theta_k\right) = 0 \tag{12.100c}$$

$$F_Z^1 \left(\frac{X}{L}, \theta_k\right) + F_{Zm} \left(\frac{X}{L}, \theta_k\right) = 0 \tag{12.100d}$$

where the bending components are taken from expressions of the form of equation (12.79), and the membrane theory parts are given by equations (8.56a–e) and (6.10a–c). Also, D_Y and $D_{\theta X}$ are expressed in terms of D_θ and D_n by equations (12.93c) and (12.93d), respectively, and F_Z is stated in terms of Q_θ and N_θ by equation (12.94c). Again, these computations may be expedited by the use of the tables in *Design of Cylindrical Concrete Roofs*.[24]

Finally, with respect to simply supported circular cylindrical shells without intermediate supports, we should assess the relevance of the rather complex analysis we have outlined. We made a start in this direction at the conclusion of Section 6.2. Now we can amplify some of these remarks, because we have, in principle, satisfied the boundary conditions on all edges.

Basically, at every cross-section, the shell must resist the statical bending moment caused by the applied loading, regardless of the theory used. For a uniform load w (force/length), the statical moment at the center line $X = L/2$ is the well-known

$$\overline{M}_X = w\frac{L^2}{8} \tag{12.101}$$

It is interesting to compare the magnitude and distribution of the stress resultant N_X over the cross-section calculated by using the bending theory of shells to the like quantity found from elementary beam theory

$$N_X = \frac{\overline{M}_X Z}{I} \qquad (12.102)$$

where I = the moment of inertia of the cross-section with unit thickness about the centroidal axis. This may be regarded as a gross measure of the relevance of the shell theory, as opposed to beam theory, calculations. In Figure 12–9(b), we show the distribution of N_X over the cross-section for a single barrel shell analyzed in *Design of Cylindrical Concrete Roofs*,[25] along with the straight-line distribution given by equation (12.102). Note that both the maximum tensile and compressive stress resultants are more than double the values computed from the linear strain, beam theory formula. It has been verified that both distributions of N_X yield practically the same statical moment $\overline{M}_X$ about the centroid of the cross-section.[25] Obviously, the analysis of an open cylindrical shell by elementary beam theory is grossly inaccurate, insofar as the elastic response is concerned.

REFERENCES

1. H. Kraus, *Thin Elastic Shells,* (New York: Wiley, 1967): 297–314.

2. S. Timoshenko and S. Woinowsky-Krieger, *Theory of Plates and Shells,* (New York: McGraw-Hill, 1959): 472–481; E. Y. W. Tsui, *Stresses in Shells of Revolution,* (Menlo Park, CA: Pacific Coast Publishers, 1968): 63.

3. C. R. Calladine, "Theory of Shell Structures," (Cambridge: Cambridge University Press 1983); and Gaussian Curvature and Shell Structures, *The Mathematics of Shell Structures* [J. A. Gregory, Ed.] (Oxford: Clarendon Press, 1986): 179–196.

4. Kraus, *Thin Elastic Shells:* 136–139.

5. Design of cylindrical concrete roofs, *ASCE Manual of Engineering Practice no. 31* (New York: American Society of Civil Engineers, 1952).

6. W. Flügge, *Stresses in Shells,* 2nd ed., (Berlin: Springer-Verlag, 1973), chap. 5 and 513–515.

7. V. V. Novozhilov, *Thin Shell Theory* [translated from 2nd Russian ed. by P. G. Lowe] (Groningen, The Netherlands: Noordhoff, 1964); 215–218.

8. V. Z. Vlasov, *General Theory of Shells and Its Application in Engineering,* NASA TTF-99 (Washington, D.C.: National Aeronautics and Space Administration, April 1964).

9. D. P. Billington, *Thin Shell Concrete Structures,* 2nd ed., (New York: McGraw-Hill, 1982): 10–21.

10. Kraus, *Thin Elastic Shells:* 221–229.

11. Ibid: 200–204.

12. Ibid: 297–314.

13. Flügge, Stresses in Shells: 217–250.

14. *Design of Cylindrical Concrete Roofs.*

15. Vlasov, *General Theory of Shells:* 359–394.

16. Ibid: 376–394.

17. P. L. Gould et al., Column supported cylindrical-conical tanks, *Journal of the Structural Division,* ASCE 102, no. ST2 (February 1976): 429–447.

18. Flügge, *Stresses in Shells:* 231–235.

19. Gould et al., *Column supported cylindrical-conical tanks:* 429–447.

20. D. P. Billington, *Thin Shell Concrete Structures,* (New York: McGraw-Hill, 1965): 171–172.

21. *Design of Cylindrical Concrete Roofs.*

22. Billington, *Thin Shell Concrete Design,* Chap. 6.

23. *Design of Cylindrical Concrete Roofs.*

24. Ibid.

25. Ibid., pp: 48–57.

EXERCISES

12.1 (a) Consider Exercise 4.8. Assume that the lower boundary is fixed and determine the bending stress resultants and couples in terms of the radius a, height H, thickness h, Young's modulus E, Poisson's ratio μ, and the unit weight γ.

   (b) Re-solve part (a), except now assume that the lower boundary is hinged.

12.2. Consider the solution for Exercises 12.1(a) and (b) with properties typical of two common engineering materials, and graph the variations of N_θ and M_X for the tank full and half-full. Locate the points of maximum stress.

Material properties

Property	Steel	Concrete	Units
E	30×10^6	3×10^6	force/area
μ	0.25	0.10	
h	0.25	10	length
H	250	250	length
a	250	250	length
γ	0.20	0.08	force/volume

For the English system, length units = inches; area unit = inches^2; volume units = inches^3; force units = pounds.

12.3. Consider the shell shown in Figure 12–5 with the following numerical values. Additionally, E, μ, and h are as listed under *steel* in Exercise 12.2.

$a = 36$	(length);	$L = 360$	(length)
$p = 100$	(force/area)		
$T_1 = 50°F$	$T_2 = 80°F$		

Compute D_n, N_θ, and M_X caused by the pressure and temperature effects separately and in combination. Locate the points of maximum stress by graphing the results.

12.4. Consider the ring-stiffened cylindrical shell shown in Figure 12–7, but with the rings located on the inside of the shell. Derive the equations to evaluate the contact force P.

12.5. Consider the tapered cylindrical shells as shown in Figure 12–10. The shell subjected to hydrostatic pressure with the unit weight of the fluid $= \gamma$.

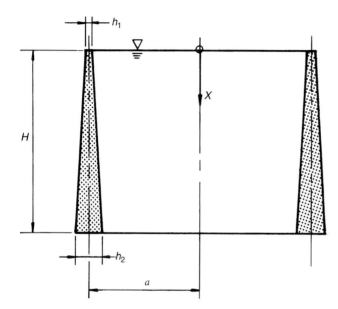

FIGURE 12–10 Exercise 12.5

(a) Derive the general solution for the displacement function, and also expressions for M_X and N_θ.

(b) If the analytical solution becomes too involved, set up a numerical solution for (a).

(c) Consider an alternative to the gradual increase in thickness, whereby the shell is constructed in two constant thickness segments.

$$h_3 = \frac{3h_1 + h_2}{4} \left(0 < X \le \frac{H}{2} \right) \text{ and } h_4 = \frac{h_1 + 3h_2}{4} \left(\frac{H}{2} < X \le H \right).$$

For this configuration, repeat the calculations in (a) and compare the results to those found for the tapered wall.

13

Bending Theory and Stability of Shells

13.1 SHELLS OF REVOLUTION

13.1.1 General

The bending of shells of revolution is a relatively complex subject from a mathematical standpoint. Also, the availability of very efficient and powerful numerically based computer programs has made many of the classical means of solving such problems somewhat archaic. Moreover, the classical solutions are developed in considerable detail in a number of popular monographs (Novozhilov,[1] Függe,[2] and Kraus[3]), and little would be accomplished here by restating these lengthy derivations or by dwelling on the small differences in the various theories, which are attributable to the necessity of introducing simplifying assumptions to facilitate analytical solutions. Instead, we explore only some of the important classical findings, and then present an energy formulation suitable for a finite element solution. In such a solution, the necessity of the simplifying assumptions accompanying the analytical solutions is greatly reduced, making the consideration of the distinctions between the various theories of little practical interest.

13.1.2 Axisymmetrical Loading: Analytical Solutions

13.1.2.1 Governing Equations. An early classical formulation of the shell-bending problem was presented by H. Reissner (the father of E. Reissner), who studied the spherical shell.[4] This "mixed" formulation, generalized by E. Meissner,[5] represents the fourth-order system by a coupled set of two second-order equations in terms of the meridional rotation $D_{\phi\theta}$ and the transverse shear resultant Q_ϕ; here, the derivation is generalized to include specifically orthotropic materials.

We begin by specializing the equilibrium equations, equations (3.24a, c, e), for the case of axisymmetric bending:

$$(R_0 N_\phi)_{,\phi} - R_\phi \cos \phi N_\theta + R_0 Q_\phi + q_\phi R_\phi R_0 = 0 \tag{13.1a}$$

$$(R_0 Q_\phi)_{,\phi} - R_0 N_\phi - R_\phi \sin \phi N_\theta + q_n R_\phi R_0 = 0 \tag{13.1b}$$

$$(R_0 M_\phi)_{,\phi} - R_\phi \cos \phi M_\theta - R_\phi R_0 Q_\phi = 0 \tag{13.1c}$$

We next combine the strain-displacement relationship for axisymmetric shells, equations (7.48a, b, d, e), and the constitutive law, equation (8.12), into a set of stress resultant and stress couple-displacement equations:

$$N_\phi = D_{or11} \left[\frac{1}{R_\phi} (D_{\phi,\phi} + D_n) \right] + D_{or12} \left[\frac{1}{R_0} (\cos \phi D_\phi + \sin \phi D_n) \right] - N_{\phi Tor} \tag{13.2a}$$

$$N_\theta = D_{or12} \left[\frac{1}{R_\phi} (D_{\phi,\phi} + D_n) \right] + D_{or22} \left[\frac{1}{R_0} (\cos \phi D_\phi + \sin \phi D_n) \right] - N_{\theta Tor} \tag{13.2b}$$

$$M_\phi = D_{or44} \left[\frac{1}{R_\phi} D_{\phi\theta,\phi} \right] + D_{or45} \left[\frac{\cos \phi}{R_0} D_{\phi\theta} \right] - M_{\phi Tor} \tag{13.2c}$$

$$M_\theta = D_{or45} \left[\frac{1}{R_\phi} D_{\phi\theta,\phi} \right] + D_{or55} \left[\frac{\cos \phi}{R_0} D_{\phi\theta} \right] - M_{\theta Tor} \tag{13.2d}$$

Symmetry of the constitutive matrix is presumed, so that $D_{or12} = D_{or21}$ and $D_{or45} = D_{or54}$. Also, as before, the thermal terms are presumed to be known at the outset.

The boundary conditions corresponding to the axisymmetric formulation are specialized from Figure 8–4 as $N_\phi = \overline{N}_\phi$ or $D_\phi = \overline{D}_\phi$; $Q_\phi = \overline{Q}_\phi$ or $D_n = \overline{D}_n$; and $M_\phi = \overline{M}_\phi$ or $D_{\phi\theta} = \overline{D}_{\phi\theta}$.

We now substitute equations (13.2c) and (13.2d) into equation (13.1c). To simplify the derivation here somewhat, the elements of the constitutive matrix $[\mathbf{D}_{or}]$ are taken as constant, but they can be generalized for shells with continuously varying material properties and/or thickness.

Continuing, we have

$$D_{\phi\theta,\phi\phi} + \frac{R_\phi}{R_0} \left(\frac{R_0}{R_\phi} \right)_{,\phi} D_{\phi\theta,\phi}$$

$$- \frac{1}{D_{or44}} \left(\frac{R_\phi}{R_0} \right) \left[D_{or45} \sin \phi + D_{or55} \frac{R_\phi \cos^2 \phi}{R_0} \right] D_{\phi\theta} - \frac{R_\phi^2}{D_{or44}} Q_\phi \tag{13.3}$$

$$= \frac{1}{D_{or44}} \left(\frac{R_\phi}{R_0} \right) [(R_0 M_{\phi Tor})_{,\phi} - R_\phi \cos \phi M_{\theta Tor}]$$

which is the first of two governing equations in $D_{\phi\theta}$ and Q_ϕ.

To obtain the second equation, we first consider equations (13.2a) and (13.2b). We form the combinations

$$\frac{R_\phi}{D_{or11}}\left[(13.2a) - \frac{D_{or12}}{D_{or22}}(13.2b)\right]$$

and

$$\frac{R_\theta}{D_{or22}}\left[(13.2b) - \frac{D_{or12}}{D_{or11}}(13.2a)\right]$$

and rewrite the expression

$$(1/R_0)(\cos\phi\, D_\phi + \sin\phi\, D_n) \text{ as } (1/R_\theta)(\cot\phi\, D_\phi + D_n)$$

to get

$$D_{\phi,\phi} + D_n = \frac{R_\phi}{D_{or11}(1 - \mu_1\mu_2)}[(N_\phi + N_{\phi Tor}) - \mu_2(N_\theta + N_{\theta Tor})] \qquad (13.4a)$$

and

$$\cot\phi\, D_\phi + D_n = \frac{R_\theta}{D_{or22}(1 - \mu_1\mu_2)}[-\mu_1(N_\phi + N_{\phi Tor}) + (N_\theta + N_{\theta Tor})] \qquad (13.4b)$$

where

$$\mu_1 = \frac{D_{or12}}{D_{or11}} \quad\text{and}\quad \mu_2 = \frac{D_{or12}}{D_{or22}} \qquad (13.4c)$$

Next, we eliminate D_n between equations (13.4a) and (13.4b) to find

$$D_{\phi,\phi} - \cot\phi\, D_\phi = \frac{1}{(1 + \mu_1\mu_2)}\left[\left(\frac{R_\phi}{D_{or11}} + \frac{\mu_1 R_\theta}{D_{or22}}\right)(N_\phi + N_{\phi Tor})\right.$$
$$\left. - \left(\frac{R_\theta}{D_{or22}} + \frac{\mu_2 R_\phi}{D_{or11}}\right)(N_\theta + N_{\theta Tor})\right] \qquad (13.5a)$$

and differentiate equation (13.4b) with respect to ϕ, which gives

$$\cot\phi\, D_{\phi,\phi} - \csc^2\phi\, D_\phi + D_{n,\phi} = \frac{1}{D_{or22}(1 - \mu_1\mu_2)}$$
$$\times\{R_\theta[-\mu_1(N_\phi + N_{\phi Tor}) \qquad (13.5b)$$
$$+ (N_\theta + N_{\theta Tor})]\}_{,\phi}$$

We then eliminate $D_{\phi,\theta}$ from equations (13.5a) and (13.5b) by taking $[(13.5b - \cot\phi\, (13.5a)]$. The lhs of this combination becomes

$$(\cot^2\phi - \csc^2\phi)D_\phi + D_{n,\phi} = D_{n,\phi} - D_\phi \qquad (13.6)$$

It is now convenient to suppress the transverse shearing strains. From equation (7.48g), note that for $\gamma_\phi = 0$,

$$D_{\phi\theta} = -\frac{1}{R_\phi}(D_{n,\phi} - D_\phi) \tag{13.7}$$

so that $-(1/R_\phi)$ [lhs (13.6)] $= D_{\phi\theta}$, which is one of our dependent variables.

To complete the derivation, the stress resultants N_ϕ and N_θ on the rhs of equation (13.5b) must be expressed in terms of the other variable Q_ϕ. This is most easily accomplished by using the overall equilibrium concept introduced in Section 4.4.1. With Q_ϕ now included, equation (4.57) generalizes to

$$\mathbf{Q}(\phi) = Q(\phi)\mathbf{t_Z} = [N_\phi(\phi)\sin\phi - Q_\phi(\phi)\cos\phi]2\pi R_0(\phi)\mathbf{t_Z} \tag{13.8}$$

where $\mathbf{Q}(\phi)$ is the resultant vertical load in the *negative* Z direction, expressed as a function of the surface loading by equation (4.59). Therefore, the magnitude of N_ϕ is

$$N_\phi = \frac{Q(\phi)}{2\pi R_0 \sin\phi} + Q_\phi \cot\phi \tag{13.9a}$$

Now, considering equation (13.1b) in view of equation (13.9a),

$$N_\theta = \frac{1}{R_\phi \sin\phi}[(R_0 Q_\phi)_{,\phi} - R_0 N_\phi + q_n R_\phi R_0]$$

$$= \frac{1}{R_\phi \sin\phi}\left[(R_0 Q_\phi)_{,\phi} - R_0 \cot\phi\, Q_\phi - \frac{Q(\phi)}{2\pi \sin\phi} + q_n R_\phi R_0\right] \tag{13.9b}$$

Equations (13.9a) and (13.9b) may be used to remove N_ϕ and N_θ from the rhs of the combined equations (13.5a) and (13.5b), in other words [(13.5b) $-$ cotϕ (13.5a)], which then becomes a second equation in $D_{\phi\theta}$ and Q_ϕ. This is tedious to carry through in the general form; consequently, at this point, the specialization for a commonly encountered case seems prudent.

13.1.2.2 Homogeneous Solution for Spherical Shells.

Consider an isotropic shell and no surface loading or thermal effects. Then, $q_n = Q(\phi) = 0$; $\mu_1 = \mu_2 = \mu$; $D_{or11} = D_{or22} = Eh/(1 - \mu^2)$; $D_{or44} = D_{or55} = D$; and $D_{or45} = \mu D$, where D has been previously defined as $Eh^3/[12(1 - \mu^2)]$. Also, for the spherical geometry, $R_\phi = R_\theta = a$ and $R_0 = a \sin\phi$. We may anticipate that the forthcoming solution will represent the influence of edge forces and moments which, in combination with a membrane theory solution for the surface loading, constitutes a complete solution to the spherical shell problem.

Now, we proceed with the aforementioned ([13.5b] $-$ cot ϕ[13.5a]) combination, which, in view of equation (13.7), gives

$$D_{\phi\theta} = \frac{1}{Eh}[(1 + \mu)\cot\phi(N_\phi - N_\theta) - N_{\theta,\phi} + \mu N_{\phi,\phi}] \tag{13.10}$$

Equations (13.9a) and (13.9b) simplify to

$$N_\phi = Q_\phi \cot\phi \tag{13.11a}$$

$$N_\theta = Q_{\phi,\phi} \tag{13.11b}$$

where we have employed the Gauss-Codazzi relationship $R_{0,\phi} = R_\phi \cos\phi$ via equation (2.51).

Substituting equations (13.11a) and (13.11b) into (13.10), we have

$$D_{\phi\theta} = -\frac{1}{Eh}[Q_{\phi,\phi\phi} + \cot\phi\, Q_{\phi,\phi} - (\cot^2\phi - \mu)Q_\phi] \qquad (13.12a)$$

The specialized form of the first of our governing equations, equation (13.3), is

$$\frac{a^2}{D}Q_\phi = D_{\phi\theta,\phi\phi} + \cot\phi D_{\phi\theta,\phi} - (\cot^2\phi + \mu)D_{\phi\theta} \qquad (13.12b)$$

Noting the similarity between the preceding two equations, we may define the operator[6]

$$\mathbf{L}(\ ) = (\ )_{,\phi\phi} + \cot\phi(\ )_{,\phi} - \cot^2\phi(\ ) \qquad (13.13)$$

whereupon the equations become

$$\mathbf{L}(Q_\phi) + \mu Q_\phi = -EhD_{\phi\theta} \qquad (13.14a)$$

$$\mathbf{L}(D_{\phi\theta}) - \mu D_{\phi\theta} = \frac{a^2}{D}Q_\phi \qquad (13.14b)$$

We substitute Q_ϕ from equation (13.14b) into (13.14a), and $D_{\phi\theta}$ from (13.14a) into (13.14b) to get, respectively,

$$\mathbf{LL}(D_{\phi\theta}) - \mu^2 D_{\phi\theta} = -\frac{a^2 Eh}{D}D_{\phi\theta} \qquad (13.15a)$$

$$\mathbf{LL}(Q_\phi) - \mu^2 Q_\phi = -\frac{a^2 Eh}{D}Q_\phi \qquad (13.15b)$$

so that the equations are uncoupled. When one variable is determined, the other may be immediately found from equations (13.14a) and (13.14b), and all other terms follow from the previous formulas.

It is of interest to rewrite equation (13.15b) in the form

$$\mathbf{LL}(Q_\phi) + 4k^4 Q_\phi = 0 \qquad (13.16a)$$

where

$$k^4 = \frac{1}{4}\left(\frac{a^2 Eh}{D} - \mu^2\right)$$

$$= 3(1 - \mu^2)\frac{a^2}{h^2} - \frac{\mu^2}{4} \qquad (13.16b)$$

The resulting equation (13.16a) is quite similar to equation (12.16a), which describes cylindrical shells. Here, of course, the operator $\mathbf{L}(\ )$ causes more complication in the eventual solution.

It is also convenient to express equation (13.16a) in complex form as[7]

$$\mathbf{L}[\mathbf{L}(Q_\phi) + 2ik^2 Q_\phi] - 2ik^2[\mathbf{L}(Q_\phi) + 2ik^2 Q_\phi] = 0 \qquad (13.17a)$$

$$\mathbf{L}[\mathbf{L}(Q_\phi) - 2ik^2 Q_\phi] + 2ik^2[\mathbf{L}(Q_\phi) - 2ik^2 Q_\phi] = 0 \qquad (13.17b)$$

which can be easily verified to be equivalent to the original equation.

Then, the solution of the two second-order equations:

$$L(Q_\phi) + 2ik^2 Q_\phi = 0 \tag{13.18a}$$

$$L(Q_\phi) - 2ik^2 Q_\phi = 0 \tag{13.18b}$$

will constitute the solution to the fourth-order equation (13.16a).

The formulation of the governing equations for other geometries follows similar, but sometimes more complicated, steps. The interested reader is referred to Novozhilov,[8] Flügge,[9] and Kraus.[10]

An exact solution to the governing equations for the spherical geometry exists. If one of the equations, for example, equation (13.18a), is fully expanded, and we have

$$Q_{\phi,\phi\phi} + \cot\phi \, Q_{\phi,\phi} - \cot^2\phi \, Q_\phi + 2ik^2 Q_\phi = 0 \tag{13.19}$$

which is a second-order linear differential equation with *variable* coefficients.

The further transformations

$$\cos^2\phi = \psi \tag{13.20a}$$

and

$$Q_\phi = \overline{Q}_\phi \sin\phi \tag{13.20b}$$

put the equation into the form

$$\overline{Q}_{\phi,\psi\psi} + \frac{1 - 5\psi}{2\psi(1 - \psi)} \overline{Q}_{\phi,\psi} - \frac{1 - 2i\psi^2}{4\psi(1 - \psi)} \overline{Q}_\phi = 0 \tag{13.21}$$

which is known as the *hypergeometric equation.* Such equations may be solved with some rather slowly converging power series expressions.[11]

13.1.2.3 Asymptotic Solutions for Spherical Shells.

One of the most powerful methods for solving the complicated differential equations that arise in the bending theory of shells is that of *asymptotic integration.* This method is developed in Kraus[12] and Novozhilov[13] and applied to both symmetrically and asymmetrically loaded shells of revolution in the latter reference. A detailed development is beyond our objectives in this book. It is sufficient to mention here that the asymptotic integration technique, as applied to second-order equations such as equation (13.19), transforms the equation into a form in which the *first derivative* of the dependent variable is eliminated.

The attractiveness of solving the bending theory equations for rotational shells in a relatively simple way has given rise to some useful approximation methods, in which it is postulated that the solutions to the homogeneous equations have the form

$$\left. \begin{array}{l} e^{\pm k\phi} \sin k\phi \\ e^{\pm k\phi} \cos k\phi \end{array} \right\} \tag{13.22}$$

Although expressions of this type can be rigorously derived using asymptotic integration, such solutions may be obtained directly by simply neglecting certain terms in the original differential equation. This is applicable for shells in which certain geometric parameters remain essentially constant over a significant portion of the shell. For example, note in equation (13.19) that the second and third terms are multiplied by $\cot \phi$ and $\cot^2 \phi$, which are relatively small when $\phi \geq \pi/4$, whereas the fourth term has a coefficient of k^2, which, from equation (13.16b), is proportional to a/h. Thus, it is reasonable to use the simplified equation

$$Q_{\phi,\phi\phi} \pm 2ik^2 Q_\phi = 0 \tag{13.23}$$

to find approximate solutions for the bending of spherical shells under edge loads. If we are interested in solutions in the vicinity of the lower boundary of spherical shells resulting from edge loading, as shown in Figure 13–1, it is only necessary that ϕ_b not be too small. This reasoning is also applicable for other shells with smoothly varying geometry such as conical shells, and is sometimes referred to as the *Geckeler approximation*.[14] Despite the lack of mathematical rigor, such "quick and dirty" solutions are often the only relatively simple analytical means available to check complex, computer-based analyses.

We may write the solution of equation (13.23) from the auxiliary equations

$$m^2 \pm 2ik^2 = 0 \tag{13.24}$$

which give

$$m = \pm(1 \pm i)k \tag{13.25}$$

so that the complete solution is

$$Q_\phi = B_1 e^{(1+i)k\phi} + B_2 e^{(1-i)k\phi} + B_3 e^{-(1+i)k\phi} + B_4 e^{-(1-i)k\phi} \tag{13.26}$$

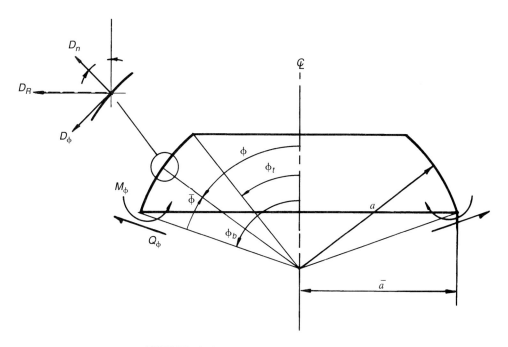

FIGURE 13–1   Edge-Loaded Spherical Shell.

where $B_1 - B_4$ are integration constants.

We now wish to redefine the integration constants so that Q_ϕ is expressed in terms of real functions. First, we use the identities

$$\left. \begin{array}{l} e^{ik\phi} = \cos k\phi + i \sin k\phi \\ e^{-ik\phi} = \cos k\phi - i \sin k\phi \end{array} \right\} \tag{13.27}$$

to rewrite equation (13.26) as

$$\begin{aligned} Q_\phi = &(B_1 + B_2)e^{k\phi} \cos k\phi + i(B_1 - B_2)e^{k\phi} \sin k\phi \\ &+ (B_3 + B_4)e^{-k\phi} \cos k\phi - i(B_3 - B_4)e^{-k\phi} \sin k\phi \end{aligned} \tag{13.28a}$$

It is convenient to express this solution as a function of the auxiliary coordinate

$$\overline{\phi} = \phi_b - \phi \tag{13.28b}$$

which we show on Figure 13–1 and introduce into the first term of equation (13.28a):

$$\begin{aligned} (B_1 + B_2)e^{k\phi} \cos k\phi &= (B_1 + B_2)e^{k(\phi_b - \overline{\phi})} \cos k(\phi_b - \overline{\phi}) \\ &= (B_1 + B_2)e^{k\phi_b}e^{-k\overline{\phi}} (\cos k\phi_b \cos k\overline{\phi} \\ &\quad + \sin k\phi_b \sin k\overline{\phi}) \\ &= [(B_1 + B_2)e^{k\phi_b} \cos k\phi_b]e^{-k\overline{\phi}} \cos k\overline{\phi} \\ &\quad + [(B_1 + B_2)e^{k\phi_b} \sin k\phi_b]e^{-k\overline{\phi}} \sin k\overline{\phi} \end{aligned} \tag{13.28c}$$

Similarly, the second term becomes

$$\begin{aligned} i(B_1 - B_2)e^{k\phi} \sin k\phi &= [i(B_1 - B_2)e^{k\phi_b} \sin k\phi_b]e^{-k\overline{\phi}} \cos k\overline{\phi} \\ &\quad - [i(B_1 - B_2)e^{k\phi_b} \cos k\phi_b]e^{-k\overline{\phi}} \sin k\overline{\phi} \end{aligned} \tag{13.28d}$$

The terms in [ ] in the preceding equations may be collected as coefficients of the variable terms $e^{-k\overline{\phi}} \cos k\overline{\phi}$ and $e^{-k\overline{\phi}} \sin k\overline{\phi}$, and can be redefined as new arbitrary constants C_1 and C_2. We retain the last two expressions in equation (13.28a) in terms of ϕ, with $(B_3 + B_4)$ and $-i(B_3 - B_4)$ redefined as C_3 and C_4 respectively, so that the revised form of the solution is

$$Q_\phi = e^{-k\overline{\phi}}(C_1 \cos k\overline{\phi} + C_2 \sin k\overline{\phi}) + e^{-k\phi}(C_3 \cos k\phi + C_4 \sin k\phi) \tag{13.29}$$

Equation (13.29) is identical in form to equation (12.20) if $\overline{\phi}$ is replaced by $\phi_b - \phi$ and the constants are redefined. This comparison gives rise to the physical interpretation of the Geckeler approximation as the replacement of the actual shell by a cylindrical shell of suitable dimensions.

Referring to the closed spherical shell or cap in Figure 4–1(a) with thickness = h, Calladine and Paskaran have shown that the Geckeler approximation is suitable provided $Z/h > 5$, where Z is measured from the pole.[15]

The foregoing transformation is an alternate means of introducing the semi-infinite concept that we utilized extensively for cylindrical shells. The solution is uncoupled into parts that decay from the boundaries $\phi = \phi_b$ and $\phi = \phi_\tau$, respectively. If the influence of the loading at one boundary does not penetrate into the region of the shell affected by the loading at the other boundary, we have a semi-infinite shell. Considering edge loading originating at $\phi = \phi_b$, we retain

$$Q_\phi(\overline{\phi}) = e^{-k\overline{\phi}}(C_1 \cos k\overline{\phi} + C_2 \sin k\overline{\phi}) \tag{13.30}$$

We now evaluate the corresponding stress resultants, couples, and displacements. In performing the required differentiations, take care to note that

$$\frac{\partial(\ )}{\partial\phi} = -\frac{\partial(\ )}{\partial\overline{\phi}} \tag{13.31}$$

From equation (13.9a), with $Q(\phi) = 0$,

$$N_\phi(\phi) = Q_\phi \cot \phi \tag{13.32a}$$

$$= e^{-k\overline{\phi}} \cot \phi(C_1 \cos k\overline{\phi} + C_2 \sin k\overline{\phi}) \tag{13.32b}$$

$$N_\phi(\overline{\phi}) = e^{-k\overline{\phi}} \cot(\phi_b - \overline{\phi})(C_1 \cos k\overline{\phi} + C_2 \sin k\overline{\phi}) \tag{13.32c}$$

Expanding equation (13.1b) with $q_n = 0$, we have

$$a(\sin \phi\, Q_\phi),_\phi - a \sin \phi\, N_\phi - a \sin \phi\, N_\theta = 0 \tag{13.33a}$$

Solving for N_θ, after replacing N_ϕ with equation (13.32a), we find

$$N_\theta(\overline{\phi}) = Q_{\phi,\phi}$$

$$= -Q_{\phi,\overline{\phi}} \tag{13.33b}$$

$$= (C_1 - C_2)ke^{-k\overline{\phi}} \cos k\overline{\phi} + (C_1 + C_2)ke^{-k\overline{\phi}} \sin k\overline{\phi}$$

Next, we turn to the rotation $D_{\phi\theta}$ which is expressed by equation (13.14a). It is consistent with our previous approximations to take $\mathbf{L}(\ ) = (\ )_{,\phi\phi}$ and also to neglect the μQ_ϕ term, whereby

$$D_{\phi\theta}(\overline{\phi}) = -\frac{1}{Eh} Q_{\phi,\phi\phi}$$

$$= -\frac{1}{Eh} Q_{\phi,\overline{\phi}\,\overline{\phi}} \tag{13.34}$$

$$= -\frac{2}{Eh} k^2 e^{-k\overline{\phi}}(C_1 \sin k\overline{\phi} - C_2 \cos k\overline{\phi})$$

The actual displacements are sometimes of interest: D_ϕ and D_n may be individually found from the simultaneous solution of equations (8.34a) and (8.34b). Of more use for our purposes is the radial displacement. D_R, which is defined in the inset of Figure 13–1 as

$$D_R = D_\phi \cos \phi + D_n \sin \phi \tag{13.35a}$$

Referring to equation (8.34b) with $R_0 (\overline{\phi}) = a\sin \overline{\phi} = \overline{a}$,

$$D_R(\overline{\phi}) = \overline{a}\varepsilon_\theta = \frac{\overline{a}}{Eh} (N_\theta - \mu N_\phi) \tag{13.35b}$$

which is easily evaluated from equations (13.32c) and (13.33b).

We now turn to the stress couples, which can be obtained from equations (13.2c) and (13.2d). Again, it is consistent with our previous arguments to neglect the second terms in these equations in comparison to the first, whereby

$$M_\phi = \frac{D}{a} D_{\phi\theta,\phi}$$ (13.36a)

which is conveniently written, using equations (13.34) and (13.16b) with μ^2 neglected, as

$$M_\phi = -\frac{a}{4k^4} Q_{\phi,\phi\phi\phi} = \frac{a}{4k^4} Q_{\phi,\overline{\phi}\,\overline{\phi}\,\overline{\phi}\,\overline{\phi}}$$ (13.36b)

and which gives

$$M_\phi(\overline{\phi}) = \frac{a}{2k} e^{-k\overline{\phi}}[(C_1 + C_2) \cos k\overline{\phi} - (C_1 - C_2) \sin k\overline{\phi}]$$ (13.36c)

Also, from equation (13.2d) with $D_{or45} = \mu$ and equation (13.36a)

$$M_\theta(\overline{\phi}) = \mu M_\phi(\overline{\phi})$$ (13.37)

We have thus derived approximate, but explicit, expressions for the forces, moments, and displacements in an edge-loaded spherical shell. From these equations, it is obvious that the assumption leading to equation (13.23)—the neglect of Q_ϕ and $Q_{\phi,\phi}$ as compared to $Q_{\phi,\phi,\phi}$—is reasonable because each subsequent differentiation gives a factor, k, that is order (a/h), in other words, $O(a/h)$. Moreover, the presence of the cot ϕ and cot^2 ϕ coefficients of $Q_{\phi,\phi}$ and Q_ϕ, respectively, in equation (13.19) further reduces the influence of the latter terms when $\phi > \pi/4$. The limits of application of the semi-infinite solution can be established using Figure 12–2, with kX replaced by $k\phi$, where k is given by equation (13.16b).

For practical application, it is helpful to represent the previous solution in the same format as we used for the edge-loaded cylindrical shell. Consider the loading shown in Figure 13–1, which is analogous to Figure 12–1 for the cylindrical shell. We take

$$M_0 = M_\phi(\phi_b)$$ (13.38a)

$$Q_0 = Q_\phi(\phi_b)$$ (13.38b)

whereupon, from equations (13.30) and (13.36c) with $\overline{\phi} = 0$,

$$Q_0 = C_1$$ (13.39a)

$$M_0 = \frac{a}{2k}(C_1 + C_2)$$ (13.39b)

from which

$$C_1 = Q_0$$ (13.40a)

$$C_2 = \frac{2k}{a} M_0 - Q_0$$ (13.40b)

A matrix of derivatives of Q_ϕ, similar to that written for D_n in Matrix 12–1, is useful for further computations.

13.1.2.4 Compound Shells.   We are now able to resume our consideration of the cylindrical shell with a spherical head, as shown in Figure 4–14 and discussed in Section 4.5. Previously, we found that the membrane theory, by itself, was inadequate, because (a) radial deformations between the two shells were incompatible; and, (b) except for the complete hemisphere, the radial component of the meridional force in the spherical shell was unbalanced. We may now fully analyze the shell by the flexibility method described in Section 8.3.1 using the notation introduced in equation (8.30).

In Figure 13–2(a), we show the pressurized shell from Figure 4–14, with the axial coordinate taken as X and the origin placed at the end of the cylinder. The respective thickness of the spherical and cylindrical segments are defined as h_1 and h_2, and the Young moduli and Poisson's ratios as E_1 and E_2 and μ_1 and μ_2. Note that $\sin \phi_1 = a_2/a_1$. We denote the characteristic constants k, defined in equation (12.16b) for the cylindrical shell and in equation (13.16b) for the spherical shell, as k_1 and k_2 respectively.

$$\left\{ \begin{matrix} \Delta_1 \\ \Delta_2 \end{matrix} \right\} = \left\{ \begin{matrix} \overline{\Delta}_1 \\ \overline{\Delta}_2 \end{matrix} \right\} + \begin{bmatrix} F_{11} & F_{12} \\ F_{21} & F_{22} \end{bmatrix} \left\{ \begin{matrix} \chi_1 \\ \chi_2 \end{matrix} \right\} \tag{13.41}$$

First, note that we are dealing with *relative* displacements and rotations which are computed from the *absolute* displacements of the two adjoining segments of the shell. Therefore, we must carefully define the sign convention for each of these quantities. Corresponding to the senses assumed for the redundants χ_1 and χ_2 in Figure 13–2(c), it is consistent to take the positive sense of the relative radial displacement as increasing, in other words, inward on the cylinder (d_{c1}) and outward on the sphere (d_{s1}). Similarly, a positive *relative* meridional rotation will *open* the internal angle between the cylinder (d_{c2}) and the sphere (d_{s2}), originally $\pi/2 + \phi_1$. These are illustrated in Figure 13–2(d) and (e). For example, if we compute the radial displacement in the cylinder D_n from one of the loading cases, and it is *positive* (outward) by the cylindrical shell convention, it takes a *negative* sign in the relative displacement expression, because it is opposite in sense to d_{c1}. On the other hand, a *positive* computed value for the radial displacement of the spherical shell D_R is consistent with d_{s1}, so that it has a *positive* sign in the relative displacement expression. Similarly, a *positive* meridional rotation $D_{X\theta}$ in the cylindrical shell may be seen from figure 7–2(a) (with $\alpha = X$ and $\beta = \theta$) to be opposite in sense to d_{c2}, so that it gets a *negative* sign. The rotation of the spherical shell is established from Figure 7–2 (a) (with $\alpha = \phi$ and $\beta = \theta$), where a *positive* meridional rotation $D_{\phi\theta}(\phi)$, corresponds to a *positive* d_{s2}.

Next, note that the compatibility condition requires that the relative radial displacement and relative meridional rotation between the two segments vanishes, or

$$\{\Delta\} = \{\Delta_1 \ \Delta_2\} = \{0\} \tag{13.42}$$

For the evaluation of the elements of $\{\overline{\Delta}\}$ and $[\mathbf{F}]$ in equation (13.41), we refer to the chart of basic solutions given in Table 13–1. Note that the cylindrical and spherical segments may each have different constant thicknesses and material properties. In the last column, the most probable algebraic sign of the computed displacement or rotation on the shell segment is assumed, and the corresponding algebraic sign of the contribution of this quantity to the relative displacement is given.

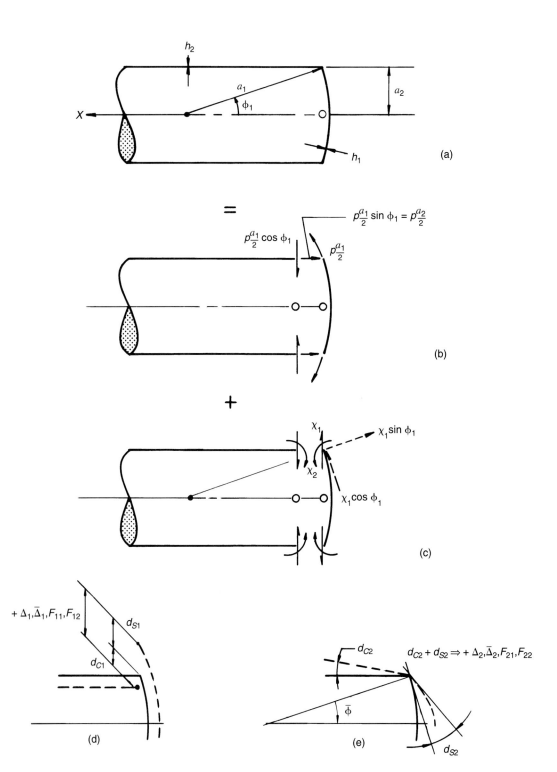

FIGURE 13–2 Compound Shell.

TABLE 13–1 Basic Solutions for a Compound Shell

Segment	Loading	Superposition Case	Compute	Equations	Contributes to	Sign Convention
Cylindrical	p	b_1	D_n $(D_{X\theta}=0)$	(12.42a) with $N_x = p(a_1/2)\sin\phi_1$ and $N_\theta = pa_2$	$\bar\Delta_1$	$+D_n \to -d_{c1}$
$a = a_2$ $h = h_2,\ \mu = \mu_2$	$Q_0 = p(a_1/2)\cos\phi_1$	b_2	D_n	matrix 12–1 row (a)	$\bar\Delta_1$	$-D_n \to +d_{c1}$
$E = E_2,\ D = D_2 = \dfrac{E_2 h_2^3}{12(1-\mu_2^2)}$	$M_0 = 0$		$D_{X\theta} = -D_{n,X}$	matrix 12–1 row (b)	$\bar\Delta_2$	$-D_{X\theta} = +d_{c2}$
$X = 0$	$Q_0 = 1$	c_1	D_n	matrix 12–1 row (a)	F_{11}	$-D_n \to +d_{c1}$
$k_2^4 = \dfrac{3(1-\mu_2^2)}{(a_2 h_2)^2}$	$M_0 = 0$		$D_{X\theta} = -D_{n,X}$	matrix 12–1 row (b)	F_{21}	$-D_{X\theta} \to +d_{c2}$
	$Q_0 = 0$	c_2	D_n	matrix 12–1 row (a)	F_{12}	$-D_n \to +d_{c1}$
$\bar k_2^4 = k_2^4 a_2^4$	$M_0 = 1$		$D_{X\theta} = -D_{n,X}$	matrix 12–1 row (b)	F_{22}	$-D_{X\theta} \to +d_{c2}$
Spherical	p	b_3	D_R $(D_{\phi\theta}=0)$	(13.35b) with $N_\phi = N_\theta = p(a_1/2)$	$\bar\Delta_1$	$+D_R \to +d_{s1}$
$R_\phi = R_\theta = a$ $a = a_1$ $h = h_1,\ \mu = \mu_1$	$Q_0 = \sin\phi_1 = a_2/a_1$ $M_0 = 0$ or $C_1 = \sin\phi_1 = a_2/a_1$	c_3	D_R	(13.35b) with N_ϕ and N_θ from (13.32c) and (13.33b)	F_{11}	$+D_R \to +d_{s1}$
$E = E_1,\ D = D_1$ $\phi = \phi_b = \phi_1$ $\bar\phi = 0$	$C_2 = -\sin\phi_1 = -a_2/a_1$		$D_{\phi\theta}$	(13.34)	F_{21}	$+D_{\phi\theta} \to +d_{s2}$
$k_1^4 = 3(1-\mu_1^2)\left(\dfrac{a_1}{h_1}\right)^2$	$Q_0 = 0$ $M_0 = 1$ or $C_1 = 0$	c_4	D_R	(13.35b) with N_ϕ and N_θ from (13.32c) and (13.33b)	F_{12}	$-D_R \to +d_{s1}$
	$C_2 = 2k_1/a_1$		$D_{\phi\theta}$	(13.34)	F_{22}	$+D_{\phi\theta} \to +d_{s2}$

Example 13.1(a)

First we calculate the displacements with the continuity constraints relaxed, $\overline{\Delta}_1$ and $\overline{\Delta}_2$.

For the cylindrical shell, we compute the normal displacement caused by the internal pressure from equation (12.42a) with $N_x = p(a_1/2)\sin \phi_1$ and $N_\theta = pa_2$ as

$$D_n(0) = p\,\frac{a_2}{E_2 h_2}\left(a_2 - \frac{\mu_2}{2}a_1 \sin \phi_1\right) = p\,\frac{a_2^2}{E_2 h_2}\left(1 - \frac{\mu_2}{2}\right) \qquad (13.43a)$$

which is a *negative* contribution to $\overline{\Delta}_1$ entered into Table 13–1 as superposition case b_1.

Also, we have the normal displacement from the unbalanced transverse shear imparted by the spherical cap to the cylinder, $Q_0 = p(a_1/2)\cos\phi_1$, which is found from Matrix 12–1, row (a), as

$$D_n(0) = -p\,\frac{a_1 \cos \phi_1}{4D_2 k_2^3} \qquad (13.43b)$$

providing a *positive* contribution to $\overline{\Delta}_1$. The accompanying rotation, neglecting transverse shearing strain, is found from Matrix 12–1, row (b), as

$$D_{x\theta}(0) = -D_{n,x} = -p\,\frac{a_1 \cos \phi_1}{4D_2 k_2^2} \qquad (13.43c)$$

which is a *positive* contribution to $\overline{\Delta}_2$. These are entered into Table 13–1 as case b_2.

For the spherical shell, we have the radial displacement caused by the internal pressure from equation (13.35b) with $N_\phi = N_\theta = p(a_1/2)$ and $\overline{a} = a_1\sin \phi_1$,

$$D_R(0) = p\,\frac{a_1^2(1 - \mu_1)\sin \phi_1}{2E_1 h_1} = p\,\frac{a_1 a_2}{2E_1 h_1}(1 - \mu_1) \qquad (13.43d)$$

giving a *positive* contribution to $\overline{\Delta}_1$ entered as case b_3. This loading does not produce a contribution to $\overline{\Delta}_2$.

Next, we compute the contribution to the flexibility influence coefficient from the unit actions on the cylindrical shell.

Solving for D_n with $Q_0 = 1$ and $M_0 = 0$ inserted into Matrix 12–1, row (a), we find

$$D_n(0) = -\frac{1}{2k_2^3 D_2} \qquad (13.44a)$$

which is a *positive* contribution to F_{11}.

The same loading in Matrix 12–1, row (b), with the use of equation (12.6d), gives

$$D_{X\theta}(0) = -D_{n,x}(0) = -\frac{1}{2k_2^2 D_2} \qquad (13.44b)$$

which is a *positive* contribution to F_{21}. The preceding quantities are entered as case c_1.

Proceeding, we repeat the computations with $Q_0 = 0$ and $M_0 = 1$, which gives

$$D_n(0) = -\frac{1}{2k_2^2 D_2} \qquad (13.44c)$$

providing a *positive* contribution to F_{12} and

$$D_{X\theta}(0) = -D_{n,x}(0) = -\frac{1}{k_2 D_2} \qquad (13.44d)$$

which is a *positive* contribution to F_{22} which are entered as case c_2.

Finally, we have the contributions to the flexibility influence coefficients from the unit actions on the spherical shell.

Solving for D_R with $Q_0 = \sin \phi_1$ and $M_0 = 0$, we find $C_1 = \sin \phi_1$, and $C_2 = -\sin \phi_1$ from equations (13.40). Then, from equations (13.32c) and (13.33b), we evaluate $N_\phi(0) = \cos \phi_1$ and $N_\theta(0) = 2k_1 \sin \phi_1$, which are inserted into equation (13.35b) to produce

$$D_R(0) = \frac{a_1 \sin \phi_1}{E_1 h_1} (2k_1 \sin \phi_1 - \mu_1 \cos \phi_1)$$

$$= \frac{a_2}{E_1 h_1} \left(\frac{2k_1 a_2}{a_1} - \mu_1 \cos \phi_1 \right)$$
(13.45a)

giving a *positive* contribution to F_{11}.

The same loading used in equation (13.34) yields

$$D_{\phi\theta}(0) = -\frac{2}{E_1 h_1} k_1^2 \sin \phi_1 = -\frac{2a_2 k_1^2}{E_1 h_1 a_1}$$
(13.45b)

which is a *negative* contribution to F_{21}. The preceding quantities are entered as case c_3.

Proceeding, we repeat the computations with $Q_0 = 0$ and $M_0 = 1$. This gives $C_1 = 0$ and $C_2 = 2(k_1/a_1)$; and $N_\phi(0) = 0$ and $N_\theta(0) = -2k_1^2/a_1$. Then, we find

$$D_R(0) = -\frac{2k_1^2}{E_1 h_1} \sin \phi_1 = -\frac{2a_2 k_1^2}{E_1 h_1 a_1}$$
(13.45c)

which is a *negative* contribution to F_{12} and

$$D_{\phi\theta}(0) = -\frac{4k_1^3}{E_1 h_1 a_1}$$
(13.45d)

which is a *negative* contribution to F_{22} which are entered as case c_4.

Collecting the results, we have from equations (13.43a), (13.43b) and (13.43d),

$$\overline{\Delta}_1 = p \left[-\frac{a_2^2}{E_2 h_2} \left(1 - \frac{\mu_2}{2} \right) + \frac{a_1 \cos \phi_1}{4D_2 k_2^3} + \frac{a_1 a_2 (1 - \mu_1)}{2E_1 h_1} \right]$$
(13.46a)

and from equation (13.43c),

$$\overline{\Delta}_2 = p \frac{a_1 \cos \phi_1}{4D_2 k_2^2}$$
(13.46b)

Also, from equations (13.44a) and (13.45a),

$$F_{11} = \frac{1}{2k_2^3 D_2} + \frac{a_2}{E_1 h_1} \left(2\frac{a_2}{a_1} k_1 - \mu_1 \cos \phi_1 \right)$$
(13.46c)

from equations (13.44b) and (13.45b),

$$F_{21} = \frac{1}{2k_2^2 D_2} - \frac{2a_2}{E_1 h_1 a_1} k_1^2$$
(13.46d)

from equations (13.44c) and (13.45c),

$$F_{12} = \frac{1}{2k_2^2 D_2} - \frac{2a_2}{E_1 h_1 a_1} k_1^2$$
(13.46e)

and, from (13.44d) and (13.45d),

$$F_{22} = \frac{1}{k_2 D_2} - \frac{4}{E_1 h_1 a_1} k_1^3 \tag{13.46f}$$

Note that $F_{21} = F_{12}$, a check of symmetry.

The equations can be put in better order for evaluation if we write the shell parameter k_2 for the cylindrical shell in the same nondimensional form as k_1 defined for the spherical shell, in other words,

$$\bar{k}_2^4 = 3(1 - \mu_2^2)\left(\frac{a_2}{h_2}\right)^2 = k_2^4 a_2^4 \tag{13.47a}$$

Then, we may restate equations (13.46a–e) as

$$\bar{\Delta}_1 = p\left[-\frac{a_2^2}{E_2 h_2}\left(1 - \frac{\mu_2}{2}\right) + \frac{a_1 a_2}{E_2 h_2}\bar{k}_2 \cos \phi_1 + \frac{a_1 a_2 (1 - \mu_1)}{2E_1 h_1}\right] \tag{13.47b}$$

$$\bar{\Delta}_2 = p \cos \phi_1 \left(\frac{a_1}{a_2}\right)\bar{k}_2^2 \tag{13.47c}$$

$$F_{11} = \frac{2a_2^2}{E_2 h_2}\bar{k}_2 + \frac{a_2}{E_1 h_1}\left(2\frac{a_2}{a_1}k_1 - \mu_1 \cos \phi_1\right) \tag{13.47d}$$

$$F_{12} = F_{21} = \frac{2}{E_2 h_2}\bar{k}_2^2 - \frac{2a_2}{E_1 h_1 a_1}k_1^2 \tag{13.47e}$$

$$F_{22} = \frac{4}{E_2 h_2 a_2}\bar{k}_2^3 - \frac{4}{E_1 h_1 a_1}k_1^3 \tag{13.47f}$$

If the materials and/or the thicknesses for the two shells are the same, the equations can be further simplified.

Once all of the terms are evaluated,

$$\begin{Bmatrix} \chi_1 \\ \chi_2 \end{Bmatrix} = -\left[\mathbf{F}^{-1}\right]\begin{Bmatrix} \bar{\Delta}_1 \\ \bar{\Delta}_2 \end{Bmatrix} \tag{13.48}$$

in view of equation (13.42). After determining χ_1 and χ_2, the stress resultants, couples, and displacements in the cylindrical and spherical segments are found as linear combinations of the solutions for the loading cases shown in Figures 13–2(b) and (c). To carry out the superposition, the relevant stress resultants from the particular solution cases (b_1, b_2, b_3) are combined with those from cases c_1 and c_3 multiplied by χ_1 and from cases c_2 and c_4 multiplied by χ_2.

Example 13–2(b)

We shall not pursue further details of the calculations here, but refer to a numerical study in Flügge.[16] There, two cases are considered, a hemispherical head and a shallower cap with $\phi_1 = 45°$. For both cases, $h_1 = h_2 = h$, $h/a = 0.01$, and $\mu = 0.3$. On Figures 13–3(a) and (b), comparative values of the meridional stress couple, M_X or M_ϕ, and the circumferential or hoop stress resultant, N_θ, are shown for the two examples.

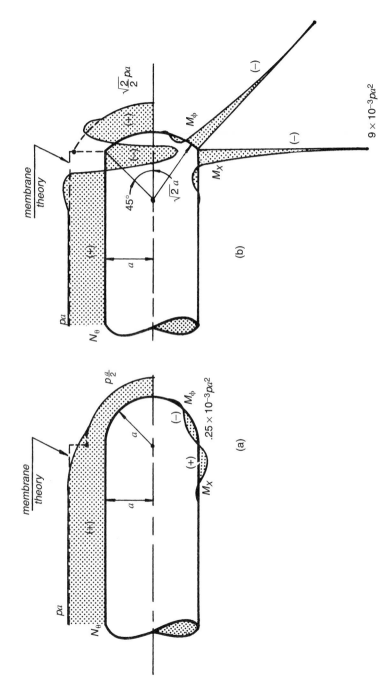

FIGURE 13–3 Stress Resultants and Moments in Compound Shells.

The meridional moments increase greatly for the discontinuous case. Similarly, for the circumferential stress resultant, the hemispherical head provides a smooth transition for the hoop stresses in the two segments; the shallow cap results in an increased magnitude of N_θ as well as a change in sense to compression, which gives rise to the possibility of circumferential buckling or wrinkling. An independent calculation gave a maximum $N_\theta = -7.8pa$ for the shallow cap.

Although the hemispherical head might seem more attractive on the basis of the preceding comparison, the fabrication advantages of shallow caps have been pointed out previously and the preferred solution is often to include a circumferential ring stiffener, as we see in Figure 2–8(r). Details regarding the design of such stiffeners and other practical considerations are found in Pirok and Wozniak.[17] When an elastic stiffener is used, it is possible to generalize the model to include the stiffener following the same reasoning we employed in Section 12.3.7. A comprehensive study of ring-stiffened cylindrical-conical shells under hydrostatic loading was performed by the author and his co-workers, and the results are reported in Gould et al.[18] and Wang and Gould.[19] In general, an adequate circumferential stiffener greatly moderates the extreme amplifications caused by a geometric discontinuity.[18]

Another possibility for reducing the stress amplification caused by a geometric discontinuity is the use of a transition segment, such as a torospherical head, discussed in Section 4.5.2 and illustrated in Figures 4–16(b) and 4–17. Ranjan and Steele[20] have examined the range of applicability of approximate analytical solutions derived from asymptotic expansions of the dependent variables to the solution of torospherical shells. From their study, one can draw some conclusions as to the optimum radius of the toroidal knuckle as a function of the cylindrical and spherical radii and the shell thickness. Also, their study suggests a lower bound approximation for the critical internal pressure corresponding to circumferential wrinkling. Although they obtain fairly accurate solutions with their analytical approach based on comparisons with numerical and experimental results, it would appear that the best present technology for an actual design case would include a finite element analysis using doubly curved rotational shell elements.

The finite element concept (FEM) is introduced in Section 13.1.4 and is widely documented in the technical literature. Results from an FEM analysis of the shell treated in Section 4.5.2, presented originally in Figure 4–19, are shown in Figure 13–4 where the smoothing of the peak membrane theory stresses is apparent. Further results for this shell, such as displacements and stress couples, are also available.[21]

13.1.3 Asymmetrical Loading: Analytical Solutions:

When the loading on a shell of revolution is not axisymmetric, the Fourier series technique that we employed extensively in Chapter 5 is again useful to separate the independent variables. The problem then reduces to the solution of a set of ordinary differential equations for each harmonic j.

General solutions for spherical and conical shells are given by Flügge.[22] These solutions are based on the classical displacement method and are exact in the sense that only quantities of $O(h^2/R^2):1$ are neglected. These solutions are valuable in studying such cases as static wind loading and discrete column supports, which we discussed in Sections 5.2.2 and 5.3.3, respectively.

$\sigma_{\theta\theta}$ middle surface

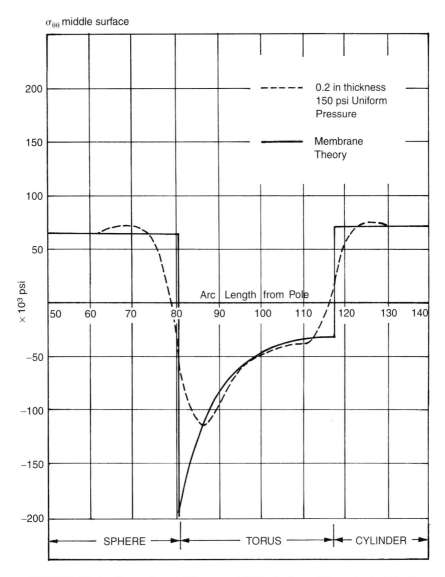

FIGURE 13–4   Circumferential Stress on Middle Surface of Torospherical Head.

For geometries other than spherical and conical and, of course, cylindrical (which we have treated separately), exact solutions are scarce.

As a general approach, the formulation of Novozhilov[23] is very attractive. He follows the classical *force* method rather than the *displacement* procedure that we have employed almost exclusively in our study of plate bending in Chapters 10 and 11 and of cylindrical shells in chapter 12. We did use a force formulation for the membrane theory in Chapter 4 and 5, but this involved only the equilibrium equations because that problem is statically determinate. In the bending theory, however, the situation is more complicated, and it is necessary to express the strain-displacement or compatibility equations, (7.52) and (7.53), in terms of the stress resultants and couples. Once this is accomplished via the constitutive law, the resulting expressions are combined with the equilibrium relations to provide a consistent set of equations in which the stress resultants and couples are the unknowns. In this context, it is interesting to note that the H. Reissner-Meissner approach presented in Section 13.1.2.1 is a combination of a *force* and a *displacement* formulation, because the dependent variables are Q_ϕ and $D_{\phi\theta}$. This is often termed a *mixed* formulation.

We may also point out an interesting feature of the Novozhilov formulation of the shell bending problem, whereby the dependent variables are grouped into complex variables. This results in a halving of the order of the eventual set of equations to be solved and thereby simplifies the algebra. This transformation was introduced in equations (13.16) through (13.18). The solution of the ensuing equations is accomplished by the asymptotic integration technique, to which we have alluded previously. Novozhilov's solution is applicable for surface loading which is described by the first two harmonics, $j = 0$ and $j = 1$. Moreover, shells with a special form of the meridian, in other words, $(1/R_\phi - 1/R_\theta)\csc^2\phi = $ constant, can be solved for the general harmonic loading $j > 1$, which suggests a substitute curve approximation for other meridional profiles. Beyond this, it is generally necessary to neglect terms of $O[j^2(h/R)]$ to achieve a solution; thus, for higher harmonics, the accuracy diminishes. However, a large number of practical problems can be described rather closely in terms of the lower harmonic load components—such as the wind loading shown on Figure 5–16—so that the Novozhilov method has wide applicability. The author has extended this approach to include cases in which terms of $O[j^2(h/R)]$ may be significant, such as in the column-supported shell problem discussed in Section 5.3.3.[24]

It is felt, however, that currently the most feasible approach to the general solution of shells of revolution under asymmetrical as well as symmetrical loading is to use an energy-based numerical solution. We develop the required energy expressions for shells of revolution in the next section.

13.1.4   Energy Formulation

13.1.4.1   General Considerations.   The total potential energy functional U_τ, defined in Section 9.4.2, is specialized here for a shell of revolution. It is convenient to consider an arbitrary segment of a shell bound by two parallel circles defined by the meridional angles ϕ_i and ϕ_{i+1}, as shown in Figure 13–5. Here, the Z-axis is oriented so that $Z = 0$ corresponds to the upper boundary of the shell. This segment may be a portion of a shell or even an entire shell. Also, ϕ_i may be 0°, which indicates a closed shell or dome. It is evident that the total potential energy for a shell composed of the assemblage of a number of such segments is the sum of the energies of all the segments. Therefore, the development, for the most part, may focus on a single general segment such as that shown in Figure 13–5. Of course, the segments adjoining external boundaries require some specialization to accommodate prescribed boundary conditions. We also should remember that the stationary condition, stated previously as equation (9.14),

$$\delta U_\tau = 0 \tag{13.49}$$

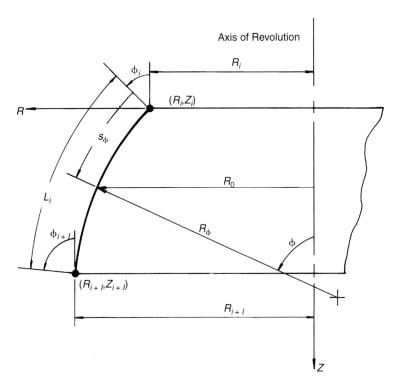

FIGURE 13–5 Shell of Revolution Finite Elements Geometry. (Reprinted with permission of American Institute of Aeronautics and Astronautics. Source: L. J. Brombolich and P. L. Gould, A high-percision curved shell finite element, *AIAA Journal,* 10, no. 6 [June 1972]: 727.)

implies a global extremum and should be applied to the entire assembled shell. Nevertheless, the numerical techniques generally applied to this problem permit almost all of the calculations to be performed at the segment or local, as opposed to the global, level.

   13.1.4.2 Geometry.   Recall from Section 2.8.2 that three possibilities were presented for the meridional coordinate: the meridional angle ϕ, the axial coordinate Z, and the meridional arc length s_ϕ. In fact, the various computations required in an actual problem sometimes involve functions of two, or even all three of these possibilities. For this topic, it is convenient to use the $R - Z$ Cartesian coordinates to locate stations on the shell, but to take the arc length as the primary dependent variable. It is convenient to define for *each* segment a nondimensional arc length coordinate

$$s = \frac{s_\phi}{L_i} \quad (0 \le s \le 1) \tag{13.50a}$$

where s_ϕ is measured from (R_i, Z_i), and where, from equation (2.50),

$$L_i = \int_{Z_i}^{Z_{i+1}} [1 + (R_{0,Z})^2]^{1/2} \, dZ \tag{13.50b}$$

We also find it convenient to express the following geometric relations that are required in the formulation as functions of Z.[25]

$$R_0(Z) = R \tag{13.51a}$$

$$R_\phi(Z) = \frac{[1 + (R_{0,Z})^2]^{3/2}}{R_{0,ZZ}} \tag{13.51b}$$

$$[\sin \phi](Z) = [1 + (R_{0,Z})^2]^{-1/2} \tag{13.51c}$$

$$[\cos \phi](Z) = R_{0,Z}[1 + (R_{0,Z})^2]^{-1/2} \tag{13.51d}$$

$$R_\theta(Z) = \frac{R_0}{\sin \phi} \tag{13.51e}$$

13.1.4.3 Fourier Series Representation. As is customary for shells of revolution, all external loading, stress resultants and couples, strains and changes in curvature, and displacements and rotations are represented by Fourier series in the circumferential variable θ. These expressions are given as equations (5.2a) and (5.2b) for the loading and in-plane stress resultants; by equations (7.50) for the strains and changes in curvature; and by equations (7.51) for the displacements and rotations. To make our discussion complete, we write the corresponding expressions for the bending stress resultants and couples:

$$\begin{Bmatrix} Q_\phi \\ Q_\theta \\ M_\phi \\ M_\theta \\ M_{\phi\theta} \end{Bmatrix} = \sum_{j=0}^{\infty} \begin{Bmatrix} Q_\phi^j \cos j\theta \\ Q_\theta^j \sin j\theta \\ M_\phi^j \cos j\theta \\ M_\theta^j \cos j\theta \\ M_{\phi\theta}^j \sin j\theta \end{Bmatrix} \tag{13.52}$$

We may then further restrict our development to typical harmonic j of segment i for which the stress resultants and couples and the strains and changes in curvature are represented by vectors $\{\mathbf{N}^j\}$ and $\{\boldsymbol{\varepsilon}^j\}$, respectively, as defined in Matrix 8–2 and equation (8.10). Typical elements of these vectors are $N_\phi^j \cos j\theta$, $\omega^j \sin j\theta$, etc. Also, we have the displacement vector $\{\boldsymbol{\Delta}_i^j\}$, defined in accordance with equation (7.51), as

$$\{\boldsymbol{\Delta}_i^j\} = \{D_\phi^j \cos j\theta \quad D_\theta^j \sin j\theta \quad D_n^j \cos j\theta \quad D_{\phi\theta}^j \cos j\theta \quad D_{\theta n}^j \sin j\theta\} \tag{13.53}$$

13.1.4.4 Strain-Displacement Relationships. If we anticipate a displacement formulation, we will have need for the strains and changes in curvature expressed as functions of the displacements and rotations. We derived such compatibility relationships for harmonic j as equations (7.52) and (7.53). Now, we may transform these equations to the nondimensional arc length variable s by noting, from Figure 13–5, that

$$R_\phi d\phi = ds_\phi = L_i ds \tag{13.54a}$$

from which we may express

$$(\ )_{,\phi} = (\ )_{,s} \frac{ds}{d\phi} \tag{13.54b}$$

as

$$(\ )_{,\phi} = (\ )_{,s} \frac{R_\phi}{L_i} \tag{13.55}$$

The resulting relations are conveniently written in the matrix form

$$\{\boldsymbol{\varepsilon}^j\} = [\mathbf{B}_i^j] \{\boldsymbol{\Delta}_i^j\} \tag{13.56}$$

where $[\mathbf{B}_i^j]$ is given by Matrix 13–1.

Matrix 13–1

$$[\mathbf{B}_i^j] = \begin{bmatrix} \dfrac{1}{L_i}(\ )_{,s} & 0 & \dfrac{1}{R_\phi} & 0 & 0 \\[2mm] \dfrac{\cos\phi}{R_0} & \dfrac{j}{R_0} & \dfrac{\sin\phi}{R_0} & 0 & 0 \\[2mm] \dfrac{-j}{R_0} & \dfrac{1}{L_i}(\ )_{,s} - \dfrac{\cos\phi}{R_0} & 0 & 0 & 0 \\[2mm] 0 & 0 & 0 & \dfrac{1}{L_i}(\ )_{,s} & 0 \\[2mm] 0 & 0 & 0 & \dfrac{\cos\phi}{R_0} & \dfrac{j}{R_0} \\[2mm] 0 & 0 & 0 & \dfrac{-j}{2R_0} & \dfrac{1}{2L_i}(\ )_{,s} - \dfrac{\cos\phi}{2R_0} \\[2mm] \dfrac{1}{R_\phi} & 0 & -\dfrac{1}{L_i}(\ )_{,s} & -1 & 0 \\[2mm] 0 & \dfrac{\sin\phi}{R_0} & \dfrac{j}{R_0} & 0 & -1 \end{bmatrix}$$

Note that $[\mathbf{B}_i^j]$ is an operator matrix and must always premultiply $\{\boldsymbol{\Delta}_i^j\}$. Modified expressions for κ_ϕ^j, κ_θ^j, and τ^j, with transverse shearing strains suppressed, are easily written but are not particularly advantageous in the formulation.

We also require a modification of the strain-displacement relationships if a closed shell is encountered. We have shown two cases in Figure 4–1 and have recorded the pole condition for both possibilities in Table 8–1. Here, the origin for the coordinate system is generally selected so that $s = 0$ when $R_0 = 0$.

To establish the pole conditions and to modify the strain-displacement relationships suitably, we simply postulate that all strains and changes in curvature remain *finite* as $s \to 0$. Consider, for example, the expression for ε_θ^j, which we write from equation (13.56) and Matrix 13–1, row (2), as

$$\varepsilon_\theta^j = \frac{\cos \phi}{R_0} D_\phi^j + \frac{j}{R_0} D_\theta^j + \frac{\sin \phi}{R_0} D_n^j \qquad (13.57)$$

For both shells shown in Figure 4–1, $R_0(0) = 0$, so that for the first term to remain finite as $s \to 0$, $(\cos \phi D_\phi^j)_{s=0} = 0$, which creates an indeterminate form. The second term will remain finite only if $(jD_\theta^j)_{s=0} = 0$, and the third term requires that $(\sin \phi\, D_n^j)_{s=0} = 0$. Hence, we have the general pole condition

$$(\cos \phi\, D_\phi^j + jD_\theta^j + \sin \phi\, D_n^j)_{s=0} = 0 \qquad (13.58)$$

for which we may investigate the individual pole conditions. We have six separate possibilities to study, consisting of two types of closed shells, $(\phi)_{s=0} = 0$ and $(\phi)_{s=0} = \phi_t$, for $j = 0$, $j = 1$, and $j > 1$. It is easiest to treat the harmonics individually.

$\quad j = 0$: For this case, equation (13.58) reduces to

$$(\cos \phi\, D_\phi^0 + \sin \phi\, D_n^0)_{s=0} = 0 \qquad (13.59a)$$

If $(\phi)_{s=0} = 0$ the second term drops out, so that no restriction is needed on D_n^j, whereas if $(\phi)_{s=0} = \phi_t$,

$$D_\phi^0(0) = 0 \qquad (13.59b)$$

as recorded in the second line, second column, of Table 8–2. Also, for the latter case, equation (13.59a) gives

$$D_\phi^0(0) = -\tan \phi_t\, D_n^0(0) \qquad (13.59c)$$

as listed in the second line, third column, of Table 8–2.

$\quad j = 1$: Here, equation (13.58) becomes

$$(\cos \phi\, D_\phi^1 + D_\theta^1 + \sin \phi\, D_n^1)_{s=0} = 0 \qquad (13.60a)$$

If $(\phi)_{s=0} = 0$,

$$D_\phi^1(0) + D_\theta^1(0) = 0 \qquad (13.60b)$$

as recorded in the third line, second column, of Table 8–2. Note that $D_n^1(0) = 0$ is also listed in the table. This arises out of consideration of γ_θ and is not required here—but neither is it contradicted. Now, if $(\phi)_{s=0} = \phi_t$, we have

$$(\cos \phi_t\, D_\phi^1 + D_\theta^1 + \sin \phi_t\, D_n^1)_{s=0} = 0 \qquad (13.61a)$$

Examining the corresponding entry in row 3, column 3, of Table 8–2, we find

$$D_\phi(0) = -\cos \phi_t\, D_\theta(0); \quad D_n(0) = -\sin \phi_t\, D_\theta(0) \qquad (13.61b)$$

which obviously cannot be completely obtained from equation (13.61a). However, equation (13.61b) does satisfy equation (13.61a), so that no contradiction is encountered.

$j > 1$: In this case,

$$D_\phi^j(0) = D_\theta^j(0) = D_n^j(0) = 0 \qquad (13.62)$$

as listed in line 4 of Table 8–2 for both types of closed shells, obviously satisfies equation (13.58), but, as before, consideration of some of the other individual relationships is required to certify the correctness of these conditions.

Proceeding with similar arguments for the other strains that have singularities at $Z = 0$—in other words, ω^j, κ_θ^j, τ^j, and γ_θ^j—the remaining entries in Table 8–2 may be verified.

Having isolated the necessary constraints on the displacements to avoid singularities, we now turn to the actual modified strain-displacement relationships. We again use ε_θ^j as given by equation (13.57) to illustrate the approach. In view of equation (13.58), equation (13.57) evaluated at $s = 0$ becomes

$$(\varepsilon_\theta^j)_{s=0} = \left[\frac{\cos\phi\, D_\phi^j + jD_\theta^j + \sin\phi\, D_n^j}{R_0} \right]_{s=0} = \frac{0}{0} \qquad (13.63)$$

which is an indeterminate form. Using L'Hospital's rule, we may investigate the $\lim_{s\to 0} \varepsilon_\theta^j$. It is convenient to differentiate with respect to ϕ, and then to convert to s using equation (13.54b). Proceeding, we have

$$\lim_{s\to 0} \varepsilon_\theta^j = \left[\frac{-\sin\phi\, D_\phi^j(L_i/R_\phi) + \cos\phi\, D_{\phi,s}^j + jD_{\theta,s}^j}{R_\phi \cos\phi\, (L_i/R_\phi)} \right]_{s\to 0}$$

$$+ \left[\frac{\cos\phi\, D_n^j(L_i/R_\phi) + \sin\phi\, D_{n,s}^j}{R_\phi \cos\phi\, (L_i/R_\phi)} \right]_{s\to 0} \qquad (13.64)$$

$$= \left[\frac{1}{L_i} D_{\phi,s}^j + \frac{j}{L_i \cos\phi} D_{\theta,s}^j + \frac{1}{R_\phi} D_n^j - \frac{\tan\phi}{R_\phi} D_\phi^j + \frac{\tan\phi}{L_i} D_{n,s}^j \right]_{s\to 0}$$

For the dome, $\phi \to 0$ as $s \to 0$ and the last two terms drop out, giving

$$\varepsilon_\theta^j(0) = \frac{1}{L_i} D_{\phi,s}^j(0) + \frac{j}{L_i} D_{\theta,s}^j(0) + \frac{1}{R_\phi(0)} D_n^j(0) \qquad (13.65)$$

In practice, when a dome is modeled by multiple segments, equation (13.65) is used for $\varepsilon_\theta^j(s)$ in the uppermost segment where $i = 0$.

To make this discussion complete, we present the entire set of modified strain-displacement equations for a *dome*.[25]

$$\{\varepsilon_0^j\} = [\mathbf{B}_0^j] \{\mathbf{\Delta}_0^j\} \qquad (13.66)$$

where rows 2, 3, 5, 6, and 8 (corresponding to ε_θ, ω, κ_ϕ, τ, and γ_θ) of $[\mathbf{B}_i^j]$ are replaced by the modified equations, such as equation (13.65), as depicted in Matrix 13–2.

Additionally, there are circumstances in which the pole cannot conveniently be located at $s = 0$ but coincides with $s = L_i$. For example, a shell which is closed at both ends will present such a situation at one end unless complete loading and geometric symmetry are present, whereupon

Matrix 13–2

$$
[\mathbf{B_0^j}] =
\begin{bmatrix}
\dfrac{1}{L_0}(\ )_{,s} & 0 & \dfrac{1}{R_\phi} & 0 & 0 \\[2ex]
\dfrac{1}{L_0}(\ )_{,s} & \dfrac{j}{L_0}(\ )_{,s} & \dfrac{1}{R_\phi} & 0 & 0 \\[2ex]
\dfrac{-j}{L_0}(\ )_{,s} & 0 & 0 & 0 & 0 \\[2ex]
0 & 0 & 0 & \dfrac{1}{L_0}(\ )_{,s} & 0 \\[2ex]
0 & 0 & 0 & \dfrac{1}{L_0}(\ )_{,s} & \dfrac{j}{L_0}(\ )_{,s} \\[2ex]
0 & 0 & 0 & \dfrac{-j}{2L_0}(\ )_{,s} & 0 \\[2ex]
\dfrac{1}{R_\phi} & 0 & \dfrac{-1}{L_0}(\ )_{,s} & -1 & 0 \\[2ex]
0 & \dfrac{1}{R_\phi} & \dfrac{-j}{L_0}(\ )_{,s} & 0 & -1
\end{bmatrix}
$$

only half the shell must be considered. Similar modified strain-displacement equations may be developed by enforcing the finite strain conditions at $s = L_i$ instead of $s = 0$.

An analogous set of equations can be developed for the case shown in Figure 4–1(b) where $\phi = \phi_t$.

Thus, we have the strain-displacement relationships given as a function of the nondimensional arc length variable s by Matrix 13–1 and equation (13.56), augmented by special equations for pole segments, Matrix 13–2, and equation (13.66.)

13.1.4.5 Total Potential Energy Functional. U_τ, as defined by equation (9.14), is the sum of U_ε, the strain energy, and U_q, the potential energy of the applied loads.

First, considering the strain energy U_ε as given by equation (9.4), we may integrate through the thickness to obtain the strain energy for segment i in harmonic j:

$$
U_{\varepsilon i}^j = \frac{1}{2}\int_0^1\int_{-\pi}^{\pi} \lfloor \mathbf{N^j} \rfloor \{\boldsymbol{\varepsilon}^j\}\, dA_i \tag{13.67a}
$$

where the differential surface area

$$
dA_i = RL_i d\theta\, ds \tag{13.67b}
$$

We proceed with the displacement formulation, generalizing for specifically orthotropic shells, by replacing $[\mathbf{N^j}]$ using equation (8.12):

$$
U_{\varepsilon i}^j = \frac{1}{2}\int_0^1\int_{-\pi}^{\pi} \Big[[\mathbf{D}_{or}]\{\boldsymbol{\varepsilon}^j\} - \{\mathbf{N_{Tor}}\} \Big]^T \{\boldsymbol{\varepsilon}^j\}\, dA_i \tag{13.68a}
$$

$$
= \frac{1}{2}\int_0^1\int_{-\pi}^{\pi} \Big(\{\boldsymbol{\varepsilon}^j\}^T [\mathbf{D}_{or}]\{\boldsymbol{\varepsilon}^j\} - \lfloor \mathbf{N_{Tor}} \rfloor \{\boldsymbol{\varepsilon}^j\} \Big) dA_i \tag{13.68b}
$$

noting that $[\mathbf{D}_{or}]$ is symmetric. Finally, we replace the strains by the displacements using Matrix 13–1 and equation (13.56), and get

$$U^j_{\varepsilon i} = \frac{1}{2} \int_0^1 \int_{-\pi}^\pi \left(\left[[\mathbf{B}_\mathbf{i}^\mathbf{j}]\{\boldsymbol{\Delta}_\mathbf{i}^\mathbf{j}\} \right]^T [\mathbf{D}_{or}] \left[[\mathbf{B}_\mathbf{i}^\mathbf{j}]\{\boldsymbol{\Delta}_\mathbf{i}^\mathbf{j}\} \right] - \lfloor \mathbf{N_{Tor}} \rfloor [\mathbf{B}_\mathbf{i}^\mathbf{j}]\{\boldsymbol{\Delta}_\mathbf{i}^\mathbf{j}\} \right) dA_i \qquad (13.69)$$

Now, we turn to the potential energy of the applied loads U_q, as defined in equations (9.10). Of the three terms contained in the general expression, equation (9.10a), which accommodate distributed, line, and concentrated loadings, only the first term and the second integral of the second term are pertinent here. Using Fourier series, line loads along the s_α or s_ϕ coordinate line, as represented by the first integral in the second term, may be included in the first term, and any concentrated loads, as represented by the third term, may be included in the second integral of the second term. Therefore, in terms of the present notation

$$U^j_{qi} = -\int_0^1 \int_{-\pi}^\pi \lfloor \boldsymbol{\Delta}_\mathbf{i}^\mathbf{j} \rfloor \{\mathbf{q}_\mathbf{i}^\mathbf{j}\} \, dA_i - \int_{-\pi}^\pi \lfloor \boldsymbol{\Delta}_\mathbf{i}^\mathbf{j}(Z_i) \rfloor \{\hat{\mathbf{q}}_\mathbf{i}^\mathbf{j}(Z_i)\} \, R(Z_i) d\theta$$

$$- \int_{-\pi}^\pi \lfloor \boldsymbol{\Delta}_\mathbf{i}^\mathbf{j}(Z_{i+1}) \rfloor \{\hat{\mathbf{q}}_\mathbf{i}^\mathbf{j}(Z_{i+1})\} R(Z_{i+1}) d\theta \qquad (13.70)$$

where $\{\mathbf{q}_\mathbf{i}^\mathbf{j}\}$, $\{\boldsymbol{\Delta}^\mathbf{j}(Z_i)\}$, and $\{\hat{\mathbf{q}}_\mathbf{i}^\mathbf{j}(Z_i)\}$ are defined corresponding to equation (9.10a). Note that in a numerical calculation, the second and the third terms of equation (13.70) would both not ordinarily appear. The second term might be included with segment $i - 1$, or the third with segment $i + 1$, depending on the computational order.

For the entire shell,

$$U^j_\tau = \sum_{i=1}^p (U^j_{\varepsilon i} + U^j_{qi}) \qquad (13.71)$$

where the shell is assumed to be composed of p segments. The principle of minimum total potential energy, equation (9.14), gives the equilibrium condition

$$\delta U^j_\tau = 0 \qquad (13.72)$$

For asymmetric loading, the complete solution is composed of the superposition of the solutions to equation (13.72) for the relevant participating harmonics.

13.1.4.6 Rayleigh-Ritz Solution.   The Rayleigh-Ritz method was presented in Section 9.4.3 as a promising numerical approach to the solution of equations such as equation (13.72). In the context of the present development for a single general harmonic j, the vector $\{\boldsymbol{\Delta}\}$, as introduced in equation (9.15), is the assemblage of the vectors $\{\boldsymbol{\Delta}_\mathbf{i}^\mathbf{j}\}$ for all of the segments comprising the shell. Recall that $\{\boldsymbol{\Delta}_\mathbf{i}^\mathbf{j}\}$ was defined in equation (13.53) as the vector of displacements for segment i. The index m introduced in equation (9.15) is equal to 5 in this case, since there are five displacement components in $\{\boldsymbol{\Delta}_\mathbf{i}^\mathbf{j}\}$, except for the $j = 0$ case, when D_θ and $D_{\theta\phi} = 0$ and m is reduced to 3. But, for the general case, there will be five *comparison* or *trial* functions of the form of equation (9.16) for each of the p segments of the shell. Polynomials of the form

$$\Delta_{ik}^j = c_{ik0}^j + c_{ik1}^j s + \cdots c_{ikl}^j s^l \cdots + c_{ikn}^j s^n \quad \begin{pmatrix} k = 1,5 \\ i = 1,p \end{pmatrix} \tag{13.73}$$

where $\Delta_{i1}^j = D_\phi^j$, $\Delta_{i2}^j = D_\theta^j$, $\Delta_{i3}^j = D_n^j$, $\Delta_{i4}^j = D_\phi^j$, and $\Delta_{i5}^j = D_\theta^j$ are quite popular for this application. It is important to note that n, the order of the approximation for each Δ_{ik}^j, may vary for each displacement k, segment i, and harmonic j.

Upon substitution of the set of $5p$ comparison functions of the form of equation (13.73) for $\{\Delta_i^j\}$ into equations (13.69) and (13.70), and the operation by $[\mathbf{B}_i^j]$ on the polynomials, the extremum problem represented by equation (13.72) is transformed into the maximum-minimum problem of the calculus

$$\frac{\partial U_\tau^j}{\partial c_{ikl}^j} = 0 \quad \begin{pmatrix} l = 0, \ldots, n \\ k = 1, \ldots, 5 \\ i = 1, \ldots, p \end{pmatrix} \tag{13.74}$$

The maximum-minimum conditions, equation (13.74), are necessary, but not sufficient, to determine the n coefficients (c_{ikl}^j), that are present for each of the five comparison functions Δ_{ik}^j in every segment $i = 1, p$. In addition to equation (13.74), *intersegment continuity conditions* recognize that the displacements are continuous along the common boundary of adjacent segments. Also, *kinematic boundary conditions* which fix the values of the corresponding displacements may be prescribed on the external boundaries. Thus, in summary, the maximum-minimum conditions, the intersegment continuity conditions, and the kinematic boundary conditions comprise the necessary and sufficient conditions to satisfy the principle of minimum total potential energy approximately. The further development of this approach is beyond our objectives here and is left to the province of the finite element method.[26]

One further point should be noted, however. Static boundary conditions are also encountered. These enter as the *natural boundary conditions* of the variational problem; this means, essentially, that if the kinematic correspondent of a certain force or moment is *not* prescribed on the boundary, that force or moment will take the value of the corresponding external loading component along the boundary. As an example, consider a typical stress resultant, N_ϕ, on given external boundary $\phi = \phi'$. If the displacement $D_\phi(\phi')$ is not prescribed *a priori*, along that boundary, then $N_\phi(\phi')$ is equal to the corresponding element of the external loading vector $\hat{q}_\phi(\phi')$. If $\hat{q}_\phi(\phi') = 0$, then $N_\phi(\phi') = 0$.

We have outlined a quite general, energy-based approach that may be developed and extended to solve many complex problems in the field of shells of revolution on a fairly routine, automated basis. This technique is developed quite completely in a companion volume,[27] and an example of the capability for analyzing a torospherical shell was shown in Figure 13–4.

13.2 SHELLS OF TRANSLATION

13.2.1 General

The general theory of shells of translation, apart from cylindrical shells, is lightly treated in contemporary English-language textbooks. Although many papers in the literature describe analysis methods for such shells, most are based on idealized boundary conditions that do not necessarily match the physical situation very closely. For example, the inclusion of flexible edge members, which are often present in reality, greatly complicates the analysis. From a practical standpoint, however, numerous translational shells with spectacularly large spans have been constructed. As an example, we may take the shell shown in Figure 2–8(a). Such major structures certainly warrant a complete investigation, beyond the scope of the membrane theory. The means for carrying out such investigations are often computer-based numerical solutions, such as finite element or finite difference techniques, or carefully executed experimental studies using physical models. The current availability and capabilities of numerical analysis procedures for translational shells are described in Schnobrich.[28] Here, we introduce two theories upon which many of the popular numerical algorithms are founded.

13.2.2 Mushtari-Donnell-Vlasov (MDV) Theory

This theory simplifies the general theory of shells in orthogonal curvilinear coordinates and is based on the assumptions that (a) the effect of the transverse shear forces Q_α and Q_β in the in-plane equilibrium equations is negligible; and, (b) the influence of the normal displacement D_n will predominate over the influences of the in-plane displacements D_α and D_β in the bending response of the shell. We have discussed the first assumption at length in connection with other derivations, and it should be acceptable provided the membrane theory boundary conditions are not grossly violated. The second assumption is most feasible as the shell becomes relatively flat, approaching a plate. Such a shell is said to be *shallow*. Shallow shells were introduced in Section 2.8.1, and we build on this geometric concept in the following paragraphs. Some quantitative criteria for a shell to be classified as shallow are also stated later.

Proceeding, we first consider the force equilibrium equations in general form, equations (3.17). We omit the transverse shear stress terms as stated in assumption (a) and take $N_{\alpha\beta} = N_{\beta\alpha} = S$:

$$(BN_\alpha)_{,\alpha} + (AS)_{,\beta} + A_{,\beta}S - B_{,\alpha}N_\beta + q_\alpha AB = 0 \tag{13.75a}$$

$$(BS)_{,\alpha} + (AN_\beta)_{,\beta} + B_{,\alpha}S - A_{,\beta}N_\alpha + q_\beta AB = 0 \tag{13.75b}$$

$$(BQ_\alpha)_{,\alpha} + (AQ_\beta)_{,\beta} - N_\alpha \frac{AB}{R_\alpha} - N_\beta \frac{AB}{R_\beta} + q_n AB = 0 \tag{13.75c}$$

We first wish to express the transverse shear forces in terms of the normal displacement D_n. With transverse shearing strains suppressed, Q_α and Q_β are given in terms of the stress couples through equations (3.22a) and (3.22b):

$$Q_\beta = \frac{1}{AB}\left[(BM_{\alpha\beta})_{,\alpha} + (AM_\beta)_{,\beta} + B_{,\alpha}M_{\beta\alpha} - A_{,\beta}M_\alpha\right] \tag{13.76a}$$

$$Q_\alpha = \frac{1}{AB}\left[(BM_\alpha)_{,\alpha} + (AM_{\beta\alpha})_{,\beta} + A_{,\beta}M_{\alpha\beta} - B_{,\alpha}M_\beta\right] \tag{13.76b}$$

Next, we apply the constitutive law, Matrix 8–2, omitting the thermal terms, and take $M_{\alpha\beta} = M_{\beta\alpha}$ to write the stress couples as functions of κ_α, κ_β, and τ:

$$M_\alpha = D(\kappa_\alpha + \mu\kappa_\beta) \tag{13.77a}$$

$$M_\beta = D(\mu\kappa_\alpha + \kappa_\beta) \tag{13.77b}$$

$$M_{\alpha\beta} = D(1 - \mu)\tau \tag{13.77c}$$

Before substituting equations (13.77a–c) into (13.76a) and (13.76b), we express the changes in curvature as functions of the displacements. These relationships are given, in general, by equations (7.47a–c), but we invoke assumption (b) and eliminate the contributions of D_α and D_β. The simplified compatibility equations are

$$\kappa_\alpha = -\frac{1}{A}\left(\frac{1}{A}D_{n,\alpha}\right)_{,\alpha} - \frac{1}{AB^2}A_{,\beta}D_{n,\beta} \tag{13.78a}$$

$$\kappa_\beta = -\frac{1}{B}\left(\frac{1}{B}D_{n,\beta}\right)_{,\beta} - \frac{1}{A^2 B}B_{,\alpha}D_{n,\alpha} \tag{13.78b}$$

$$\tau = -\frac{1}{AB}\left(D_{n,\alpha\beta} - \frac{1}{A}A_{,\beta}D_{n,\alpha} - \frac{1}{B}B_{,\alpha}D_{n,\beta}\right) \tag{13.78c}$$

By comparison to equations (7.55), these relationships are essentially those for a flat plate. The next step is to substitute equations (13.78a–c) into equations (13.77a–c) and then into equations (13.76a) and (13.76b). The algebra is considerable, and we defer to Novozhilov[29] to write

$$Q_\alpha = -D\frac{1}{A}(\nabla^2 D_n)_{,\alpha} \tag{13.79a}$$

$$Q_\beta = -D\frac{1}{B}(\nabla^2 D_n)_{,\beta} \tag{13.79b}$$

where the Laplacian, in general terms, is given by

$$\nabla^2(\ ) = \frac{1}{AB}\left\{\left[\frac{B}{A}(\ )_{,\alpha}\right]_{,\alpha} + \left[\frac{A}{B}(\ )_{,\beta}\right]_{,\beta}\right\} \tag{13.79c}$$

Finally, substituting equations (13.79a) and (13.79b) into equation (13.75c) and dividing through by AB, we obtain

$$\frac{-D}{AB}\left\{\left[\frac{B}{A}(\nabla^2 D_n)_{,\alpha}\right]_{,\alpha} + \left[\frac{A}{B}(\nabla^2 D_n)_{,\beta}\right]_{,\beta}\right\} - \frac{N_\alpha}{R_\alpha} - \frac{N_\beta}{R_\beta} + q_n = 0 \tag{13.80}$$

or, in view of equation (13.79c),

$$D\nabla^4 D_n + \frac{N_\alpha}{R_\alpha} + \frac{N_\beta}{R_\beta} - q_n = 0 \tag{13.81}$$

Equation (13.81) is very similar to the equilibrium equation for a plate, augmented by the contribution of the extensional stress resultants to the resistance to the normal loading q_n. Thus, we have equations (13.75a) and (13.75b), together with equation (13.81) and the requisite boundary conditions, constituting the shell theory which was apparently derived independently by Mushtari in the Soviet Union and Donnell in the United States.

The governing equations at this stage of the development are somewhat similar to those encountered in Section 11.3.5 following the derivation of equation (11.114). Equations (11.72) and (11.114) formed a coupled set of equations for the finite displacement theory of plates. The dependent variables in those equations were the *normal displacement* and the *in-plane stress resultants; that is* essentially the same situation we now have with equations (13.75a) and (13.75b) and equation (13.81). Therefore, the introduction of a stress function $\mathcal{F}$, defined such that

$$N_\alpha = \frac{1}{B}\left(\frac{1}{B}\,\mathcal{F}_{,\beta}\right)_{,\beta} + \frac{1}{AB}B_{,\alpha}\frac{1}{A}\mathcal{F}_{,\alpha} \tag{13.82a}$$

$$N_\beta = \frac{1}{A}\left(\frac{1}{A}\,\mathcal{F}_{,\alpha}\right)_{,\alpha} + \frac{1}{AB}A_{,\beta}\frac{1}{B}\mathcal{F}_{,\beta} \tag{13.82b}$$

$$S = -\frac{1}{AB}\left(\mathcal{F}_{,\alpha\beta} - \frac{1}{A}A_{,\beta}\mathcal{F}_{,\alpha} - \frac{1}{B}B_{,\alpha}\mathcal{F}_{,\beta}\right) \tag{13.82c}$$

is quite logical.[30]

The steps in the back substitution are tedious and, basically, are similar to our previous exercises in the development of the von Kármán equations in Section 11.3.5. We omit the details and follow the procedure suggested by Vlasov.[31] Also, we consider the special case in which the in-plane components of the surface loading q_α and q_β are equal to 0. The resulting equations are

$$\frac{1}{A^2B}\left[\left(\frac{1}{A}B_{,\alpha}\right)_{,\alpha} + \left(\frac{1}{B}A_{,\beta}\right)_{,\beta}\right]\mathcal{F}_{,\alpha} = -\frac{1}{R_\alpha R_\beta}\frac{1}{A}\mathcal{F}_{,\alpha} = 0 \tag{13.83a}$$

$$\frac{1}{AB^2}\left[\left(\frac{1}{A}B_{,\alpha}\right)_{,\alpha} + \left(\frac{1}{B}A_{,\beta}\right)_{,\beta}\right]\mathcal{F}_{,\beta} = -\frac{1}{R_\alpha R_\beta}\frac{1}{B}\mathcal{F}_{,\beta} = 0 \tag{13.83b}$$

and

$$D\nabla^4 D_n + \overline{\nabla}^2\mathcal{F} - q_n = 0 \tag{13.84a}$$

where

$$\overline{\nabla}^2(\ ) = \frac{1}{AB}\left\{\left[\frac{1}{R_\beta}\frac{B}{A}(\ )_{,\alpha}\right]_{,\alpha} + \left[\frac{1}{R_\alpha}\frac{A}{B}(\ )_{,\beta}\right]_{,\beta}\right\} \tag{13.84b}$$

The equalities in equations (13.83a) and (13.83b) have been obtained by invoking equation (2.37), the third Gauss-Codazzi relationship; the second term in equation (13.84a) is expressed in terms of the operator $\overline{\nabla}^2(\ )$ by using the first two Gauss-Codazzi relations, equations (2.35) and (2.36).

It may now be argued that for certain classes of problems, equations (13.83a) and (13.83b) may be considered to be satisfied. First, for shells of zero Gaussian curvature, such as cylindrical or conical shells, equations (13.83a) and (13.83b) are *identically* satisfied. Also, if we think of shells that are relatively flat or shallow, the Gaussian curvature will be small. Therefore the coefficients of $\mathcal{F}_{,\alpha}$ and $\mathcal{F}_{,\beta}$ in equations (13.83a) and (13.83b) will be an order of magnitude *smaller* than the coefficients of the same terms in equations (13.84), so that the former equations can be regarded to be *approximately* satisfied. A third class of problems for which the first two equations may be ignored are those cases in which the stress resultants are rapidly varying functions of α and β. Then, the higher derivatives of $\mathcal{F}$ predominate, and the contribution of the terms associated with $\mathcal{F}_{,\alpha}$ and $\mathcal{F}_{,\beta}$ will be relatively small as compared to those multiplying $\mathcal{F}_{,\alpha\alpha}$ and $\mathcal{F}_{,\beta\beta}$. Similar reasoning was used to motivate the Geckeler approximation in Section 13.1.2.3.

We require a second relationship between D_n and $\mathcal{F}$. Since equations (13.83) and (13.84) are a statically indeterminate system of equilibrium equations, an appropriate set of compatibility equations is sought. This accomplishment is attributed to Vlasov by Novozhilov[32] and requires two general steps.

The first operation is the derivation of the compatibility equations for the deformation of the middle surface in terms of strains and curvatures only. These equations are analogous to the St. Venant equations of the theory of elasticity and may be established by following the same steps employed in the derivation of the Gauss-Codazzi relations in Section 2.6, if the tangent vectors to the *deformed* middle surface are used in place of the tangent vectors to the *undeformed* middle surface. Specifically, $\mathbf{t}_{n,\alpha}$ and $\mathbf{t}_{n,\beta}$ in equation (2.33) are replaced by $\mathbf{r}'_{,\alpha}$ *and* $\mathbf{r}'_{,\beta}$, as given by equations (7.9) and (7.11), respectively. Then the identical steps are carried out on these tangent vectors to derive the compatibility equations, which may be viewed as Gauss-Codazzi relations for the *deformed* middle surface.[31]

The second operation required is the substitution of equations (13.78a–c) for the changes in curvature; the replacement of the in-plane strains by the corresponding stress resultants using Matrix 8–2; and the subsequent introduction of the stress function $\mathcal{F}$ via equations (13.82a–c).

The final result, after eliminating terms similar to those previously dropped in equations (13.83a) and (13.83b), is

$$\frac{1}{Eh}\nabla^4\mathcal{F} - \overline{\nabla}^2 D_n = 0 \tag{13.85}$$

Thus, the solution of the coupled equations (13.84a) and (13.85) is sufficient to describe the complete pattern of stress and deformation within the limitations of the approximations introduced in the course of the derivation. Novozhilov observed that these equations form the basis for the investigation of many practical problems.[33] Also, explicit solutions are provided by Vlasov[34] and are based on some further approximations: (a) the shell is described by Cartesian coordinates, with

$$A = B \simeq 1 \tag{13.86}$$

and, (b) R_α and R_β are taken as constant (average values) and are given by

$$R_\alpha(X,Y) = \overline{R}_X \tag{13.87}$$

$$R_\beta(X,Y) = \overline{R}_Y \tag{13.88}$$

These assumptions are specifically tied to the shallow shell notion and, according to Vlasov, can generally be utilized for shells that cover rectangular plan areas if the rise is no more than 1/5 of the smaller side of the rectangle.

The introduction of the simplifications given in equations (13.86) to (13.88) into the previous equations reduces the operators $\nabla^2(\ )$ and $\overline{\nabla}^2(\ )$ to

$$\nabla^2(\ ) = (\ )_{,XX} + (\ )_{,YY} \tag{13.89}$$

$$\overline{\nabla}^2(\ ) = \frac{1}{\overline{R}_Y}(\ )_{,XX} + \frac{1}{\overline{R}_X}(\ )_{,YY} \tag{13.90}$$

where $\overline{R}_Y$ and $\overline{R}_X$ are average values of the radii of curvature along the respective coordinate lines. The solution then proceeds with the introduction of a potential function G, such that

$$D_n = \nabla^4 G \tag{13.91a}$$

and

$$\mathscr{F} = Eh\overline{\nabla}^2 G \tag{13.91b}$$

When introduced into equation (13.84a), equations (13.91) give[35]

$$\nabla^8 G + \frac{12(1 - \mu^2)}{h^2}\overline{\nabla}^4 G - \frac{q_n}{D} = 0 \tag{13.92}$$

There is a marked similarity between equation (13.92) and equation (12.75), which governs the general bending of cylindrical shells, and the solution strategies are similar. Once G is found, $\mathscr{F}$ is determined from equation (13.91b). The interested reader is referred to Vlasov[36] for the complete development of this solution. As an example, the bending of an elliptic paraboloid shell of a similar configuration to that shown on Figure 6–15 can be investigated using equation (13.92). Also, these equations can be readily generalized to the study of the instability of certain shells, as we see later.

13.2.3 Theory of Shallow Shells

The theory of shallow shells is roughly equivalent to the simplification of the MDV theory that is facilitated by equations (13.87) and (13.88) and has been widely used for shell analysis. The approach is to employ some simplifying assumptions at the outset to derive the governing equations in a form which is particularly suited for shell roofs covering rectangular plan areas. This theory has also been used to analyze shells that become *locally* shallow when the original shell is divided into finite segments or elements—a procedure to which we have often alluded in our previous discussions.

Refer to the basic geometric relationships introduced in Section 2.8.1. Recalling the assumption that $(Z_{,X})^2$ and $(Z_{,Y})^2$ may be neglected in comparison to unity, we may evaluate the principal radii of curvature from equation (2.29)

$$R_X = -\frac{1}{Z_{,XX}} \tag{13.93a}$$

Similarly,

$$R_Y = -\frac{1}{Z_{,YY}} \tag{13.93b}$$

The expressions for the extensional strains ε_X and ε_Y are obtained directly from equation (7.46a) and (7.46b) as

$$\varepsilon_X = D_{X,X} - Z_{,XX}D_n \tag{13.94a}$$

and

$$\varepsilon_Y = D_{Y,Y} - Z_{,YY}D_n \tag{13.94b}$$

The equation for the in-plane shearing strain could similarly be written from equation (7.46c). However, it is more correct to use a somewhat different form that arises because the Cartesian coordinates are not orthogonal when projected on the surface. The expanded form of ω is[37]

$$\omega = D_{Y,X} + D_{X,Y} - 2Z_{,XY}D_n \tag{13.95}$$

The term $Z_{,XY} = -(1/R_{XY})$ may be called the twisting curvature and was encountered previously in Section 6.3 when planar $X - Y$ coordinates were also used. Additionally, we have the expressions for $\bar{\kappa}_\alpha$, $\bar{\kappa}_\beta$, and τ, which with $A = B = 1$, reduce to

$$\kappa_X = -D_{n,XX} \tag{13.96a}$$

$$\kappa_Y = -D_{n,YY} \tag{13.96b}$$

$$\tau = -D_{n,XY} \tag{13.96c}$$

Equations (13.94) to (13.96) are the strain-displacement or compatibility equations for this theory.

We now proceed as in the previous section. The analogue to the normal equilibrium equation, (13.84a), is[38]

$$D\nabla^4 D_n - (Z_{,YY}\mathscr{F}_{,XX} - 2Z_{,XY}\mathscr{F}_{,XY} + Z_{,XX}\mathscr{F}_{,YY}) - q_n = 0 \tag{13.97}$$

where the stress function is related to the stress resultants by the simplified form of equations (13.82a–c),

$$N_X = \mathscr{F}_{,YY} \tag{13.98a}$$

$$N_Y = \mathscr{F}_{,XX} \tag{13.98b}$$

$$S = -\mathscr{F}_{,XY} \tag{13.98c}$$

Similarly, an analogous equation to equation (13.85) is developed. This equation is written as[39]

$$\frac{1}{Eh}\nabla^4\mathscr{F} + (Z_{,YY}D_{n,XX} - 2Z_{,XY}D_{n,XY} + Z_{,XX}D_{n,YY}) = EhP \tag{13.99a}$$

where P is derived from the applied surface loading as

$$P = (1 + \mu)(P_{X,YY} + P_{Y,XX}) - \mu\nabla^2(P_X + P_Y) \tag{13.99b}$$

$$P_X(X,Y) = \int q_X \, dX \tag{13.99c}$$

$$P_Y(X,Y) = \int q_Y \, dY \tag{13.99d}$$

and

$$\nabla^2(\ ) = (\ )_{,XX} + (\ )_{,YY} \tag{13.99e}$$

P_X and P_Y in equations (13.99c) and (13.99d) are cumulative in-plane forces and are set to zero on boundaries that are unrestrained in the X and Y directions, respectively. Note that this theory is slightly complicated by starting with nonorthogonal coordinates, but retains the generality of including the in-plane components of the surface loading q_X and q_Y. For most shells encountered in practice, the MDV theory and the theory of shallow shells are fairly equivalent and both are sometimes referred to as *shallow shell theories*. From an applications standpoint, the most significant difference is probably in the nature of the external boundaries. If the boundaries are coincident with the coordinate lines of the middle surface, the general form of the Mushtari-Donnell-Vlasov theory is appropriate; if the boundaries fall along the Cartesian coordinate lines, the theory of shallow shells, or the even simpler specialization of the MDV theory, seems to be more attractive.

13.2.4   Strain Energy for Shallow Shells

The energy expressions and principles derived in Chapter 9 are directly adaptable to shells of translation, but we have not carried out the details of the specializations based on the theories presented in the last two sections. However, it is of general interest to examine the expression for strain energy, equation (9.6), using the simplified strain-displacement relationship, equations (13.94–13.96). With the transverse shearing strains neglected, term [**I**] of equation (9.7) becomes

$$
\begin{aligned}
[\mathbf{I}] &= \varepsilon_x^2 + \varepsilon_Y^2 + 2\mu\varepsilon_x\varepsilon_Y + \frac{1-\mu}{2}\,\omega^2 \\[2mm]
&= \left(D_{X,X} - \frac{1}{Z_{,XX}} D_n\right)^2 + \left(D_{Y,Y} - \frac{1}{Z_{,YY}} D_n\right)^2 \\[2mm]
&\quad + 2\mu\left(D_{X,X} - \frac{1}{Z_{,XX}} D_n\right)\left(D_{Y,Y} - \frac{1}{Z_{,YY}} D_n\right) \\[2mm]
&\quad + \frac{1-\mu}{2}\left(D_{Y,X} + D_{X,Y} - 2Z_{,XY}D_n\right)^2
\end{aligned}
\tag{13.100}
$$

Term [**II**] is dropped in the linear theory, and

$$
\begin{aligned}
[\mathbf{III}] &= (\kappa_X + \kappa_Y)^2 - 2(1-\mu)(\kappa_X\kappa_Y - \tau^2) \\
&= (D_{n,XX} + D_{n,YY})^2 - 2(1-\mu)(D_{n,XX}D_{n,YY} - D_{n,XY}^2)
\end{aligned}
\tag{13.101}
$$

The resulting expression,

$$
U_\varepsilon = \frac{E}{2(1-\mu^2)} \int_S \left\{ h[\mathbf{I}] + \frac{h^3}{12}[\mathbf{III}] \right\} dX\,dY \tag{13.102}
$$

is frequently encountered in the literature. The thermal terms, as defined in equation (9.6), can easily be added if required.

13.3 INSTABILITY AND FINITE DEFORMATIONS

13.3.1 General

Local instabilities, characterized by displacements of comparatively small wave length similar to those occurring in plate structures, may be significant in thin-walled vessels where compressive stresses are present. This mode of instability is delineated from general instability, which could correspond to the buckling of an entire stack or tower as a beam-column. Because the amplitudes encountered in local instabilities are comparatively small, this phenomenon can be analyzed in many cases by using the simplifications of the shallow shell theories presented in the previous sections.

There are several possible approaches to the *elastic* instability analysis of shells. These are illustrated schematically on Figure 13–6, where we have plotted the normal displacement, D_n, against a loading parameter, $\bar{N}$, which represents some combination of the in-plane loading.

First, if we can arrive at a general solution for the normal displacement in the presence of transverse and in-plane loading, and then we observe that certain discrete values of $\bar{N}$ cause the displacements to become excessively large, we term these values of $\bar{N}$ the *critical* loads for a *perfect* shell, $\bar{N}_{cr}$. This is illustrated by the curves labeled one in Figure 13–6, with the various branches corresponding to different magnitudes of the known transverse loading. This approach was used for the study of plate instability in Section 11.3.3. Note that the same $\bar{N}_{cr}$ is obtained regardless of the transverse loading, which we call q_n for shells.

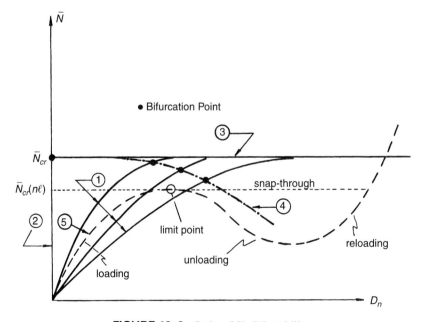

FIGURE 13–6 Paths of Shell Instability.

The second approach follows from the realization that $\overline{N}_{cr}$ is *independent* of q_n. This suggests that the transverse loading can be ignored in the stability analysis, which means that for $\overline{N} < \overline{N}_{cr}$, $D_n = 0$. This is indicated by curve 2, which is the special case of 1 for $q_n = 0$. Then, at $N = \overline{N}_{cr}$, the displacements become unbounded as before, following path 3. This approach is attractive because the complete solution for D_n as a function of q_n and $\overline{N}$ is *not* required. Rather, the states of equilibrium just prior to and just after instability can be examined without concern for the remainder of the load-deflection curve. At the transition point $(\overline{N}_{cr}, 0)$, a *bifurcation* of equilibrium is said to occur, because there are at least two possible states of equilibrium, $D_n = 0$ and $D_n \to \infty$. This approach is, of course, familiar from the classical solution of the Euler column buckling problem. In regard to the unbounded displacements following bifurcation, curve 3, a more sophisticated nonlinear analysis may yield an unstable (descending) post-buckling path, curve 4.

Although it is implied by curve 2 that the equilibrium path prior to buckling is linear, buckling of shells may also be preceded by a nonlinear prebuckling response, such as a curve 1.[40] The bifurcation would then occur at the intersection of curves 1 and 4.[41]

A third approach which should be mentioned is that of using *geometrically nonlinear* strain-displacement relationships to determine the complete load-deflection curve—unloading as well as loading. As illustrated by curve 5, a critical load $\overline{N}_{cr(nl)}$, which occurs at a limit point, may be found somewhat below $\overline{N}_{cr}$. A classical illustration is an axially compressed conical shell, such as that shown in Figure 4–13, where the axial shortening and accompanying rotational deformation would cause an immediate degradation of the stiffness.[40] Then, the curve would show unloading, and perhaps reloading, as other elements of resistance are mobilized by the finite deformations. If this latter resistance is adequate, the load may eventually surpass $\overline{N}_{cr(nl)}$, and even $\overline{N}_{cr}$, provided the material capacity is not exceeded. An actual shell cannot unload in most cases, so there may be a sudden transition from the *loading* to the *reloading* branches of the curve at $\overline{N} = \overline{N}_{cr(nl)}$, with a corresponding jump in D_n. This is termed *snap-through* buckling and is characteristic of shallow arches and shells.

Shells with initial imperfections follow a path such as 5. As the magnitude of the imperfections decreases, the descending branch of curve 5 approaches curve 4.

Note that all three of the general approaches discussed have been predicated on elastic buckling. If material nonlinearities are present for $\overline{N} \leq \overline{N}_{cr}$, then additional complications arise. Sometimes these can be handled conveniently by the tangent modulus theory.

13.3.2 Equilibrium Equation Method

As we mentioned in our discussion of the bifurcation of equilibrium approach to elastic instability in the previous section, we need only consider the immediate neighborhood of the transition point $(\overline{N}_{cr}, 0)$. Up to this point, the shell will be regarded as being subject to a membrane state of stress N_{X0}, N_{Y0}, and S_0.[42] At the transition point, the greatly increased normal displacements result in the development of normal components of these in-plane stress resultants. Such normal components may be treated as surface loading in the $\mathbf{t_n}$ direction, because the starting point for the analysis is the state of stress N_{X0}, N_{Y0}, and S_0. With this in mind, the normal components of the membrane stress resultants are used for q_n in equations (13.81) and (13.84a) if the MDV theory is used; in equation (13.92) if the simplified version of the

MDV theory is used; and in equation (13.97) if we choose the theory of shallow shells. It now remains to determine the normal components of the membrane stress resultants.

Refer to Figure 11–3, where the normal components of the in-plane stress resultants for a plate are apparent. (The analysis corresponding to this figure is contained in Section 11.3.1.) Because the net normal force caused by any single stress resultant is the *difference* of the normal projections at the two ends of the differential segment, that analysis is valid for shells as well. Following equation (11.70), with the transverse shear forces and in-plane loads omitted, we have for Cartesian coordinates

$$q_{n1} = -N_{X0}D_{XY,X} - N_{Y0}D_{YX,Y} - N_{XY0}(D_{YX,X} + D_{XY,Y}) \tag{13.103}$$

With transverse shearing strains suppressed and $N_{XY0} = S_0$, and with D_X and D_Y neglected in comparison with D_n in the Cartesian coordinate version of equations (7.46g) and (7.46h), in the spirit of the shallow shell assumptions, we have

$$\left.\begin{aligned} D_{XY} &= -D_{n,X} \\ D_{YX} &= -D_{n,Y} \end{aligned}\right\} \tag{13.104}$$

so that

$$q_{n1} = N_{X0}D_{n,XX} + N_{Y0}D_{n,YY} + 2S_0D_{n,XY} \tag{13.105a}$$

which is sometimes written as

$$q_{n1} = -[N_{X0}\kappa_X + N_{Y0}\kappa_Y + 2S_0\tau] \tag{13.105b}$$

in view of equations (13.96). For orthogonal curvilinear coordinates, the same analysis is valid, and

$$q_{n1} = -[N_{\alpha0}\kappa_\alpha + N_{\beta0}\kappa_\beta + 2S_0\tau] \tag{13.105c}$$

We proceed with the simplest of the three theories, the MDV theory in Cartesian coordinates. The governing equation is written in terms of the potential function G from equation (13.92) and q_n is taken as q_{n1}, as given by equation (13.105a):[43]

$$\nabla^8 G + \frac{12(1-\mu^2)}{h^2}\overline{\nabla}^4 G = \frac{1}{D}(N_{X0}D_{n,XX} + N_{Y0}D_{n,YY} + 2S_0D_{n,XY}) \tag{13.106a}$$

where again

$$\nabla^2(\ ) = (\ )_{,XX} + (\ )_{,YY} \tag{13.106b}$$

$$\overline{\nabla}^2(\ ) = \frac{1}{R_y}(\ )_{,XX} + \frac{1}{R_x}(\ )_{,YY} \tag{13.106c}$$

and G is related to D_n and $\mathscr{F}$ by

$$D_n = \nabla^4 G \tag{13.106d}$$

and

$$\mathscr{F} = Eh\overline{\nabla}^2 G \tag{13.106e}$$

If we write the rhs of equation (13.106a) in terms of G, we have

$$\nabla^8 G + \frac{12(1 - \mu^2)}{h^2} \overline{\nabla}^4 G$$

$$= \frac{1}{D} [N_{X0}(\nabla^4 G)_{,XX} + N_{Y0}(\nabla^4 G)_{,YY} + 2S_0(\nabla^4 G)_{,XY}]$$

(13.107)

13.3.3 Buckling of Pressurized Spherical Shells

Vlasov has investigated the stability of pressurized spherical shells by using equations (13.106a–e) and (13.107).[44] For a sphere of radius a,

$$\overline{\nabla}^2(\ ) = \frac{1}{a} \nabla^2(\ )$$

(13.108)

The membrane theory of axisymmetrically loaded spherical shells has been presented in Section 4.3.2.2. For a uniform internal suction or external pressure p, $q_n = -p$, equations (4.27) and (4.28) give

$$N_\phi = N_\theta = -\frac{pa}{2}$$

(13.109)

from which the prebuckling stresses

$$N_{X0} = N_{Y0} = -\frac{pa}{2}$$

(13.110a)

and

$$S_0 = 0$$

(13.110b)

Then the rhs of equation (13.107) becomes

$$-\frac{pa}{2D} [(\nabla^4 G)_{,XX} + (\nabla^4 G)_{,YY}] = -\frac{pa}{2D} \nabla^6 G$$

(13.111)

Now, in view of equations (13.109–13.111), equation (13.107) reduces to

$$\nabla^8 G + \frac{12(1 - \mu^2)}{h^2 a^2} \nabla^4 G + \frac{pa}{2D} \nabla^6 G = 0$$

(13.112)

We seek solutions of the form

$$\nabla^2 G = \lambda G$$

(13.113)

Where λ is a characteristic or eigenvalue related to $p = p_{cr}$, and obtain

$$\lambda^2 \left[\lambda^2 + \frac{p_{cr} a}{2D} \lambda + \frac{12(1 - \mu^2)}{h^2 a^2} \right] = 0$$

(13.114)

The trivial solution $\lambda = 0$ is discarded, and we have remaining the quadratic equation

$$\lambda^2 + \frac{p_{cr} a}{2D} \lambda + \frac{12(1 - \mu^2)}{h^2 a^2} = 0$$

(13.115)

which we write in terms of p_{cr} as

$$p_{cr} = -\frac{2D}{a}\left[\lambda + \frac{12(1-\mu^2)}{h^2a^2\lambda}\right] \tag{13.116}$$

It is initially unclear as to which value of λ corresponds to the minimum buckling load.

We may proceed by invoking the minimization condition to find the lowest value of p_{cr}, p_{cr1}. This condition is

$$p_{cr,\lambda} = 0 \tag{13.117a}$$

from which

$$1 - \frac{12(1-\mu^2)}{h^2a^2\lambda^2} = 0 \tag{13.117b}$$

or

$$\lambda = \pm\frac{2\sqrt{[3(1-\mu^2)]}}{ha} \tag{13.117c}$$

The negative root of λ is meaningful, because it results in a positive p_{cr}, which has already been defined as an internal suction or external pressure.

With

$$\lambda = \frac{-2\sqrt{[3(1-\mu^2)]}}{ha} \tag{13.118}$$

$$p_{cr1} = \frac{2E}{\sqrt{[3(1-\mu^2)]}}\left(\frac{h}{a}\right)^2 \tag{13.119}$$

is the lowest buckling pressure for a spherical shell. This equation is applicable for complete spheres, as well as for spherical shells that have close to ideal membrane boundary conditions. A direct treatment of spherical shell buckling in polar coordinates is developed in Brush and Almroth.[45]

Note that we have arrived at the lowest critical load, *without* first finding a general solution for the normal displacement D_n. Subsequently we may investigate the normal displacement D_n, which is the *mode shape* of the buckle.

If we return to equation (13.106d),

$$D_n = \nabla^4 G$$
$$= \nabla^2(\lambda G) \tag{13.120}$$
$$= \lambda\nabla^2 G$$

The potential function G, in turn, is found from equation (13.113)

$$\nabla^2 G - \lambda G = 0 \tag{13.121}$$

where λ is given by equation (13.118). The solutions of equation (13.121) are the eigenfunctions of the problem. Recall, however, that we have used the simplest form of the MDV theory based on Cartesian coordinates. Consequently, solutions of equations (13.120) and

(13.121) that satisfy the kinematic boundary conditions are only feasible if the shell covers a rectangular plan area. For boundaries coincident with the coordinate lines, the original form of the theory, equations (13.84a), (13.84b), and (13.85), should be used in place of the subsequently simplified equation (13.92). This is carried out in some detail in Vlasov,[46] where it is shown that the lowest critical pressure only differs from p_{cr1}, as computed from equation (13.119), by a term of $O[\mu(h/a)]:1$. Nevertheless, the more rigorous solution provides a basis for determining the eigenfunctions and subsequent displacement or mode shapes. In many applications, only p_{cr1} is of interest, and the solution obtained from equation (13.119) is sufficient to estimate the buckling pressure. Note that, in this regard, quite large factors of safety against *elastic* buckling—perhaps three to five or even more—are commonly specified for the design of thin shells. Because the bifurcation analysis accounts for neither initial imperfections nor nonlinear behavior, the critical pressure obtained theoretically is very difficult to approach in reality, even in carefully controlled experiments,[47] and a very ample factor of safety is prudent unless a more sophisticated analysis can be accomplished. A means for introducing additional refinements into the analysis is described in Section 13.3.6. However, many *design* procedures are based on the *elastic* critical load, reduced substantially by a "knockdown factor."

13.3.4   Energy Method for Cylindrical Shells

From an energy standpoint, the transition between the prebuckled and postbuckled states may be represented by the following stability criterion[48]: (a) there is no bending prior to the onset of buckling, so that the total strain energy is due to the in-plane stress resultants; (b) at the onset of instability, there are additional contributions to the strain energy due to middle surface straining and bending; and, (c) the increase in strain energy as buckling occurs must be equal to the work done by the external loading and by the components of the in-plane forces that act through *normal* displacements. The latter source of "external" work is analogous to the load component q_{n1} introduced in section 13.3.2 and is not present during infinitesimal deformations. We illustrate this for a cylindrical shell under a uniform axial ring loading, as shown in Figure 13–7.

We first consider the strains and charges in curvature prior to and following buckling.

In the prebuckled state, which we designated by the subscript 0, we assume that there is no bending. There is no normal load acting, so that the third equilibrium equation, (12.5b), with $Q_X = q_{n0} = 0$, gives $N_{\theta 0} = 0$. We now refer to the constitutive law, equations (8.10) with $\alpha = X$, $\beta = \theta$, and no thermal terms.

The axial strain just before buckling is found from the first row of Matrix 8–2 with $N_\alpha = N_X = -\bar{N}_{cr}$ as

$$\varepsilon_{X0} = -\frac{\bar{N}_{cr}}{Eh} \tag{13.122a}$$

The corresponding circumferential strain follows from the second row of Matrix 8–2 as

$$\varepsilon_{\theta 0} = -\mu \varepsilon_{X0} \tag{13.122b}$$

As a result of buckling, the normal displacement becomes D_n producing circumferential strain $\varepsilon_{\theta 1}$. This strain is proportional to D_n through equation (12.6b), so that

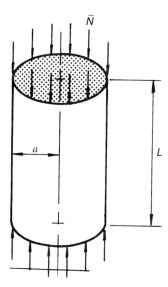

FIGURE 13–7   Axially Loaded Cylindrical Shell.

$$\varepsilon_{\theta 1} = \frac{D_n}{a} \tag{13.123}$$

Therefore, the total circumferential strain after the shell has buckled is

$$\varepsilon_\theta = \varepsilon_{\theta 0} + \varepsilon_{\theta 1}$$

$$= \frac{D_n}{a} - \mu\varepsilon_{X0} \tag{13.124}$$

To find the corresponding meridional strain, we now refer to the first equation of Matrix 8–2 with $N_\alpha = N_X = -\overline{N}_{cr}$:

$$-\overline{N}_{cr} = \frac{Eh}{1 - \mu^2}(\varepsilon_X + \mu\varepsilon_\theta) \tag{13.125a}$$

from which

$$\varepsilon_X = -\frac{(1 - \mu^2)}{Eh}\overline{N}_{cr} - \mu\varepsilon_\theta \tag{13.125b}$$

Substituting for $\overline{N}_{cr}/Eh$ from equation (13.122a) and for ε_θ from equation (13.124), we find

$$\varepsilon_X = (1 - \mu^2)\varepsilon_{X0} - \mu\left(\frac{D_n}{a} - \mu\varepsilon_{X0}\right)$$

$$= \varepsilon_{X0} - \mu\frac{D_n}{a} \tag{13.126}$$

Also, we have the change in curvature after buckling

$$\kappa_X = -D_{n,XX} \tag{13.127}$$

Of course, $\kappa_X = 0$ prior to buckling.

Now, we may evaluate the change in strain energy during buckling. We may specialize equation (9.6) for the case of cylindrical shells with axisymmetric loading to

$$U_\varepsilon = \frac{Ea\pi}{1 - \mu^2} \int_0^L \left\{ h(\varepsilon_X^2 + \varepsilon_\theta^2 + 2\mu\varepsilon_X\varepsilon_\theta) + \frac{h^2}{12}\kappa_X^2 \right\} dX \qquad (13.128)$$

The change in strain energy is given by

$$\Delta U_\varepsilon = U_\varepsilon(\varepsilon_X, \varepsilon_\theta, \kappa_X) - U_{\varepsilon 0}(\varepsilon_{X0}, \varepsilon_{\theta 0}, 0) \qquad (13.129)$$

We substitute equations (13.126), (13.124), and (13.127), respectively, for ε_X, ε_θ, and κ_X into equation (13.128) to compute U_ε; also, using equations (13.122a) and (13.122b), we take ε_{X0} and $-\mu\varepsilon_{X0}$ for ε_X and ε_θ, respectively, to evaluate $U_{\varepsilon 0}$. Equation (13.129) then becomes

$$\Delta U_\varepsilon = a\pi \int_0^L \left[Eh\left(\frac{D_n}{a}\right)^2 - 2\mu Eh\varepsilon_{X0}\left(\frac{D_n}{a}\right) + D(D_{n,XX})^2 \right] dX \qquad (13.130)$$

Next, we consider the work done by the external loading and the normal components of the in-plane forces during buckling. For the external loading, $\overline{N} = \overline{N}_{cr}$ and

$$\Delta U_{q\overline{N}} = 2\pi a \int_0^L -\overline{N}_{cr}(\varepsilon_X - \varepsilon_{X0})dX$$

$$= 2\mu\pi\overline{N}_{cr} \int_0^L D_n dX \qquad (13.131)$$

in view of equation (13.126). Note that there is no 1/2 coefficient in this expression, because the full $\overline{N}_{cr}$ is already present prior to the onset of buckling. For the normal components of the in-plane forces, refer to the loading q_{n1} given by equation (13.105a). The work done over the circumference by q_{n1} acting through D_n is

$$\Delta U_{qn} = \frac{1}{2} \int_0^L (q_{n1}D_n)(2\pi a \, dX) \qquad (13.132a)$$

In equation (13.105a) for q_{n1}, we take $N_{X0} = -\overline{N}_{cr}$ and $N_{Y0} = S_0 = 0$, whereupon equation (13.132a) becomes

$$\Delta U_{qn} = -\int_0^L \pi a\overline{N}_{cr}D_{n,XX}D_n dX \qquad (13.132b)$$

Adding equations (13.131) and (13.132b), we have

$$\Delta U_q = \Delta U_{q\overline{N}} + \Delta U_{qn}$$

$$= 2\pi\overline{N}_{cr}\left\{ \mu \int_0^L \left[D_n - \frac{a}{2} D_{n,XX}D_n \right] dX \right\} \qquad (13.133)$$

Finally, equating equations (13.130) and (13.133) completes the application of the stability criterion:

$$\Delta U_\varepsilon = \Delta U_q \qquad (13.134)$$

These equations can be treated most easily with a Rayleigh-Ritz-type approximation for D_n. We provide an example in the following section.

13.3.5 Buckling of Axially Loaded Cylindrical Shells

We now investigate the buckling of the shell shown in Figure 13–7, assuming that the buckled form is axisymmetrical. We use the energy criterion, equation (13.134), and assume a one-term Rayleigh-Ritz-type solution for D_n

$$D_n = C \sin m\pi \frac{X}{L} \tag{13.135}$$

Substituting equation (13.135) into ΔU_ε and ΔU_q as given by equations (13.130) and (13.133), respectively, gives

$$\Delta U_\varepsilon = a\pi \left\{ \frac{EhC^2L}{2a^2} - \frac{2\mu Eh\varepsilon_{X0}C}{a} \int_0^L \sin m\pi \frac{X}{L} dX + D \frac{m^4\pi^4C^2}{2L^3} \right\} \tag{13.136}$$

and

$$\Delta U_q = 2\pi \overline{N}_{cr} \left\{ \mu C \int_0^L \sin m\pi \frac{X}{L} dX + \frac{a}{4} \frac{m^2\pi^2}{L} C^2 \right\} \tag{13.137}$$

After substituting equations (13.136) and (13.137) into equation (13.134), and noting from equation (13.122a) that $\varepsilon_{X0} = (-N_{cr}/Eh)$, we find the coefficients of the C terms cancelling and the coefficients of C^2 producing

$$\overline{N}_{cr} = D \left[\frac{m^2\pi^2}{L^2} + \frac{Eh}{a^2D} \frac{L^2}{m^2\pi^2} \right] \tag{13.138}$$

The first term on the r.h.s of equation (13.138) corresponds to plate buckling, equation (11.95), while the second term represents a restoring effect due to membrane stretching.[49] The minimum $\overline{N}_{cr}$ is found from

$$\overline{N}_{cr,m} = 0 \tag{13.139a}$$

giving

$$\frac{m\pi}{L} = \sqrt[4]{\left(\frac{Eh}{a^2D} \right)} \tag{13.139b}$$

and

$$\overline{N}_{cr1} = 2D \sqrt{\left(\frac{Eh}{a^2D} \right)}$$

$$= \frac{Eh^2}{a\sqrt{[3(1 - \mu^2)]}} \tag{13.140a}$$

For $\mu = 0.3$

$$\overline{N}_{cr} = 0.605 \, Eh^2/a \tag{13.140b}$$

or, in terms of stress

$$\sigma_{cr} \approx 0.6 \, Eh/a \tag{13.140c}$$

which is a well-known and widely used formula.

The length of the buckled waves is given, in terms of the characteristic $\sqrt{ah}$ term noted in Section 12.3.2, by

$$\frac{L}{m} = \pi \sqrt[4]{\left(\frac{a^2 D}{Eh}\right)} \simeq 1.72 \sqrt{(ah)} \tag{13.141}$$

It is obvious from equation 13.138 that N_{cr} is relatively insensitive to m, similar to plate buckling as illustrated in Figure 11–4. Thus, circumferential stiffeners alone would not significantly increase the buckling strength of cylindrical shells under axial load.[50] On the other hand, properly spaced meridional stiffeners would be effective in increasing the bucking load. Essentially, referring to Figure 8–1, the effective thickness in equations (13.140) would be increased from h to h_X.

The identical result may be obtained by solving the equilibrium equation, equation (13.107).

Various other specific solutions based on the two approaches developed in this chapter are available in the literature for spherical and cylindrical shells. Using either the differential equation or the energy approach the solutions are quite simple and may be used for other geometries with some judicious geometric approximations (see Vlasov[51] and Flügge[52]). Also, an extensive array of more complex situations, including body forces, fluid pressure, torsion, orthotropic shells, and initial imperfections, is treated in Brush and Almroth.[53]

In summary, however, to the exposition of the linear theory of shell stability, we restate the comment of Flügge[52] "And, after all, there is only a loose correlation between the actual collapse of a shell and the buckling load obtained from a linear theory."

13.3.6 Finite Deformation of Shallow Shells

Recall from Section 11.3.5 that the introduction of the geometrically nonlinear strain-displacement relations, (11.109a–c), lead to the von Kármán equations, describing finite deformations for plates. Vlasov has derived a similar set of equations for shallow shells.[54]

In this case, the nonlinear strain-displacement relations are the same as for plates except that $\bar{\varepsilon}_X$ and $\bar{\varepsilon}_Y$ each have an additional term proportional to the radii of curvature, as we may verify from equations (7.46a) and (7.46b). Therefore, the appropriate equations in Cartesian coordinates are

$$\bar{\varepsilon}_X = D_{X,X} + \frac{D_n}{R_X} + \frac{1}{2}(D_{n,X})^2 \tag{13.142a}$$

$$\bar{\varepsilon}_Y = D_{Y,Y} + \frac{D_n}{R_Y} + \frac{1}{2}(D_{n,Y})^2 \tag{13.142b}$$

$$\bar{\omega} = D_{Y,X} + D_{X,Y} + D_{n,X}D_{n,Y} \tag{13.142c}$$

If we follow the identical steps leading to equation (11.116), we obtain[55]

$$\nabla^4 \mathcal{F} = Eh\left[(D_{n,XY})^2 - D_{n,XX}D_{n,YY} - \frac{1}{R_X}D_{n,YY} - \frac{1}{R_Y}D_{n,XX}\right] \tag{13.143}$$

as the generalized form of that compatibility equation. As before, $\mathcal{F}$ is the stress function defined by equations (11.115).

Now considering the second equation, which was derived for plates as equation (11.117), note that this equation followed from equation (11.72), which is the finite displacement version of the basic biharmonic equation for plate bending, equation (10.12). Following the same reasoning, the generalization of equation (10.12) to shallow shell theory is equation (13.81), where the term

$$\frac{N_\alpha}{R_\alpha} + \frac{N_\beta}{R_\beta}$$

is added to the lhs. This term enters into the shallow shell equivalent of equation (11.72) as

$$\frac{N_X}{R_X} + \frac{N_Y}{R_Y}$$

and subsequently into the shallow shell form of equation (11.117) as

$$-\left(\frac{1}{R_X}\mathcal{F}_{,YY} + \frac{1}{R_Y}\mathcal{F}_{,XX}\right)$$

on the rhs. Finally, we have the generalization of equilibrium equation (11.117):

$$\nabla^4 D_n = \frac{1}{D}\left[q_n + \mathcal{F}_{,YY}D_{n,XX} + \mathcal{F}_{,XX}D_{n,YY} - 2\mathcal{F}_{,XY}D_{n,XY}\right.$$
$$\left. - \frac{1}{R_X}\mathcal{F}_{,YY} - \frac{1}{R_Y}\mathcal{F}_{,XX}\right] \tag{13.144}$$

Equations (13.143) and (13.144) constitute the generalized von Kármán equations, which form a finite deformation theory for shallow shells. In these equations, the in-plane stress resultants N_X and N_Y are assumed as positive. The corresponding plate equations (11.116) and (11.117), of course, follow for $R_X = R_Y = \infty$; also, for a cylindrical shell of radius a, $R_X = \infty$ and $R_Y = a$. Similarly, for a spherical shell with radius a, $R_X = R_Y = a$.

We have noted in the previous section that the elastic buckling load based on infinitesimal deformation theory, as presented in Sections 13.3.1 to 13.3.5, may only be a rough indication of the load at which an actual shell may buckle. We have also suggested that a more elaborate analysis may provide theoretical results which are in reasonable agreement with physical experiments. Such an analysis has been provided by two of the most distinguished mechanicians and engineers of the twentieth century, T. von Kármán and H. S. Tsien, in their study of the postbuckling behavior of cylindrical shells based on the finite deformation equations presented in this section.[56] They used a Rayleigh-Ritz approach and assumed an approximate expression for D_n, selecting the unspecified parameters based on minimization of the potential energy functional and on some test results. This analysis provides a strain energy expression similar to equation (13.136), modified by a factor proportional to the buckled normal deflection to thickness ratio D_n/h. For $D_n/h = 0$, the critical load is given by equation (13.140a); as D_n/h increases, the critical load decreases rapidly to only about 1/3 of this value.[57]

A frequently used equation for the buckling of axially loaded cylindrical shells based on their analysis is

$$\overline{N}_{cr1} = \frac{Eh^2}{3a\sqrt{[3(1-\mu^2)]}}$$
$$\simeq 0.2\frac{Eh^2}{a} \tag{13.145}$$

implying a "knockdown factor" of 3 when compared to equation (13.140b).

By way of confirming this theory, an experimental study of open aluminum cylindrical shells with simply supported, uniformly compressed ends and free longitudinal edges by Yang and Guralnick gave measured buckling loads that correlated quite well with numerical predictions based on equation (13.140).[58] It was observed that the correlation between analytically predicted and experimentally measured buckling loads was considerably better than similar comparisons for closed cylindrical shells. It was also noted that open shells have a theoretical critical stress of only 10% to 15% of the corresponding values for comparable closed shells in the range of parameters investigated. A relatively complete compilation of the rather extensive literature in this area is contained in the study.

13.3.7 Postbuckling and Imperfection Sensitivity

Inasmuch as the theoretically determined buckling load calculated by a small deflection theory is rarely attained or even approached in experiments, attention has been focused on the sources of discrepancy such as (a) the influence of boundary conditions; (b) the presence of prebuckling deformations caused by edge constraints; and, (c) the effect of initial imperfections.[59] It is generally assumed that the initial imperfections are the most influential contributor.

The effect of initial imperfections is closely related to the load-displacement response following the onset of buckling, the *postbuckling path.* Although the calculation of the secondary equilibrium path is beyond our present scope, we may observe some possibilities from Figure 13–6. Curve 4 has a negative slope and is said to be *unstable,* whereas a system which follows curve 5 and snaps through to the positive slope branch is said to be *stable* in the postbuckling region. The possibilities for slightly imperfect shells are indicated more clearly in Figure 13–8,[58] where three cases are shown. Between the cases of stable symmetric postbuckling (**I**) and unstable symmetric postbuckling (**III**) is the case of unsymmetric postbuckling (**II**) which may follow a stable or unstable course.

Structures which behave as case **I** are said to be *imperfection-insensitive.* For this case, the classical buckling load is a useful measure, whereas for the other cases, a more refined calculation is usually necessary. Imperfection-sensitive structures may be prone to a more violent buckling failure and sudden collapse, so a careful analysis is warranted. The seminal contribution in the field of postbuckling analysis and imperfection sensitivity is the theory of Koiter.[59,60]

Rotational shells subject to axial compression are generally highly imperfection sensitive; in the case of plates or shallow roof shells which can develop a secondary mode of resistance, such as cantenary or tension field action, buckling is relatively insensitive to initial imperfections.

Shells may also buckle under external normal pressure or internal suction. Spherical shells under such loading are generally imperfection-sensitive. More generally, Wittek[61] considered free-ended rotational shells of varying Gaussian curvature, in other words, an ellipsoid (+), a cylinder (0), and a hyperboloid (−), under an axisymmetrical normal pressure. In Figure 13–9, results are shown for the critical pressure λ as a function of R_1/R_2, where $R_1 = R_0(\phi = \pi/2)$ which is proportional to R_θ, and $R_2 = R_\phi$. Three different computations were performed.

The upper curve, λ_{cr}, is the small deformation (linear) critical load, whereas the lower curve, λ_b, is a reduced critical load based only on the bending portion of the strain energy,

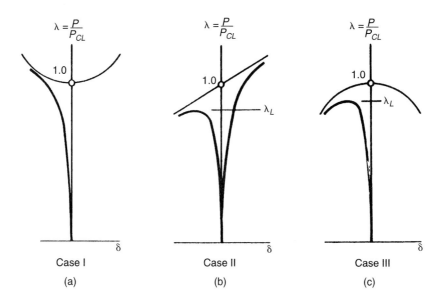

(a) Case I: Stable Postbuckling.
(b) Case II: Unsymmetric Postbuckling
(c) Case III: Unstable Postbuckling

FIGURE 13–8   Postbuckling Paths for Slightly Imperfect Structures (a) Case I: Stable Postbuckling, (b) Case II: Unsymmetric Postbuckling, (c) Case III: Unstable Postbuckling.

roughly equivalent to neglecting the group-I terms in equation (9.7). The latter calculation is claimed to yield a lower bound solution to the linear buckling value. The third computation, a nonlinear analysis indicated by the plotted points, will be discussed subsequently.

Buckling may be idealized as a shift in the load resistance mechanism from a stiff membrane mode, associated with the group-I terms in equation (9.7), to a more flexible bending mode. Accordingly, the initial imperfections would be primarily acted upon by the in-plane stress resultants. It follows that this sensitivity would be measured by the difference of the curves, $\lambda_{cr} - \lambda_b = \lambda_m$. This difference obviously reduces with decreasing Gaussian curvature and is quantified by the factor α_g. In the negative curvature region, the points where the curves intersect correspond to geometrical combinations for which pure bending or inextensional deformations, discussed in Section 5.2.1 and 7.6.2, may occur.

Between the curves for λ_{cr} and λ_b, the plotted points indicate values calculated by a nonlinear analysis. The difference between these values and the lower curve is regarded as confirmation of λ_b as the lower bound, not only for the linear but also for the nonlinear critical load.

Shells must also be investigated for buckling caused by asymmetric normal pressure, such as the wind load depicted on Figure 5–18(a) for a spherical shell and on Figure 5–15 for

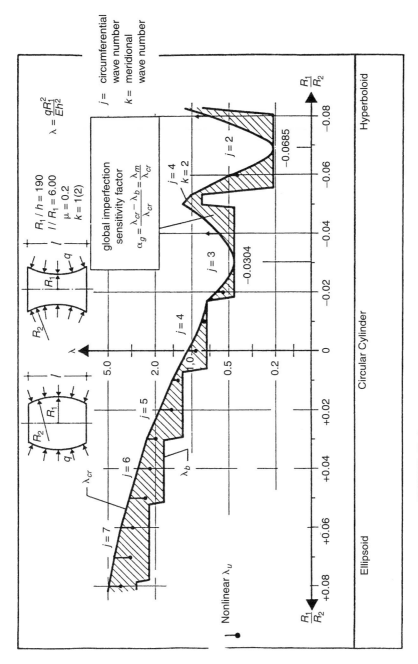

FIGURE 13-9 Global Imperfection Sensitivity of a Rotational Shell.

a cylinder or hyperboloid. We expect buckling to initiate in regions of high compression under those loading conditions.

For the spherical shell, the resultants plotted in Figure 5–18(b) indicate that there are regions along the windward meridian where both N_ϕ and N_θ are compressive. This is true for the simplified as well as for the realistic pressure distribution. The biaxial state of compression tends to increase the imperfection sensitivity.

In contrast, for the cylindrical shell considered in Section 5.3.4, the circumferential stress resultant N_θ is proportional to the load, equation (5.65b), and thus would be compressive along the windward meridian, $\theta = 0°$. However, the meridional stress, equation (5.65a), would be tensile in this region as a result of overturning and would tend to counter a circumferentially initiated buckle. Similarly, in the separation zone, $\theta \simeq 70°$, the tendency for a meridionally initiated buckle caused by a compressive N_ϕ is countered by a tensile value of N_θ. The same behavior is true for hyperboloidal shells. Such shells, where the compression is only uniaxial and perhaps opposed in the perpendicular direction, may be regarded as imperfection-insensitive for that particular loading. For an axial or self-load, a state of biaxial compression may be present in the same shell, for example, Figure 4–8, and the shell would be more sensitive to imperfections.

REFERENCES

1. V. V. Novozhilov, *Thin Shell Theory* [translated from 2nd Russian ed. by P. G. Lowe] (Groningen, The Netherlands: Noordhoff, 1964), chapter 4.

2. W. Flügge, *Stresses in Shells,* 2nd ed., (Berlin: Springer-Verlag, 1973), chap. 6.

3. H. Kraus, *Thin Elastic Shells,* (New York: Wiley, 1967), chapters 5, 6, 7.

4. H. Reissner, *Spannungen in Kugelschalen,* (Leipzig: Muller-Breslau-Fetschrift, 1912):181–193.

5. E. Meissner, Des elastizitatsproblem für dünne schalen von ringflächen-, kugel-, and kegelform, *Physik Zeit.* 14 (1913):343–349.

6. Flügge, *Stresses in Shells:* 326–344.

7. Ibid.

8. Novozhilov, *Thin Shell Theory,* chaps. 3–4.

9. Flügge, *Stresses in Shells,* chap. 6.

10. Kraus, *Thin Elastic Shells,* chaps. 5, 6, and 7.

11. Flügge, *Stresses in Shells:* 326–334.

12. Kraus, *Thin Elastic Shells:* 271–283.

13. Novozhilov, *Thin Shell Theory:* 315–319.

14. Kraus, *Thin Elastic Shells:* 261–271.

15. C. R. Calladine *Theory of Shell Structures,* (Cambridge: Cambridge University Press, 1983):371.

16. Flügge, *Stresses in Shells:* 341–351.

17. J. N. Pirok and R. S. Wozniak, Steel tanks, in *Structural Engineering Handbook,* E. H. Gaylord, Jr., and C. N. Gaylord, eds. (New York: McGraw-Hill, 1968), chap. 23.

18. P. L. Gould, S. K. Sen, R. S. C. Wang, H. Suryontomo and R. D. Lowrey, Column supported cylindrical-conical tanks. *Journal of the Structural Division,* ASCE 102, no. ST2 (February 1976): 429–447.

19. R. S. C. Wang and P. L. Gould, Continuously supported cylindrical-conical tanks, *Journal of the Structural Division, ASCE* 100, no. ST10 (October 1974): 2037–2052.

20. G. V. Ranjan and C. R. Steele, Analysis of torosphencal pressure vessels, *Journal of the Engineering Mechanics Division, ASCE* 102, no. EM4 (August 1976): 643–657.

21. P. L. Gould, J. S. Lin, and J. M. Rotter, Linear stress analysis of torospherical head, *Journal of Engineering Mechanics,* ASCE III, no. 10 (October 1985): 1296–1300.

22. Flügge, *Stresses in Shells:* 386–413.

23. Novozhilov, *Thin Shell Theory,* chap. 4.

24. P. L. Gould and S. L. Lee, Hyperboloids of revolution supported on columns, *Journal of the Engineering Mechanics Division, ASCE* 95, no. EM5 (October 1969): 1083–1100; P. L. Gould and S. L. Lee, Column-supported hyperboloids under wind load, (Zurich, Switzerland: Publications, International Association for Bridge and Structural Engineering, 1971), vol 31-11: 47–64.

25. L. J. Brombolich and P. L. Gould, Finite element analysis of shells of revolution by minimization of the potential energy functional, Proceedings of the Symposium on Applications of Finite Element Methods in Civil Engineering (Nashville, Tenn.: Vanderbilt University, 1969: 279–307.

26. Ibid and L. J. Brombolich and P. L. Gould, A high-precision curved shell finite element, *Synoptic, AIAA Journal 10,* no. 6 (June 1972): 727–728.

27. P. L. Gould, *Finite Element Analysis of Shells of Revolution,* (London: Pitman, 1985).

28. W. C. Schnobrich, Analysis of hyperbolic paraboloid shells, *Concrete Thin Shells,* publication SP-28 (Detroit: American Concrete Institute, 1971): 275–311; Different methods of numerical analysis of structures, Proc. of Symposium on Practical Aspects in the Computation of Shell and Spatial Structures, (K. U. Leuven, Belgium, July 1986).

29. Novozhilov, *Thin Shell Theory:* 88–94.

30. Ibid.

31. Vlasov, *General Theory of Shells:* 345–346.

32. Novozhilov, *Thin Shell Theory:* 92.

33. Ibid.

34. Vlasov, *General Theory of Shells:* 495–514.

35. Ibid.

36. Ibid.

37. Novozhilov, *Thin Shell Theory:* 94–99.

38. Ibid.

39. Ibid.

40. B. O. Almroth and J. H. Starnes, The computer in shell stability analysis, *Journal of the Engineering Mechanics Division, ASCE* 101, no. EM6 (December 1975): 873–888.

41. D. Bushnell, Static collapse: A survey of methods and modes of behavior, *Finite Elements in Analysis and Design* 1, (1985), pp. 165–205.

42. Vlasov, *General Theory of Shells:* 521–525.

43. Ibid.

44. Ibid., 525–529.

45. D. O. Brush and B. O. Almroth, *Buckling of Bars, Plates, and Shells,* (New York: McGraw-Hill, 1975): 142–258.

46. Vlasov, *General Theory of Shells:* 525–529.

47. S. Timoshenko and J. Gere, *Theory of Elastic Stability,* (New York: McGraw-Hill, 1961): 468–473.

48. Ibid., *457–519.*

49. Calladine, *Theory of Shell Structures:* 482–488.

50. Ibid., 535–542.

51. Vlasov, *General Theory of Shells:* 529–538.

52. Flügge, *Stresses in Shells,* iv and chap. 7.

53. Brush and Almroth, *Buckling of Bars, Plates, and Shells.*

54. Vlasov, *General Theory of Shells:* 538–543.

55. Ibid.

56. T. von Kármán and H. S. Tsien, The buckling of thin cylindrical shells under elastic compression, *Journal of Aeronautical Sciences* 8 (1941):303.

57. Timoshenko and Gere, *Theory of Elastic Stability:* 471–472.

58. T. H. Yang and S. A. Guralnick, Buckling of axially loaded open shells, *Journal of the Engineering Mechanics Division, ASCE* 102, no. EM2 (April 1976): 199–211.

59. H. Kolisnik and C. Tahiani, A survey of methods of analysis of stiffened shell structures, Civil Engineering Research Report No. 85-3, Royal Military College of Canada (August 1985).

60. W. T. Koiter, Over de stabiliteit van het elastisch evenwicht, Dissertation Delft, 1945, English Translation NASA TTF-10, 833 (1967).

61. U. Wittek, Beitrag zum tragverhalten der strukturen bei endlichen verformungen unter besonderer beachtung des nachbeulmechanismus dünner flächentragwerke, *Mitteilung* Nr. 80-1, Technical Reports, Institut für Konstruktiven Ingenieurbau, Ruhr-Universität Bochum (May 1980).

62. V. S. Kelkar and R. T. Sewell, *Fundamentals of the Analysis and Design of Shell Structures,* (Englewood Cliffs, N.J.: Prentice-Hall, 1987): 115–116.

63. Ibid: 127.

EXERCISES

13.1. Continue the solution of the compound shell shown in Figure 13–2 from equation (13.48) and verify the key points on the graphs shown in Figure 13–3(a). Use $h_1 = h_2 = h$, $h/a = 0.01$, and $\mu = 0.3$. Assume reasonable values for any additional properties required.

13.2. Verify the pole conditions for a shell of revolution listed in Table 8–2.

13.3. Derive $[\mathbf{B}_0^j]$, Matrix 13–2, from $[\mathbf{B}_1^j]$, Matrix 13–1.

13.4. Derive a matrix analogous to $[\mathbf{B}_0^j]$ for the closed shell shown in Figure 4–1(b).

13.5. Specialize the MDV and the shallow shell theories for cylindrical shells.

13.6. Investigate the bending theory solution under gravity loading for the elliptic paraboloid shown in Figure 6–15. Assume that the shell is shallow.

13.7. Compute the buckling pressure for the spherical shell treated in Section 13.3.3 using the energy method.

13.8. Compute the buckling load for the cylindrical shall treated in Section 13.3.5 using the equilibrium equation method.

13.9. Consider a closed cylindrical shell of length L and radius a and derive an expression for the critical uniform internal suction.

13.10. Compute the value of the critical density of material ρ (force/volume) for which a hemispherical shell of thickness h and radius a would buckle under self-weight.

13.11. Consider the cylindrical shell shown in Figure 13–2, but with a flat circular plate end. Determine the meridional moment and radial force at the junction. Take the plate thickness as twice the shell thickness h and assume reasonable values for any other properties required.

13.12. Try to obtain the solution to Exercise 13.11 as the degenerate case of the spherical head considered in Exercise 13.1

13.13. A conical tank is filled with water and is supported at its mid-height by a ring support, as shown in Figure 13–10. The tank has a circular slab at its base. Plot the variations of membrane stress resultants in the tank due to (a) self-weight and (b) water load. Discuss design problems which arise at the ring support and at the base slab. Hint: See the discussion at the end of Section 12.3.2 and also Kelkar and Sewell.[62] Also, see Section 11.2.6.

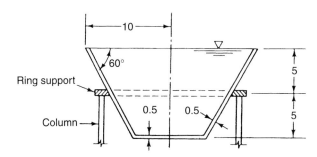

FIGURE 13–10   Exercise 13.13. Conical Water Tank Supported at Its Midheight by a Ring Beam. (Adapted from Kelkar and Sewell.)

13.14. The shell shown in Figure 13–11 is to be analyzed by the Flexibility Method. Outline the solution to the problem listing all of the pertinent equation in generic form.[63]

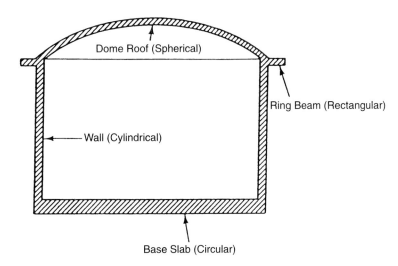

FIGURE 13–11   Exercise 13.14. Wall/dome/plate Junction to be Analyzed by the Flexibility Method. (Adapted from Kelkar and Sewell.)

13.15 Consider Exercise 4.1, which refers to Figure 4–2. Now assume that the base is hinged rather than free to displace in the normal direction and perform a complete analysis using an approximate bending solution. If the algebra becomes unwieldy, assume reasonable values for the material and geometric parameters.

14
Conclusion

14.1 GENERAL

The preceding chapters have treated the analysis of thin elastic plates and shells from a unified viewpoint, insofar as possible. Specializations to specific forms, for example, membrane shells, flat plates—were introduced in a logical sequence after the general elements of the theory were set forth. The emphasis was on explaining the mechanics of surface structures, with ample examples to illustrate the most important aspects. In general, relatively simple and idealized cases, which could readily be solved analytically, were used to illustrate the salient points. Fairly complete compilations of known analytical solutions for both plates[1] and thin shells[2] are available elsewhere to supplement the examples given in the text. The solution of more complex structures is relegated to the domain of numerical analysis, and several citations to the literature in this area were provided. Although the calculations for those situations may be more involved, such structures generally reflect the same basic behavioral characteristics as the relatively simple illustrations contained in this book.

14.2 PROPORTIONING

One aspect of the shell design process that has not been extensively addressed in this text is *proportioning*. Rather, the emphasis here has been on *analysis*. In a practical sense, proportioning is vital, and it is our hope that interest in this topic has been aroused. It is clear that the specific form of a surface structure is mainly determined by the anticipated function: flat plates for carrying pedestrian and vehicular traffic, shells of revolution for containment, and shallow shells for long, clear-span roofs. Also, the construction materials—generally metals and/or concrete—play a very important role in the proportioning process.

Although the proportioning of surface structures is obviously very much related to the anticipated function and the material of construction, it is possible to offer a few general comments on this topic. We consider separately (a) regions of the structure where the stresses are primarily in-plane or membrane; and, (b) regions where there is significant bending action.

In case (a), direct tensile stresses should be resisted entirely by reinforcing steel in concrete shells, or by providing ample wall thicknesses for metal shells. Regions with direct compressive stresses are generally controlled by stability requirements, particularly in structures formed from metal plates.

For case (b), the bending moments or stress couples may be resisted by considering a concrete section with reinforcement near the surfaces to act as a wide flexural member. However, the relatively small effective depth available may complicate the provision of ample reinforcing steel. For thin metal structures, it is often impractical to resist high local bending stresses with the basic shell thickness, so that stiffening members are attached. The stiffeners may also be effective in increasing the buckling strength.

Beyond these brief generalizations, the proportioning of thin surface structures is a somewhat subjective and always challenging branch of structural design which has become somewhat specialized for various structural forms and materials.

For thin concrete shells, the Building Code Requirements for Reinforced Concrete (ACI 318) give some general guidelines.[3] The shell thickness and the reinforcement area and spacing should satisfy the conditions of strength and serviceability which prevail for all concrete structures. Additionally, thin shells must be investigated for general and local instability. Often, the shell thickness is determined by the latter consideration. A shell may also require gradual thickening to resist stress concentrations due to abrupt changes in surface geometry, for example, along the ridge and valley lines of an HP shell such as shown in Figure 2–8(b). From an international perspective, the state-of-the-art on reinforcing concrete shells, along with a proposed unified approach for plates and slabs as well as shells, is presented in a recent paper by Laurenco and Figueiras.[4]

Reinforcement is required to resist tensile stresses, as well as to control shrinkage and thermal cracking. The bars which resist the in-plane stress resultants should be placed in two or more directions and should ideally be oriented in the general directions of the principal tensile stresses, especially in regions of high tension. Often, however, this is impractical, so that the required resistance in every direction must be provided by an orthogonal mesh. In providing reinforcement to replace tensile capacity of the concrete, considerations should be given to the strains as well as the stresses. Limiting the strains by using lower stress levels in the reinforcement is thought to minimize cracking in the adjacent concrete. Thus, the use of high-strength reinforcing bars is not advantageous. Also, smaller bars spaced closer together are generally preferred to an equivalent amount of larger diameter bars, both for crack control and increased cover.

A representative situation is illustrated in Figure 14–1, where the various types of reinforcement required in an umbrella shell, similar to those shown in Figure 2–8(b), are shown.

FIGURE 14–1   Double-Layered Roof Shell Under Construction.

The orthogonal grid for the shell proper, the additional shell steel in the vicinity of the columns, and the edge beam reinforcement are difficult to accommodate in a relatively thin section, and careful planning and supervision are necessary to ensure a sound, crack-free structure.

Reinforcement to resist stress couples should be placed near both faces even though moment reversal is not indicated by the calculations, since the bending may vary rapidly along the surface. The two layers may also include the membrane reinforcement for the in-plane tensile stresses, or there may be three or more layers. The provision of adequate clearance and cover may necessitate increasing the shell thickness in such cases.

At the junction of the shell and the edge or supporting member, the computed stresses in the shell reinforcement should be developed by anchorage within or beyond the width of the member. In turn, the edge members must be proportioned to resist the forces imparted by the shell. Often an adjacent portion of the shell can be assumed to act with the member, or the edge member may be formed by a smooth transition from the shell cross section. Frequently in concrete shells, post-tensioning is employed to improve the efficiency of these members. However, the anchor zone for the tendons may require special reinforcement to resist the stresses from the concentrated reactions.

For steel tanks, J. N. Pirok and R. S. Wozniak[5] have treated several aspects of proportioning. They discuss the calculation of the plate thickness in the tensile regions with regard to efficiency of the welded joints. In regions of compression, the proper consideration of stability effects on the allowable stress is elaborated. In junction regions (rings and knuckles), guidelines are provided for determining the portion of the shell wall that might be effective as a ring beam, acting alone or with an additional structural member. Stiffeners and opening reinforcement are also discussed briefly. The tank shown in Figure 2–8(r) serves as a representative example of these concepts.

Beyond the rather general treatments elaborated in the preceding paragraphs, there are a number of authoritative works for concrete cooling towers,[6] steel tanks,[7] pressure vessels,[8] bins,[9,10] and cylinders on column supports.[11] Also a number of unusual concrete tanks and other reinforced concrete shell and plates are analyzed and partially detailed in a fairly recent volume on shell structures.[12]

14.3 FUTURE APPLICATIONS OF THIN SHELLS

At the conclusion of a book which focuses on a rather specialized class of structures, it seems appropriate to comment on future applications of such constructions. Shell structures continue to be widely used in shipbuilding; in the offshore, aerospace, and automotive industries; and for pressure vessels, grain silos, and similar containments. However, one previously prominent application, roof shells of reinforced concrete, have been built far less frequently in the last quarter of the twentieth century than in the immediately preceding decades. This has been attributed to the increased availability of cheaper prefabricated systems.[13]

One of the world's most distinguished engineers, J. Schlaich, has recently considered this issue.[13] Schlaich eloquently states:

1. Shells are the most honest structures.
2. He, who cares about the shape of his structures, needs the shells.
3. He, who cares about the genuine use of materials, does not want to miss the concrete shells.

This being the case, he concludes[13] that "he, who cares about *beautiful* and *genuine* structures, cannot accept the fact that concrete shells are slowly disappearing." He conjectures that with the imaginative application of modern falsework techniques, such as pneumatic forms, and new materials, such as fiber concrete, light and versatile concrete shells may again be competitive, especially if some premium is allowed for quality.

Newspaper editor Horace Greeley, a renowned observer of the American scene, wrote in 1860: "I think concrete walls, rightly made, will last a thousand years."[14] The same is surely true of concrete shells, and it is a challenge to the present and future generations of structural engineers to see that such structures are indeed "rightly made."

The vision of Greeley and the optimism of Schlaich seem to be fulfilled by several new projects which exploit the considerable attributes of shells while controlling the costs of forming curved surfaces. In Figure 14–2, a double layered roof shell is shown under construction. This structure is designed to protect the only B-52 bomber on display in Europe, at a museum in England. Because the nose requires less clearance than the tail, the obvious choice of a cylindrical barrel shell, discussed in Chapter 13, is improved by taking a "bite" out of a torus, as shown in Figure 14–3. This segment is exploded into units of constant curvature formed by inverted precast T-beams topped by precast concrete panels. In turn, the units are stitched together by cast-in-place concrete and are supported on twisting edge

FIGURE 14–2   Double-Layered Roof Shell Under Construction.

Museum Roof Matamorphosis

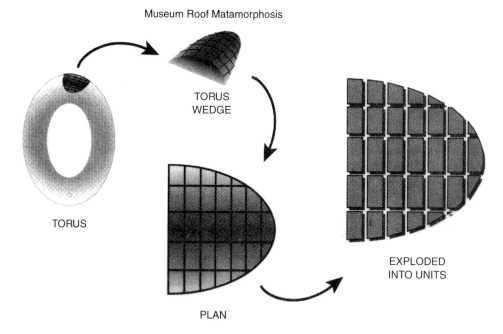

FIGURE 14–3   Torus Section Used for Museum Roof Under Construction.[15]
(Reprinted by permission of McGraw-Hill.)

FIGURE 14–4   Completed Museum Roof.

beams.[15] The completed structure, Figure 14–4, seems to demonstrate the advantages of both traditional and contemporary thin shell construction.

REFERENCES

1. R. Szilard, *Theory and Analysis of Plates,* (Englewood Cliffs, NJ: Prentice-Hall, 1974).

2. E. H. Baker, L. Kovalevsky, and F. L. Rish, *Structural Analysis of Shells,* (New York: McGraw-Hill, 1972).

3. Building code requirements for reinforced concrete (ACI 318-95) with Commentary (ACI 318 R-95) (Detroit: American Concrete Institute, 1995), chap. 19.

4. P.B. Lourenco and J. A. Figueiras, Solution for the design of reinforced concrete plates and shells, *Journal of Structural Engineering,* ASCE, Vol. 121, No. 5, May 1995: 815–823.

5. J. N. Pirok and R. Wozniak, Steel tanks, in *Structural Engineering Handbook,* E. H. Gaylord, Jr., and C. N. Gaylord, eds., (New York: McGraw-Hill, 1968), sec. 23.

6. Reinforced concrete cooling tower shells: Practice and commentary, *Journal of the ACI,* Title 81-52, 81, no. 6 (November–December 1984): 623–631.

7. "Recommended Rules for Design and Construction of Large, Welded, Low Pressure Storage Tanks," API Standard 620 American Petroleum Institute, 1966, "Welded Steel Tanks for Oil Stor-

age," API Standard 650, American Petroleum Institute, 1973; "AWWA Standard for Steel Tanks—Standpipes, Reserving and Elevated Tanks—for Water Storage," AWWA D100-67, American Water Works Association, or AWS D5.2-67, American Welding Society.

8. "Rules for Construction of Pressure Vessels" div. 1, 1971 ea., ASME Boiler and Pressure Vessel Code, sec. VIII, American Society of Mechanical Engineers, 1971; "Alternative Rules for Pressure Vessels," div. 2, 1971 ed., ASME Boiler and Pressure Vessel Code, sec. VIII, American Society of Mechanical Engineers, 1971.

9. N. S. Trahair, A. Abel, P. Ansourian, H. M. Irvine, J. M. Rotter, *Structural Design of Steel Bins for Bulk Solids,* (Sydney: Australian Institute of Steel Construction, 1983).

10. Design and construction of welded steel bins, AWRA Technical Note 14, Australian Welding Research Association (Milsons Point, Australia, December 1974).

11. V. S. Kelkar and R. T. Sewell, *Fundamentals of the Analysis and Design of Shell Structures* (Englewood Cliffs, NJ: Prentice-Hall, 1987).

12. W. Guggenberger, Proposal for design rules of axially loaded steel cylinders on local supports, Advances in Steel Structures, VII, *Proceedings of the International Conference on Advances in Steel Structures* (Pergamon Press, Dec. 1966): 1225–1230.

13. J. Schlaich, Do concrete shells have a future? *Bulletin of the IASS,* no. 89 (December 1985); 38–46.

14. *St. Louis Construction News & Review,* vol. 18, no. 4 (April 15, 1987): 36.

15. Putting U.S. Warplane Under Wraps in U.K. *Engineering News-Record,* (McGraw-Hill, New York, July 8, 1996), pp. 34–36.

Index